Applications of Synthetic Biology in Health, Energy, and Environment

Muhammad Arshad
International Islamic University, Islamabad, Pakistan

A volume in the Advances in Bioinformatics and Biomedical Engineering (ABBE) Book Series

IGI Global
PUBLISHER of TIMELY KNOWLEDGE

Published in the United States of America by
IGI Global
Medical Information Science Reference (an imprint of IGI Global)
701 E. Chocolate Avenue
Hershey PA, USA 17033
Tel: 717-533-8845
Fax: 717-533-8661
E-mail: cust@igi-global.com
Web site: http://www.igi-global.com

Library of Congress Cataloging-in-Publication Data

Names: Arshad, Muhammad, DATE- editor.
Title: Applications of synthetic biology in health, energy, and environment / Muhammad Arshad.
Description: Hershey, PA : Engineering Science Reference, [2023] | Includes bibliographical references and index. | Summary: "The book deliberates on principles and advancement of synthetic biology and their translation in the field of health & disease, energy, environment"-- Provided by publisher.
Identifiers: LCCN 2022045607 (print) | LCCN 2022045608 (ebook) | ISBN 9781668465776 (hardcover) | ISBN 9781668465783 (ebook)
Subjects: LCSH: Biomedical engineering. | Biomass energy.
Classification: LCC R856 .A657 2023 (print) | LCC R856 (ebook) | DDC 610.285--dc23/eng/20221115
LC record available at https://lccn.loc.gov/2022045607
LC ebook record available at https://lccn.loc.gov/2022045608

This book is published in the IGI Global book series Advances in Bioinformatics and Biomedical Engineering (ABBE) (ISSN: 2327-7033; eISSN: 2327-7041)

British Cataloguing in Publication Data
A Cataloguing in Publication record for this book is available from the British Library.

Advances in Bioinformatics and Biomedical Engineering (ABBE) Book Series

Ahmad Taher Azar
College of Computer & Information Sciences, Prince Sultan University, Riyadh, Saudi Arabia & Faculty of Computers and Artificial Intelligence, Benha University, Benha, Egypt

ISSN:2327-7033
EISSN:2327-7041

Mission

The fields of biology and medicine are constantly changing as research evolves and novel engineering applications and methods of data analysis are developed. Continued research in the areas of bioinformatics and biomedical engineering is essential to continuing to advance the available knowledge and tools available to medical and healthcare professionals.

The **Advances in Bioinformatics and Biomedical Engineering (ABBE) Book Series** publishes research on all areas of bioinformatics and bioengineering including the development and testing of new computational methods, the management and analysis of biological data, and the implementation of novel engineering applications in all areas of medicine and biology. Through showcasing the latest in bioinformatics and biomedical engineering research, ABBE aims to be an essential resource for healthcare and medical professionals.

Coverage

- Biomedical Sensors
- Protein Structure
- Genetics
- Artificial Organs
- Rehabilitation Engineering
- Applied Biosystems
- Biomechanical Engineering
- Structural Biology
- Computational Biology
- Neural Engineering

The Advances in Bioinformatics and Biomedical Engineering (ABBE) Book Series (ISSN 2327-7033) is published by IGI Global, 701 E. Chocolate Avenue, Hershey, PA 17033-1240, USA, www.igi-global.com. This series is composed of titles available for purchase individually, each title is edited to be contextually exclusive from any other title within the series. For pricing and ordering information please visit http://www.igi-global.com/book-series/advances-bioinformatics-biomedical-engineering/73671. Postmaster: Send all address changes to above address.

Titles in this Series

For a list of additional titles in this series, please visit: www.igi-global.com/book-series

Isolation, Characterization, and Therapeutic Applications of Natural Bioactive Compounds
Ajeet Singh (ICAR-Indian Institute of Wheat and Barley Research, India)
Medical Information Science Reference • © 2022 • 401pp • H/C (ISBN: 9781668473375) • US $345.00

AI-Enabled Smart Healthcare Using Biomedical Signals
Rahul Kumar Chaurasiya (Maulana Azad National Institute of Technology, Bhopal, India) Dheeraj Agrawal (Maulana Azad National Institute of Technology, Bhopal, India) and Ram Bilas Pachori (Indian Institute of Technology, Indore, India)
Medical Information Science Reference • © 2022 • 322pp • H/C (ISBN: 9781668439470) • US $435.00

Advancements in Controlled Drug Delivery Systems
Shekhar Verma (University College of Pharmacy, Pandit Deendayal Upadhyay Memorial Health Science, India & Ayush University of Chattisgarh, India) and Santosh Kumar Verma (Yulin University, China)
Medical Information Science Reference • © 2022 • 395pp • H/C (ISBN: 9781799889083) • US $360.00

Polymer-Based Functional Materials for Biomedical Applications
Nithin Kundachira Subramani (The National Institute of Engineering, India) and D.V. Gowda (JSS College of Pharmacy, India)
Engineering Science Reference • © 2022 • 320pp • H/C (ISBN: 9781799871989) • US $265.00

Emerging Developments and Applications of Low Temperature Plasma
Aamir Shahzad (Government College University, Faisalabad, Pakistan) and Maogang He (Xi'an Jiaotong University, China)
Engineering Science Reference • © 2022 • 239pp • H/C (ISBN: 9781799883982) • US $250.00

Biomedical Computing for Breast Cancer Detection and Diagnosis
Wellington Pinheiro dos Santos (Universidade Federal de Pernambuco, Brazil) Washington Wagner Azevedo da Silva (Universidade Federal de Pernambuco, Brazil) and Maira Araujo de Santana (Universidade Federal de Pernambuco, Brazil)
Medical Information Science Reference • © 2021 • 357pp • H/C (ISBN: 9781799834564) • US $325.00

IGI Global
PUBLISHER of TIMELY KNOWLEDGE

701 East Chocolate Avenue, Hershey, PA 17033, USA
Tel: 717-533-8845 x100 • Fax: 717-533-8661
E-Mail: cust@igi-global.com • www.igi-global.com

Dedicated to my son, Muhammad Shayan (Late), whose memory lives in our hearts forever.

Table of Contents

Detailed Table of Contents

in cell-free platforms, such as microfluidics, machine learning, CRISPR-mediated genome editing, etc. Developments in cell-free platforms satisfy many of the requirements of preparatory biochemical research, clinical purposes, diagnostic devices, industrial application, biosensors, etc. that offer exciting opportunities to transform synthetic biology and biotechnology. Therefore, this chapter aims to elaborate on the rapidly advancing field of cell-free platforms.

Assistive robotics and synthetic biology have revolutionized the medical field by offering personalized interventions and precise exercises in physical therapy and rehabilitation. This synergy accelerates patient recovery and improves functional outcomes. Synthetic biology techniques contribute to neural repair and regeneration, providing innovative solutions for neurological disorders and brain injuries. Robotic prosthetic limbs with synthetic sensory feedback revolutionize amputees' lives, offering enhanced motor control and reducing pain. The collaboration has led to breakthroughs in personalized gene therapies, smart drug delivery systems, and brain-computer interfaces.

Deep learning techniques have revolutionized biomedical applications such as arrhythmia detection, analysis of cardiac sensed signals, and cell-free synthetic biology. This chapter provides an introduction to each topic, discusses the role of deep learning techniques, highlights recent advancements, and explores the challenges and future directions in these areas. It also provides an introduction to cardiac sensed signals and their relevance in cardiology and explores the role of deep learning in analyzing and interpreting cardiac signals. Cell-free synthetic biology is also discussed, including applications in gene circuit design and metabolic engineering. Deep learning techniques are increasingly being used in biomedical applications and are important for personalized medicine, explainability and interpretability, real-time monitoring and intervention, adaptive optimization, and integration with other technologies.

Jay Prakash Maurya, Samrat Ashok Technological Institute, India
Manish Manoria, Bansal Institute of Research Technology and Science, India
Sunil Joshi, Samrat Ashok Technological Institute, India

Arrhythmia is a general type of cardiac disease in persons between 30-40 years of age. Cardiac system in human body generates electrical pulses that can be captured and plotted through electrical system called ECG. Computer-aided diagnosis system (CADS) is a good approach to help the healthcare field for early, regular, and accurate diagnosis and treatment plan during critical care conditions. Deep learning-based CADS can helps in critical condition for more quick diagnosis and treatment in countries where doctor ratio is comparatively low. With the help of machine learning (ML) algorithm, intervariable relationships may be used for prediction. However, machine learning algorithms are also limited due to its datasets availability, established framework, and clinician unfamiliarity. This chapter aims to provide an idea of arrhythmia and CADS approach using cascaded deep learning model of CNN, LSTM, GRU, and RNN. The chapter focuses on techniques used in past years, comparative studies, and direction of research as future improvements in respective fields.

Fatma Ece Altinisik Kaya, Marmara University, Turkey & Eberhard Karl University of Tübingen, Germany
Fatma Gizem Avci, Üsküdar University, Turkey

Synthetic biology is an engineering approach that designs biological systems to have new functions for useful purposes. Rational cell design is crucial for synthetic biology and can be considered a strategy that includes increasing the production efficiency of a cell or developing a new product by changing or adding specific biochemical reactions and regulations using modern metabolic engineering techniques. Metabolic flux and control analyses with models created in silico following cellular genome mapping have also contributed significantly to rational cell design. Models make the prediction of behaviors possible without performing a vast number of experiments when biological systems are modified. The modeling of microbial cells for the biosynthesis of high-value products is gaining increasing attention in biotechnology. In this chapter, the main focus will be on the concept of the genome scale metabolic modeling of microbial cells in biotechnological applications with an emphasis on biofuels and pharmaceuticals.

R. Kelmani Chandrakanth, Gulbarga University, India
Sahanapriya Gubbewad, Gulbarga University, India
Ajay Kumar Oli, Shri Dharmasthala Manjunatheshwara University, Dharwad, India

Molecular pharming is a method that combines the use of plants and biotechnological techniques for the production of proteins and bioactive metabolites which are important to support human health and industries. Using genetic engineering techniques, transgenic plants are developed. These plants have applications in the field of agriculture, biofuel, chemical industries, and pharmaceutical industries. These transgenic crop plants with limited facilities can synthesize a large amount of biomass at a low cost, through which a novel source of molecular medicines like plasma proteins, antibodies, vaccines, and growth factors can be isolated. Plant molecular pharming can be used as an alternative to mammalian cells, yeast, or bacterial culture systems for the production of pharmaceutical proteins. This also helps in studying and understanding the mechanism of expression of genes in plants. This chapter deals with the current strategies and approaches used in plant molecular pharming including the strategies, applications, limitations, and biosafety considerations.

A number of disorders can be treated promisingly with therapeutic proteins. Pharmaceutical corporations are focused on protein-based research in their pursuit for cutting-edge, efficient treatments. The manufacturing and application of therapeutic proteins have recently advanced significantly as a consequence of the enormous attention given by researchers. Biotechnological developments have accelerated and streamlined the manufacturing of therapeutically important proteins to fight a variety of terminal illnesses. Although protein-based treatments have assumed a prominent role in drug research and development and improved human safety, a number of difficulties still exist, including problems with safety, protein stability, and immunogenicity and degradation. The authors made an effort to address all potential contributing aspects to the creation of therapeutic proteins as well as any obstacles that researchers may be facing at the moment.

Polycystic ovary syndrome (PCOS) is the most common endocrine disorder in females of reproductive age. It is one of the main causes of infertility in females with a prevalence of 5-10%. PCOS is characterized by Oligo or anovulation, hyperandrogenism, and polycystic ovaries. PCOS is also associated with obesity, type 2 diabetes, hyperinsulinemia, insulin resistance, lipid abnormalities, and cardiovascular diseases. The early diagnosis of PCOS is complicated and complex due to the overlapping symptoms of PCOS. Genetic variance, epigenetic changes, and disturbed lifestyles lead to the development of pathophysiological disturbances and disturbed lifestyles in women. At the molecular level, different proteins and molecular and signaling pathways are involved in disease progression. Proteomics is advantageous in linking both genotype and phenotype and contributing to biomarker discovery. Investigating the molecular mechanism and proteins underlying PCOS is imperative to gain insight into pathophysiology of PCOS and formulate novel diagnostic and treatment strategies.

More than 65% of the elderly population globally suffers from dementia. The term "smart drug" refers to any drug substance that positively impacts mental skills. Nootropics are agents that are used for boosting the performance of the brain and to increase one's concentration. These are also called cognitive enhancers as these can enhance memory, creativity, awareness, and attention. Natural sources, as well as synthetic approaches, both, are being employed by researchers worldwide for developing smart drugs. Smart drugs are intended to enhance academic, social, as well as career performance. These drugs work by boosting the oxygen supply and nutrition to the brain and are useful for treating various neurodegenerative disorders. Several mechanistic pharmacological approaches have been explored for the development of smart drugs and enhancement of brain activity, the most prominent ones being glutamatergic, dopaminergic, serotonergic, and cholinergic pathways. This chapter reviews the various approaches for developing synthetic smart drugs and their applications and limitations.

Chapter 11
Molecular Level Analogy and Approaches of Gibson Assembly in Genome-Wide Analysis 212
Venkata Satya Sai Nagasri Abhishek Chembuli, Northeastern University, USA
Ajaykumar Chittipolu, Osmania University, India
Haritha Katasani, Northeastern University, USA

DNA assembly has always been a vital foundational technology for synthetic biology and other bioengineering processes. The development of restriction digestion and ligation methods has played a significant role in developing better DNA assembly methods with higher fidelity, efficiency, and modularity. Daniel Gibson discovered an exonuclease-based robust cloning method to assemble the DNA in the correct sequence known as the Gibson assembly. This method has been widely adopted as a major workhouse for synthetic biology experiments that have provided a new platform for the rapid cloning of multiple DNA fragments into the vector with a minimal time frame under isothermal conditions utilizing an overhang generated by 5' exonuclease, polymerase, and a DNA ligase. The Gibson assembly technique is widely applied in assembling single fragments, multiple fragments, site-directed mutagenesis, shotgun cloning, and creating library construct. Several advancements were observed in terms of developing minimal cells and synthesizing the cells and combining them with CRISPR technology.

Chapter 12
Integrated Machine Learning Approach for Healthcare Disabilities Tracing and Detection:
Applied Intelligent Techniques 235
Raghuram Bhukya, Department of Computer Science and Engineering, Kakatiya Institute of Technologey and Science, India
M. Sangeetha, Department of Data Science and Business Systems, SRM Institute of Science and Technology, India
Dler Salih Hasan, College of Science, University of Salahaddin, Erbil, Iraq
L. Raja, Department of Electronics and Communication Engineering, Sri Eshwar College of Engineering, India
Abdul Hannan, Department of Computer Science and Engineering, Koneru Lakshmaiah Education Foundation, India

The convergence of internet of things (IoT) technology and machine learning has paved the way for innovative solutions in the healthcare domain. This chapter presents an integrated approach that leverages IoT devices and advanced machine learning techniques to trace and detect healthcare-related disabilities. By continuously monitoring physiological signals and activity patterns, the proposed system enables

Foreword

The dawn of the 21st century has witnessed an astonishing convergence of science, technology, and innovation that has propelled humanity towards unprecedented advancements in the field of biology. Among these groundbreaking developments, none has captured the imagination and potential for transformative change quite like Synthetic Biology. It is with great pleasure and anticipation that I introduce this remarkable volume, *Applications of Synthetic Biology in Health, Energy, and Environment*. As I go through the book, it becomes abundantly clear that we are embarking on a journey through the cutting-edge landscape of scientific discovery and innovation, one that holds the promise of transforming our world in profound ways. The field of synthetic biology has undergone a remarkable evolution, and this book serves as a testament to the incredible progress made thus far. The editor, with foresight and dedication, has brought together a diverse group of experts and scholars who have shared their insights and expertise across a wide range of topics. One of the striking aspects of this book is its interdisciplinary nature. The collaborative efforts of researchers from various fields, including biotechnology, computer science, artificial intelligence, and environmental science, showcase the power of convergence. It is at these crossroads that some of the most exciting breakthroughs are occurring, and this book is a testament to the potential that arises when different disciplines come together in pursuit of common goals.

Synthetic Biology, often described as the "engineering of biology," leverages the principles of biology, chemistry, and engineering to design and construct novel biological systems, pathways, and organisms. The contributors to this volume, leading experts and pioneers in their respective domains, unravel the remarkable potential of Synthetic Biology in revolutionizing healthcare, sustainable energy production, and environmental stewardship. The editor rightly points out that we are in the midst of a scientific revolution, an era where the fusion of genetic engineering techniques with fields such as genomics, proteomics, systems biology, and computational biology has given birth to the field of Synthetic Biology. It is a field that challenges the boundaries of what is possible and envisions a future where the building blocks of life can be manipulated and engineered to address some of the most pressing challenges facing humanity.

In the pages that follow, readers will embark on a journey through the dynamic landscape of Synthetic Biology—a realm where scientists, engineers, and visionaries collaborate to redefine the boundaries of what is possible. This edited collection stands as a testament to the remarkable strides made in this field, offering a comprehensive exploration of its multifaceted applications in some of the most critical sectors of our global society: health, energy, and the environment. In the realm of health, from genetic engineering to deep learning, from bioremediation to smart drugs, each chapter offers a unique perspective on the boundless possibilities that cell-free synthetic biology presents. The chapters in this section illuminate the ingenious ways in which synthetic cell-free approaches and artificial intelligence are poised to transform our understanding of diseases, drug development, and medical interventions. The energy sector stands at the precipice of transformation, and this volume highlights how Synthetic Biology is a driving force behind this shift. From biofuels to microbial factories and machine learning approaches, these chapters present a compelling narrative of how biology's toolbox can provide sustainable solutions to our ever-growing energy demands. Environmental sustainability has never been more crucial, and the third section of this book delves into the applications of Synthetic Biology in ecological restoration, pollution remediation, and conservation. The stories of synthetic organisms engineered to restore damaged ecosystems and mitigate the effects of climate change are both inspiring and thought provoking. As we delve deeper into the myriad of possibilities that Synthetic Biology offers, it is essential to recognize the ethical, safety, and regulatory challenges that accompany this powerful technology. The contributors to this volume are acutely aware of these concerns and offer a balanced perspective on the responsibilities that come with wielding the power to engineer life itself.

In *Applications of Synthetic Biology in Health, Energy, and Environment*, the intersection of science, technology, and human creativity takes center stage. Whether you are a seasoned scientist, a curious student, a policymaker, or an engaged citizen, this book invites you to explore a future where biology becomes a versatile tool for addressing some of our planet's most pressing challenges. I am confident that this collection will ignite your curiosity, provoke your imagination, and inspire you to contemplate the limitless potential of Synthetic Biology. I extend my heartfelt congratulations to the editors and authors for their outstanding contributions to this field, and I hope that this book serves as a beacon guiding us towards a future where science and innovation harmonize with the health of our planet and the well-being of all its inhabitants.

In closing, I am excited to see how the insights and innovations within these pages will inspire new ideas, drive innovation, and deepen our understanding of the boundless potential that Synthetic Biology offers. This book will be a great aid to researchers and teachers. The future is bright, the possibilities are limitless, and it is my hope that this book serves as a guiding light on the path to a brighter, more sustainable future. The opportunities are greater than the problems.

Best wishes,

Johar Ali
UIT University, Karachi, Pakistan

Johar Ali, *upon completing his Bachelor of Science, was awarded a full scholarship to earn his Master's from the University of Missouri Columbia, USA. Then he was offered a Research/Teaching Assistant position at the University of Missouri, USA to pursue his Ph. D. After completing his Ph.D., Dr. Ali joined Washington University, School of Medicine to work in the areas of Bioinformatics and Biotechnology and worked in the high-profile human genome project. Dr. Ali was subsequently offered a job in BC Cancer Agency, Vancouver BC, Canada where he led a team of scientists in the area of bioinformatics and ran several multimillion dollars projects to completion. Later, working as a Team Leader in the area of cancer genomics at Ontario Institute for Cancer Research for 6.5 years, Dr. Ali was involved in setting up the genome center from designing, planning, and executing a variety of genomics-related projects. He was also instrumental in setting up the first genome center in Pakistan. He also served as a member at PARC, Ministry of National Food Security and Research. Currently, he is working as a Vice Chancellor at UIT University in Karachi, as well as Editor-in-Chief of the journal "Current Trend in OMICS". Dr. Ali has published in Science, Nature, BMC, and PLOS with 700+ impact points. Dr. Ali is the recipient of numerous international awards including excellence in teaching award at the University of Missouri, Columbia, MO USA, certificate of excellence at an international meeting held in Pennsylvania USA, Kempster memorial scholarship USA, and F21C summer fellowship in the USA. Dr. Ali is also the recipient of the "Spirit of Detroit Award for outstanding leadership" in Michigan, USA.*

Preface

As editor of this comprehensive reference book, I am delighted to present a compilation of cutting-edge research and insights into the dynamic realm of synthetic biology. The rapid progression of genetic engineering techniques, combined with the interdisciplinary convergence of fields such as genomics, proteomics, systems biology, and computational biology, has ushered in the era of synthetic biology. This field holds immense promise in reshaping our approach to health, energy, environment, and beyond.

ORIGINS AND GROWTH OF SYNTHETIC BIOLOGY

In this era of unparalleled scientific advancement, the principles of synthetic biology have taken center stage. This field has evolved into a multidisciplinary playground for innovators, researchers, and problem solvers. By ingeniously repurposing and redesigning biological systems, synthetic biology offers new horizons for tackling complex life processes. This book serves as a testament to the progress made and the uncharted territories yet to be explored.

NAVIGATING THE BOOK

The chapters of this edited volume delve into the core principles and breakthroughs of synthetic biology while elucidating their profound impact on health, disease, energy, and the environment. Designed to cater to a diverse audience including students, researchers, industrialists, policymakers, and stakeholders, this book endeavors to demystify the complexities of synthetic biology and bring its advancements to the forefront.

TOPICS EXPLORED

From DNA synthesis techniques, both chemical and enzymatic, to the synthesis of unnatural nucleotides, the book spans an impressive array of topics. Chapters explore the frontiers of genome editing, metabolic network remodeling, synthetic cells for industrial enzymes and therapeutics, and the production of synthetic molecules for environmental cleanup. The applications extend to the development of biosensors for health monitoring across species and the innovation of synthetic cells for energy production.

ENVISIONING THE FUTURE

As we traverse the pages of this book, we encounter the visions of tomorrow's world. Bio-remediation utilizing synthetic systems, molecular pharming, pathway engineering, synthetic promoters, and intricate synthetic circuits are explored in detail. The possibilities extend to smart drugs, soft living materials, and ethical considerations, opening a dialogue on responsible innovation and intellectual property rights.

A CALL TO ACTION

I extend an invitation to experts, scholars, and enthusiasts to contribute to the ever-evolving landscape of synthetic biology. This book is more than a collection of chapters; it is a platform for shared knowledge, insights, and inspirations that will shape the trajectory of this field. I hope that this volume serves as a catalyst for further exploration, collaboration, and realization of the limitless potential that synthetic biology offers.

As embark on this intellectual journey, I express my gratitude to all contributors who have dedicated their expertise to make this book a reality. Together, we strive to unlock the full promise of synthetic biology for the betterment of humanity and our planet.

CHAPTER OVERVIEW

From Cells to Cell-Free Platforms: A Comprehensive Overview

Mohammad Uzair, Hammad Qaiser, Muhammad Asad Ghuffran, and Asif Mir provide a comprehensive overview of the transition from cellular to cell-free platforms in synthetic biology. This chapter serves as a foundational guide, detailing principles, historical background, preparation methods, and applications of cell-free systems. By highlighting the advantages of cell-free approaches and the key factors influencing their efficiency, the authors offer insights into how cell-free systems are reshaping biotechnology research and applications.

Cell-free Synthetic Biology: The Novel Approach Towards Biotechnological Applications

In this chapter, authors Hammad Qaiser, Mohammad Uzair, Abida Arshad, Jabbar Zaman Khan Khattak and Shahid Bashir explore the groundbreaking advances in cell-free synthetic biology. They shed light on how cell-free workflows, which employ purified enzymes and cell extracts, have revolutionized biomanufacturing. With a particular focus on cell-free protein synthesis processes, the authors delve into the creation and analysis of biosynthetic pathways, as well as the synthesis of therapeutics. Recent technological breakthroughs in microfluidics, machine learning, and CRISPR-mediated genome editing have propelled the development of high-yielding cell-free platforms. By highlighting applications in preparatory biochemical research, clinical purposes, diagnostics, industrial applications, and biosensors, this chapter underscores the transformative potential of cell-free synthetic biology.

Assistive Robotics and Synthetic Biology in Medical Applications

Authors Dhiresh S. Shastri, Narhar K. Patil, Dler Salih Hasan, Tushar S Nikumbh, and Naresh Babanrao Chaudhari discuss the transformative impact of assistive robotics and synthetic biology in medical applications. This chapter delves into how the synergy between these disciplines has revolutionized patient recovery and functional outcomes in physical therapy and rehabilitation. The authors explore synthetic biology's contributions to neural repair and regeneration, as well as how robotic prosthetic limbs with synthetic sensory feedback are enhancing amputees' lives. By highlighting breakthroughs in personalized gene therapies, smart drug delivery systems, and brain-computer interfaces, the chapter demonstrates how these technologies are reshaping healthcare.

Deep Learning Techniques Revolutionizing Biomedical Applications: Arrhythmia Detection, Cardiac Sensed Signals, and Cell-Free Synthetic Biology

Pranjit Das, D. Vetrithangam, G. Hari Krishnan, Ashok Kumar, and R. Vijaya Kumar Reddy highlight the transformative role of deep learning techniques in biomedical applications. This chapter focuses on the revolutionary impact of deep learning in arrhythmia detection, cardiac signal analysis, and cell-free synthetic biology. The authors offer an introduction to each topic, emphasizing the pivotal role of deep learning techniques. By exploring recent advancements, challenges, and future directions, the chapter highlights how deep learning is reshaping diagnostics, patient care, and biotechnology.

Arrhythmia Detection and Classification on Cardiac Sensed Signals Using Deep Learning Techniques

Authors Jay Maurya, Manish Manoria, and Sunil Joshi delve into the critical field of arrhythmia detection using deep learning techniques. This chapter provides a comprehensive overview of how deep learning methods, specifically cascaded models like CNN, LSTM, GRU, and RNN, are applied to analyze and interpret cardiac sensed signals. The authors emphasize the potential of these techniques for early and accurate arrhythmia diagnosis, offering a glimpse into the future of data-driven cardiology and patient care.

Modeling the Microbial Cells for Biotechnological Applications

Fatma Ece Altinisik Kaya and Fatma Avci delve into the critical role of modeling microbial cells for biotechnological applications. This chapter underscores how synthetic biology's rational cell design, anchored in metabolic engineering and in silico modeling, has revolutionized the development of new functions and products. The authors explore the significance of genome-scale metabolic modeling in biotechnology, with a focus on biofuels and pharmaceuticals. By examining the integration of modeling techniques and their implications, the chapter offers insights into the role of synthetic biology in advancing biotechnological innovation.

Molecular Pharming

Kelmani R., Sahanapriya Gubbewad, and Ajay Kumar Oli elucidate the cutting-edge field of molecular pharming, which merges plant utilization and biotechnological techniques for the production of bio-active metabolites and proteins vital for human health and industries. Genetic engineering paves the way for transgenic plants with applications in agriculture, biofuels, and pharmaceuticals. This chapter delves into the strategies, applications, limitations, and biosafety considerations associated with plant molecular pharming. The potential of these transgenic crop plants to synthesize abundant biomass at low cost, thus providing a novel source of molecular medicines like antibodies, vaccines, and growth factors, is explored in detail.

New Trends in Developing Therapeutic Proteins by Synthetic Cells on Diseased Tissues

Authors Chittipolu Ajaykumar, Ravinder Doneti, Pasha Akbar, and Venkata Satya Sai Nagasri Abhishek Chembuli explore the expanding landscape of therapeutic protein development through synthetic cell applications. This chapter spotlights the significance of therapeutic proteins and how biotechnological advancements enable their production for combating various terminal illnesses. The authors underscore the advantages of protein-based treatments while acknowledging challenges such as safety, stability, and immunogenicity. By reviewing current strategies, applications, and obstacles, the chapter provides insights into the ongoing efforts to harness synthetic cell systems for therapeutic protein generation.

Pathomechanisms of Proteins in Polycystic Ovary Syndrome (PCOS)

Apoorva Jain and Ajay Oli delve into the intricate molecular mechanisms and protein pathways underpinning the pathophysiology of Polycystic Ovary Syndrome (PCOS). This chapter sheds light on PCOS, a prevalent endocrine disorder, affecting females of reproductive age, emphasizing its genetic and epigenetic basis. The authors underscore how genetic variance, epigenetic changes, and lifestyle factors contribute to PCOS's complex pathogenesis. The role of proteomics in uncovering the links between genotype and phenotype, aiding in biomarker discovery, and informing diagnostic and therapeutic strategies is also discussed.

Smart Drugs in Cognitive Disorders

Authors Jatin Phogat, Alka Bali, and Nishika Kapoor delve into the exciting realm of smart drugs and their applications in cognitive disorders. This chapter highlights the potential of nootropics—cognitive enhancers—to boost brain performance, memory, and awareness. The authors explore both natural and synthetic approaches to smart drug development, discussing their role in neurological disorders and brain injuries. By highlighting the mechanistic pharmacological pathways and various challenges, this chapter provides a comprehensive overview of the evolving landscape of smart drug research.

Molecular Level Analogy and Approaches of Gibson Assembly in Genome-Wide Analysis

Authors Venkata Satya Sai Nagasri Abhishek Chembuli, Ajaykumar Chittipolu, and Haritha Katasani delve into the foundational technology of DNA assembly in synthetic biology: Gibson Assembly. This chapter details the evolution from traditional methods to exonuclease-based robust cloning techniques. The authors explain the versatility of the Gibson Assembly method, highlighting its application in single fragment assembly, site-directed mutagenesis, library constructs, and more. The chapter sheds light on how this method's advancements facilitate the rapid construction of complex DNA constructs with minimal effort and time.

Integrated Machine Learning Approach for Healthcare Disabilities Tracing and Detection: Applied Intelligent Techniques

Raghuram Bhukya, M Sangeetha, Dler Salih Hasan, L Raja, and Abdul Hannan explore the synergy between IoT and machine learning for healthcare disability tracing and detection. This chapter demonstrates how IoT-enabled devices and advanced machine-learning techniques are revolutionizing healthcare by providing proactive solutions for disability tracing. The authors delve into data collection, preprocessing, feature extraction, and application of machine learning algorithms. The chapter discusses potential benefits, challenges, and ethical considerations, offering insights into the integration of IoT and machine learning for improved healthcare outcomes.

Biofuel Production, Bioremediation, and the Frontier of Synthetic Cell-Free Processes: IoT-AI Integration

Nalla Bhanu Teja, Sivakumar Karthikeyan, S Bharathi, Rayavarapu Bhavani Sankar, and Rohi Prasad delve into the intersection of biofuel production, bioremediation, and synthetic cell-free processes. This chapter underscores how these domains intertwine to promote sustainability and innovation. By examining the potential of IoT-AI integration, the authors emphasize its role in optimizing biofuel production and bioremediation processes. The chapter explores the integration of smart technologies, shedding light on how IoT and AI could propel the fields forward while considering ethical and environmental implications.

Energy Monitoring and Control in the Smart Grid: Integrated Intelligent IoT and ANFIS

Authors Kavitha C.R, Varalatchoumy M, Mithuna HR, Bharathi K, Geethalakshmi N M, and Sampath Boopathi explore the integration of intelligent technologies for energy monitoring and control in the smart grid. This chapter focuses on the convergence of IoT and Adaptive Neuro-Fuzzy Inference System (ANFIS) techniques to optimize energy usage and enable demand response. By highlighting the benefits and challenges of these technologies, the authors emphasize their potential to revolutionize energy management, enhance efficiency, and contribute to sustainability.

Hospital Waste Management Using IoT and Deep Learning: Enhanced Efficiency and Sustainability

R.E. Ugandar, U. Rahamathunnisa, S. Sajithra, M. Beulah Viji Christiana, Basanta Kumar Palai, and Sampath Boopathi offer a comprehensive exploration of the integration of Internet of Things (IoT) and Deep Learning (DL) technologies to revolutionize hospital waste management. With a focus on waste volumes, environmental concerns, and real-time monitoring, this chapter delves into IoT-DL integration's potential benefits and applications. It investigates how IoT facilitates real-time monitoring of waste parameters, while advanced DL techniques enhance waste collection efficiency, segregation accuracy, and data-driven decision-making. The chapter highlights the challenges, ethical considerations, and sustainability implications inherent in this innovative approach.

Bioremediation of Pharma Pollutants Using Synthetic Systems

Meghmala Waghmode, Aparna Gunjal, Manisha Gadekar, and Dipali Mali provide insights into the use of synthetic biology for the bioremediation of pharmaceutical pollutants. The authors highlight how synthetic systems can contribute to gene mining, enzyme discovery, and biodegradation pathways. This chapter explores the challenges associated with in-situ remediation using cell-based systems and presents a comprehensive approach to the bioremediation of pharmaceutical waste using synthetic biology techniques.

Synthetic Biology for Wastewater to Energy Conversion: IoT and AI Approaches

Authors N. Venkateswaran, S. Senthil Kumar, G. Diwakar, D. Gnanasangeetha, and S. Boopathy explore the innovative application of synthetic biology for wastewater-to-energy conversion. This chapter explores how synthetic systems and bioremediation can be harnessed to address environmental challenges. The authors emphasize the potential of integrating IoT and AI to optimize these processes and contribute to a sustainable and energy-efficient future. By providing case studies and insights, the chapter highlights the transformative impact of these technologies on wastewater treatment and energy generation.

SUMMARY

In concluding this edited reference book, we are presented with a mosaic of chapters that collectively represent the forefront of biotechnological advancements and innovations. The multidisciplinary nature of these chapters, authored by esteemed experts globally, underscores the intricate interplay between fields such as synthetic biology, IoT, deep learning, and more. These interdisciplinary intersections not only reflect the state of the art but also point towards the exciting trajectories ahead.

As an editor, I am inspired by the depth and breadth of knowledge encapsulated within these chapters. The diverse range of topics covered, from cell-free synthetic biology to assistive robotics, from smart drugs to wastewater-to-energy conversion, attests to the dynamism and versatility of contemporary biotechnology. The collaborative efforts and insights shared by the experts display the collective commitment to addressing complex challenges and ushering in transformative solutions.

Throughout these chapters, a common thread emerges—the convergence of cutting-edge technologies and scientific inquiry with the pressing concerns of our times. Climate change, healthcare disparities, environmental pollution, and more find solutions in the creative amalgamation of synthetic biology, IoT, AI, and other intelligent systems. This interplay underlines the inherent synergy between scientific progress and societal impact.

The chapters, written in the distinctive voices of experts, weave a narrative of discovery, innovation, and exploration. From unraveling the mysteries of disease pathomechanisms to harnessing cell-free systems for biomanufacturing, each chapter contributes a unique facet to the grand tapestry of biotechnological advancements. These chapters collectively exemplify how research and application in these domains are shaping the present and future of human well-being.

As an editor, I extend my heartfelt gratitude to all the contributors who have shared their knowledge, insights, and experiences in this volume. Their dedication and expertise have brought depth and substance to this compilation. I hope that readers—students, researchers, professionals, and curious minds alike—find inspiration and valuable insights within these pages that spark new ideas, drive innovation, and deepen their understanding of the potential at the nexus of biotechnology and intelligent systems.

The journey encapsulated in these chapters is a testament to human ingenuity, curiosity, and our unceasing quest to unlock the mysteries of life and the universe. The fusion of science, technology, and human endeavor that resonates through these pages embodies the essence of exploration, inquiry, and progress at the heart of biotechnology. I look forward with excitement to future breakthroughs and collaborations that will continue to shape the landscape of biotechnological applications and their impact on society.

With sincere appreciation,

Muhammad Arshad
International Islamic University, Islamabad, Pakistan

Acknowledgment

The completion of this book, *Applications of Synthetic Biology in Health, Energy, and Environment*, has been made possible through the dedication and collaboration of many individuals and organizations. I would like to express my heartfelt gratitude to all those who have contributed to this project.

First and foremost, I extend my special thanks to all the contributors who generously shared their expertise and insights to make this book a reality. Your valuable contributions have enriched the content and depth of this work, and I am truly grateful for your commitment to advancing the field of synthetic biology.

I am also deeply appreciative of Prof. Dr. Johar Ali, Vice Chancellor of UIT University (formerly Usman Institute of Technology), for his invaluable contribution to this book by writing the foreword. His expert perspective has added significant value and credibility to the book, and I am honored to have his endorsement.

My sincere thanks go to IGI Global for providing us with the opportunity to publish this edition. Their support and guidance throughout the publishing process have been instrumental in bringing this book to fruition. I would like to extend a special acknowledgment to Ms. Nina Eddinger for her direct involvement and efforts in making this book a reality.

Furthermore, I want to express my gratitude to all those who have been directly involved in supporting this book during its entire journey. This includes individuals who provided reviews, valuable comments, assistance with dissemination, and guidance. Your contributions have played a crucial role in shaping the content and quality of this book.

In conclusion, this book is the result of collective effort, dedication, and collaboration. I am immensely grateful to everyone who has been part of this endeavor, and without your contributions, this book would not have been possible. It is my hope that this work will serve as a valuable resource for researchers, professionals, and all those interested in the dynamic field of synthetic biology.

Chapter 1
From Cells to Cell-Free Platforms:
A Comprehensive Overview

Mohammad Uzair
https://orcid.org/0000-0003-2039-0870
International Islamic University, Islamabad, Pakistan

Hammad Qaiser
International Islamic University, Islamabad, Pakistan

Muhammad Asad Ghuffran
International Islamic University, Islamabad, Pakistan

Asif Mir
International Islamic University, Islamabad, Pakistan

ABSTRACT

Cell-free systems are powerful tools in biotechnology, allowing researchers to study and manipulate biological processes outside of intact cells. This chapter provides a comprehensive introduction to cell-free systems, covering their principles, historical background, preparation methods, and various approaches used in research and applications. It emphasizes the advantages of cell-free systems, their historical development, and the different methods for preparing cell extracts or purified components. The chapter explores approaches like extracts-based cell-free systems and the PURE system, discussing their methodologies and applications. It also highlights the key components and factors influencing the efficiency of cell-free systems. Therefore, this chapter serves as a foundation for understanding the potential of cell-free systems in biotechnology research.

DOI: 10.4018/978-1-6684-6577-6.ch001

INTRODUCTION

Cells have molecular machinery to make proteins, which are essential tools for cellular differentiation and growth. DNA contains genetic instructions needed to make different proteins. Biotechnological tools allow transferring the specific gene from one organism to another, i.e., bacterial cells, which are cultivated on an artificial medium and used to isolate the target proteins. For instance, *Escherichia coli* and *Saccharomyces cerevisiae* are mostly used for the production of recombinant human insulin. Moreover, other protein-based medicines, such as vaccines or antibodies, have also been developed using recombinant gene technology for therapeutic application in humans. During in vivo recombinant protein expression, once DNA is transferred, bacterial cells are cultured in liquid culture, at either pilot scale or industrial scale. The desired proteins are isolated from the cells through a complex process, protein purification, and utilized for therapeutic purposes. Some other cellular pathways can be utilized to carry out biochemical reactions to develop various products, including biofuels, bioplastics, etc. However, cell-based systems are highly complex and cannot be used directly due to the maintenance of cell viability and the controlled environment. In vivo recombinant technology for protein manufacturing also has drawbacks, including protein aggregation, degradation, and DNA loss, as well as several time-consuming experimental procedures, such as DNA cloning in the vector, DNA transformation, and overexpression of the target protein (Yoshihiro Shimizu et al., 2006). With the help of recent developments in synthetic biology, it is now possible to design and create biological systems that have the capacity to carry out intricate molecular processes at all levels, from the molecular to the general, to extracellular (Laohakunakorn, 2020). In contrast, another strategy is the use of a Cell-free System (CFS), in which molecular processes are, activated efficiently using crude cellular extracts, instead of whole cells, supplemented with multitudes of accessory ingredients, including amino acids, transcription factors, cofactors, translational regulators, nucleotides, and energy substrates (ATP, GTP). CFS, generally known as Cell-free gene expression (CFE), is an in vitro technique that retains protein synthesis and other native biochemical reactions independently from an intact cell. Hence, simplifying the complicated interconnections encountered when working in an intact cell (Kelwick et al., 2020; Koo et al., 2020). A purified bacterial RNA polymerase is employed in the first step of CFE for mRNA synthesis, whereas a cellular extract supplies the ribosomes and every other necessary element required for protein expression. *E. coli*, yeast, wheat germ, rabbit reticulocytes, insect cells, Chinese hamster ovary (CHO) cells, and even human cell cultures are just a few of the many different cell-free systems that have been developed (Zemella et al., 2015). CFE offers an unparalleled high degree of controllability and simplicity over the biochemical reactions for gene expression and metabolism by overcoming the restriction of maintaining cell viability.

CFS can prototype cellular function and utilized as a key research technique in biology and biotechnology for more than 120 years. Furthermore, over the past two decades, the multitude of cell-free platforms have undergone tremendous development, and increases in the yield of bioactive recombinant DNA (rDNA) proteins have led to the application of these systems for the in vitro expression of proteins. Recent development in cell-free platforms has to speed up the use of this technology, which can satisfy many of the requirements of biochemical research, clinical purposes, paper-based diagnostic devices, industrial application, biosensors, etc. Therefore, this chapter aims to elaborate on the rapidly advancing field of cell-free platforms. This chapter describes cell-free systems obtained from different model organisms, with their advantages and limitations.

HISTORICAL BACKGROUND

Cell-free systems have been used for more than a century to clarify fundamental ideas underlying biological systems. CFS approaches have for history served as a fundamental methodology to explore molecular pathways through the construction and reconstruction of cellular processes. In 1897, Eduard Buchner prepared the first extract-based cell-free system from yeast for fermenting sugar instead of live yeast cells (Buchner, 1897). Buchner was awarded Nobel Prize in 1907 for his groundbreaking research. Following the finding of the structure of DNA by Watson and Crick, the research interest shifted towards molecular biology and central dogma, i.e., transcription and translation. The initial step towards understanding the process of enzymatic DNA replication was directly assessed by polymerase activity in cell-free extracts of *Xenopus* and *E. coli* (Blow & Laskey, 1986; Fuller et al., 1981). Studies on *Saccharomyces cerevisiae* (yeast) lysates, as an in vitro translation assay, discovered the role of 5' cap and 3'poly(A) tail on mRNA in translation initiation through internal ribosomes (Iizuka & Sarnow, 1997; Preiss & Hentze, 1998). Later, to know more about transcription mechanism, nucleotides translation into amino acids, or protein synthesis, in 1948, Zamecnik reported the first application of cell-free protein synthesis (CFPS) using liver cell extracts from rats to study protein synthesis (Zamecnik et al., 1948). Nirenberg and Matthaei (1961) develop *E. coli* cell-extract and supplement with combinations of RNA, discovering the first genetic codon (3 nucleotides per amino acid), UUU, encoding phenylalanine, as well as the mechanism of mRNA translation into amino acids and proteins. Marshall Nirenberg received Nobel Prize in 1968 for this experiment, i.e., breaking the genetic code and describing how it operates in protein synthesis using *E. coli* cell-free extract. This experiment was one of the first to demonstrate that a cell-free system contains molecular machinery that carries out transcription and translation without intact cells. Various biological processes, from metabolism to central dogma (DNA→ RNA→ Protein) of molecular biology, were discovered in the early years of the discovery of cell-free systems (Laohakunakorn, 2020). The Krebs cycle and decoding the information in DNA are a few examples of novel discoveries that used cell lysate or in vitro protein translation. Moreover, cell-free systems also identify the role of ATP and GTP in protein synthesis, tRNA, amino acids, ribosomes, etc. (Zamecnik et al., 1948). The synthesis of the enzyme beta-galactosidase was also reported using a microbial cell-free system (DeVries & Zubay, 1967; Zubay, 1973). Application of CFS includes the study of biological and chemical samples, pharmaceutical proteins, metabolites, and enzymes, and the production of gene and protein libraries. By enabling revolutionary techniques, these discoveries provide fascinating prospects for fundamentally revolutionizing molecular and synthetic biology.

CELL-FREE SYSTEM PREPARATION

CFS have open nature, i.e., the absence of any physical barrier (such as cell wall or cell membrane) that allows the modification and direct control over the function and outcome of the gene network (Didovyk et al., 2017). The absence of a cell wall or cell membrane allows free access to the biochemical reactions for probing the synthesis reaction. All resources are directed toward expressing the desired protein because it is not necessary to keep cells alive. Because the gene of interest is expressed in the bulk of the protein generated, this property facilitates a simpler purification procedure (Chiba et al., 2021). Therefore, CFS facilitates in vitro gene expression or amplification (i.e., PCR reaction), known as CFPS, without selective markers or cell-based cloning procedures. CFS facilitates molecular methods to the model-driven design to decipher the hidden information in DNA by accelerating design-build-test-run steps through rapid, predictable, and robust prototyping platforms for synthetic biology and synthetic gene regulatory network (Laohakunakorn, 2020). In most cases, the function of protein synthesis from cell-free systems of different origins is the same. Crude extract prepared from centrifuging cultured cells contains all necessary components, including proteins, enzymes, ribosomes, etc. The cell-free extract can be prepared by collecting supernatant of lysates of different cells centrifuged at 30,000× g, known as S30 extract. (Brookwell et al., 2021). Zubay (1973) reported the S30 extract protocol. Later in 2006, S12 extract was also reported by centrifugation at 12,000× g (T. W. Kim et al., 2006). In addition to extraction, the crude extract is supplemented with nucleotides, amino acids, and polymerases to synthesize desired protein. Adding nucleotide or DNA template, known as a coupled reaction, RNA polymerase transcripts the sequence, and the ribosomes translate the desired amino acid sequence for de novo protein production. The energy source is also necessary to proceed with the reaction (Rosenblum & Cooperman, 2014; Zemella et al., 2015). Energy consumption and supply is the challenging task associated with protein yield and reaction lifetime. Therefore, consumable substrate, including ATP and amino acids, facilitates ADP phosphorylation in the CFS reaction. The cell-free extract is usually supplemented with phosphoenolpyruvate (PEP), glucose, or glucose-6-phosphate. However, these ATP resources have some limitations, i.e., degradation of PEP into pyruvate and inorganic phosphate, glucose/glucose-6-phosphate results in a lower production rate. The buildup of inorganic phosphate causes the chelation of magnesium ions (Mg^{2+}). Mg^{2+}, an essential metal ion cofactor in enzymatic systems, is converted to magnesium ions by inorganic phosphate. The reduction of Mg^{2+} causes the inhibition of protein synthesis. Moreover, the accumulation of inorganic phosphate also depletes ATP concentration in the CFS milieu. Fructose-1, 6-bisphosphate was also implemented in the CFS reaction, which reduced the cost of produced protein (D. M. Kim & Swartz, 2001; Zemella et al., 2015). Moreover, creatine, phosphates, and kinases could substitute energy sources for CFS reactions. ATP regeneration could be a potential approach for continuous protein synthesis reactions. The utilization of inorganic phosphates by high-energy substrates facilitates the increment of reaction lifetime, including pyruvate or acetyl phosphate. However, when using pyruvate, glucose oxidation might occur, which causes a pH decline. Appropriate buffers, such as Bis-Tris buffer, could be utilized to maintain pH and increase the protein yield. The CFS approach is a cost-effective technique for the high-yield production of the desired protein, for instance, from milligram quantities per 1ml reaction volume to grams of proteins per liter (Garenne et al., 2021). Moreover, DNA templates, dNTPs, salts (e.g., initiation factors, sigma factors, and elongation factors), canonical/non-canonical amino acids, transcription and translation factors, polyamines (spermidine), chaperones/chaperonins, enzymes, metabolic cofactors (coenzyme A), reducing agents, microsomes, mRNAs, and tRNAs are also added in CFS reaction mixture in adequate amounts to carry out the process of transcription and translation for protein synthesis, efficiently.

CFPS systems vary significantly in many ways, including the quality and quantity of proteins, due to the origin, i.e., the organism from which they are prepared. Several eukaryotic and prokaryotic cell-free extracts have been developed in different reaction approaches, such as classical batch reactions or dialysis CF format that includes continuous exchange cell-free (CECF), and continuous-flow cell-free (CFCF) reactions (Zemella et al., 2015). For instance, an *E. coli* cell-free system with endogenous transcription and translation components yields up to 2.3mg/ml of a reporter protein in a batch mode reaction (Caschera & Noireaux, 2014). A significantly higher yield of bispecific antibodies, about 1300 mg/ml, using *E. coli* cell-free extract in a batch-mode reaction was also reported (Xu et al., 2015). Using a CECF reaction based on lysates Chinese hamster ovary (CHO) cells, 980mg/ml of epidermal growth factor receptor (EGFR) was obtained (Lena Thoring et al., 2017). The batch format is the commonly practiced CFS reaction, occurring in one reaction vessel. However, the accumulation of by-products could inhibit the synthesis of in vitro protein synthesis (Smolskaya et al., 2020). In contrast, the dialysis CF format avoids the accumulation of by-products passively by dialysis (CECF) or actively by a flow (CFCF), separating substrates supply, synthesized proteins, and by-product removal. The dialysis CF system, therefore, has a sustained reaction lifetime with a higher protein yield (Pardee, 2018). A semi-continuous format has also been developed, using a dialysis membrane reactor that extended reaction rate of a continuous-flow system (D. M. Kim & Choi, 1996). Hence, CFPS is a novel, inexpensive, and scalable approach for synthesizing pharmaceutical proteins. Moreover, CFS reactions could be coupled or uncoupled formats, distinguished by transcription and translation conditions. Coupled CFS reaction is a one-pot reaction, i.e., a DNA template is added, and transcription and translation occur in a single tube, whereas in an uncoupled reaction, mRNA purification is done and supplied to the CFS reaction for the translation process (Gregorio et al., 2019).

The cell-free extracts require freeze-dring to avoid degradation of the proteins, ribosomes, and the polymerases inside the cell-free system if left for a long time at room temperature. Once the extract is stored, it only needs water containing desired DNA template (DNA or mRNA), and the reactions will occur and give the desired protein product. Lyophilization exerts the potential benefit of long-term stability at storage above freezing. When necessary, freeze-dried cell-free (FD-CF) systems can then be turned on by simply adding water. FD-CF reactions have also been used for synthetic biology educational kits (Huang et al., 2018; Stark et al., 2018). In addition to being non-living, CFS is a cellular mimic, having cytoplasmic lysate, functional metabolic pathways, lyophilized cell-free gene circuits, and other essential isolates.

CELL-FREE SYSTEM APPROACHES

CFS can be obtained from either prokaryotes (e.g., *E. coli*) or eukaryotes (i.e., plants, fungi, mammals), and exerts different applications and drawbacks regarding CFPS approaches (Table 1). The extract of *E. coli* is widely accepted as a first-choice model organism in cell-free systems for recombinant protein expression. Various approaches have been developed for the cell-free synthesis of protein and peptides due to the simplicity, fast growth, and cost-effectiveness of *E. coli* culture and protein extraction (Smolskaya et al., 2020). Various other sources have been utilized for developing cell-free platforms, such as protozoans parasites (*Plasmodium falciparum*, *Perkinsus marinus*, and *Leishmania tarentolae*), yeast (*Pichia pastoris* and *Saccharomyces cerevisiae*), alga (*Chromera velia*), plants (wheat-germ and tobacco BY-2) insects (*Spodoptera frugiperda*, *Spodoptera frugiperda* 21), mammals (rabbits reticulocytes, HeLa cells, human K562 cells, human HEK293 cell, Chinese hamster ovary cells), etc. (Ezure et al., 2010; Fernández-Robledo & Vasta, 2010; Zemella et al., 2015).

The CFPS reaction is assembled in a test tube, i.e., DNA template, amino acids, energy source, and TXTL machinery, to initiate essential reactions (TXTL) for synthesizing required proteins, such as therapeutic proteins, biosensors, antibodies, etc. Protein synthesis might occur in different formats; (a) The batch format utilizes one vesicle comprising both reaction medium (RM) and feeding solution (FS). (b) Continuous-flow CF methodology: FS is permanently pumped in RM through the dialysis membrane (DM); the newly synthesized proteins or by-products are removed from the RM by passing through the DM with other cut-off pores, allowing for target protein collection. (c) Continuous-exchange CF is based on dialysis through the membrane, allowing for constant substrate addition and by-product removal from the RM. (d) Bilayer format: RM and FD represent two different liquid phases; the exchange and prolonged protein synthesis occur at the phase border or interphase (IF). Reaction components include ribosomes and RNA polymerase (when necessary). The feeding buffer includes amino acids, energy, etc.

Figure 1. Schematic representation of components, designs, and formats of in vivo CFPS system

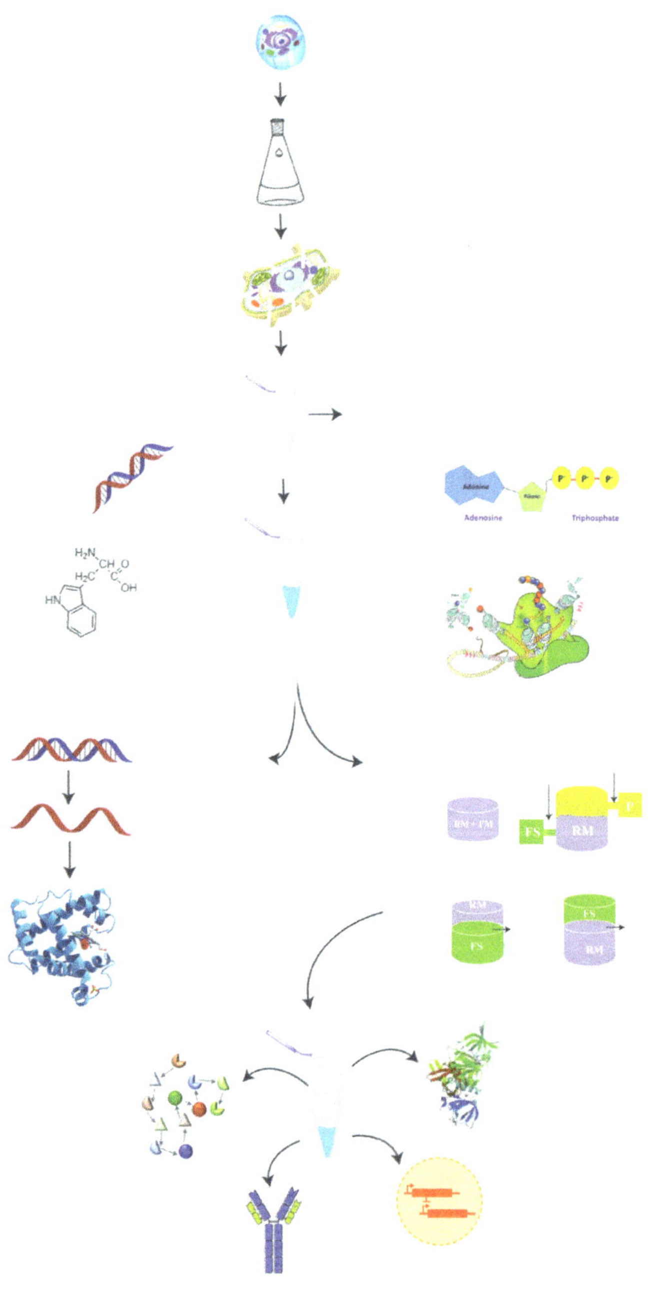

E. coli-Based CFS

The extract of *E. coli* is one of the earliest and most widely accepted as a first-choice model organism in cell-free systems for recombinant protein expression. Various approaches have been developed for the cell-free synthesis of protein and peptides due to the simplicity, fast growth, and cost-effectiveness of *E. coli* culture and protein extraction (Smolskaya et al., 2020) (Table. 1). In *E. coli* cell-free settings, post-translational modifications for functional proteins are severely constrained since only a few alterations are possible. The *E. coil*-based cell-free platforms have undergone modifications to increase protein yields created from scratch. *E. coli* has been successfully employed to produce peptides and protein subunits at a low cost. Many strains have undergone genetic modification to aid in bioprocessing and industrial-scale fermentation. Several biotech companies created and commercialised cell-free systems based on *E. coli*, including Invitrogen, Qiagen, Promega, Sigma-Aldrich/Merck, Biotechrabbit, and others. Cell-free protein synthesis based on *E. coli* facilitates the synthesis of difficult proteins. NEBExpress Cell-free E. coli Protein Synthesis System (NEB), Thermo Scientific 1-Step Human in vitro Protein Expression Kits, and ALiCE® Mini Kit are only a handful of the things that have been made available for purchase (Sigma)(Ezure et al., 2010; Fernández-Robledo & Vasta, 2010; Zemella et al., 2015). However, *E. coli's* inability to perform post-translational modifications observed in eukaryotic proteins and the absence of a secretion mechanism for the rapid release of protein into the culture medium are its principal drawbacks as an expression system (Arya et al., 2008; Makrides, 1996).

Archaea-Based CFS

To investigate the impact of various antibiotics on the translational machinery of archaeal cells, the first archaeal cell extracts were demonstrated (Elhardt & Böck, 1982). The study of ribosomal subunit gathering in cell extracts generated from sulfur-dependent thermophilic Archaea provided new visions into the translation mechanism. (Londei et al., 1986). The translationally active extracts were obtained from the extremely thermophilic *Sulfolobus solfataricus* (Ruggero et al., 1993) and the hyperthermophilic *Thermococcus kodakaraensis* (Endoh et al., 2006) for a coupled CFPS reaction operating at high temperatures. Archaeal cell extract-based systems are crucial for the synthesis of thermostable proteins, which may require high translation temperatures (up to 80 °C) for optimal protein folding (Table. 1). High temperatures may also reduce the translation hindrance caused by mRNA secondary structures in coupled transcription-translation events (Zemella et al., 2015).

Virus-Based CFS

Baculoviruses are regarded as being harmless to humans and not contagious. For diagnostic purposes, the baculovirus-based cell-free protein expression approach has been utilised to create G-protein coupled receptors, autoantigens, cytochromes, and herpes simplex virus capsid proteins (Condreay & Kost, 2007). In contrast to mammalian cells, the baculovirus expression method has several significant difficulties regarding the structure of recombinant proteins (Condreay & Kost, 2007; Hu, 2005). Additionally, the simultaneous protein expression and baculovirus spread methods are time-taking, labor-intensive, and expensive to operate (Table. 1). Compared to yeasts or E. coli, the growth and culture media are more expensive because of their complexity (Arya et al., 2008).

Protozoans-Based CFS

Tetrahymena pyriformis (Renoux & Rohmer, 1986), *Plasmodium falciparum* (Mudeppa, 2007), *Paramecium aurelia, Paramecium tetraurelia* (Yano et al., 2020), *Toxoplasma gondii* (Luk et al., 2008) are one of the protozoan species whose CFS facilitates analyzing the integration efficacy of amino acids dependent on buffers, energy, pH, ions, and other chemicals. Because translation and transcription start routes differ amongst organisms, standard in vivo protein synthesis methods were largely ineffective for parasite protein expression (Fernández-Robledo & Vasta, 2010). Protozoans-based CFS is typically less expensive to cultivate and scaleable compared to CFS from other model organisms. In addition, the proteins synthesized are more soluble than those produced from *E. coli* lysates (Kovtun et al., 2011) (Table. 1). *Leishmania tarentolae* extracts have several advantages, including improved translation initiation and the availability of identical splice leader sequences on all endogenous mRNAs (Zemella et al., 2015).

Amoeba-Based CFS

The soil amoeba *Dictyostelium discoideum* is a non-mammalian model organism that has been extensively studied for the expression of the eukaryotic protein, due to reliability, reproducibility, and controllability to automation. *D. discoideum* facilitates post-transcriptional modifications, i.e., incorporate N- and O-linked phosphoglycosylation to the recombinant protein, similar to mammalian or other higher eukaryote systems. Antithrombin III, Insulin-like growth factor-binding protein 6, GLUT1, and other recombinant therapeutic proteins have been produced in massive quantities and at significantly lower costs from *D. discoideum*. Continuous growing of *D. discoideum* in a bioreactor with cell holding via microfiltration and low space velocity increased cell density in suspension culture. Relevant progress has been made in the use of the *D. discoideum* expression system for the excessive production of medicinal significant proteins for the novel drug development investigation (Arya et al., 2008).

Yeast-Based CFS

Hansenulla polymorpha, Saccharomyces cerevisiae, and *Pichia pastoris* yeast extracts are excellent prospects for manufacturing FDA-approved therapeutic proteins such as hepatitis B antigen, insulin, and glucagon. Saccharomyces cerevisiae promotes high-yielding protein synthesis while also ensuring effective protein folding (Gellissen & Hollenberg, 1997; Gellissen et al., 2005). On simple chemically specified conditions, yeast may multiply quickly to high cell densities and is compatible with fermentation operations. Yeasts are safe for genetic engineering because they are food-grade organisms and do not release endotoxins or contain viruses. Their capacity to secrete the desired protein is a further benefit, resulting in better recovery of pure protein from broth at much lower costs (Arya et al., 2008). CFS from yeast extracts can also help establish post-translational modifications such as glycosylation. As a result, there are some fundamental limitations of his CFPS system in eukaryotes. It overcomes the problems of low protein production (compared to E. coli), expensive extract preparation, and small reaction size (Zamecnik et al., 1948).

Wheat Germ-Based CFS

The most popular and commercially available eukaryotic CFPS systems are those based on wheat-germ extract (WGE). WGE-based CFS approach offers benefits over the *E. coli* system for the production of some complex protein types and can produce post-translational modifications not present in bacteria (Fogeron et al., 2021). Eukaryotic CFPS procedures, on the other hand, are often more expensive, take more time to prepare the extract, and generate less protein in batch processing. In terms of protein yields, WGE is the most productive approach (Khambhati et al., 2019) (Table. 1). WGE typically yields between a few hundred micrograms and milligrammes of recombinant protein per millilitre reaction, depending on the protein and format. WGE are wheat seed embryos that have been isolated (Gregorio et al., 2019). Although WGE is the most effective method for producing proteins, it is not always appropriate for post-translational modifications like glycosylation (Chen & Kashina, 2021). Wide-spectrum expression of eukaryotic proteins has frequently been accomplished by WGE. WGE has subpar tools for genetic modification (Dong & Ronald, 2019). The lengthy and difficult extract preparation process for WGE results in a low yield of cell extract.

Tobacco BY 2-Based CFS

Based on tobacco BY-2 cells, another plant-based eukaryotic system has been developed over the past 15 years. Tobacco cells are only exposed for 4 to 5 hours in order to produce a translationally active lysate, as opposed to the 4 to 5 days needed for the manufacture of wheat-germ extract (Buntru et al., 2014). There aren't many assessments of the tobacco system. The preliminary findings pointed to a system that might challenge the wheat-germ system (Endo, 2021) (Table. 1). A preliminary examination of post-translational changes revealed that glycosylation and disulfide bond formation are likely to occur, leading to the creation of a fully functioning antibody (Fung & Liu, 2018). To obtain a reliable protein manufacturing system, additional optimization processes, scale-up, and functional study of more complicated proteins are necessary (Gräslund et al., 2008).

Insect Cells-Based CFS

According to reports, several tens of micrograms of protein per milliliter response can be obtained from ICE, which is typically made from *Spodoptera frugiperda* cells. ICE offers an advantage over RRL in terms of glycosylation because core glycosylation does not required the insertion of mitochondrial membranes, which have been proven to provide the categorization and enzymes required for appropriate post-translational alterations in RRL (Bekri, 2006). It is necessary to independently purify these microsomal membranes before including them in the CFPS process. It is undesirable to add this extra processing step. As a result, ICE is becoming the CFPS platform with the quickest growth (Rolf et al., 2019). In ICE, cell culturing is costly, time-consuming, and provides subpar instruments for genetic manipulation (Table. 1).

Rabbit Reticulocytes-Based CFS

Protein yields in CFS reactions based on rabbit reticulocytes lysate (RRL) are frequently reported to be several to tens of micrograms per milligram response, which is nearly a two order of magnitude lower. RRL employs an RNA fragment from the encephalomyocarditis virus (Carlson et al., 2012). A technique for direct creation of target proteins without background synthesis of indigenous proteins was developed using rabbit reticulocyte lysate (Anastasina et al., 2014). The lysate was treated with micrococcal nuclease, an endo-exonuclease that selectively digests single-stranded nucleic acids, in order to get rid of endogenous mRNA (Adams et al., 1992). The activity of the enzyme can be reduced by EGTA since it needs Ca^{2+} ions to function. The relatively poor protein output of systems for protein combination dependent RRL is a drawback (Trachsel et al., 1978) (Table. 1). It's worth noting that a recent study discovered that adding viral translation enhancers like mRNA and proteins components like internal ribosome binding sites improved protein output by a factor of more than 10 (Au & Jan, 2014).

CHO Cells-Based CFS

The most commonly used mammalian cell line for in vivo synthesis of complex therapeutic proteins is Chinese Hamster Ovary (CHO) cells. The efficient and inexpensive production of stable and diverse proteins using his CHO lysate-based CFPS system is promising (Lakshmanan et al., 2019). The platform was developed by gently disturbing cultivated CHO cells and includes translationally active mitochondrial structures arising from the ER (Schloßhauer et al., 2022). As a result, the cell-free CHO system facilitates post-translational modifications such as glycosylation of protein targets and co-translational incorporation of protein complexes into biological membranes (Table. 1). Advanced membrane protein characterization can be done directly with inserted membrane proteins (L. Thoring et al., 2016). CHO-lysate CFPS reactions are currently carried out in the linked transcription-translation mode, enabling for rapid and effective protein synthesis as a basis for masterpieces screening applications (Khambhati et al., 2019). To solve the fundamental problem of most eukaryotes cell-free systems, which really is poor protein output due to restricted translational initiation, internal ribosomal entry sites (IRES) have already been employed to begin translation (Marques et al., 2022).

Human Cells/-Based CFS

Currently, a variety of cultivated human cell lines are accessible for both academic study and commercial protein production. Cell lines that come from virtually every type of human tissue have been modified for in vivo culture (Dumont et al., 2016). Thus, it appears possible to create a number of innovative cell-free systems that provide the best circumstances for the creation of human proteins (Table. 1). HEK293 cells and HeLa cells have been used to create cell-free systems depend on human cells in culture (Abaandou et al., 2021). The natural codon usage of the human CFPS systems, which makes it easier to synthesize high-molecular-weight human proteins, is one of their main advantages (Zemella et al., 2015). The creation of whole viruses and virus-like particles in extracts generated from cultured human cells is a potential approach. For instance, a platform was created to study the processes of viral replication before anti-viral medications were created (Dolskiy et al., 2020). Picorna viruses, the poliovirus, and the encephalomyocarditis virus have all been studied using human cell-free systems.

Table 1. Comparison of different CFS systems (Arya et al., 2008; Carlson et al., 2012; Fernández-Robledo & Vasta, 2010; Zemella et al., 2015)

Source of Cell-Free System		Advantages	Disadvantages	Applications
Group	Organism			
Prokaryote	*Escherichia coli*	• Well-established genetics • Clearly elucidated biochemical knowledge allows the transformation easily through genetic engineering • Simple and cost-effective cultivation and fast cell growth and extract preparation • Inexpensive growth media • Multiple cloning vectors, purification tags, and host strains • Low cost energy source • High protein synthesis yield • High rate of protein synthesis • Rapid growth at high densities • Able to fold complex proteins • Suitable for large-scale production	• No endogenous membrane structures for the production of integral membrane proteins • Limited post-translational modifications • Only native prokaryotic chaperones available: eukaryotic proteins might not be correctly folded • Substitutions or changes to amino acids • Heterogeneous goods • Endotoxin contamination • Simplicity in functional organelle targeting • There are no FDA-approved SOPs.	• The use of non-canonical amino acids, • Protein engineering for human medicines, modified enzymes, protein polymers, and ribosome display methods • Large-scale synthesis (reaction volume up to 100 L) • Commercialization of antibody-drug conjugates
	Archea (*Sulfolobus solfataricus, Thermococcus kodakaraensis*)	• Extreme synthesis conditions, such as high temperature, resulting in less translation hindrance by secondary structures in mRNA • Producing appropriately folded thermostable proteins • AT rich genome • GPI anchored proteins		• Synthesis of thermostable proteins
Virus	Baculovirus expression vector system	• Fundamental cell culture equipment • Minimal safety precautions • Extra-large inserts (up to 15 Kb) • Post-translational alterations	• Labor-intensive and costly • Time taking (up to 3 weeks to obtain enough viruses) • No easy agreeable for automation • Glycosylation patterns distinct from the native parasite protein	• Development of viral vaccinations and gene therapy vectors • Nine BEVS-derived products have been approved for use in humans (Cervarix®, Provenge®, Glybera®, and Flublok®) and in animals (Porcilis® Pesti, BAYOVAC CSF E2®, Circumvent® PCV, Ingelvac CircoFLEX®, and Porcilis® PCV).
Eukaryote (Protozoan)	*Tetrahymena pyriformis, Crithidia oncopelti, Paramecium aurelia Entamoeba histolytica*	• Less costly cultivation and lysate preparation • Good scalability • High solubility of manufactured proteins • Increased initiation of translation by adding of a special RNA sequence	• Types of post-translational modifications not well known • Less used or recognized • Low yield	• High-throughput analysis by PCR-based CFPS
Eukaryote (Amoeba)	*Dictyostelium discoideum*	• Easy culture at high densities • Eukaryotic organelle complexity for functional studies • Multiple cloning vectors • Easy genetic modifications • Easy mutant selection • Fast revival spores • Post-translational modification similar to higher eukaryotes	• Preferential codon usage • Still under development • Not commercially available • No FDA-approved SOPs • Glycosylation pattern may be different from human native protein • Cultivation and fermentation processes to be optimized for industrial scale production	• Expression of protein is reliable, reproducible and amendable to automation • Ideal production system for recombinant subunit vaccines

continues on following page

Table 1. Continued

Source of Cell-Free System		Advantages	Disadvantages	Applications
Group	Organism			
Eukaryote (Fungal)	Yeast	• Ability to perform post-translational modifications like glycosylation • Simple and fast cultivation of cells for lysate preparation • Well-known in vivo system, so established methods for cell engineering available • Inexpensive • Multiple cloning vectors, purification tags, and host strains • Suitable for large-scale production • Secretion into culture media • Trafficking through the secretory pathway favors disulphide bounds formation and protein folding • Easy genetic manipulation	• Low protein yield derived from codon bias, depletion of precursors and energy, proteolysis • No mammalian-like post-translational modifications • Glycosylation patterns different from those in the native parasite protein • Hyper-glycosylation and hypermanosylation • Proteolytic degradation • Unfolded proteins shooting down the secretory pathway • Truncated proteins	• Production of virus like particles for anti-viral drug research 2. Production of bioethanol and (S)-l-acetoxyalkan-2-ol in cell-free bioreactors
Eukaryote (Plants)	Wheat germ	• Highly productive cell-free system, so high yield of complex proteins • Systems available for synthesis of disulfide-bridged proteins • Correct folding of many protein types, so high solubility of proteins • Well-known system • Wide-spectrum expression of eukaryotic proteins has been achieved repeatedly • Sophisticated high-throughput method for proteomics	• Laborious and expensive lysate preparation • Limited post-translational modifications possible • No endogenous membrane structures • Low protein yield compare to prokaryotic and wheat germ systems • Poor genetic modification tools • Extract preparation is lengthy and complex	• Production of malaria proteins to characterize novel vaccine candidates • On-chip protein synthesis • High-throughput applications • Production of monoclonal antibodies against GPCRs • Investigation of translational processes (conformation analysis of ribosomes)
	Tobacco BY-2	• Fast and easy lysate preparation procedure • Glycosylation and disulfide-bridge formation are possible • Yield comparable to wheat germ extracts	• Limited evaluations of the system are available • Contain endogenous amino acids, so difficult to obtain good protein yield	• Novel cell-free system • High potential for future applications
Eukaryote (Insect)	Insect cell	• Easy cell breakage and quick preparation of extract • Eukaryotic-specific post-translational modifications (glycosylation, disulfide-bridge formation, lipidation, signal peptide cleavage phosphorylation) • Signal sequence processing • Endogenous microsomes are available • Direct synthesis and integration of membrane proteins	• Cell cultivation is expensive and time-consuming • Poor genetic modification tools • Insect cells grow slowly • Expensive media • Improper folding of protein at late stages • Complex glycosylation not possible	• GUV formation for membrane protein model to study biological processes • Engineering of proteins with integration of non-canonical amino acids • Automated production of membrane proteins

continues on following page

Table 1. Continued

Source of Cell-Free System		Advantages	Disadvantages	Applications
Group	**Organism**			
Eukaryotes (Mammalian)	Rabbit reticulocytes	• Well-established system • Mammalian system • Easy cell breakage and quick preparation of extract • Eukaryotic-specific post-translational modifications • Protein synthesis in the presences of supplemented, heterogeneous microsomes possible (e.g., canine pancreas microsomes)	• Low protein yield • Post-translational modifications only possible by supplementation with exogenous microsomes • Treatment of living animals required for lysate preparation • Complex anipulation of animal tissue required • Narrow spectrum of proteins expressed to date • High background of endogenous globin mRNAs and abundance of RNase M • Poor genetic modification tools • Low protein synthesis yields • Complex nutrient requirement	• Protein microarray technologies • Protein–molecule interaction studies • Display technologies • Screening technologies
	CHO cells extract	• Well-known and characterized cell line (often used for pharmaceutical in vivo protein production) • Contain endogenous microsomal structures • Mammalian post-translational modifications • Direct production of membrane proteins • Increase in protein yield by IRES-mediated translation initiation	• Low protein yield compared to prokaryotic cell-free systems • High cultivation cost • Robust cell line	• The system contains translocationally active endogenous microsomes, enabling the co-translational incorporation of membrane proteins into biological membranes • FDA approved SOPs
	Human cells	• Optimal environment for correct folding and assembly of human proteins • Contain endogenous microsomal structures • Human-like post-translational modifications possible • Adapted codon usage to facilitate synthesis of high-molecular-weight proteins	• Low protein yield compared to prokaryotic cell-free systems • High cultivation cost • Laborious cultivation technologies are necessary (sensitive cells)	• Investigation of viral replication mechanisms • Development of drugs

PURE System

Mathematical modelling and the progress of reconstructed cell-free systems comprised of an energy solution and enzymes blended in a specified mixture are also commercially available. Lysates are far more sophisticated than reconstituted systems. To translate difficult-to-express proteins, for example, widely used versions based on the PURE (Protein Expression Using Recombinant Elements) technology are designed. The PURE system, which was reconstituted from E. coli, is the first synthetic cell-free system, with purified cell machinery supplemented with amino acids, buffers, cofactors, energy sources (ATP, GTP, UTP TP, creatine phosphate, potassium glutamate, magnesium acetate), enzymes kinases, nucleotides, phosphatases, ribosomes, salts, translation factors, template DNA, tRNAs, aminoacy (Cui et al., 2022). In 2001, transcription and translation were demonstrated using PURE system (Y. Shimizu et al., 2001). PURE system has precise and free of proteases composition, making it considerably significant. PURE system provides high controllability (i.e. the composition of reaction mixture can be adjusted accordingly) and programmability (i.e. easy to manipulate of translation) (Cui et al., 2022; Gregorio et al., 2019; Ohashi et al., 2010). Release factor one (RF1) is a protein truncator that frequently causes premature protein termination at the stop codon, amber (UAG), in cell extracts. By employing UAG, the decrease of RF1 produced a highly effective integration of amino acids into the protein. It is possible to introduce post-translational changes at specific locations in a protein by using non-canonical amino acids (Hong et al., 2014). Although PURE systems can be developed in lab, commercial PURE systems are also available, such as Magic PURE system (Creative Biolabs), PURExpress (NEB), and PUREfrex 2.0 (GeneFrontier). The PURE system, which can easily assess the effects of pharmaceuticals on particular pathways in addition to manufacturing a variety of therapeutic compounds, is incredibly useful in the prototyping of drugs without protein purification (Cui et al., 2022). PURE systems are commonly used for the synthesis of increasingly complex proteins due to their excellent settings and low background protein interference. Because it is a pure system, the PURE system provides a simplified framework for modelling biological phenomena linked to transcription and translation dynamics. Furthermore, the PURE system has the potential to be employed as a biosensor due to its compatibility with colorimetric reporter enzymes, allowing for fast visual detection of a variety of target analytes (Sharpes et al., 2021). However, compared to *E. coli* extract-based cell-free protein synthesis reactions, PURE cell-free methods often give lower protein yields.

CONCLUSION

Difficult-to-express proteins pose special challenges to standard protein synthesis processes (cell-based expression). Cell-free protein synthesis (CFPS) has the potential to overcome fundamental problems of cell-based expression. The potential of CFPS continues to expand thanks to the development of many cellular extracts and techniques that stimulate the formation of complex proteins, including those with diverse post-translational modification needs and functions. Membranes, difficult-to-synthesize proteins (antibodies, large proteins, ice-structure proteins, and metalloproteins), post-translational modifications, high-throughput expression, generation of virus-like particles, large-scale synthesis, and other applications are examples of these. (Research on the translation machinery, genetic code expansion, metabolic engineering, and genetic circuitry.) Because of its open system, lack of reliance on living cells, lack of regeneration, and ability to focus all system energy on targeted protein synthesis. Secondly, CFPS has outstanding in vivo protein expression. Additional work is required to develop CFE, but past and current efforts have made the technology more widely available. Moreover, synthetic biology CFE systems already provide design guidelines and technical approaches developed in the D-test, as well as practical solutions to manufacturing, diagnostic, and teaching problems. A number of engineering strategies are used to improve CFS performance and increase high-throughput protein yields. CFE is a fascinating platform technology that explores complex biological problems and impacts nearly every area of our daily lives.

In conclusion, cell-free synthetic biology is a promising, highly flexible and sustainable biotechnology approach that can aid in the industrialization of sustainable treatments and materials. It is important that synthetic biology fosters interdisciplinary collaboration with designers and materials scientists during the development phase. If current trends continue, a new frontier for the development of cell-free materials and an expanding bioeconomy will emerge. They understanding the fundamental principles and historical context of cell-free systems, researchers can harness their potential to address critical challenges in biotechnology, pharmaceutical development, and synthetic biology. This chapter serves as a foundational introduction, providing a comprehensive overview of the concept, historical background, and diverse approaches employed in cell-free systems. It paves the way for further exploration and advancements in this rapidly evolving field of biotechnology research.

ACKNOWLEDGMENT

This chapter received no specific grant from any funding agency in the public, commercial, or not-for-profit sectors.

REFERENCES

Abaandou, L., Quan, D., & Shiloach, J. (2021). Affecting HEK293 Cell Growth and Production Performance by Modifying the Expression of Specific Genes. *Cells*, *10*(7), 1667. Advance online publication. doi:10.3390/cells10071667 PMID:34359846

Adams, R. L., Knowler, J. T., & Leader, D. P. (1992). Degradation and modification of nucleic acids. In *The biochemistry of the nucleic acids* (pp. 97–133). Springer. doi:10.1007/978-94-011-2290-0_4

Anastasina, M., Terenin, I., Butcher, S. J., & Kainov, D. E. (2014). A technique to increase protein yield in a rabbit reticulocyte lysate translation system. *BioTechniques*, *56*(1), 36–39. doi:10.2144/000114125 PMID:24447137

Arya, R., Bhattacharya, A., & Saini, K. S. (2008). Dictyostelium discoideum—A promising expression system for the production of eukaryotic proteins. *The FASEB Journal*, *22*(12), 4055–4066. doi:10.1096/fj.08-110544 PMID:18714070

Au, H. H., & Jan, E. (2014). Novel viral translation strategies. *Wiley Interdisciplinary Reviews. RNA*, *5*(6), 779–801. doi:10.1002/wrna.1246 PMID:25045163

Bekri, S. (2006). Importance of glycosylation in enzyme replacement therapy. *Fabry disease: Perspectives from 5 years of FOS.*

Blow, J. J., & Laskey, R. A. (1986). Initiation of DNA replication in nuclei and purified DNA by a cell-free extract of Xenopus eggs. *Cell, 47*(4), 577–587. doi:10.1016/0092-8674(86)90622-7 PMID:3779837

Brookwell, A., Oza, J. P., & Caschera, F. (2021). Biotechnology Applications of Cell-Free Expression Systems. *Life (Basel, Switzerland), 11*(12), 1367. Advance online publication. doi:10.3390/life11121367 PMID:34947898

Buchner, E. (1897). Alcoholic fermentation without yeast cells. *Cornish-Bowden*, 25-31.

Buntru, M., Vogel, S., Spiegel, H., & Schillberg, S. (2014). Tobacco BY-2 cell-free lysate: An alternative and highly-productive plant-based in vitro translation system. *BMC Biotechnology*, *14*(1), 37. doi:10.1186/1472-6750-14-37 PMID:24886601

Carlson, E. D., Gan, R., Hodgman, C. E., & Jewett, M. C. (2012). Cell-free protein synthesis: Applications come of age. *Biotechnology Advances*, *30*(5), 1185–1194. doi:10.1016/j.biotechadv.2011.09.016 PMID:22008973

Caschera, F., & Noireaux, V. (2014). Synthesis of 2.3 mg/ml of protein with an all Escherichia coli cell-free transcription-translation system. *Biochimie*, *99*, 162–168. doi:10.1016/j.biochi.2013.11.025 PMID:24326247

Chen, L., & Kashina, A. (2021). Post-translational modifications of the protein termini. *Frontiers in Cell and Developmental Biology*, 9. PMID:34395449

Chiba, C. H., Knirsch, M. C., Azzoni, A. R., Moreira, A. R., & Stephano, M. A. (2021). Cell-free protein synthesis: Advances on production process for biopharmaceuticals and immunobiological products. *BioTechniques*, *70*(2), 126–133. doi:10.2144/btn-2020-0155 PMID:33467890

Condreay, J. P., & Kost, T. A. (2007). Baculovirus expression vectors for insect and mammalian cells. *Current Drug Targets*, *8*(10), 1126–1131. doi:10.2174/138945007782151351 PMID:17979672

Cui, Y., Chen, X., Wang, Z., & Lu, Y. (2022). Cell-Free PURE System: Evolution and Achievements. *BioDesign Research*, *9847014*. Advance online publication. doi:10.34133/2022/9847014

DeVries, J. K., & Zubay, G. (1967). DNA-directed peptide synthesis. II. The synthesis of the alpha-fragment of the enzyme beta-galactosidase. *Proceedings of the National Academy of Sciences of the United States of America*, *57*(4), 1010–1012. doi:10.1073/pnas.57.4.1010 PMID:5342418

Didovyk, A., Tonooka, T., Tsimring, L., & Hasty, J. (2017). Rapid and Scalable Preparation of Bacterial Lysates for Cell-Free Gene Expression. *ACS Synthetic Biology*, *6*(12), 2198–2208. doi:10.1021/acssynbio.7b00253 PMID:28795570

Dolskiy, A. A., Grishchenko, I. V., & Yudkin, D. V. (2020). Cell Cultures for Virology: Usability, Advantages, and Prospects. *International Journal of Molecular Sciences*, *21*(21), 7978. Advance online publication. doi:10.3390/ijms21217978 PMID:33121109

Dong, O. X., & Ronald, P. C. (2019). Genetic Engineering for Disease Resistance in Plants: Recent Progress and Future Perspectives. *Plant Physiology*, *180*(1), 26–38. doi:10.1104/pp.18.01224 PMID:30867331

Dumont, J., Euwart, D., Mei, B., Estes, S., & Kshirsagar, R. (2016). Human cell lines for biopharmaceutical manufacturing: History, status, and future perspectives. *Critical Reviews in Biotechnology*, *36*(6), 1110–1122. doi:10.3109/07388551.2015.1084266 PMID:26383226

Elhardt, D., & Böck, A. (1982). An in vitro polypeptide synthesizing system from methanogenic bacteria: Sensitivity to antibiotics. *Molecular & General Genetics*, *188*(1), 128–134. doi:10.1007/BF00333006

Endo, Y. (2021). Development of a cell-free protein synthesis system for practical use. *Proceedings of the Japan Academy. Series B, Physical and Biological Sciences*, *97*(5), 261–276. doi:10.2183/pjab.97.015 PMID:33980755

Endoh, T., Kanai, T., Sato, Y. T., Liu, D. V., Yoshikawa, K., Atomi, H., & Imanaka, T. (2006). Cell-free protein synthesis at high temperatures using the lysate of a hyperthermophile. *Journal of Biotechnology*, *126*(2), 186–195. doi:10.1016/j.jbiotec.2006.04.010 PMID:16730835

Ezure, T., Suzuki, T., Shikata, M., Ito, M., & Ando, E. (2010). A cell-free protein synthesis system from insect cells. *Methods in Molecular Biology (Clifton, N.J.)*, *607*, 31–42. doi:10.1007/978-1-60327-331-2_4 PMID:20204846

Fernández-Robledo, J. A., & Vasta, G. R. (2010). Production of recombinant proteins from protozoan parasites. *Trends in Parasitology*, *26*(5), 244–254. doi:10.1016/j.pt.2010.02.004 PMID:20189877

Fogeron, M.-L., Lecoq, L., Cole, L., Harbers, M., & Böckmann, A. (2021). Easy synthesis of complex biomolecular assemblies: Wheat germ cell-free protein expression in structural biology. *Frontiers in Molecular Biosciences*, *8*, 639587. doi:10.3389/fmolb.2021.639587 PMID:33842544

Fuller, R. S., Kaguni, J. M., & Kornberg, A. (1981). Enzymatic replication of the origin of the Escherichia coli chromosome. *Proceedings of the National Academy of Sciences of the United States of America*, *78*(12), 7370–7374. doi:10.1073/pnas.78.12.7370 PMID:6278471

Fung, T. S., & Liu, D. X. (2018). Post-translational modifications of coronavirus proteins: Roles and function. *Future Virology*, *13*(6), 405–430. doi:10.2217/fvl-2018-0008 PMID:32201497

Garenne, D., Haines, M. C., Romantseva, E. F., Freemont, P., Strychalski, E. A., & Noireaux, V. (2021). Cell-free gene expression. *Nature Reviews. Methods Primers*, *1*(1), 49. doi:10.103843586-021-00046-x

Gellissen, G., & Hollenberg, C. P. (1997). Application of yeasts in gene expression studies: A comparison of Saccharomyces cerevisiae, Hansenula polymorpha and Kluyveromyces lactis- a review. *Gene*, *190*(1), 87–97. doi:10.1016/S0378-1119(97)00020-6 PMID:9185853

Gellissen, G., Kunze, G., Gaillardin, C., Cregg, J. M., Berardi, E., Veenhuis, M., & van der Klei, I. (2005). New yeast expression platforms based on methylotrophic Hansenula polymorpha and Pichia pastoris and on dimorphic Arxula adeninivorans and Yarrowia lipolytica– A comparison. *FEMS Yeast Research*, *5*(11), 1079–1096. doi:10.1016/j.femsyr.2005.06.004 PMID:16144775

Gräslund, S., Nordlund, P., Weigelt, J., Hallberg, B. M., Bray, J., Gileadi, O., ... Gunsalus, K. C. (2008). Protein production and purification. *Nature Methods*, *5*(2), 135–146. doi:10.1038/nmeth.f.202 PMID:18235434

Gregorio, N. E., Levine, M. Z., & Oza, J. P. (2019). A User's Guide to Cell-Free Protein Synthesis. *Methods and Protocols*, *2*(1), 24. Advance online publication. doi:10.3390/mps2010024 PMID:31164605

Hong, S. H., Kwon, Y. C., & Jewett, M. C. (2014). Non-standard amino acid incorporation into proteins using Escherichia coli cell-free protein synthesis. *Frontiers in Chemistry*, *2*, 34. doi:10.3389/fchem.2014.00034 PMID:24959531

Hu, Y. C. (2005). Baculovirus as a highly efficient expression vector in insect and mammalian cells. *Acta Pharmacologica Sinica*, *26*(4), 405–416. doi:10.1111/j.1745-7254.2005.00078.x PMID:15780188

Huang, A., Nguyen, P. Q., Stark, J. C., Takahashi, M. K., Donghia, N., Ferrante, T., Dy, A. J., Hsu, K. J., Dubner, R. S., Pardee, K., Jewett, M. C., & Collins, J. J. (2018). BioBits™ Explorer: A modular synthetic biology education kit. *Science Advances*, *4*(8), eaat5105. Advance online publication. doi:10.1126ciadv.aat5105 PMID:30083608

Iizuka, N., & Sarnow, P. (1997). Translation-competent extracts from Saccharomyces cerevisiae: Effects of L-A RNA, 5′ cap, and 3′ poly(A) tail on translational efficiency of mRNAs. *Methods (San Diego, Calif.)*, *11*(4), 353–360. doi:10.1006/meth.1996.0433 PMID:9126550

Kelwick, R. J. R., Webb, A. J., & Freemont, P. S. (2020). Biological Materials: The Next Frontier for Cell-Free Synthetic Biology. *Frontiers in Bioengineering and Biotechnology*, *8*, 399. doi:10.3389/fbioe.2020.00399 PMID:32478045

Khambhati, K., Bhattacharjee, G., Gohil, N., Braddick, D., Kulkarni, V., & Singh, V. (2019). Exploring the potential of cell-free protein synthesis for extending the abilities of biological systems. *Frontiers in Bioengineering and Biotechnology*, *7*, 248. doi:10.3389/fbioe.2019.00248 PMID:31681738

Kim, D. M., & Choi, C. Y. (1996). A semicontinuous prokaryotic coupled transcription/translation system using a dialysis membrane. *Biotechnology Progress*, *12*(5), 645–649. doi:10.1021/bp960052l PMID:8879155

Kim, D. M., & Swartz, J. R. (2001). Regeneration of adenosine triphosphate from glycolytic intermediates for cell-free protein synthesis. *Biotechnology and Bioengineering*, *74*(4), 309–316. doi:10.1002/bit.1121 PMID:11410855

Kim, T. W., Keum, J. W., Oh, I. S., Choi, C. Y., Park, C. G., & Kim, D. M. (2006). Simple procedures for the construction of a robust and cost-effective cell-free protein synthesis system. *Journal of Biotechnology*, *126*(4), 554–561. doi:10.1016/j.jbiotec.2006.05.014 PMID:16797767

Koo, J., Yang, J., & Park, H. (2020). Cell-free systems: Recent advances and future outlook. *Biotechnology and Bioprocess Engineering; BBE*, *25*(6), 955–961. doi:10.100712257-020-0013-x

Kovtun, O., Mureev, S., Jung, W., Kubala, M. H., Johnston, W., & Alexandrov, K. (2011). Leishmania cell-free protein expression system. *Methods (San Diego, Calif.)*, *55*(1), 58–64. doi:10.1016/j.ymeth.2011.06.006 PMID:21704167

Lakshmanan, M., Kok, Y. J., Lee, A. P., Kyriakopoulos, S., Lim, H. L., Teo, G., Poh, S. L., Tang, W. Q., Hong, J., Tan, A. H.-M., Bi, X., Ho, Y. S., Zhang, P., Ng, S. K., & Lee, D. Y. (2019). Multi-omics profiling of CHO parental hosts reveals cell line-specific variations in bioprocessing traits. *Biotechnology and Bioengineering*, *116*(9), 2117–2129. doi:10.1002/bit.27014 PMID:31066037

Laohakunakorn, N. (2020). Cell-Free Systems: A Proving Ground for Rational Biodesign. *Frontiers in Bioengineering and Biotechnology*, *8*, 788. Advance online publication. doi:10.3389/fbioe.2020.00788 PMID:32793570

Londei, P., Altamura, S., Cammarano, P., & Petrucci, L. (1986). Differential features of ribosomes and of poly(U)-programmed cell-free systems derived from sulphur-dependent archaebacterial species. *European Journal of Biochemistry*, *157*(3), 455–462. doi:10.1111/j.1432-1033.1986.tb09689.x PMID:3087750

Luk, F. C., Johnson, T. M., & Beckers, C. J. J. (2008). N-linked glycosylation of proteins in the protozoan parasite Toxoplasma gondii. *Molecular and Biochemical Parasitology*, *157*(2), 169–178. doi:10.1016/j.molbiopara.2007.10.012 PMID:18096254

Makrides, S. C. (1996). Strategies for achieving high-level expression of genes in Escherichia coli. *Microbiological Reviews*, *60*(3), 512–538. doi:10.1128/mr.60.3.512-538.1996 PMID:8840785

Marques, R., Lacerda, R., & Romão, L. (2022). Internal Ribosome Entry Site (IRES)-Mediated Translation and Its Potential for Novel mRNA-Based Therapy Development. *Biomedicines*, *10*(8), 1865. Advance online publication. doi:10.3390/biomedicines10081865 PMID:36009412

Mudeppa, D. G., Pang, C. K. T., Tsuboi, T., Endo, Y., Buckner, F. S., Varani, G., & Rathod, P. K. (2007). Cell-free production of functional Plasmodium falciparum dihydrofolate reductase-thymi-dylate synthase. *Molecular and Biochemical Parasitology*, *151*(2), 216–219. doi:10.1016/j.molbiopara.2006.10.016 PMID:17157934

Nirenberg, M. W., & Matthaei, J. H. (1961). The dependence of cell-free protein synthesis in E. coli upon naturally occurring or synthetic polyribonucleotides. *Proceedings of the National Academy of Sciences of the United States of America*, *47*(10), 1588–1602. doi:10.1073/pnas.47.10.1588 PMID:14479932

Ohashi, H., Kanamori, T., Shimizu, Y., & Ueda, T. (2010). A highly controllable reconstituted cell-free system—A breakthrough in protein synthesis research. *Current Pharmaceutical Biotechnology*, *11*(3), 267–271. doi:10.2174/138920110791111889 PMID:20210745

Pardee, K. (2018). Perspective: Solidifying the impact of cell-free synthetic biology through lyophilization. *Biochemical Engineering Journal*, *138*, 91–97. doi:10.1016/j.bej.2018.07.008 PMID:30740032

Preiss, T., & Hentze, M. W. (1998). Dual function of the messenger RNA cap structure in poly(A)-tail-promoted translation in yeast. *Nature*, *392*(6675), 516–520. doi:10.1038/33192 PMID:9548259

Renoux, J. M., & Rohmer, M. (1986). Enzymatic cyclization of all-trans pentaprenyl and hexaprenyl methyl ethers by a cell-free system from the protozoon Tetrahymena pyriformis: The biosynthesis of scalarane and polycyclohexaprenyl derivatives. *European Journal of Biochemistry*, *155*(1), 125–132. doi:10.1111/j.1432-1033.1986.tb09467.x PMID:3081345

Rolf, J., Rosenthal, K., & Lütz, S. J. C. (2019). Application of cell-free protein synthesis for faster biocatalyst development. *Catalysts*, *9*(2), 190. doi:10.3390/catal9020190

Rosenblum, G., & Cooperman, B. S. (2014). Engine out of the chassis: Cell-free protein synthesis and its uses. *FEBS Letters*, *588*(2), 261–268. doi:10.1016/j.febslet.2013.10.016 PMID:24161673

Ruggero, D., Creti, R., & Londei, P. (1993). In vitro translation of archaeal natural mRNAs at high temperature. *FEMS Microbiology Letters*, *107*(1), 89–94. doi:10.1111/j.1574-6968.1993.tb06009.x

Schloßhauer, J. L., Cavak, N., Zemella, A., Thoring, L., & Kubick, S. (2022). Cell Engineering and Cultivation of Chinese Hamster Ovary Cells for the Development of Orthogonal Eukaryotic Cell-free Translation Systems. *Frontiers in Molecular Biosciences*, *9*, 832379. doi:10.3389/fmolb.2022.832379 PMID:35586195

Sharpes, C. E., McManus, J. B., Blum, S. M., Mgboji, G. E., & Lux, M. W. (2021). Assessment of Colorimetric Reporter Enzymes in the PURE System. *ACS Synthetic Biology*, *10*(11), 3205–3208. doi:10.1021/acssynbio.1c00360 PMID:34723497

Shimizu, Y., Inoue, A., Tomari, Y., Suzuki, T., Yokogawa, T., Nishikawa, K., & Ueda, T. (2001). Cell-free translation reconstituted with purified components. *Nature Biotechnology*, *19*(8), 751–755. doi:10.1038/90802 PMID:11479568

Shimizu, Y., Kuruma, Y., Ying, B.-W., Umekage, S., & Ueda, T. (2006). Cell-free translation systems for protein engineering. *The FEBS Journal*, *273*(18), 4133–4140. doi:10.1111/j.1742-4658.2006.05431.x PMID:16930131

Smolskaya, S., Logashina, Y. A., & Andreev, Y. A. (2020). Escherichia coli Extract-Based Cell-Free Expression System as an Alternative for Difficult-to-Obtain Protein Biosynthesis. *International Journal of Molecular Sciences*, *21*(3), 928. Advance online publication. doi:10.3390/ijms21030928 PMID:32023820

Stark, J. C., Huang, A., Nguyen, P. Q., Dubner, R. S., Hsu, K. J., Ferrante, T. C., Anderson, M., Kanapskyte, A., Mucha, Q., Packett, J. S., Patel, P., Patel, R., Qaq, D., Zondor, T., Burke, J., Martinez, T., Miller-Berry, A., Puppala, A., Reichert, K., ... Jewett, M. C. (2018). BioBits™ Bright: A fluorescent synthetic biology education kit. *Science Advances*, *4*(8), eaat5107. Advance online publication. doi:10.1126ciadv.aat5107 PMID:30083609

Thoring, L., Dondapati, S. K., Stech, M., Wüstenhagen, D. A., & Kubick, S. (2017). High-yield production of "difficult-to-express" proteins in a continuous exchange cell-free system based on CHO cell lysates. *Scientific Reports*, *7*(1), 11710. doi:10.103841598-017-12188-8 PMID:28916746

Thoring, L., Wüstenhagen, D. A., Borowiak, M., Stech, M., Sonnabend, A., & Kubick, S. (2016). Cell-Free Systems Based on CHO Cell Lysates: Optimization Strategies, Synthesis of "Difficult-to-Express" Proteins and Future Perspectives. *PLoS One*, *11*(9), e0163670. doi:10.1371/journal.pone.0163670 PMID:27684475

Trachsel, H., Ranu, R. S., & London, I. M. (1978). Regulation of protein synthesis in rabbit reticulocyte lysates: Purification and characterization of heme-reversible translational inhibitor. *Proceedings of the National Academy of Sciences of the United States of America*, *75*(8), 3654–3658. doi:10.1073/pnas.75.8.3654 PMID:278981

Xu, Y., Lee, J., Tran, C., Heibeck, T. H., Wang, W. D., Yang, J., Stafford, R. L., Steiner, A. R., Sato, A. K., Hallam, T. J., & Yin, G. (2015). Production of bispecific antibodies in "knobs-into-holes" using a cell-free expression system. *mAbs*, *7*(1), 231–242. doi:10.4161/19420862.2015.989013 PMID:25427258

Yano, D., Funadani, R., Uda, K., Matsuoka, T., & Suzuki, T. (2020). Role of arginine kinase in Paramecium tetraurelia (Ciliophora, Peniculida): Subcellular localization of AK3 and phosphoarginine shuttle system in cilia. *European Journal of Protistology*, *74*, 125705. doi:10.1016/j.ejop.2020.125705 PMID:32464434

Zamecnik, P. C., & Frantz, I. D. Jr. (1948). Incorporation in vitro of radioactive carbon from carboxyl-labeled dl-alanine and glycine into proteins of normal and malignant rat livers. *The Journal of Biological Chemistry*, *175*(1), 299–314. doi:10.1016/S0021-9258(18)57260-4 PMID:18873305

Zemella, A., Thoring, L., Hoffmeister, C., & Kubick, S. (2015). Cell-Free Protein Synthesis: Pros and Cons of Prokaryotic and Eukaryotic Systems. *ChemBioChem*, *16*(17), 2420–2431. doi:10.1002/cbic.201500340 PMID:26478227

Zubay, G. (1973). In vitro synthesis of protein in microbial systems. *Annual Review of Genetics*, *7*(1), 267–287. doi:10.1146/annurev.ge.07.120173.001411 PMID:4593305

KEY TERMS AND DEFINITIONS

Artificial Cell: An engineered particle that mimics one or many functions of a biological cell. Artificial cells are biological or polymeric membranes, enclosing biologically active materials from cell-free system. As such, nanoparticles, liposomes, polymersomes, microcapsules and a number of other particles have qualified as artificial cells.

Cell Extract: A lysed the cell of interest having cell walls, DNA genome, and other debris centrifuged out, with the remains including ribosomes, aminoacyl-tRNA synthetases, translation initiation and elongation factors, nucleases, etc.

Cell-Free Protein Synthesis: CFPS, or in vitro protein synthesis, is the production of proteins using biological machinery in a cell-free system, i.e., without the use of living cells.

Freeze-Dried Cell-Free Systems: A platform that retains not only the basic protein synthesis ability of living cells but also abiotic, aseptic and portable characteristics.

Post-Translation Modifications: The changes that are made to proteins after synthesis, typically mediated by enzymes.

Protein Yield: Protein secreted by a cell or a number of cells.

PURE System: The protein synthesis using recombinant elements (PURE) system reconstitutes the *Escherichia coli* translational machinery with fully recombinant proteins, with the exception of ribosomes and tRNAs.

Synthetic Biology: A multidisciplinary area of research that involves the application of engineering principles to biology.

Chapter 2
Cell-Free Synthetic Biology:
The Novel Approach Towards the Biotechnological Applications

Hammad Qaiser
International Islamic University, Islamabad, Pakistan

Mohammad Uzair
https://orcid.org/0000-0003-2039-0870
International Islamic University, Islamabad, Pakistan

Abida Arshad
Pir Mehr Ali Shah Arid Agriculture University, Rawalpindi, Pakistan

Jabar Zaman Khan Khattak
International Islamic University, Islamabad, Pakistan

Shahid Bashir
King Fahad Specialist Hospital, Dammam, Saudi Arabia

ABSTRACT

Innovations in sustainable biomanufacturing are made possible by cell-free synthetic biology advances. Cell-free workflows frequently combine purified enzymes with cell extracts for cell-free protein synthesis processes to create and analyze biosynthetic pathways and synthesize therapeutics. The development of cell-free platforms, high-yielding, cell-free gene expression systems, and multiplexed strategies for quickly evaluating biological design have all been made possible by recent technological advances in cell-free platforms, such as microfluidics, machine learning, CRISPR-mediated genome editing, etc. Developments in cell-free platforms satisfy many of the requirements of preparatory biochemical research, clinical purposes, diagnostic devices, industrial application, biosensors, etc. that offer exciting opportunities to transform synthetic biology and biotechnology. Therefore, this chapter aims to elaborate on the rapidly advancing field of cell-free platforms.

DOI: 10.4018/978-1-6684-6577-6.ch002

INTRODUCTION

Cell-free systems (CFS) have emerged as powerful tools in biotechnology, enabling the study and manipulation of biological processes outside the traditional boundaries of intact cells. By harnessing the essential components and machinery of cellular systems, cell-free systems offer unique advantages, including increased control, rapid prototyping, and simplified manipulation of biological processes (Figure 1). CFS offer several advantages for synthetic biology and biotechnology applications, including synthesis and production of vaccines, antibiotics, antibodies, proteins, enzymes, and other biomaterials. Therefore, this chapter focuses on the cell-free formats for protein synthesis and recent engineering and technological intervention advancing cell-free platforms, including CRISPR, microfluidics, and artificial cell. This chapter also highlights the necessary supplements for the cell-free synthesis of valuable proteins for their biomedical application.

Figure 1. Advantages of cell-free system over cell-based platforms

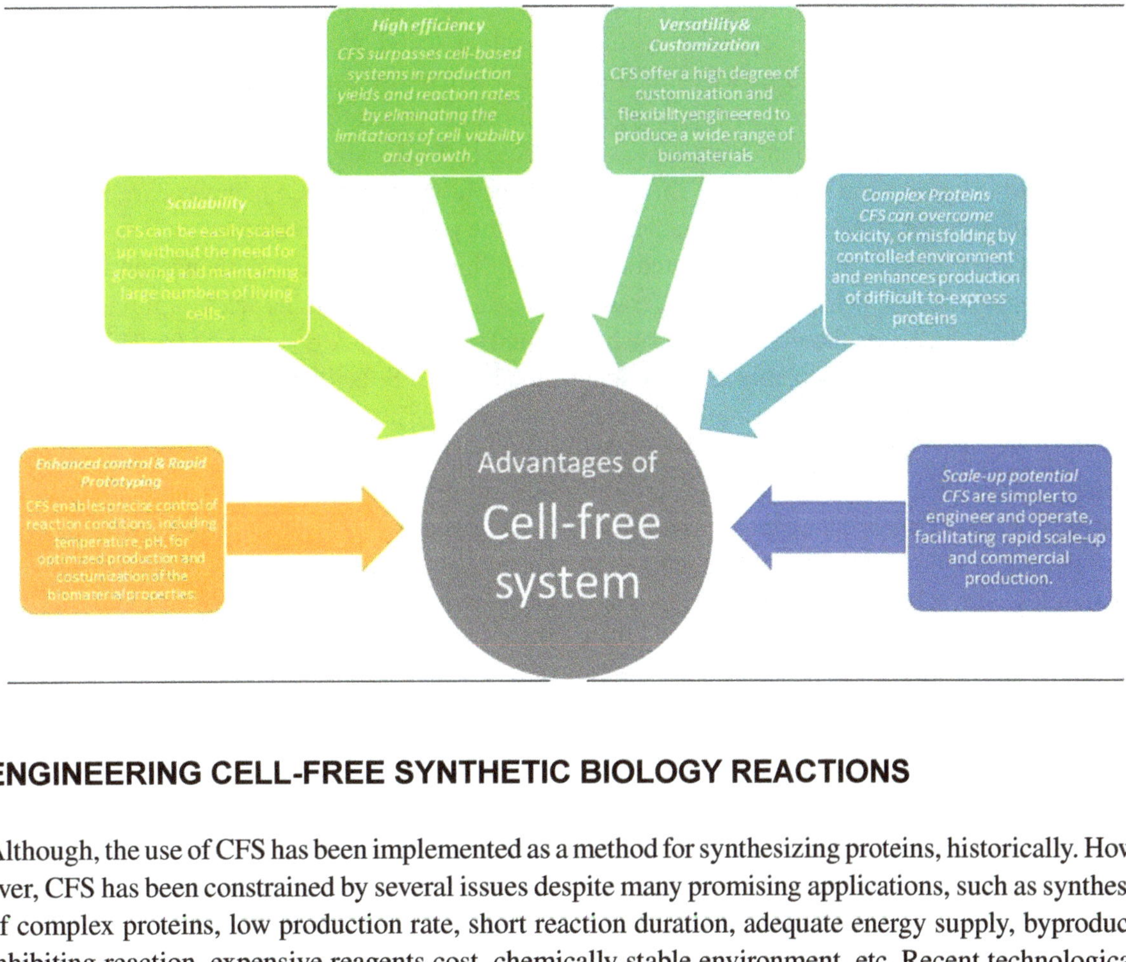

ENGINEERING CELL-FREE SYNTHETIC BIOLOGY REACTIONS

Although, the use of CFS has been implemented as a method for synthesizing proteins, historically. However, CFS has been constrained by several issues despite many promising applications, such as synthesis of complex proteins, low production rate, short reaction duration, adequate energy supply, byproduct-inhibiting reaction, expensive reagents cost, chemically stable environment, etc. Recent technological, technical, or engineering advances have addressed such limitations of CFS approaches, resulting in reduced cost and high production rates at the manufacturing scale. Changes in extract preparation procedures in recent years have resulted in more durable extracts, the ability to activate central metabolism for cell-free protein synthesis (CFPS) energy supply, and the ability to improve production scale. Cell extract is the most critical component of the CFPS reaction because CFPS systems use a group of catalytic proteins derived from cell lysates. Cell extract composition is affected by the growing medium, lysis procedure, and processing conditions (Carlson et al., 2012). For CFPS, proteins in the crude lysate and the composition of cell extract significantly affect protein synthesis and productivity. The cell extract depends on the experimental procedure, such as growth media, lysis method, and homeostatic conditions. Therefore, modifying the composition of cell extract through emerging engineering and technological approaches.

Energy Regeneration in Cell-Free System

The cellular and molecular factors utilized during the reaction in CFS and their depletion lead to the early cessation of protein synthesis. For instance, phosphatases in the cell extract degrade PEP, the most widely used energy source for ATP regeneration, into pyruvate and inorganic phosphates. The accumulation of phosphates inhibits protein synthesis (D.-M. Kim & Swartz, 2000). However, acetyl phosphate was also utilized in CFPS, which facilitates ATP regeneration. Acetyl phosphate degradation also produces inorganic phosphate; however, recycled to synthesize another acetyl phosphate molecule utilizing oxygen. The utilization of oxygen forms a by-product, H_2O_2, which is also degraded by endogenous catalase activity. However, dissolved oxygen depletion would limit the generation of acetyl phosphate (D. M. Kim & Swartz, 1999). In addition, 3-phosphoglyceric acid (3-PGA) supplemented CFS achieve higher protein yield (Shin & Noireaux, 2010). Therefore, various in vitro synthetic pathways are involved in energy production for protein synthesis, rather than only the phosphorylation pathway.

During CFPS, certain amino acids depleted significantly, including arginine, cysteine, and tryptophan. The addition of chloramphenicol acetyltransferase, magnesium, PEP, and amino acids (arginine, cysteine, and tryptophan) increases the yield of the batch reaction (D.-M. Kim & Swartz, 2000). The unnatural amino acids and mischarge suppressor tRNA supplemented to CFS facilitate codon-anticodon interactions and rewriting of the genetic circuit (Shimizu et al., 2006). The bacteriophage RNA polymerases, when introduced in CFS, also improves the transcription-translation system, increasing protein yield. The plasmid and ions concentration (K^+, Mg^{2+}, etc.) also significantly affect protein production in the CFPS approach. Instead of nucleoside triphosphates (NTPs) as nucleotide sources, the substitution of nucleoside monophosphates (NMPs) enables the stability of the nucleotide pool during CFPS reaction and increases the proteins yield (Calhoun & Swartz, 2005). The supplementation of molecular chaperones, the crucial proteins for the correct folding of proteins, in the CFS facilitates the increment of the protein synthesis rate. In the reconstituted CFS, incorporating redox-dependent chaperone protein disulfide isomerase and disulfide oxidoreductase ensures the correct disulfide bonding in complex proteins under oxidative conditions (Shimizu et al., 2006). Therefore, cytoplasmic mimicking components, strain engineering, and energy substrate. The progress in technical renaissance enables monitoring the kinetic progress of processes by evaluating expression levels of mRNA and proteins, as well as real-time metabolomics analysis (Laohakunakorn, 2020; M. Vilkhovoy et al., 2019). Technological advances in CFSs highlight their benefit for the biotechnological applications of desired protein synthesis, particularly those that are complex or pose difficulties for cell-based systems.

Encapsulated Cell-Free System: An Artificial Cell Approach

The artificial cells based on cell-free systems may result in the expansion of innovative CFS, serving as a system to study biological procedures. To make a cell-like bioreactor, for example, an E. coli cell-free expression system is contained in a phospholipid vesicle (Noireaux & Libchaber, 2004). There is a lot of effort to enclose cell-free reactions in different compartments, such as liposomes, polymersomes, and droplets, to add crowding and organization into CFS and replicate more realistic reaction circumstances (Laohakunakorn, 2020). There are currently a variety of approaches that might be used to create compartmentalized CFS, such as vesicles, microfluidic devices, and water-in-oil emulsions. Molecular machinery can also be encapsulated in phospholipid vesicles through encapsulation or extrusion. Each of the alternatives presented has both advantages and disadvantages. Despite their high encapsulation efficiency, water-in-oil emulsions cannot adequately imitate the permeability barrier of live cells (Spencer et al., 2013). An artificial cell's membrane border, which encapsulates CFS, serves as a selective barrier and a matrix for membrane proteins to act as anchoring sites and support functions for internal processes (Cho & Lu, 2020). Phospholipid vesicles are considerably utilized for encapsulating CFPS reactions due to their similarity with natural membranes. In addition, cell-free phospholipid production has been established by CFS-encapsulated genes supplemented with acyl-coenzyme A (acyl-CoA) as a fatty acid (FA) source (Blanken et al., 2020). Therefore, if the enzymes needed for lipid manufacturing are formed inside the cell, artificial cells can synthesize their phospholipids (Peruzzi et al., 2022; Scott et al., 2016). Recently, Eto et al. (2022) constructed a lipid-synthesizing artificial cell for in vitro fatty acid synthesis by reconstituting FA-biding enzymes, acyl carrier protein, and thioesters.

Moreover, encapsulating CFS can introduce the cell membrane properties, such as molecular containment and selective permeability, improving sensing capabilities (Boyd et al., 2022). Cell-mimetic membranes are potential tools or investigating lipid-protein interactions and developing membrane-based products. However, due to a reduction in internal volume brought on by a modification in osmotic pressure, the form of the membrane vesicle is altered regardless of internal metabolic processes (Eto et al., 2022). Another fascinating compartmental border material that is replacing or substituting lipids is polymers. They are the second most common cellular mimic after liposomes because of their substantial structural and formation similarities to liposomes. Polymersomes may have thicker membranes than liposomes, which might be helpful in applications needing kinetic or thermodynamic stability (Cho & Lu, 2020; Schmid-Dannert & López-Gallego, 2019). When creating a multicompartmental structure, hydrogel-based ACs are preferable to liposomes or polymersomes. Hydrogels are a relatively new approach for compartmentalizing CFS that can provide a considerably more distinct spatial compartmental microenvironment due to their ability to immobilize (Chen et al., 2020; Cho & Lu, 2020).

Machine Learning Approach in Cell-Free System

Machine learning and automation could be integrated with CFPS for metabolic engineering, prototyping, and biomanufacturing, resulting in enhanced effectiveness with optimal configuration. CFS may utilize microfluidic volumes to their maximum potential, and standard operating procedures could work automatically (Brookwell et al., 2021). Additionally, computer modeling may be crucial for a CFS inside a compartment to efficiently balance substrate flux, translation, and folding parameters and speed up process optimization. More accurate descriptions and predictions of individual artificial cell behavior will be made possible by increasingly advanced modeling, resulting in the construction of more complicated model artificial cell systems (Ajjolli Nagaraja et al., 2020). Despite the well-known interpretability issues, parametric or non-parametric models constructed from data using machine learning can be similarly or even more precisely predicted (Doshi-Velez & Kim, 2017). Another advantage of cell-free platforms is that machine-learning algorithms may considerably aid in refining the system for more effective prototyping and high-throughput experimentation. Even now, in CFS, a liquid-handled robotic workstation paired with high-throughput trials has been demonstrated to boost protein synthesis yields by integrating automation with deep learning optimization (Caschera et al., 2011). A similar optimization method might improve genetic circuit and metabolic pathway prototypes (Jiang et al., 2018). Likewise, iSAT (*in vitro* integrated synthesis, assembly, and translation), a more complicated CFPS method for ribosome construction, was evaluated (Caschera et al., 2018). An evolutionary learning technique that combines stochastic exploration with data structure by constructing a prediction model can be sampled as virtual experimentation to improve CFPS (Caschera et al., 2011). An ensemble artificial neural network model was repeatedly trained using an evolutionary design of experiments approach to optimize the circumstances for CFPS (Caschera et al., 2018), improving the yield of ~4000 CFPS reactions by 34-folds (Borkowski et al., 2020). The development of a metabolic pathway utilizing CFPS was also optimized using a data-driven strategy based on an artificial neural network, increasing a 20-fold hemoglobin synthesis in vivo (Karim et al., 2020). Recently, Galaxy-SynBioCAD portal designs and engineers end-to-end metabolic pathways in host organisms using different computational approaches for automated DNA assembly and strain transformation (Hérisson et al., 2022). The machine-learning algorithm improved the CFPS reaction designs after it understood the outcomes of the CFPS reaction. Although the machine learning technique successfully increases protein yield in CFPS, further research might improve genetic circuits and metabolic pathways (Jiang et al., 2018). Furthermore, the standardization and automation of workflows in CFS is ideally adapted to enable machine learning-based experimentation for efficient system design. Machine learning and automation, especially in CFS; demonstrate tremendous promise for speeding experimental and industrial procedures.

Microfluidics and Cell-Free System

Cells can be enclosed within the created droplets in synthetic biology, or the droplet itself can be used as a cell architecture. As a result, microfluidic devices are frequently employed in biological and chemical studies (Damiati et al., 2018). Protein synthesis and characterization are advancing in parallel with synthetic biology through a combination of microfluidic technology and CFPS methods. The intricate channel design of microfluidic devices promotes mixing, provides a steady supply of nutrients, such as nucleotides, amino acids, and ATP, and separates by-products and essential proteins (Damiati et al., 2018; Khnouf et al., 2011). A microfluidic device with 96 units each expressing a protein enables CFPS through sequential exchange, increasing protein expression. For example, β-galactosidase expression increased 87-fold (5.5 g/L) (Khnouf et al., 2010). A microfluidic chip was created by assembling a series of components. Each component contains an internal cavity and an external cavity. The cell-free medium, used in the inner wells, supplied ribosomes and other molecular components required for protein synthesis. The outer well, which was concentric with the inner well, served as a nutrient reservoir. A nanoporous membrane was placed between the inner and outer wells to preserve the produced proteins and remove reaction by-products. Microfluidic tubing connects these two wells to provide fresh nutrients for longer reaction times and higher expression yields (Mei et al., 2010). CFPS with microfluidics has many advantages, including low reagent cost (using 125-800 times less reagent than commercially available protein expression systems), simultaneous production of the same protein under various situations to enhance the interpretation process, and simultaneous creation of proteins for protein-rich expression with high yield (Khnouf et al., 2009).

CRISPR-Mediated Genome-Editing Approach for Cell-Free Protein Synthesis

Another technological implication in CFS is Clustered Regularly-Interspaced Short Palindromic Repeats (CRISPR). CRISPR technology can program CRISPR-associated enzymes (or Cas) to cut or cleave a specific sequence of DNA. CRISPR is used for genome editing due to its ability to cut and cleave specific DNA sequences, making CRISPR significant for disease diagnostics. The cell-free biosensors based on toehold switch and CRISPR-Cas can detect viral RNA or genomic DNA of particular diseases. For instance, when CRISPR-based cleavage occurs, some quenched reporter molecule causes a color change or fluorescence that would report the presence of pathogenic DNA in that sample (Kaminski et al., 2021). This can be applied to test patients for cancer screening, bacterial infection, antibiotic resistance, and viral infection. The CRISPR-based cell-free technology has been applied to make a rapid diagnostic test for COVID-19 (Ebrahimi et al., 2022; Soh et al., 2022). A CRISPR test can detect the presence of genes from the SARS-CoV-2 virus in a patient sample. This test occurs in 30 minutes in a test strip and has a very easy readout result. The bands become if the virus is present and no band demonstrates the absence of the virus (Javalkote et al., 2022). In just 3 months after the SARS-CoV-2 virus was sequenced, CRISPR-based cell-free rapid diagnostic was developed, which is remarkable and detects the virus efficiently. Moreover, CRISPR/Cas9-based module enables the distinction of Zika genotypes with the single-base resolution, increasing the diagnostic capabilities (W.-Q. Liu et al., 2019b). In CRISPR biosensor systems incorporating Cas effectors such as Cas9, Cas12a, and Cas13a, several shear cleavage principles have been identified. When the CRISPR-Cas13a system identifies a particle capable of instantaneously cleaving its target, lateral disruption of the Cas13-crRNA complex is activated, cleaving adjacent non-target RNAs. The quenched fluorescent RNA is cleaved in response to produce a fluorescent signal as a reporter (Gootenberg et al., 2017; Myhrvold et al., 2018; Zhang et al., 2020).

BIOTECHNOLOGICAL APPLICATIONS OF CELL-FREE SYSTEM

Biosynthetic Pathways

The interplay between synthetic biology and CFS has several possibilities. As a result, the metabolic pathways required for synthesizing essential chemicals (drugs, tastes, odors, etc.) have been transferred from poorly suited organisms to heterologous hosts, such as yeast (G.-M. Lin et al., 2019). This CFS-based method has proven highly effective, combining genes from many sources to form specific signaling pathways. Enzyme-based catalysis enables stereoselectivity in aqueous, low-energy processes such as green chemistry (Wu et al., 2021). It is possible to synthesize a wide range of chemicals from simple precursors using naturally occurring methods (Daletos & Stephanopoulos, 2020).

During microbial bioproduction, normal cell metabolism and growth deplete a large portion of the input, making the overall process less efficient. Cell-free synthetic biology addresses such issues, for instance, CFS has shown efficacy in enzyme discovery that can be used for biosynthetic purposes (Tinafar et al., 2019). Enzymes and their homologs can be quickly tested for efficiency without the time-consuming steps (such as plasmid assembly and transformation) required in cell-based assays. This technique can be used for simple route prototyping and automatic multiple shuffling of complex route components (Currin et al., 2021). Cell-free prototyping allows the use of enzymes determined as linear structures (DNA or RNA). It is also possible to assess substrate selection without purifying the enzyme (Garenne et al., 2021).

Enzymes and signaling pathways found in CFS are often reintroduced into cells for research purposes. However, there is an increasing need to use CFS directly as a manufacturing medium. Profitable efforts (such as Sutro and Greenlight) have already shown that CFS can improve protein and RNA production economics (White et al., 2010). As a result, similar approaches can deliver a reliable source of high-quality small molecules, allowing biological manufacturing without the requirement for toxicity controls, complicated separation methods, or interaction with host strains (Atanasov et al., 2021). These systems have gained popularity because they enable rapid prototyping (Ali et al., 2016).

Biomaterial Production

CFS allows the production of complex biomolecules, such as proteins, peptides, and nucleic acids, without the need for intact living cells. This enables precise control over the synthesis process, resulting in enhanced yields and reduced production times. Additionally, CFS can be easily manipulated and optimized to produce biomaterials with desired properties, such as specific functionalities or tailored structures (S. Liu et al., 2023). CFS offer several advantages over traditional cell-based approaches, including improved control, scalability, and efficiency in biomaterial production.

Another key application of CFS in biomaterial production is the assembly of functional biomaterials (Michael Vilkhovoy et al., 2020). These systems provide a controlled environment where biomolecular components can be assembled into higher-order structures, such as nanoparticles, nanofibers, or hydrogels (Z. Li et al., 2022). CFS offer advantages in terms of scalability, reproducibility, and ease of customization, making them ideal for generating biomaterials with complex architectures and tailored properties. Furthermore, CFS have been utilized in the synthesis of hybrid biomaterials by combining biological components with synthetic materials. For instance, CFS can be engineered to incorporate non-natural amino acids or other non-biological building blocks into the synthesized biomaterials, expanding the range of available materials and functionalities (Shimizu et al., 2006). This versatility allows for the development of innovative biomaterials with unique properties, such as improved mechanical strength, enhanced biocompatibility, or advanced drug delivery capabilities.

Biosensors Applications

Molecular detection underpins almost every living activity involving the pairing of nucleic acids and bases, giving the main tenets a unique syntax. Scientists have long sought to integrate these activities into cell-free in vitro environments to better understand and apply the molecular mechanisms underlying diagnostics and chemical detection (B. Nguyen et al., 2009). One of the fruits of this effort is the polymerase chain reaction (PCR). PCR is a critical technique in most molecular genetics research laboratories, including medical testing. Decentralized and mobile diagnostics that can be used quickly in the field, such as during transferrable disease epidemics or in agriculture, are becoming increasingly popular (H. Y. Liu et al., 2019a). Sensing technologies, on the other hand, have traditionally been restricted to large-city labs due to the specific equipment and knowledge required (Buchan & Ledeboer, 2014). The cell-free biosensor workflow can be divided into two parts: Generation of a readable signal by expression of a reporter gene and detection and response of analytes using specific discrimination mechanisms. First, once analytes are delivered, specific chemical structures and linkages govern the analyte identification process in cell-free systems. Second, the downstream reporter gene is triggered and translated into a reporter protein in a cell-free environment. (Zhang et al., 2020).

Due to their biological safety and stability, freeze-dried cell-free (FD-CF) systems offer a unique molecular environment that meets the unmet demand for ubiquitous and low-cost sensing. The transcriptional and translational properties of CFS make it a genetic circuit-based sensor with high sensitivity and specificity for detecting minute substances and nucleic acids (Mehrotra, 2016). Many biosensors and circuits developed for cell-based applications are equally applicable to non-cellular contexts. These include ring oscillators, transcription cascades, logic gates, negative feedback loops, and various conventional switches (such as TetO- and Lacl-based systems) (Naresh & Lee, 2021). Many other diagnostic technologies require temperature-controlled environments and cold-chain logistics. However, FD-CF units can be shipped and serviced at room temperature (Tinafar et al., 2019). FD-CF reaction on an adsorbent showed that small reactions (1–2 L) can be easily carried out in this medium (e.g. paper) (Hua, 2019). Paper-based cell-free technology can be used outside the laboratory in a controlled, biosafe environment. Paper-based FD-CF protein synthesis and development of a toe-switch regulator to generate the bioluminescent reporter NanoLuc can provide a visible bioluminescent signal in response to SARS-CoV-2 RNA in saliva samples (Hunt et al., 2022).

Moreover, FD-CF supplemented with a toehold switch acts as a kind of riboregulator (Green et al., 2014). Toe switches have been used in paper-based FD-CF methods for the early identification of antibiotic-resistance genes and strain-specific recognition of the Ebola virus. These switches can find virtually any sequence of interest (Pardee, 2018). However, the device required the sensitivity to identify RNA levels in patient samples (Vidic et al., 2017). This sensitivity limit can be overcome by performing a thermal amplification step before cell-free reactions (Woo et al., 2020). A one-pot isothermal ligation and transcription assay for fluorescence-based diagnostics detect RNA from pathogens and viruses (including SARS-CoV-2) with high specificity and sensitivity (Woo et al., 2020). When the Zika virus outbreak was discovered in Brazil in early 2016, improved embodiments allowed the FD-CF toe sensor to detect all Zika virus strains worldwide in viral plasma at clinically relevant levels (Garcia Serpa Osorio-de-Castro et al., 2017). In addition, the first CRISPR-based in vitro diagnostic approach made it possible to distinguish between viral genotypes with single-base-pair accuracy (e.g., American Zika and African Zika strains)(Abudayyeh et al., 2020). Genetic circuits encoding biosensors and reporter proteins are often designed in silico before they are manufactured chemically. Target analytes are extracted and amplified in certain situations while patient or environmental samples are being collected. CFS is complemented with target analytes and gene circuits. Diagnostics can be used on the front lines of epidemics. Since CFS can be lyophilized onto test tubes or paper for diagnostic purposes, it seems likely that these reactions also permeate fabrics (P. Q. Nguyen et al., 2021). For example, lab coats and facemasks may incorporate acellular diagnostic sensors. If healthcare workers or other professionals are exposed to the virus, the diagnostic device will change color to instantly identify who has been exposed to the infection and protect those on the front lines (Lu, 2022). Although this wearable technology is still a work in progress, self-free systems are an active and interesting research area that will sooner or later become a reality.

CFS can be used as a biosensor to identify contamination/quality of environmental samples such as drinking water. A CFS-based biosensor, known as RNA exit sensor activated by ligand induction

(ROSALIND), uses an RNA sequence that acts as a switch (Jung et al., 2020). Without contamination, the switch remains off, and there is no reaction. However, when contaminants (such as lead) are present in water, RNA molecules wrap around the contaminant, resulting in a reporter enzyme that catalyzes a chemical reaction that produces a yellow molecule, reporting the presence of the contaminant (Jung et al., 2020). A variety of biosensors can detect various chemical contaminants or additives in water, antibiotics, metals such as lead, and fluoride in toothpaste (Soni et al., 2022).

Modification of Proteins and Codon Tables

Many protein-based medicines rely on particular control of natural or artificial alterations in the peptide sequence. One of the most attractive applications for such alterations is the production of antibody-drug conjugates (ADCs), which are rapidly achieving popularity as a new class of cancer medication (Peters & Brown, 2015). Conventional conjugation methods produce diverse labeled antibodies because they rely on random conjugation to a large number of amino acid side chains. Current studies show that site-specific conjugation can enhance the pharmacological effects of ADCs (Axup et al., 2012). Unnatural amino acids can be used for site-specific conjugation. Co-translational incorporation of over 100 different unnatural amino acids has been verified in vivo, allowing a large number of alterations (Leisle et al., 2015). Many of these modifications, such as orientation-directed immobilization and site-specific functionalization (phosphorylation, drug conjugation, PEGylation, etc.), have proven useful in a variety of cell-free applications (Shadish & DeForest, 2020).

CFS methods offer better control and flexibility in protein modification by avoiding harmfulness and penetrability limitations associated with cell-based techniques. Stop codons are often reused to incorporate synthetic amino acids into cell-based procedures, reducing the negative impact of recoding on cell survival (Jenkins, 2007). The complete codon table could potentially be reprogrammed in a cell-free environment, allowing the construction of entirely new codon tables and the incorporation of synthetic amino acids (Nowak et al., 2021).

The latter, if performed consistently, can help protect intellectual property. DNA sequences become obscured and ineffective outside of a cell-free environment. This complex code makes it difficult to replicate the patented design (Cook-Deegan & Heaney, 2010). Because of codon obfuscation, identifying DNA sequences that may be utilized maliciously could become extremely difficult. Companies that synthesize DNA, for example, would find it far more difficult to check out DNA sequences that may be exploited for illegal reasons (e.g., bioterrorism) (Van Aken & Hammond, 2003). Recent studies have shown that adding unnatural base pairs to the four-letter genetic alphabet can also expand the codon table. As a result, CFS-made proteins may contain an infinite number of synthetic amino acids (Jefferson et al., 2014).

CFS can also be used to naturally modify proteins. An example is the fixing of sugars (ie, glycans) commonly known as glycosylation. Inappropriate glycosylation can decrease the potency and circulating half-life of many beneficial proteins, so efficient glycosylation is generally essential for the successful manufacture of many pharmaceuticals. (Freeze & Elbein, 2009). Some CFS have a natural potential for glycosylation. However, they are usually restricted to glycan structures that are naturally formed by the cell type supplying the lysate (H. Li & d'Anjou, 2009). Furthermore, glycosylation typically demands the recapitulation of the parent cell's protein trafficking pathways in these systems. As a result, there has lately been increased attention on artificial glycosylation routes in CFS (Reynders et al., 2011).

Manufacturing of Therapeutics

An important area of CFS research is the biological manufacture of pharmaceuticals and other protein-based reagents. Remarkable possibilities for synthesizing a wide diversity of molecules, from metabolites to biopolymers, have emerged in natural biological systems (Hodgman & Jewett, 2012). These processes can be combined in a finely tuned approach to allow cell-free protein expression systems to generate molecules of interest. Several potential agents have been developed through cell-free expression, including proteins, enzymes, antibodies, vaccines, nanomaterials, RNA, and RNA-based therapeutics (Kai et al., 2013). CFS play a vital role in the manufacturing of therapeutics by enabling efficient production, rapid prototyping, customization, and scalability. These systems offer advantages in terms of speed, flexibility, and access to complex protein products, making them a valuable tool in the development and production of various therapeutic molecules. By using cell extracts that contain the necessary cellular machinery, such as ribosomes, tRNAs, and amino acids, protein synthesis can be carried out in a controlled and scalable manner. This approach allows for the production of complex proteins, including antibodies, enzymes, and growth factors. Furthermore, through directed evolution or rational design approaches, enzymes can be modified to enhance their activity, stability, or substrate specificity. CFS provide a rapid and high-throughput platform for screening and characterizing these engineered enzymes. CFS are employed in metabolic engineering approaches to optimize and produce small molecules with therapeutic applications. By reconstituting specific metabolic pathways in a cell-free environment, researchers can manipulate and enhance the production of valuable compounds, such as antibiotics, anticancer drugs, or biofuels. In addition, by using cell extracts that contain specific cellular components or pathways of interest, the efficacy and toxicity of potential drug candidates can be evaluated in a simplified and controlled environment. CFS are employed in the production of therapeutic RNA molecules, such as mRNA and small interfering RNA (siRNA). By using cell extracts, synthetic RNA templates can be translated into functional RNA molecules, which can then be utilized for gene therapy, RNA interference (RNAi), or vaccine development.

Vaccines

Cell-free biomanufacturing is particularly well-suitable for vaccine manufacture as it can be rapidly scaled up in reply to public health crises. Many recombinant vaccines, including those against botulism, diphtheria, and anthrax, have been effectively expressed in CFS (Sheridan, 2005). CFS have gained significant attention in the production of vaccines due to their ability to rapidly synthesize viral proteins or nucleic acids, allowing for a more streamlined and scalable vaccine manufacturing process. CFS play a crucial role in the production of mRNA-based vaccines, such as the COVID-19 vaccines developed by Pfizer-BioNTech and Moderna. In this approach, a synthetic mRNA encoding the viral antigen of interest is synthesized using cell-free systems. The mRNA is then encapsulated in lipid nanoparticles and administered as the vaccine. CFS allow for efficient and large-scale production of mRNA, enabling the rapid response to emerging infectious diseases. CFS are utilized in the production of protein subunit vaccines, which involve synthesizing and purifying specific viral proteins that serve as antigens. For example, the hepatitis B vaccine is produced using recombinant DNA technology. The gene encoding the viral surface antigen is inserted into a plasmid, which is then used in a CFS to produce the viral protein. The purified protein is then used as a vaccine to elicit an immune response. VLP vaccines mimic the structure of viruses without the genetic material, making them safe and effective for vaccination.

CFS are used to produce the viral structural proteins that self-assemble into VLPs. The HPV (human papillomavirus) vaccine is an example of a VLP vaccine. In this case, CFS are employed to express and assemble the viral capsid proteins into VLPs, which are then used as the vaccine. CFS are utilized in the production of influenza vaccines, particularly those based on recombinant hemagglutinin (HA) proteins. HA is a major surface antigen of the influenza virus, and its production in cell-based systems can be challenging. CFS offer an alternative approach, allowing for the efficient production of HA proteins that can be used as a vaccine component. The use of CFPS in vaccine production offers several advantages, including rapid and scalable manufacturing, flexibility in the production of different vaccine types, and the ability to respond quickly to emerging infectious diseases. These systems enable the production of viral proteins or nucleic acids that are critical for vaccine development, facilitating the production of safe and effective vaccines. Commercialization of CFS-based immunizations is expected to accelerate in the coming years, as many of these agents require microgram doses.

Antibiotics

Antimicrobial resistance is recognized as a serious global health threat, causing 23,000 deaths and 2 million illnesses each year in the United States alone. As a result, some organizations are focusing on cell-free synthesis of antimicrobial compounds such as antimicrobial proteins and small molecule medicines (Ventola, 2015). Additionally, numerous labs have demonstrated how well CFS expresses phages. Furthermore, multiple laboratories have revealed successful CFS-based expression of phages. Because of the increasing prevalence of known antibiotic resistance, phage therapy is being considered as a viable substitute for existing antibiotic regimens (D. M. Lin et al., 2017). The use of phages has also been successful as a therapeutic technique for various plant diseases, and certain phages are now commercialized for general consumption. Manufacturing these new antimicrobials on CFS-based platforms could be a key weapon in the fight against antimicrobial resistance and contribute to improving global food security (Vu & Oh, 2020).

FUTURE OF BIOTECHNOLOGY RESEARCH DIRECTIONS WITH CELL-FREE SYSTEM

The landscape of synthetic biology is changing with the mixture of cell-free systems and a variety of genetically programmable technical tools, leading to impressive in vitro platforms. With wearable diagnostics and on-site drug manufacturing, these platforms can decentralize healthcare (Pardee, 2018). They also hold significant potential for centralized, effective manufacture of high-value supplies. Cell-free synthetic biology technologies will undoubtedly push the boundaries of synthetic biology and biotechnology, yielding a wide range of innovative and unexpected results (Wang & Zhang, 2019). We anticipate that the field will now continue to evolve and merge with other planned systems. It is possible to have controlled interactions with nanomaterials as well as interactions with various customized enzymes. We are interested to see how CFS can further bioengineering integration with electronics, computers, and machine learning technologies.

Synthetic biology seeks to improve current industrial processes by addressing problems such as low yields and low proportions, as well as current practices that ultimately affect ecosystems through pollution activities. In each of these cases, it is prudent to consider alternative approaches, such as new alternative routes to manufacturing multiple value-added product lines have been discovered. CFPS systems have a great scientific impact that can advance research in many areas of high-throughput manufacturing by focusing on the synthesis of essential products such as proteins, pharmaceuticals, and viruses/virus-like particles. CFPS systems have proven to be quite effective in terms of creating high expression levels, purity, and output, as well as allowing for the easier integration of tagged amino acids, attributes that allow for better protein structure analysis (Fogeron et al., 2021; Khambhati et al., 2019).

The future of biotechnology research with cell-free systems holds tremendous potential for advancing scientific knowledge, enabling novel applications, and addressing complex challenges. Here are some key directions and potential developments in this field:

1. *Improved Protein Production:* CFS offers advantages over traditional cell-based methods, such as faster production, simpler purification, and the ability to produce complex proteins. Future research will focus on enhancing protein synthesis efficiency, increasing yields, and developing methods for post-translational modifications to produce fully functional proteins with desired properties.
2. *Synthetic Biology and Genetic Circuit Engineering*: CFS provide a controlled environment for the construction and testing of synthetic gene circuits and genetic networks. Future research will explore the design and implementation of more complex and programmable biological systems, allowing for precise manipulation of cellular processes, bioengineering, and biocomputing applications.
3. *Pharmaceutical Development:* CFS has the potential to revolutionize drug discovery and development. Future research will involve the use of cell-free systems for high-throughput screening of drug candidates, studying drug-target interactions, and optimizing therapeutic protein production. Cell-free systems can also be used to produce personalized medicines tailored to individual patient needs.
4. *Artificial Organelles and Cell Mimics:* CFS can be utilized to recreate and study the functions of organelles or cellular compartments. Future research will focus on developing artificial organelles and cell mimics that can perform specific cellular functions, allowing for the understanding of complex cellular processes and the development of novel biotechnological applications.
5. *Biomanufacturing and Industrial Applications:* CFS offers great potential for biomanufacturing due to its scalability, ease of operation, and ability to produce complex biomolecules. Future research will aim to optimize cell-free systems for large-scale production of biofuels, pharmaceuticals, industrial enzymes, and other valuable compounds, leading to more sustainable and efficient bioproduction processes.
6. *Nanotechnology and Biomaterials:* The integration of CFS with nanotechnology holds promise for the development of functional nanomaterials and nano-devices. Future research will explore the synthesis of bio-inspired materials, the fabrication of nanodevices for biosensing and drug delivery, and the integration of cell-free systems with synthetic biology to create bio-hybrid systems with enhanced capabilities.
7. *Environmental and Agricultural Applications:* CFS can be utilized for environmental monitoring, detection of pollutants, and biosensing applications. Future research will focus on developing cell-free biosensors with high sensitivity and specificity for detecting various environmental contaminants. Additionally, cell-free systems can be employed for the production of sustainable agricultural products, such as biopesticides and biostimulants, contributing to eco-friendly and efficient agricultural practices.

8. *Cell-Free Medicine:* CFS holds promise for advancing personalized medicine approaches. Future research will explore the use of cell-free systems for rapid and on-demand production of therapeutics, including vaccines, antibodies, and gene therapies, tailored to individual patient's needs, improving treatment outcomes and reducing costs.

CONCLUSION

Cell-free systems offer exciting possibilities for scientific research and technological advancement. By providing a versatile platform for the study of biological processes outside the confines of intact cells, cell-free systems unlock new insights and capabilities. The future of biotechnology research with cell-free systems is incredibly exciting and holds immense potential. Advancements in protein production, synthetic biology, pharmaceutical development, biomanufacturing, nanotechnology, and personalized medicine are expected to significantly affect various sectors, leading to the development of innovative solutions, sustainable practices, and improved human health. Continued research and development in this field will pave the way for transformative discoveries and applications in biotechnology.

REFERENCES

Ajjolli Nagaraja, A., Charton, P., Cadet, X., Fontaine, N., Delsaut, M., Wiltschi, B., Voit, A., Offmann, B., Damour, C., Grondin-Perez, B., & Cadet, F. (2020). A Machine Learning Approach for Efficient Selection of Enzyme Concentrations and Its Application for Flux Optimization. *Catalysts*, *10*(3), 291. doi:10.3390/catal10030291

Ali, A., Zafar, H., Zia, M., Ul Haq, I., Phull, A. R., Ali, J. S., & Hussain, A. (2016). Synthesis, characterization, applications, and challenges of iron oxide nanoparticles. *Nanotechnology, Science and Applications*, *9*, 49–67. doi:10.2147/NSA.S99986 PMID:27578966

Atanasov, A. G., Zotchev, S. B., Dirsch, V. M., Orhan, I. E., Banach, M., Rollinger, J. M., & … . (2021). Natural products in drug discovery: Advances and opportunities. *Nature Reviews. Drug Discovery*, *20*(3), 200–216. doi:10.103841573-020-00114-z PMID:33510482

Axup, J. Y., Bajjuri, K. M., Ritland, M., Hutchins, B. M., Kim, C. H., Kazane, S. A., ... Stafin, K. (2012). Synthesis of site-specific antibody-drug conjugates using unnatural amino acids. *Proceedings of the National Academy of Sciences of the United States of America*, *109*(40), 16101–16106. doi:10.1073/pnas.1211023109 PMID:22988081

Blanken, D., Foschepoth, D., Serrão, A. C., & Danelon, C. (2020). Genetically controlled membrane synthesis in liposomes. *Nature Communications*, *11*(1), 1–13. doi:10.103841467-020-17863-5 PMID:32859896

Borkowski, O., Koch, M., Zettor, A., Pandi, A., Batista, A. C., Soudier, P., & Faulon, J. L. (2020). Large scale active-learning-guided exploration for in vitro protein production optimization. *Nature Communications*, *11*(1), 1872. doi:10.103841467-020-15798-5 PMID:32312991

BoydM. A.ThavarajahW.LucksJ. B.KamatN. P. (2022). Robust and tunable performance of a cell-free biosensor encapsulated in lipid vesicles. bioRxiv, 2022.2003.2002.482665. doi:10.1101/2022.03.02.482665

Brookwell, A., Oza, J. P., & Caschera, F. (2021). Biotechnology Applications of Cell-Free Expression Systems. *Life (Basel, Switzerland)*, *11*(12), 1367. Advance online publication. doi:10.3390/life11121367 PMID:34947898

Buchan, B. W., & Ledeboer, N. A. (2014). Emerging technologies for the clinical microbiology laboratory. *Clinical Microbiology Reviews*, *27*(4), 783–822. doi:10.1128/CMR.00003-14 PMID:25278575

Calhoun, K. A., & Swartz, J. R. (2005). An Economical Method for Cell-Free Protein Synthesis using Glucose and Nucleoside Monophosphates. *Biotechnology Progress*, *21*(4), 1146–1153. doi:10.1021/bp050052y PMID:16080695

Carlson, E. D., Gan, R., Hodgman, C. E., & Jewett, M. C. (2012). Cell-free protein synthesis: Applications come of age. *Biotechnology Advances*, *30*(5), 1185–1194. doi:10.1016/j.biotechadv.2011.09.016 PMID:22008973

Caschera, F., Bedau, M. A., Buchanan, A., Cawse, J., de Lucrezia, D., Gazzola, G., Hanczyc, M. M., & Packard, N. H. (2011). Coping with complexity: Machine learning optimization of cell-free protein synthesis. *Biotechnology and Bioengineering*, *108*(9), 2218–2228. doi:10.1002/bit.23178 PMID:21520017

Caschera, F., Karim, A. S., Gazzola, G., d'Aquino, A. E., Packard, N. H., & Jewett, M. C. (2018). High-Throughput Optimization Cycle of a Cell-Free Ribosome Assembly and Protein Synthesis System. *ACS Synthetic Biology*, *7*(12), 2841–2853. doi:10.1021/acssynbio.8b00276 PMID:30354075

Chen, X., Sun, Q., & Lu, Y. (2020). Creating a locally crowded environment with nanoclay hydrogels for cell-free biosynthesis. *Soft Matter*, *16*(22), 5132–5138. doi:10.1039/D0SM00636J PMID:32478769

Cho, E., & Lu, Y. (2020). Compartmentalizing cell-free systems: Toward creating life-like artificial cells and beyond. *ACS Synthetic Biology*, *9*(11), 2881–2901. doi:10.1021/acssynbio.0c00433 PMID:33095011

Cook-Deegan, R., & Heaney, C. (2010). Patents in genomics and human genetics. *Annual Review of Genomics and Human Genetics*, *11*(1), 383–425. doi:10.1146/annurev-genom-082509-141811 PMID:20590431

Currin, A., Parker, S., Robinson, C. J., Takano, E., Scrutton, N. S., & Breitling, R. (2021). The evolving art of creating genetic diversity: From directed evolution to synthetic biology. *Biotechnology Advances*, *50*, 107762. doi:10.1016/j.biotechadv.2021.107762 PMID:34000294

Daletos, G., & Stephanopoulos, G. (2020). Protein engineering strategies for microbial production of isoprenoids. *Metabolic Engineering Communications*, *11*, e00129. doi:10.1016/j.mec.2020.e00129 PMID:32612930

Damiati, S., Mhanna, R., Kodzius, R., & Ehmoser, E.-K. (2018). Cell-Free Approaches in Synthetic Biology Utilizing Microfluidics. *Genes*, *9*(3), 144. Advance online publication. doi:10.3390/genes9030144 PMID:29509709

Doshi-Velez, F., & Kim, B. (2017). Towards a rigorous science of interpretable machine learning. *arXiv preprint arXiv.08608*.

Ebrahimi, S., Khanbabaei, H., Abbasi, S., Fani, M., Soltani, S., Zandi, M., & Najafimemar, Z. (2022). CRISPR-Cas System: A Promising Diagnostic Tool for Covid-19. *Avicenna Journal of Medical Biotechnology*, *14*(1), 3–9. doi:10.18502/ajmb.v14i1.8165 PMID:35509363

Eto, S., Matsumura, R., Shimane, Y., Fujimi, M., Berhanu, S., Kasama, T., & Kuruma, Y. (2022). Phospholipid synthesis inside phospholipid membrane vesicles. *Communications Biology*, *5*(1), 1016. doi:10.103842003-022-03999-1 PMID:36167778

Fogeron, M.-L., Lecoq, L., Cole, L., Harbers, M., & Böckmann, A. (2021). Easy synthesis of complex biomolecular assemblies: Wheat germ cell-free protein expression in structural biology. *Frontiers in Molecular Biosciences*, *8*, 639587. doi:10.3389/fmolb.2021.639587 PMID:33842544

Freeze, H. H., & Elbein, A. D. (2009). Glycosylation precursors. Essentials of Glycobiology.

Garcia Serpa Osorio-de-Castro, C., Silva Miranda, E., Machado de Freitas, C., Rochel de Camargo, K. Jr, & Cranmer, H. H. (2017). The Zika Virus Outbreak in Brazil: Knowledge Gaps and Challenges for Risk Reduction. *American Journal of Public Health*, *107*(6), 960–965. doi:10.2105/AJPH.2017.303705 PMID:28426311

Garenne, D., Haines, M. C., Romantseva, E. F., Freemont, P., Strychalski, E. A., & Noireaux, V. (2021). Cell-free gene expression. *Nature Reviews. Methods Primers*, *1*(1), 49. doi:10.103843586-021-00046-x

Gootenberg, J. S., Abudayyeh, O. O., Lee, J. W., Essletzbichler, P., Dy, A. J., Joung, J., Verdine, V., Donghia, N., Daringer, N. M., Freije, C. A., Myhrvold, C., Bhattacharyya, R. P., Livny, J., Regev, A., Koonin, E. V., Hung, D. T., Sabeti, P. C., Collins, J. J., & Zhang, F. (2017). Nucleic acid detection with CRISPR-Cas13a/C2c2. *Science*, *356*(6336), 438–442. doi:10.1126cience.aam9321 PMID:28408723

Green, A. A., Silver, P. A., Collins, J. J., & Yin, P. (2014). Toehold switches: De-novo-designed regulators of gene expression. *Cell*, *159*(4), 925–939. doi:10.1016/j.cell.2014.10.002 PMID:25417166

Hérisson, J., Duigou, T., du Lac, M., Bazi-Kabbaj, K., Sabeti Azad, M., Buldum, G., Telle, O., El Moubayed, Y., Carbonell, P., Swainston, N., Zulkower, V., Kushwaha, M., Baldwin, G. S., & Faulon, J. L. (2022). The automated Galaxy-SynBioCAD pipeline for synthetic biology design and engineering. *Nature Communications*, *13*(1), 5082. doi:10.103841467-022-32661-x PMID:36038542

Hodgman, C. E., & Jewett, M. C. (2012). Cell-free synthetic biology: Thinking outside the cell. *Metabolic Engineering*, *14*(3), 261–269. doi:10.1016/j.ymben.2011.09.002 PMID:21946161

Hua, S. (2019). Physiological and pharmaceutical considerations for rectal drug formulations. *Frontiers in Pharmacology*, *10*, 1196. doi:10.3389/fphar.2019.01196 PMID:31680970

Hunt, J. P., Zhao, E. L., Free, T. J., Soltani, M., Warr, C. A., Benedict, A. B., Takahashi, M. K., Griffitts, J. S., Pitt, W. G., & Bundy, B. C. (2022). Towards detection of SARS-CoV-2 RNA in human saliva: A paper-based cell-free toehold switch biosensor with a visual bioluminescent output. *New Biotechnology*, *66*, 53–60. doi:10.1016/j.nbt.2021.09.002 PMID:34555549

Javalkote, V. S., Kancharla, N., Bhadra, B., Shukla, M., Soni, B., Sapre, A., Goodin, M., Bandyopadhyay, A., & Dasgupta, S. (2022). CRISPR-based assays for rapid detection of SARS-CoV-2. *Methods (San Diego, Calif.)*, *203*, 594–603. doi:10.1016/j.ymeth.2020.10.003 PMID:33045362

Jefferson, C., Lentzos, F., & Marris, C. (2014). Synthetic biology and biosecurity: Challenging the "myths". *Frontiers in Public Health*, *2*, 115. doi:10.3389/fpubh.2014.00115 PMID:25191649

Jenkins, N. (2007). Modifications of therapeutic proteins: Challenges and prospects. *Cytotechnology*, *53*(1-3), 121–125. doi:10.100710616-007-9075-2 PMID:19003198

Jiang, L., Zhao, J., Lian, J., & Xu, Z. (2018). Cell-free protein synthesis enabled rapid prototyping for metabolic engineering and synthetic biology. *Synthetic and Systems Biotechnology*, *3*(2), 90–96. doi:10.1016/j.synbio.2018.02.003 PMID:29900421

Jung, J. K., Alam, K. K., Verosloff, M. S., Capdevila, D. A., Desmau, M., Clauer, P. R., Lee, J. W., Nguyen, P. Q., Pastén, P. A., Matiasek, S. J., Gaillard, J.-F., Giedroc, D. P., Collins, J. J., & Lucks, J. B. (2020). Cell-free biosensors for rapid detection of water contaminants. *Nature Biotechnology*, *38*(12), 1451–1459. doi:10.103841587-020-0571-7 PMID:32632301

Kai, L., Dötsch, V., Kaldenhoff, R., & Bernhard, F. (2013). Artificial environments for the co-translational stabilization of cell-free expressed proteins. *PLoS One*, *8*(2), e56637. doi:10.1371/journal.pone.0056637 PMID:23451062

Kaminski, M. M., Abudayyeh, O. O., Gootenberg, J. S., Zhang, F., & Collins, J. J. (2021). CRISPR-based diagnostics. *Nature Biomedical Engineering*, *5*(7), 643–656. doi:10.103841551-021-00760-7 PMID:34272525

Karim, A. S., Dudley, Q. M., Juminaga, A., Yuan, Y., Crowe, S. A., Heggestad, J. T., Garg, S., Abdalla, T., Grubbe, W. S., Rasor, B. J., Coar, D. N., Torculas, M., Krein, M., Liew, F. M., Quattlebaum, A., Jensen, R. O., Stuart, J. A., Simpson, S. D., Köpke, M., & Jewett, M. C. (2020). In vitro prototyping and rapid optimization of biosynthetic enzymes for cell design. *Nature Chemical Biology*, *16*(8), 912–919. doi:10.103841589-020-0559-0 PMID:32541965

Khambhati, K., Bhattacharjee, G., Gohil, N., Braddick, D., Kulkarni, V., & Singh, V. (2019). Exploring the potential of cell-free protein synthesis for extending the abilities of biological systems. *Frontiers in Bioengineering and Biotechnology*, *7*, 248. doi:10.3389/fbioe.2019.00248 PMID:31681738

Khnouf, R., Beebe, D. J., & Fan, Z. H. (2009). Cell-free protein expression in a microchannel array with passive pumping. *Lab on a Chip*, *9*(1), 56–61. doi:10.1039/B808034H PMID:19209336

Khnouf, R., Chapman, B. D., & Fan, Z. H. (2011). Fabrication optimization of a miniaturized array device for cell-free protein synthesis. *Electrophoresis*, *32*(22), 3101–3107. doi:10.1002/elps.201100376 PMID:22038694

Khnouf, R., Olivero, D., Jin, S., & Fan, Z. H. (2010). Miniaturized fluid array for high-throughput protein expression. *Biotechnology Progress*, *26*(6), 1590–1596. doi:10.1002/btpr.474 PMID:20661923

Kim, D. M., & Swartz, J. R. (1999). Prolonging cell-free protein synthesis with a novel ATP regeneration system. *Biotechnology and Bioengineering*, *66*(3), 180–188. doi:10.1002/(SICI)1097-0290(1999)66:3<180::AID-BIT6>3.0.CO;2-S PMID:10577472

Kim, D.-M., & Swartz, J. R. (2000). Prolonging Cell-Free Protein Synthesis by Selective Reagent Additions. *Biotechnology Progress*, *16*(3), 385–390. doi:10.1021/bp000031y PMID:10835240

Laohakunakorn, N. (2020). Cell-Free Systems: A Proving Ground for Rational Biodesign. *Frontiers in Bioengineering and Biotechnology*, *8*, 788. Advance online publication. doi:10.3389/fbioe.2020.00788 PMID:32793570

Leisle, L., Valiyaveetil, F., Mehl, R. A., & Ahern, C. A. (2015). Incorporation of non-canonical amino acids. *Novel Chemical Tools to Study Ion Channel Biology*, 119-151.

Li, H., & d'Anjou, M. (2009). Pharmacological significance of glycosylation in therapeutic proteins. *Current Opinion in Biotechnology*, *20*(6), 678–684. doi:10.1016/j.copbio.2009.10.009 PMID:19892545

Li, Z., Li, Y., Lin, X., Cui, Y., Wang, T., Dong, J., & Lu, Y. (2022). Supramolecular protein assembly in cell-free protein synthesis system. *Bioresources and Bioprocessing*, *9*(1), 28. doi:10.118640643-022-00520-8

Lin, D. M., Koskella, B., & Lin, H. C. (2017). Phage therapy: An alternative to antibiotics in the age of multi-drug resistance. *World Journal of Gastrointestinal Pharmacology and Therapeutics*, *8*(3), 162. doi:10.4292/wjgpt.v8.i3.162 PMID:28828194

Lin, G.-M., Warden-Rothman, R., & Voigt, C. A. (2019). Retrosynthetic design of metabolic pathways to chemicals not found in nature. *Current Opinion in Systems Biology*, *14*, 82–107. doi:10.1016/j.coisb.2019.04.004

Liu, H. Y., Hopping, G. C., Vaidyanathan, U., Ronquillo, Y. C., Hoopes, P. C., & Moshirfar, M. (2019a). Polymerase Chain Reaction and Its Application in the Diagnosis of Infectious Keratitis. *Medical Hypothesis, Discovery and Innovation in Ophthalmology*, *8*(3), 152–155. PMID:31598517

Liu, S., Yu, J. M., Gan, Y. C., Qiu, X. Z., Gao, Z. C., Wang, H., Chen, S.-X., Xiong, Y., Liu, G.-H., Lin, S.-E., McCarthy, A., John, J. V., Wei, D.-X., & Hou, H. H. (2023). Biomimetic natural biomaterials for tissue engineering and regenerative medicine: New biosynthesis methods, recent advances, and emerging applications. *Military Medical Research*, *10*(1), 16. doi:10.118640779-023-00448-w PMID:36978167

Liu, W.-Q., Zhang, L., Chen, M., & Li, J. (2019b). Cell-free protein synthesis: Recent advances in bacterial extract sources and expanded applications. *Biochemical Engineering Journal*, *141*, 182–189. doi:10.1016/j.bej.2018.10.023

Lu, Y. (2022). Textile-embedded cell-free biosensors. *Nature Biomedical Engineering*, *6*(3), 225–226. doi:10.103841551-022-00869-3 PMID:35314799

Mehrotra, P. (2016). Biosensors and their applications - A review. *Journal of Oral Biology and Craniofacial Research*, *6*(2), 153–159. doi:10.1016/j.jobcr.2015.12.002 PMID:27195214

Mei, Q., Khnouf, R., Simon, A., & Fan, Z. H. (2010). Protein synthesis in a device with nanoporous membranes and microchannels. *Lab on a Chip*, *10*(19), 2541–2545. doi:10.1039/c005233g PMID:20730191

Myhrvold, C., Freije, C. A., Gootenberg, J. S., Abudayyeh, O. O., Metsky, H. C., Durbin, A. F., Kellner, M. J., Tan, A. L., Paul, L. M., Parham, L. A., Garcia, K. F., Barnes, K. G., Chak, B., Mondini, A., Nogueira, M. L., Isern, S., Michael, S. F., Lorenzana, I., Yozwiak, N. L., ... Sabeti, P. C. (2018). Field-deployable viral diagnostics using CRISPR-Cas13. *Science*, *360*(6387), 444–448. doi:10.1126cience.aas8836 PMID:29700266

Naresh, V., & Lee, N. (2021). A review on biosensors and recent development of nanostructured materials-enabled biosensors. *Sensors (Basel)*, *21*(4), 1109. doi:10.339021041109 PMID:33562639

Nguyen, B., Neidle, S., & Wilson, W. D. (2009). A role for water molecules in DNA-ligand minor groove recognition. *Accounts of Chemical Research*, *42*(1), 11–21. doi:10.1021/ar800016q PMID:18798655

Nguyen, P. Q., Soenksen, L. R., Donghia, N. M., Angenent-Mari, N. M., de Puig, H., Huang, A., Lee, R., Slomovic, S., Galbersanini, T., Lansberry, G., Sallum, H. M., Zhao, E. M., Niemi, J. B., & Collins, J. J. (2021). Wearable materials with embedded synthetic biology sensors for biomolecule detection. *Nature Biotechnology*, *39*(11), 1366–1374. doi:10.103841587-021-00950-3 PMID:34183860

Noireaux, V., & Libchaber, A. (2004). A vesicle bioreactor as a step toward an artificial cell assembly. *Proceedings of the National Academy of Sciences of the United States of America*, *101*(51), 17669–17674. doi:10.1073/pnas.0408236101 PMID:15591347

Nowak, K., Błażej, P., Wnetrzak, M., Mackiewicz, D., & Mackiewicz, P. (2021). Some theoretical aspects of reprogramming the standard genetic code. *Genetics*, *218*(1), iyab040. doi:10.1093/genetics/iyab040 PMID:33711098

Pardee, K. (2018). Perspective: Solidifying the impact of cell-free synthetic biology through lyophilization. *Biochemical Engineering Journal*, *138*, 91–97. doi:10.1016/j.bej.2018.07.008 PMID:30740032

Peruzzi, J., Steinkuehler, J., Vu, T., Lu, P., Baker, D., & Kamat, N. (2022). Organizing cell-free expressed membrane proteins in synthetic membranes using lipid-protein interactions. *Biophysical Journal*, *121*(3), 313a. doi:10.1016/j.bpj.2021.11.1194

Peters, C., & Brown, S. (2015). Antibody–drug conjugates as novel anti-cancer chemotherapeutics. *Bioscience Reports*, *35*(4), e00225. doi:10.1042/BSR20150089 PMID:26182432

Reynders, E., Foulquier, F., Annaert, W., & Matthijs, G. (2011). How Golgi glycosylation meets and needs trafficking: The case of the COG complex. *Glycobiology*, *21*(7), 853–863. doi:10.1093/glycob/cwq179 PMID:21112967

Schmid-Dannert, C., & López-Gallego, F. (2019). Advances and opportunities for the design of self-sufficient and spatially organized cell-free biocatalytic systems. *Current Opinion in Chemical Biology*, *49*, 97–104. doi:10.1016/j.cbpa.2018.11.021 PMID:30551093

Scott, A., Noga, M. J., de Graaf, P., Westerlaken, I., Yildirim, E., & Danelon, C. (2016). Cell-free phospholipid biosynthesis by gene-encoded enzymes reconstituted in liposomes. *PLoS One*, *11*(10), e0163058. doi:10.1371/journal.pone.0163058 PMID:27711229

Shadish, J. A., & DeForest, C. A. (2020). Site-Selective Protein Modification: From Functionalized Proteins to Functional Biomaterials. *Matter*, *2*(1), 50–77. doi:10.1016/j.matt.2019.11.011

Sheridan, C. (2005). The business of making vaccines. *Nature Biotechnology*, *23*(11), 1359–1366. doi:10.1038/nbt1105-1359 PMID:16273061

Shimizu, Y., Kuruma, Y., Ying, B.-W., Umekage, S., & Ueda, T. (2006). Cell-free translation systems for protein engineering. *The FEBS Journal*, *273*(18), 4133–4140. doi:10.1111/j.1742-4658.2006.05431.x PMID:16930131

Shin, J., & Noireaux, V. (2010). Efficient cell-free expression with the endogenous E. Coli RNA polymerase and sigma factor 70. *Journal of Biological Engineering*, *4*(1), 8. doi:10.1186/1754-1611-4-8 PMID:20576148

Soh, J. H., Balleza, E., Abdul Rahim, M. N., Chan, H.-M., Mohd Ali, S., Chuah, J. K. C., Edris, S., Atef, A., Bahieldin, A., Ying, J. Y., & Sabir, J. S. M. (2022). CRISPR-based systems for sensitive and rapid on-site COVID-19 diagnostics. *Trends in Biotechnology*, *40*(11), 1346–1360. doi:10.1016/j.tibtech.2022.06.002 PMID:35871983

Soni, T., Vivekanand, V., & Pareek, N. (2022). Biosensors as an effective tool for detection of emerging water and wastewater pollutants. In I. Haq, A. S. Kalamdhad, & M. P. Shah (Eds.), *Biodegradation and Detoxification of Micropollutants in Industrial Wastewater* (pp. 39–54). Elsevier. doi:10.1016/B978-0-323-88507-2.00010-5

Spencer, A. C., Torre, P., & Mansy, S. S. (2013). The encapsulation of cell-free transcription and translation machinery in vesicles for the construction of cellular mimics. *Journal of Visualized Experiments*, (80), e51304. PMID:24192867

Tinafar, A., Jaenes, K., & Pardee, K. (2019). Synthetic Biology Goes Cell-Free. *BMC Biology*, *17*(1), 64. doi:10.118612915-019-0685-x PMID:31395057

Van Aken, J., & Hammond, E. (2003). Genetic engineering and biological weapons: New technologies, desires and threats from biological research. *EMBO Reports*, *4*(S1), S57–S60. doi:10.1038j.embor.embor860 PMID:12789409

Ventola, C. L. (2015). The antibiotic resistance crisis: part 1: causes and threats. *P&T*, *40*(4), 277. PMID:25859123

Vidic, J., Manzano, M., Chang, C.-M., & Jaffrezic-Renault, N. (2017). Advanced biosensors for detection of pathogens related to livestock and poultry. *Veterinary Research*, *48*(1), 11. doi:10.118613567-017-0418-5 PMID:28222780

Vilkhovoy, M., Adhikari, A., Vadhin, S., & Varner, J. D. (2020). The Evolution of Cell Free Biomanufacturing. *Processes (Basel, Switzerland)*, *8*(6), 675. Advance online publication. doi:10.3390/pr8060675

Vilkhovoy, M., Dai, D., Vadhin, S., Adhikari, A., & Varner, J. D. (2019). Absolute Quantification of Cell-Free Protein Synthesis Metabolism by Reversed-Phase Liquid Chromatography-Mass Spectrometry. *Journal of Visualized Experiments*, (152). Advance online publication. doi:10.3791/60329-v PMID:31710042

Vu, N. T., & Oh, C.-S. (2020). Bacteriophage usage for bacterial disease management and diagnosis in plants. *The Plant Pathology Journal*, *36*(3), 204–217. doi:10.5423/PPJ.RW.04.2020.0074 PMID:32547337

Wang, F., & Zhang, W. (2019). Synthetic biology: Recent progress, biosafety and biosecurity concerns, and possible solutions. *Journal of Biosafety and Biosecurity*, *1*(1), 22–30. doi:10.1016/j.jobb.2018.12.003

White, A., Light, A., Hughen, R., Bateman, L., Martins, T., Hill, H., & Light, K. (2010). Severity of symptom flare after moderate exercise is linked to cytokine activity in chronic fatigue syndrome. *Psychophysiology*, *47*, 615–624. doi:10.1111/j.1469-8986.2010.00978.x PMID:20230500

Woo, C. H., Jang, S., Shin, G., Jung, G. Y., & Lee, J. W. (2020). Sensitive fluorescence detection of SARS-CoV-2 RNA in clinical samples via one-pot isothermal ligation and transcription. *Nature Biomedical Engineering*, *4*(12), 1168–1179. doi:10.103841551-020-00617-5 PMID:32948855

Wu, S., Snajdrova, R., Moore, J. C., Baldenius, K., & Bornscheuer, U. T. (2021). Biocatalysis: Enzymatic Synthesis for Industrial Applications. *Angewandte Chemie International Edition*, *60*(1), 88–119. doi:10.1002/anie.202006648 PMID:32558088

Zhang, L., Guo, W., & Lu, Y. (2020). Advances in Cell-Free Biosensors: Principle, Mechanism, and Applications. *Biotechnology Journal*, *15*(9), 2000187. doi:10.1002/biot.202000187 PMID:32667120

KEY TERMS AND DEFINITIONS

Artificial Cell: An engineered particle that mimics one or many functions of a biological cell. Artificial cells are biological or polymeric membranes, enclosing biologically active materials from cell-free system. As such, nanoparticles, liposomes, polymersomes, microcapsules and a number of other particles have qualified as artificial cells.

Biomanufacturing: The production of valuable compounds, such as pharmaceuticals, biofuels, and industrial enzymes, using living organisms or cell-free systems.

Cell-Free Protein Synthesis: CFPS, or in vitro protein synthesis, is the production of proteins using biological machinery in a cell-free system, i.e., without the use of living cells.

CRISPR: Clustered Regularly Interspaced Short Palindromic Repeats, a gene-editing technology that allows for precise modifications of DNA sequences, enabling targeted genetic modifications and potential therapeutic applications.

Genetic Circuit: A network of genes and regulatory elements that interact with each other to produce specific biological functions or behaviors, analogous to electronic circuits, enabling programmable control of gene expression and cellular behavior.

Machine Learning: A field of artificial intelligence that focuses on the development of algorithms and statistical models that enable computer systems to learn and improve from data without explicit programming.

Microfluidics: The science and technology of manipulating fluids and controlling their behavior at the microscale, typically in devices with channels and chambers on a small scale.

Synthetic Biology: A multidisciplinary area of research that involves the application of engineering principles to biology.

Chapter 3
Assistive Robotics and Synthetic Biology in Medical Applications

Dhiresh S. Shastri
https://orcid.org/0009-0000-6302-2002
Department of Mechanical Engineering, Dr. Vishwanath Karad MIT World Peace University, India

Narhar K. Patil
Department of Mechanical Engineering, Dr. Vishwanath Karad MIT World Peace University, India

Dler Salih Hasan
https://orcid.org/0009-0008-3212-5509
Department of Computer Science and Information Technology, College of Science, University of Salahaddin, Iraq

Tushar S. Nikumbh
Department of Mechanical Engineering, Dr. Vishwanath Karad MIT World Peace University, India

Naresh Babanrao Chaudhari
Department of Mechanical Engineering, Dr. Vishwanath Karad MIT World Peace University, India

ABSTRACT

Assistive robotics and synthetic biology have revolutionized the medical field by offering personalized interventions and precise exercises in physical therapy and rehabilitation. This synergy accelerates patient recovery and improves functional outcomes. Synthetic biology techniques contribute to neural repair and regeneration, providing innovative solutions for neurological disorders and brain injuries. Robotic prosthetic limbs with synthetic sensory feedback revolutionize amputees' lives, offering enhanced motor control and reducing pain. The collaboration has led to breakthroughs in personalized gene therapies, smart drug delivery systems, and brain-computer interfaces.

DOI: 10.4018/978-1-6684-6577-6.ch003

INTRODUCTION

In recent years, the fields of assistive robotics and synthetic biology have witnessed significant advancements that hold immense promise for revolutionizing medical applications. Assistive robotics involves the design and development of robotic systems that interact with humans to provide support, enhance capabilities, and improve quality of life. On the other hand, synthetic biology is an interdisciplinary branch of science that combines principles from biology, engineering, and computer science to design and construct novel biological components, systems, and organisms (Anitha, R, et al., 2023).

The convergence of these two cutting-edge disciplines has given rise to a new era of medical innovation, where robotic technologies are seamlessly integrated with biological systems to address a wide array of healthcare challenges. From drug delivery to surgical procedures, rehabilitation, and disease detection, the synergistic collaboration between assistive robotics and synthetic biology offers a powerful toolbox to tackle complex medical problems (Boopathi, 2022; Boopathi et al., 2022; Kannan et al., 2022; Saravanan et al., 2022). In this book chapter, we aim to explore the potential applications of assistive robotics and synthetic biology in the medical field. We will delve into the fundamental concepts of each discipline, examining the tools, techniques, and methodologies that underpin their respective domains. By understanding these foundations, we can better grasp the transformative impact that the amalgamation of assistive robotics and synthetic biology can have on healthcare (Anitha, R, et al., 2023).

Throughout this chapter, we will investigate the collaborative efforts and novel approaches that merge the precision and adaptability of robotic systems with the programmability and versatility of synthetic biological components. We will explore the use of robotics for drug delivery, the creation of bioengineered tissues and organs, and the integration of genetic circuitry for disease detection and treatment. Moreover, we will delve into the potential applications of assistive robotics in surgery and rehabilitation, where the marriage of robotics and synthetic biology can potentially reshape the landscape of medical procedures and patient care (Boopathi et al., 2023).

However, with these advancements come unique challenges and ethical considerations. We will also explore the human-robot interaction aspect, emphasizing the importance of designing intuitive and safe user interfaces to foster seamless integration. Additionally, we will address the ethical implications surrounding the use of these technologies, ensuring that the principles of beneficence, non-maleficence, autonomy, and justice are upheld throughout the development and deployment of these innovations (Rahamathunnisa, Sudhakar, Murugan, Thivaharan S., et al., 2023).

The convergence of assistive robotics and synthetic biology represents a transformative shift in the medical landscape. This chapter aims to shed light on the exciting possibilities that lie ahead, as well as the responsibilities and considerations that accompany such transformative technologies. By embracing interdisciplinary collaboration and upholding ethical standards, we can harness the full potential of assistive robotics and synthetic biology to enhance patient outcomes, improve healthcare accessibility, and pave the way for a healthier future (Boopathi, 2023c; Ramudu et al., 2023). Assistive robotics refers to the integration of robotic systems into medical practices to assist and augment the capabilities of healthcare professionals, patients, and individuals with physical or cognitive impairments. These robots are designed to perform various tasks, such as surgical procedures, rehabilitation exercises, drug delivery, and diagnostic support, among others. By combining advanced sensors, actuators, artificial intelligence, and human-robot interaction interfaces, assistive robotics offers the potential to improve patient outcomes, enhance precision, and increase the efficiency of medical interventions (Boopathi, 2023a).

Robotic-assisted surgery allows for enhanced precision, reduced invasiveness, and improved dexterity during complex procedures, leading to shorter recovery times and better patient outcomes. Assistive robots are used in physical therapy and neurorehabilitation to help patients regain mobility, strength, and functionality after injuries or neurological disorders. Robotic systems can be employed for precise and targeted drug delivery, ensuring optimal dosage and minimizing side effects (Pramila et al., 2023; Rahamathunnisa, Sudhakar, Murugan, Thivaharan, et al., 2023a; T. Senthil et al., 2023). Robots integrated with artificial intelligence can aid in medical imaging, diagnosis, and analysis, providing more accurate and timely results. Synthetic Biology in Medicine: Synthetic biology is a multidisciplinary field that combines principles from biology, engineering, and computer science to design and construct new biological systems, devices, or organisms with novel functionalities (Koshariya, Khatoon, et al., 2023; Maheswari et al., 2023; Mohanty, Jothi, et al., 2023a). In medicine, synthetic biology offers revolutionary approaches for disease diagnosis, treatment, and therapeutics development. By reprogramming living organisms or building artificial biological systems, synthetic biology enables the creation of designer drugs, engineered tissues, and gene circuits that can sense and respond to specific disease markers (T. S. Senthil et al., 2023).

- *Engineered Biological Systems*: Synthetic biology techniques are used to design and construct engineered biological systems, such as gene circuits and cellular pathways, for therapeutic applications.
- *Tissue Engineering and Regenerative Medicine*: Synthetic biology is employed to create bioengineered tissues and organs, fostering advancements in regenerative medicine and transplantation.
- *Drug Development*: Synthetic biology facilitates the production of therapeutic agents, such as vaccines and gene therapies, for targeted treatment of diseases.
- *Disease Detection and Diagnosis*: Genetic circuits and biosensors are engineered to detect specific biomarkers, enabling early and accurate disease diagnosis.

Synergy in Medical Applications: The integration of assistive robotics and synthetic biology in medicine presents a synergistic approach to address complex medical challenges. By combining the adaptability and precision of robotic systems with the programmability and versatility of synthetic biological components, researchers and medical practitioners can create innovative solutions that were previously unimaginable. This collaboration opens up new possibilities for personalized medicine, minimally invasive procedures, and improved patient care (T. S. Senthil et al., 2023). Moreover, we will explore the ethical considerations and future directions of this transformative field, highlighting the importance of responsible innovation to maximize the benefits of assistive robotics and synthetic biology while minimizing potential risks. The integration of assistive robotics and synthetic biology holds significant importance for medical advancements (Boopathi et al., 2023; Pramila et al., 2023; Reddy et al., 2023; Sengeni et al., 2023). This convergence offers a range of benefits that can revolutionize healthcare and improve patient outcomes in various ways:

- *Precision and Personalization*: The integration of robotics and synthetic biology enables precise and personalized medical interventions. Robotic systems provide enhanced accuracy and control during surgical procedures, while synthetic biology allows for the design of personalized therapeutics tailored to an individual's genetic makeup.

- *Minimally Invasive Procedures*: By combining robotics with synthetic biology, medical procedures can become less invasive, reducing patient trauma, pain, and recovery time. Minimally invasive techniques lead to shorter hospital stays and quicker return to daily activities.
- *Targeted Drug Delivery*: Assistive robots can be used to precisely deliver drugs to specific areas of the body, minimizing side effects and improving drug efficacy. Synthetic biology contributes to the development of novel drug-delivery vehicles, enhancing the targeting and release of therapeutic agents.
- *Rehabilitation and Recovery*: Assistive robots in rehabilitation settings help patients regain mobility and function after injuries or surgeries. Synthetic biology can aid in tissue engineering and regenerative medicine, promoting faster healing and tissue repair.
- *Early Disease Detection*: Synthetic biology allows for the engineering of genetic circuits and biosensors that detect disease markers at early stages, enabling timely intervention and improving patient outcomes.
- *Research and Drug Development*: The integration of robotics and synthetic biology expedites research processes, facilitating the screening of large datasets and testing of drug candidates. This accelerates drug development and increases the potential for finding novel treatments.
- *Accessibility and Affordability*: The use of robotic systems can automate certain medical procedures, reducing the burden on healthcare professionals and making treatments more accessible and affordable for patients.
- *Data-driven Healthcare*: The integration of robotics and synthetic biology generates vast amounts of data, enabling data-driven decision-making in healthcare. This data can be utilized to optimize treatments, predict patient outcomes, and improve overall healthcare practices.
- *Collaboration and Innovation*: The convergence of assistive robotics and synthetic biology fosters interdisciplinary collaboration between experts from various fields, leading to new ideas and innovative solutions for medical challenges.
- *Futuristic Possibilities*: The integration of these technologies paves the way for futuristic medical advancements, such as biohybrid devices, self-regulating therapies, and personalized nanorobotics, which were once confined to science fiction.

Despite these advantages, the integration of assistive robotics and synthetic biology also comes with ethical and safety considerations, such as data privacy, human-robot interaction, and the potential misuse of technology. Responsible development and careful consideration of ethical principles are crucial to harness the full potential of this convergence while ensuring patient safety and well-being (T. S. Senthil et al., 2023).

ASSISTIVE ROBOTICS AND ITS ROLE IN HEALTHCARE

Assistive robotics is a rapidly evolving field that involves the integration of robotic systems into various aspects of healthcare to provide support, assistance, and improved functionality for patients, caregivers, and healthcare professionals (Boopathi et al., 2023; Pramila et al., 2023; Reddy et al., 2023; Sengeni et al., 2023). These robotic devices are designed to interact with humans in a safe and intuitive manner, enhancing their capabilities and addressing challenges in medical settings. The ultimate goal of assistive robotics is to improve patient outcomes, enhance the quality of care, and optimize the efficiency of medical interventions (Kagadis et al., 2013).

The Role of Assistive Robotics in Healthcare

- *Precision and Accuracy*: Robotic systems offer a level of precision and accuracy that surpasses human capabilities, making them ideal for performing delicate and complex medical procedures with minimal errors.
- *Minimally Invasive Interventions*: Many assistive robots enable minimally invasive procedures, reducing patient trauma, pain, and recovery time while promoting faster healing and shorter hospital stays.
- *Rehabilitation and Physical Therapy*: Assistive robotics plays a crucial role in rehabilitation and physical therapy, helping patients regain mobility, strength, and independence after injuries, surgeries, or neurological disorders.
- *Elderly Care*: Robotic devices can assist elderly individuals with daily activities, such as lifting and transferring, promoting independence and relieving caregivers' burden.
- *Diagnostic Support*: Robotics integrated with artificial intelligence can aid in medical imaging and diagnostics, providing faster and more accurate analysis of medical data.
- *Drug Delivery*: Robotic systems can be utilized for precise and targeted drug delivery, ensuring optimal dosage and reducing side effects.

Types of Assistive Robots and Their Applications in Medicine

- *Surgical Robots*: Surgical robots are one of the most prominent applications of assistive robotics in medicine. They assist surgeons during complex procedures, providing greater dexterity, stability, and magnification, leading to more precise surgeries and improved patient outcomes.
- *Rehabilitation Robots*: These robots are used in physical therapy and neurorehabilitation to assist patients in relearning motor skills and regaining functional independence after neurological injuries or disorders.
- *Exoskeletons*: Exoskeletons are wearable robotic devices designed to support and augment human movements. They are particularly useful for individuals with mobility impairments, helping them walk or perform daily activities with greater ease.
- *Telepresence Robots*: Telepresence robots enable remote communication and interaction between healthcare professionals and patients. They allow physicians to remotely assess and monitor patients, especially in situations where physical presence is challenging.
- *Robotic Prosthetics*: Robotic prosthetics are artificial limbs that mimic natural movements, providing greater functionality and quality of life for individuals with limb loss.

Challenges and Opportunities

While assistive robotics holds immense potential for healthcare advancement, several challenges must be addressed for successful implementation (Seifert et al., 2021):

- *Cost and Accessibility*: The cost of developing and deploying assistive robots can be substantial, making them less accessible to smaller healthcare facilities and individuals with limited financial resources.
- *Safety and Reliability*: Ensuring the safety and reliability of assistive robots is crucial to gain the trust of both patients and healthcare professionals. Robotic devices must undergo rigorous testing and validation to minimize the risk of errors or malfunctions.
- *Ethical and Legal Considerations*: The use of assistive robots raises ethical concerns, such as privacy, data security, and the potential replacement of human caregivers. Clear ethical guidelines and legal frameworks are needed to address these issues.
- *Human-Robot Interaction*: Designing intuitive and user-friendly human-robot interaction interfaces is essential for seamless integration and user acceptance. Improving the user experience will increase the adoption of assistive robots in healthcare settings.
- *Training and Education*: Proper training of healthcare professionals and caregivers is necessary to effectively utilize assistive robotics. A comprehensive understanding of the technology and its applications ensures its safe and efficient use.

Opportunities for implementing assistive robotics

- *Advancing Medical Procedures*: Assistive robots can enhance surgical precision, leading to better outcomes in complex surgeries (Anitha, Komala, et al., 2023; Boopathi, Umareddy, et al., 2023; Boopathi & Davim, 2023b, 2023a; Subha, Inbamalar, Komala, et al., 2023).
- *Improving Patient Rehabilitation*: Robotic devices can accelerate recovery and improve functional outcomes in rehabilitation settings.
- *Enhancing Elderly Care*: Robots can assist in elderly care, supporting daily activities and promoting independence.
- *Enabling Remote Healthcare*: Telepresence robots enable remote consultations and assessments, increasing healthcare accessibility in remote or underserved areas.
- *Research and Innovation*: Assistive robotics presents opportunities for interdisciplinary research and innovation, leading to novel technologies and medical breakthroughs.

EXPLORING SYNTHETIC BIOLOGY

Synthetic biology is a multidisciplinary field that combines principles from biology, engineering, and computer science to design and construct new biological systems, devices, or organisms with novel functionalities (Hanumanthakari et al., 2023b; Reddy et al., 2023; Sathiyaseelan et al., 2023; Sengeni et al., 2023). It aims to engineer living organisms or build artificial biological components to perform specific tasks or produce desired products. Synthetic biology provides a systematic and engineering-based approach to manipulate biological systems, offering immense potential in various fields, including medicine (Alam et al., 2019).

Fundamentals of Synthetic Biology and Its Potential in Medicine

1. *Genetic Engineering*: At the core of synthetic biology lies genetic engineering, which involves the deliberate modification of an organism's DNA to achieve specific traits or functions. This process enables scientists to create new genetic circuits, modify existing pathways, or introduce novel genes, which can be harnessed for medical applications.
2. *DNA Synthesis and Assembly*: Synthetic biology relies on DNA synthesis and assembly techniques to build genetic constructs or entire DNA sequences from scratch. Advances in DNA synthesis have made it possible to create custom-designed genetic components and circuits for medical purposes.
3. *Genetic Circuitry*: Synthetic biologists create genetic circuits, similar to electronic circuits, to control and regulate cellular processes. These circuits can sense specific molecules, respond to environmental cues, or trigger targeted therapeutic actions, making them valuable tools for medical interventions.
4. *Gene Editing and CRISPR*: Gene editing technologies, especially CRISPR-Cas9, have revolutionized synthetic biology. CRISPR allows precise and targeted modifications of DNA, opening up new possibilities for gene therapies and disease treatment.
5. *Biosecurity and Ethics*: As synthetic biology becomes more powerful, concerns regarding biosecurity and ethical considerations arise. Ensuring responsible use and addressing potential risks are essential for the safe and ethical application of synthetic biology in medicine.

Genetic Engineering and Its Applications in Healthcare

1. *Gene Therapy*: Genetic engineering facilitates gene therapy, a promising approach to treat genetic disorders by replacing or repairing defective genes. Synthetic biology tools are used to introduce functional genes into patients, aiming to correct the underlying genetic cause of the disease.
2. *Engineered Cells for Therapeutics*: Synthetic biology enables the engineering of cells to produce therapeutic proteins, antibodies, or other molecules for medical use. Engineered cells can serve as factories for producing personalized medicines or targeted therapies.
3. *Synthetic Vaccines*: Synthetic biology is used to design and construct synthetic vaccines against infectious diseases. These vaccines can be more easily and rapidly developed to address emerging pathogens or vaccine-resistant strains.
4. *Biosensors and Diagnostics*: Synthetic biology allows the creation of biosensors that can detect specific disease markers or environmental toxins. These biosensors have applications in medical diagnostics and environmental monitoring.

Bioinformatics and Computational Approaches for Advancing Synthetic Biology

1. *Genome Design and Optimization*: Bioinformatics tools aid in the design and optimization of synthetic genomes, ensuring their functionality and stability in the target organism.
2. *Gene Prediction and Design*: Computational methods are used to predict gene function and design gene sequences that exhibit desired characteristics.
3. *Data Analysis and Modeling*: Synthetic biology relies on data analysis and modeling to understand complex biological systems, predict the behavior of genetic circuits, and optimize their performance.

4. *Standardization and Automation*: Bioinformatics plays a crucial role in standardizing genetic parts and sequences, promoting the exchange of biological data, and automating genetic design processes.
5. *Synthetic Biology Software*: Several software tools and platforms have been developed to assist synthetic biologists in designing and simulating genetic circuits, facilitating the development and implementation of synthetic biology applications.

Synthetic biology integration in medical applications offers significant potential for personalized therapies, regenerative medicine, and disease detection. Interdisciplinary collaboration among biologists, engineers, computer scientists, and healthcare professionals is crucial for harnessing the full potential of synthetic biology, transforming medicine and improving patient care (Javaid & Haleem, n.d.).

THE SYNERGY OF ASSISTIVE ROBOTICS AND SYNTHETIC BIOLOGY

The convergence of assistive robotics and synthetic biology presents a powerful synergy that can significantly impact various aspects of healthcare and medical research (Figure 1). By combining the precision, adaptability, and advanced capabilities of robotic systems with the programmability, versatility, and self-regulation of synthetic biological components, researchers and healthcare professionals can achieve novel and transformative applications (Domakonda et al., 2022; Gunasekaran et al., 2022; Jeevanantham et al., 2022). This collaboration unlocks new possibilities for personalized medicine, advanced therapeutics, and cutting-edge medical interventions that were once considered beyond reach (Z. Li et al., 2018; Wisniewski et al., 2020).

Figure 1. Synthetic biology research versus medical robotics research

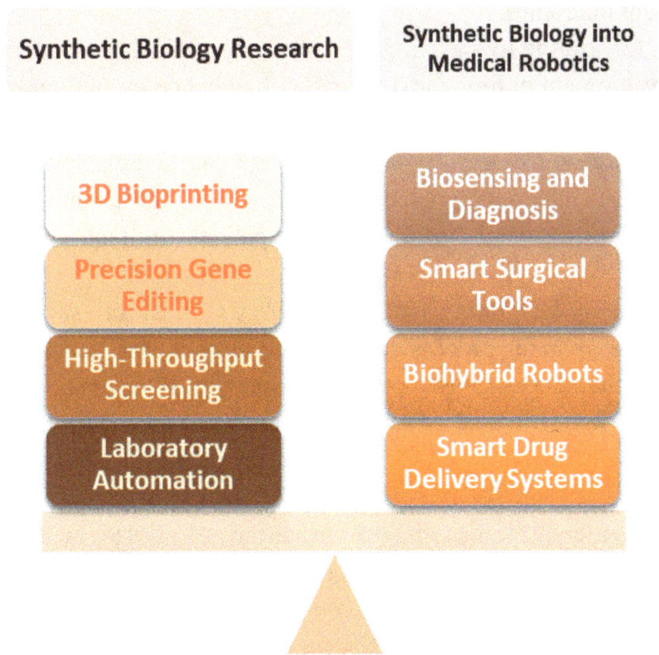

Leveraging Robotics to Enhance Synthetic Biology Research

- *Laboratory Automation*: Robotics can be utilized to automate laborious and repetitive tasks involved in synthetic biology research, such as DNA synthesis, cloning, and screening of genetic constructs. This not only saves time and resources but also improves the accuracy and consistency of experimental results.
- *High-Throughput Screening*: Assistive robots enable high-throughput screening of vast libraries of genetic constructs, accelerating the discovery of functional biological components and circuits. This significantly enhances the efficiency of synthetic biology research and the identification of potential candidates for medical applications.
- *Precision Gene Editing*: Robotics can aid in precise and targeted gene editing techniques, such as CRISPR-Cas9, ensuring accurate modifications in genetic sequences. This level of precision is crucial for developing safe and effective gene therapies.
- *3D Bioprinting*: Combining robotics with synthetic biology allows for 3D bioprinting of tissues and organs using bioink containing living cells and biomaterials. This has immense implications for regenerative medicine and transplantation.
- *Data Analysis and Modeling*: Robotics integrated with bioinformatics and computational approaches can process and analyze vast amounts of biological data generated from synthetic biology experiments. This data-driven approach helps researchers make informed decisions and gain insights into complex biological systems.

Integrating Synthetic Biology into Medical Robotics for Improved Functionality

- *Smart Drug Delivery Systems*: By incorporating synthetic biology components into drug delivery robots, it is possible to create smart systems that respond to specific disease markers or physiological conditions. These systems can release drugs at precise locations and controlled rates, optimizing treatment outcomes.
- *Biohybrid Robots*: Integrating living cells or synthetic biological components into robotic systems can lead to the development of biohybrid robots with enhanced functionalities. These robots may have self-regulatory capabilities or the ability to adapt to changing environments.
- *Customized Rehabilitation Devices*: Synthetic biology can be employed to engineer cells or tissues that promote tissue repair and regeneration. Integrating these components into rehabilitation robots can enhance the effectiveness of therapies for patients with musculoskeletal or neurological disorders.
- *Smart Surgical Tools*: By integrating synthetic biology circuits into surgical robots, it is possible to enhance the decision-making process during procedures. These smart tools can respond to real-time data and adjust their actions based on patient-specific conditions.
- *Biosensing and Diagnosis*: Medical robots equipped with synthetic biology biosensors can detect disease biomarkers or pathogens more accurately, enabling faster and more precise diagnoses.

Assistive robotics and synthetic biology integration has the potential to revolutionize medical applications, including personalized medicine and advanced diagnostics. This synergy will shape healthcare's future, improve patient outcomes, and open new avenues for innovation.

ASSISTIVE ROBOTICS FOR DRUG DELIVERY

Research Focus: Robotic Systems for Precise and Targeted Drug Delivery

Drug delivery is crucial in medical treatment, and assistive robotics offers a promising solution for improved accuracy, control, and customization. This research aims to explore designing and integrating robotic systems into drug delivery processes to enhance patient outcomes and optimize treatment strategies (Deheyab et al., 2022; Paul et al., 2013).

- *Robotic Systems for Drug Formulation*: Assistive robotics can play a significant role in the formulation and preparation of drugs, especially for personalized medicine. Researchers are investigating how robotic systems can automate and optimize drug compounding, enabling precise dosing and reducing human errors in the preparation of complex medications.
- *Smart Drug Delivery Devices*: The development of smart drug delivery devices involves integrating robotics with advanced sensors and actuators to provide real-time feedback and adaptability. These devices can sense physiological cues or disease markers to release drugs at specific times and locations within the body, optimizing therapeutic outcomes.
- *Nanorobotics for Drug Delivery*: Nanorobotics, a field that combines nanotechnology and robotics, holds immense potential for targeted drug delivery at the cellular or molecular level. Researchers are exploring how nanorobots can navigate through the bloodstream and deliver drugs directly to diseased cells, minimizing off-target effects and improving treatment efficacy.
- *Implantable Drug Delivery Systems*: Robotic systems are being explored to design and implement implantable drug delivery devices. These devices can release drugs over extended periods, eliminating the need for frequent administrations and improving patient compliance.
- *Robotic-assisted Surgical Drug Delivery*: Assistive robotics can aid surgeons in delivering drugs directly to specific tissues or organs during surgical procedures. This approach reduces the risk of complications and ensures precise drug placement in critical areas.

Synthetic Biology Approaches for Developing Therapeutic Agents

Synthetic biology provides innovative tools and techniques to engineer biological systems for therapeutic applications. This chapter explores how synthetic biology approaches can be employed to develop novel therapeutic agents, including engineered cells, gene therapies, and synthetic proteins (Maier et al., 2019).

- *Engineered Cells for Therapeutic Protein Production*: Synthetic biology allows the engineering of cells to produce therapeutic proteins with enhanced properties. Researchers are modifying cellular pathways to improve protein expression, stability, and functionality, paving the way for more effective treatments.
- *Gene Therapies Using Synthetic Biological Components*: Gene therapy holds great promise for treating genetic disorders and other diseases. Synthetic biology techniques, such as CRISPR-Cas9, enable precise gene editing and gene insertion, providing potential cures for previously untreatable conditions.

- *Synthetic Peptides and Antibodies*: Synthetic biology offers the ability to design and produce synthetic peptides and antibodies with unique properties. These molecules can be tailored for specific disease targets, leading to more targeted and effective therapies.
- *Synthetic Vaccine Development*: Synthetic biology plays a crucial role in developing novel vaccines against infectious diseases. Researchers are using synthetic components to design vaccines that elicit a specific immune response, enhancing protection against pathogens.
- *Biohybrid Therapeutics*: Synthetic biology allows for the creation of biohybrid therapeutics, where living cells are combined with synthetic components to create novel therapeutic agents with enhanced functionality and adaptability.

Researchers are combining assistive robotics and synthetic biology to develop therapeutic agents, aiming to revolutionize medical treatment strategies, improve drug efficacy, reduce side effects, and enhance patient care. This interdisciplinary approach holds great promise for precision medicine and targeted therapies.

BIOENGINEERED TISSUE AND ORGAN CONSTRUCTS

Figure 2. Bioengineered tissue and organ constructs

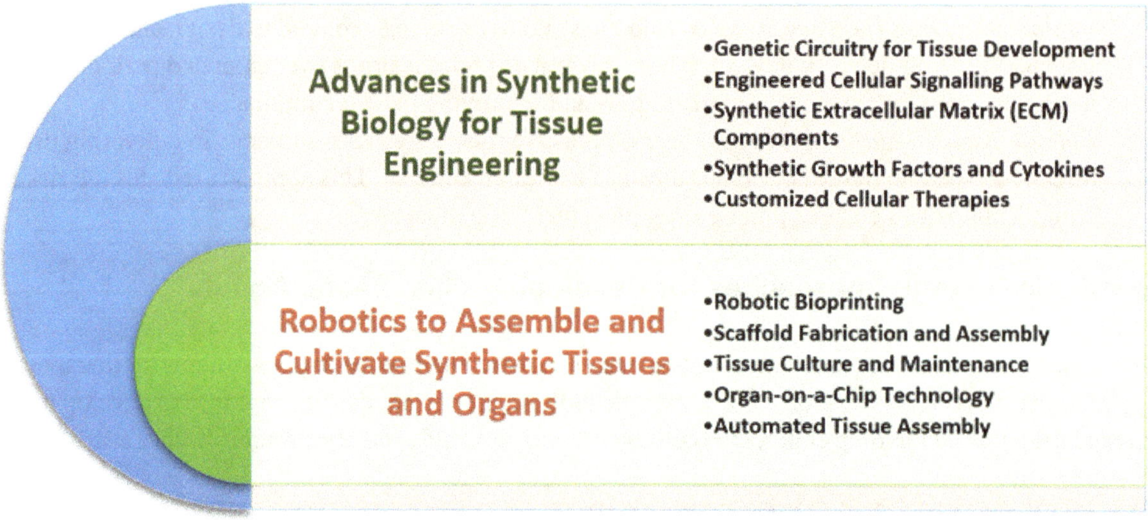

Utilizing Robotics to Assemble and Cultivate Synthetic Tissues and Organs

Bioengineered tissues and organs hold immense potential for addressing the critical shortage of donor organs for transplantation and providing personalized regenerative therapies (Figure 2). Robotics, when combined with tissue engineering techniques, offers a powerful approach to fabricate, assemble, and cultivate complex tissue constructs with precision and reproducibility. This research area focuses on how robotics can be leveraged to enhance tissue engineering processes, leading to the development of functional bioengineered tissues and organs (Klangsin & Harding, 1998).

- *Robotic Bioprinting*: Robotic bioprinting involves the precise deposition of bioinks, containing living cells and biomaterials, to create three-dimensional tissue structures layer by layer. Researchers are exploring how robotic bioprinters can produce intricate tissue architectures, mimicking the native tissue organization for improved functionality.
- *Scaffold Fabrication and Assembly*: Robotics can automate the fabrication and assembly of scaffolds, which serve as the structural framework for tissue growth. Robotic systems enable the production of scaffolds with precisely controlled porosity, mechanical properties, and spatial arrangements, ensuring optimal cell growth and tissue development.
- *Tissue Culture and Maintenance*: Robotic systems equipped with advanced sensing capabilities can monitor and control the environmental conditions within tissue culture chambers. This real-time feedback allows for precise regulation of nutrient supply, oxygen levels, and mechanical forces, promoting tissue maturation and functionality.
- *Organ-on-a-Chip Technology*: Utilizing robotics, organ-on-a-chip devices can be engineered to replicate the physiological functions of specific organs. These microfluidic systems enable high-throughput drug testing and disease modeling, providing valuable insights for drug development and personalized medicine.
- *Automated Tissue Assembly*: Robotics can automate the process of assembling multiple tissue constructs to create larger and more complex tissue structures or organoids. This approach streamlines the tissue engineering process and enables the construction of organ-like structures for transplantation and research purposes.

Advances in Synthetic Biology for Tissue Engineering

Synthetic biology offers transformative tools to enhance tissue engineering by providing new ways to design and control cellular behaviours. Through genetic engineering and synthetic biological components, researchers can influence cell functions and create bioengineered tissues with enhanced properties and functionalities (Boopathi et al., 2023).

- *Genetic Circuitry for Tissue Development*: Synthetic biology allows the design of genetic circuits that regulate cell behaviour, proliferation, and differentiation. By incorporating these circuits into cells used for tissue engineering, researchers can precisely control tissue development and maturation.
- *Engineered Cellular Signalling Pathways*: Synthetic biology techniques enable the engineering of cellular signalling pathways to promote tissue growth, vascularization, and integration with host tissues. These engineered pathways enhance tissue functionality and integration after transplantation.
- *Synthetic Extracellular Matrix (ECM) Components*: Synthetic biology enables the production of bioengineered ECM components with tailor-made properties. These synthetic ECMs can be incorporated into tissue constructs to improve their mechanical strength, biocompatibility, and regenerative potential.

- *Synthetic Growth Factors and Cytokines*: Synthetic biology allows for the production of custom-designed growth factors and cytokines that promote tissue regeneration and repair. These synthetic molecules can be incorporated into tissue constructs to enhance cellular proliferation and differentiation.
- *Customized Cellular Therapies*: Synthetic biology enables the engineering of cells with specific properties, such as improved survival, proliferation, or immune-modulatory capabilities. These customized cellular therapies can enhance tissue engineering approaches and facilitate successful tissue transplantation.

Researchers are integrating robotics with synthetic biology to create bioengineered tissues and organs, addressing medical challenges like organ transplantation, tissue repair, and disease modelling. This interdisciplinary approach contributes to regenerative medicine and personalized healthcare.

GENETIC CIRCUITRY FOR DISEASE DETECTION AND TREATMENT

Genetic circuitry, a key component of synthetic biology, offers innovative solutions for disease detection and treatment. By engineering genetic circuits within living cells, researchers can create biological systems capable of sensing specific disease biomarkers and triggering targeted therapeutic responses (Selvakumar, Adithe, et al., 2023; Selvakumar, Shankar, et al., 2023; Sengeni et al., 2023; T. Senthil et al., 2023; Subha, Inbamalar, Komala, et al., 2023). This research focus explores the potential of genetic circuitry for disease detection and treatment, with a particular emphasis on utilizing robotic systems for biomarker detection and applying synthetic biology to develop genetic therapies for genetic diseases (Sathish et al., 2023).

- *Genetic Biosensors*: Genetic biosensors are engineered genetic circuits that can detect specific biomolecules or disease biomarkers. Researchers are developing biosensors that respond to various disease-related molecules, such as proteins, nucleic acids, or metabolites. Integrating these biosensors with robotic systems allows for high-throughput and automated biomarker detection in clinical samples.
- *Robotic-assisted Diagnostic Devices*: Robotics can be employed to design and build diagnostic devices that incorporate genetic circuitry for disease detection. These devices can be portable, user-friendly, and capable of rapid and accurate diagnosis, enabling early detection and timely treatment.
- *Programmable Therapeutic Responses*: Genetic circuits can be programmed to respond to disease biomarkers by producing therapeutic molecules or activating immune responses. By integrating these circuits with robotic drug delivery systems, researchers can develop smart therapies that selectively target diseased cells while sparing healthy tissues.
- *Precision Medicine*: Genetic circuitry offers the potential for personalized medicine, where therapies can be tailored to an individual's genetic makeup and disease profile. Robotic systems can aid in the automation and customization of genetic therapies, ensuring patient-specific treatments for improved efficacy.

Robotic Systems for Detecting Disease Biomarkers

- *Lab-on-a-Chip Devices*: Robotic systems can automate lab-on-a-chip devices, which integrate multiple biochemical and biological assays on a small chip. These devices can be used for rapid and high-throughput biomarker detection, facilitating point-of-care diagnostics.
- *Automated Sample Handling*: Robotics can automate sample handling processes, such as sample preparation, extraction, and purification. This ensures consistency and reduces human error in biomarker detection workflows.
- *High-throughput Screening*: Assistive robotics enables high-throughput screening of large sample libraries, accelerating the discovery of disease biomarkers and potential therapeutic targets.
- *Real-time Monitoring*: Robotics can be integrated with real-time monitoring systems to track biomarker levels continuously. This continuous monitoring provides valuable data for disease diagnosis and treatment assessment.

Synthetic Biology Applications for Genetic Disease Therapies

- *Gene Replacement Therapy*: Synthetic biology enables the design and delivery of therapeutic genes to replace defective or mutated genes responsible for genetic diseases. Gene replacement therapy aims to restore normal cellular functions and alleviate disease symptoms.
- *Gene Editing*: Advances in synthetic biology have led to powerful gene editing technologies, such as CRISPR-Cas9, that can precisely modify or correct disease-causing genetic mutations. Gene editing holds tremendous potential for treating a wide range of genetic disorders.
- *RNA Therapies*: Synthetic biology allows for the engineering of RNA molecules with specific functions, such as RNA interference (RNAi) or antisense RNA. These RNA-based therapies can target disease-causing genes or regulate gene expression for therapeutic purposes.
- *Synthetic Gene Regulation*: Synthetic biology offers the ability to design genetic circuits that regulate the expression of specific genes. These synthetic regulatory circuits can be utilized to control gene expression in response to disease biomarkers or environmental cues.

The integration of genetic circuitry, robotic systems, and synthetic biology offers potential for advancing disease detection and treatment strategies. By combining genetic engineering and robotics, researchers can create smart diagnostic devices, programmable therapeutic responses, and personalized genetic therapies, improving patient outcomes and transforming healthcare. This interdisciplinary approach enables precision medicine and targeted therapies for various diseases, including genetic disorders.

ASSISTIVE ROBOTICS IN SURGERY AND MINIMALLY INVASIVE PROCEDURES

Robotics in Surgical Procedures: Enhancing Precision and Safety

Robotic systems have revolutionized surgical procedures by providing surgeons with enhanced precision and dexterity. With the use of robotic arms and advanced imaging, surgeons can perform complex operations with increased accuracy and reduced invasiveness, leading to shorter recovery times and minimized risk of complications. Robotic-assisted surgeries also enable better visualization of the surgical site and allow for more precise maneuvers, making them particularly valuable for delicate and intricate procedures (Rahamathunnisa, Sudhakar, Murugan, Thivaharan S., et al., 2023; Sathish et al., 2023; Subha, Inbamalar, R, et al., 2023).

Synthetic Biology Contributions to Minimally Invasive Surgeries

Synthetic biology plays a vital role in advancing minimally invasive surgical techniques by providing innovative tools for tissue engineering and regenerative medicine. By utilizing synthetic biological components, such as engineered tissues, biomaterials, and drug delivery systems, surgeons can achieve better tissue repair and regeneration. Additionally, synthetic biology enables the development of smart surgical tools that respond to specific cues within the body, enhancing the efficacy and safety of minimally invasive procedures. The integration of robotics and synthetic biology holds the potential to transform surgical practices, offering a new frontier in precision medicine and patient-centered care (Jiang et al., 2018; F. Li et al., 2018).

Assistive Robots for Physical Therapy and Rehabilitation

Physical therapy and rehabilitation play a critical role in helping individuals recover from injuries, surgeries, or neurological conditions and regain functional independence. The integration of assistive robots in these processes has emerged as a revolutionary approach, offering enhanced support, personalized treatment, and accelerated recovery (Boopathi, Alqahtani, et al., 2023a; Gowri et al., 2023a; Hanumanthakari et al., 2023b; Maguluri et al., 2023a). This article explores the use of assistive robots in physical therapy and rehabilitation, highlighting their benefits, applications, and the future potential they hold in transforming healthcare and improving patient outcomes (Carbonell et al., 2019; Tinafar et al., 2019).

The Role of Assistive Robots in Physical Therapy: Assistive robots in physical therapy serve as valuable tools for both patients and healthcare professionals. These robots are designed to interact with patients in a safe and intuitive manner, providing various therapeutic interventions and exercises to promote healing and functional recovery. Key roles of assistive robots in physical therapy include:

- *Gait Training*: Robots equipped with exoskeletons or wearable devices can assist individuals with gait impairments, helping them regain mobility and improve walking patterns.
- *Muscle Strengthening*: Robots provide controlled resistance and support for muscle strengthening exercises, helping patients rebuild strength after injuries or surgeries.
- *Range of Motion Exercises*: Robots aid in performing passive or active range of motion exercises, assisting patients in regaining flexibility and joint mobility.
- *Neurological Rehabilitation*: Robotics can facilitate repetitive and task-specific exercises, which are crucial for neurorehabilitation in conditions such as stroke or spinal cord injury.
- *Functional Training*: Assistive robots allow patients to practice functional movements, such as reaching or grasping, to improve activities of daily living.

Benefits of Assistive Robots in Rehabilitation: The integration of assistive robots in rehabilitation offers several advantages that enhance the overall therapeutic process:

- *Personalization*: Assistive robots can be programmed to adapt to individual patient needs, providing personalized treatment plans and tracking progress over time.
- *Repetition and Consistency*: Robots offer consistent and repetitive training, which is essential for neuroplasticity and skill acquisition during rehabilitation.
- *Real-time Feedback*: Robots can provide real-time feedback on patients' performance, enabling them to make adjustments and improvements during therapy sessions.
- *Reduced Strain on Therapists*: By automating certain aspects of therapy, assistive robots help reduce the physical strain on therapists, allowing them to focus on more specialized interventions.
- *Motivation and Engagement*: The interactive and engaging nature of assistive robots can motivate patients to actively participate in therapy, leading to better adherence and outcomes.

Applications of Assistive Robots in Physical Therapy: Assistive robots have found applications in various fields of physical therapy and rehabilitation:

- *Stroke Rehabilitation*: Robots can assist in retraining upper limb movements and gait patterns in stroke survivors, promoting functional recovery and independence.
- *Spinal Cord Injury*: Robots are used to support standing and walking exercises for individuals with spinal cord injuries, aiding in muscle re-education and bone density maintenance.
- *Orthopedic Rehabilitation*: Assistive robots play a role in post-surgery rehabilitation, helping patients regain strength and range of motion in affected joints.
- *Pediatric Rehabilitation*: Robots are employed in pediatric therapy to make rehabilitation more enjoyable for children and facilitate their development and recovery.
- *Elderly Care*: Assistive robots can support the elderly in maintaining mobility and performing daily activities, contributing to a higher quality of life.

Challenges and Considerations: While assistive robots offer significant potential, there are challenges that need to be addressed for their wider adoption:

- *Cost*: The initial investment and maintenance costs of robotic systems can be substantial, limiting their availability in certain healthcare settings.
- *Human-Robot Interaction*: Ensuring a seamless and safe interaction between patients and robots requires careful design and consideration of human factors.
- *Customization*: Each patient's rehabilitation needs are unique, requiring robots with a high degree of flexibility and adaptability to individual cases.
- *Skill Transfer*: Therapeutic gains made with assistive robots need to transfer to real-world functional improvements for patients to maintain independence outside of therapy sessions.

Synthetic Biology Techniques for Neural Repair and Recovery

Neural repair and recovery are critical challenges in addressing neurological disorders, brain injuries, and degenerative conditions. Synthetic biology offers innovative approaches to facilitate neural repair, regeneration, and functional recovery. This section explores the applications of synthetic biology in neural repair and recovery, focusing on its potential to revolutionize the treatment of neurological conditions and enhance the understanding of brain function (Baumschlager & Khammash, 2021; Nguyen et al., 2023; Schuergers et al., 2017a).

- **Neural Tissue Engineering**: Synthetic biology techniques enable the design and fabrication of neural tissue constructs with biomaterials and engineered cells. These constructs can be used as grafts to repair damaged neural tissue or as models to study neural development and disease. The bioengineered neural tissue offers a platform to investigate the interactions between neurons and glial cells, contributing to the understanding of neural regeneration.
- **Gene Therapy for Neurological Disorders**: Gene therapy, a powerful application of synthetic biology, holds significant promise for treating genetic neurological disorders. Researchers can use synthetic genetic circuits to introduce functional genes into affected neurons, replacing or compensating for defective genes. Additionally, gene editing technologies, such as CRISPR-Cas9, offer the potential to correct genetic mutations responsible for neurological conditions.
- **Optogenetics and Neural Modulation**: Optogenetics, a synthetic biology technique, involves the use of light-sensitive proteins to control neural activity with precision. By introducing optogenetic constructs into specific neural circuits, researchers can manipulate neural firing and study the neural basis of behavior. This technique has therapeutic potential in modulating neural networks to treat conditions like epilepsy, depression, and Parkinson's disease.
- **Neural Regeneration and Growth Factors**: Synthetic biology allows the engineering of cells to produce growth factors and cytokines that promote neural regeneration and repair. Engineered cells can be transplanted into the damaged brain, releasing therapeutic factors that support neurogenesis and axonal growth. This approach is particularly relevant for traumatic brain injuries and neurodegenerative diseases.
- **Brain-Computer Interfaces (BCIs):** BCIs are innovative devices that interface with the brain, translating neural signals into external commands or feedback. Synthetic biology plays a role in developing biohybrid BCIs, where living neural cells are incorporated into the device to enhance biocompatibility and neural integration. BCIs have significant potential for restoring communication and motor control in individuals with severe neurological impairments.

The synergy of assistive robots in physical therapy and synthetic biology techniques for neural repair represents a transformative frontier in healthcare. Assistive robots offer precise and personalized therapeutic interventions, accelerating rehabilitation and improving functional outcomes. Synthetic biology opens new avenues for neural repair and recovery, providing novel approaches to treat neurological conditions and advance our understanding of the brain. By combining these two fields, researchers and clinicians can revolutionize healthcare, bringing us closer to the realization of regenerative therapies and enhanced quality of life for patients with neurological disorders.

REAL TIME CASE STUDY

Robotic Prosthetic Limbs with Enhanced Sensory Feedback for Amputees

The collaboration between assistive robotics and synthetic biology has led to ground breaking advancements in the field of medicine, particularly in the development of advanced prosthetic limbs for amputees. This real-time case study highlights the impactful applications and achievements resulting from the integration of assistive robotics and synthetic biology in the design of robotic prosthetic limbs with enhanced sensory feedback (Schuergers et al., 2017b; Singh, 2014; Sun et al., 2013).

Amputation due to trauma, congenital conditions, or disease can severely impact an individual's quality of life and functional independence. Traditional prosthetic limbs, although offering mobility, often lack sensory feedback, leading to challenges in performing precise movements and interacting with the environment. The collaboration between assistive robotics and synthetic biology has sought to address these limitations and revolutionize prosthetic limb technology.

Applications of Assistive Robotics and Synthetic Biology

- *Robotic Limb Design and Control*: Assistive robotics plays a crucial role in the design and development of robotic prosthetic limbs. Advanced robotic arms and legs are equipped with sensors and actuators that mimic the natural movements of human limbs. Synthetic biology techniques contribute to the design of biocompatible materials and engineered cells that can integrate with the human body, ensuring a seamless connection between the robotic limb and the user.
- *Sensory Feedback Integration*: One of the most significant achievements in this collaboration is the integration of synthetic biology techniques to provide sensory feedback to amputees. By incorporating engineered sensory cells, such as optogenetic sensors, into the prosthetic limb, researchers have enabled real-time feedback to the user's brain. This feedback allows amputees to experience touch, temperature, and pressure, enhancing their ability to interact with objects and their environment more naturally.
- *Neural Interface and Brain-Computer Integration*: The collaboration between assistive robotics and synthetic biology has enabled the development of advanced neural interfaces for prosthetic limbs. By incorporating synthetic neural circuits and optogenetic components, researchers have created brain-computer interfaces (BCIs) that facilitate seamless communication between the user's brain and the robotic limb. This breakthrough enables intuitive control of the prosthetic limb, making it an extension of the user's body.

Impactful Collaborations and Achievements

- *Enhanced Motor Control and Precision*: The integration of synthetic biology in the design of robotic prosthetic limbs has led to significant improvements in motor control and precision. Amputees using these advanced prosthetic limbs report smoother and more natural movements, allowing them to perform intricate tasks with greater ease.
- *Restoring Sensory Perception*: The incorporation of synthetic sensory feedback has had a profound impact on amputees' daily lives. With the ability to sense touch and pressure through the prosthetic limb, users have reported a restored sense of embodiment and increased confidence in their movements.
- *Reduced Phantom Limb Pain*: Phantom limb pain is a common issue faced by amputees, wherein they experience pain or discomfort in the missing limb. The sensory feedback provided by the synthetic biology-engineered prosthetic limb has shown promising results in reducing phantom limb pain, leading to improved overall well-being and mental health for amputees.

- *Enhanced Rehabilitation and Adaptation*: The collaborative efforts between assistive robotics and synthetic biology have streamlined the rehabilitation process for amputees. The intuitive control and sensory feedback from the robotic limb have accelerated the adaptation period, allowing users to integrate the prosthetic limb seamlessly into their daily routines.

The real-time case study of robotic prosthetic limbs with sensory feedback demonstrates the transformative impact of assistive robotics and synthetic biology in medicine.

FUTURE PERSPECTIVES AND CHALLENGES

Assistive robotics and synthetic biology have already made significant strides in medicine, but the field continues to evolve with exciting future prospects. However, there are several challenges that need to be addressed to fully realize the potential of this collaboration (French, 2019; Nguyen et al., 2023; Schuergers et al., 2017a).

Emerging Trends and Potential Breakthroughs in the Field

- Advanced brain-computer interfaces and synthetic biology could revolutionize neuroprosthetics, enabling seamless communication between the brain and external devices, enabling individuals with neurological conditions to control robotic limbs or communicate through thought.
- Personalized therapies in synthetic biology offer potential for gene and cellular treatments for various diseases. Advancements in robotics and automation could enable efficient and safe delivery of these therapies to patients.
- Robotics and synthetic biology enable smart drug delivery systems targeting diseased cells, reducing side effects and improving treatment effectiveness.

Addressing Challenges and Limitations for Wider Adoption: Robotic systems and synthetic biology techniques face barriers to widespread adoption due to their high initial cost. Research and development are needed to make these technologies more affordable and accessible. Ethical considerations, particularly in gene editing and brain-computer interfaces, must be addressed. Regulatory approval is crucial for safety and efficacy before widespread implementation (Gowri et al., 2023b; Maguluri et al., 2023b; Mohanty, Venkateswaran, et al., 2023; T. S. Senthil et al., 2023).

Assistive robotics and synthetic biology are transforming medicine, enabling personalized treatments, improved patient outcomes, and enhanced quality of life for individuals with disabilities and neurological conditions (Boopathi, 2023b; Mohanty, Jothi, et al., 2023b). This field has already demonstrated remarkable achievements in physical therapy, neural repair, and prosthetic technology. The collaboration highlights the benefits of assistive robots in physical therapy and rehabilitation, providing personalized interventions for improved patient mobility and functional outcomes. Synthetic biology contributes to neural repair and recovery, paving the way for advanced brain-computer interfaces and neuroprosthetics. Robotic prosthetic limbs with sensory feedback significantly impact amputees, offering enhanced motor control, reduced pain, and improved well-being (Boopathi, Alqahtani, et al., 2023b; Hanumanthakari et al., 2023a; Hema et al., 2023; Koshariya, Kalaiyarasi, et al., 2023; Rahamathunnisa, Sudhakar, Murugan, Thivaharan, et al., 2023b).

Assistive robotics and synthetic biology in medicine hold great promise for groundbreaking breakthroughs in neuroprosthetics, personalized therapies, and smart drug delivery. However, addressing challenges like cost, accessibility, and ethical considerations is crucial for unlocking their full potential. The integration of these technologies is shaping the future of medicine, bringing innovative solutions to physical therapy, rehabilitation, neural repair, and prosthetic technology. As research and development continue, these technologies will improve patient care, foster independence, and revolutionize healthcare for generations to come. The journey towards a more inclusive and transformative healthcare system is well underway, driven by the synergy of assistive robotics and synthetic biology in medicine.

CONCLUSION

The collaboration between assistive robotics and synthetic biology in medicine is a ground-breaking convergence of cutting-edge technologies, revolutionizing patient care and treatment outcomes. This synergistic approach has shown significant advancements in physical therapy, rehabilitation, neural repair, and prosthetic technology, with promising future prospects.

The collaboration between assistive robotics and synthetic biology has led to several key benefits, including precision and personalization in robotics, neural repair and regeneration for neurological disorders and brain injuries, innovative therapies in medicine, and enhanced prosthetic technology. Synthetic biology techniques have opened new avenues for neural repair and regeneration, restoring sensory feedback, improving motor control, and reducing phantom limb pain in amputees. Additionally, the integration of synthetic biology in medicine has led to the development of gene therapies, cellular treatments, and smart drug delivery systems, offering targeted and customized approaches to diseases at the molecular level. The collaboration has also improved motor control and quality of life for amputees, providing a sense of embodiment and restored functionality.

Challenges like cost, accessibility, and ethical considerations persist in the adoption of assistive robotics and synthetic biology in medicine. Overcoming these obstacles is crucial for the full realization of their potential. Emerging trends like neuroprosthetics, personalized therapies, and brain-computer interfaces hold promise for groundbreaking healthcare breakthroughs. Continued research and development will unlock their transformative potential, paving the way for a more inclusive and advanced healthcare system.

ABBREVIATIONS

BCI - Brain-Computer Interface

DNA - Deoxyribonucleic Acid

CRISPR-Cas9 - Clustered Regularly Interspaced Short Palindromic Repeats-CRISPR associated protein 9

ECM - Extracellular Matrix

RNA - Ribonucleic Acid

RNAi - RNA Interference

BCIs - Brain-Computer Interfaces

REFERENCES

Alam, I., Alam, G., Ayub, S., & Siddiqui, A. A. (2019). Assessment of bio-medical waste management in different hospitals in Aligarh city. *Advances in Waste Management: Select Proceedings of Recycle, 2016*, 501–510. doi:10.1007/978-981-13-0215-2_36

Anitha, C., R, K. C., Vivekanand, C. V., Lalitha, S. D., Boopathi, S., & R, Revathi. (2023, February). Artificial Intelligence driven security model for Internet of Medical Things (IoMT). *IEEE Explore*. doi:10.1109/ICIPTM57143.2023.10117713

Anitha, C., Komala, C., Vivekanand, C. V., Lalitha, S., & Boopathi, S. (2023). Artificial Intelligence driven security model for Internet of Medical Things (IoMT). *IEEE Explore*, 1–7.

Baumschlager, A., & Khammash, M. (2021). Synthetic Biological Approaches for Optogenetics and Tools for Transcriptional Light-Control in Bacteria. *Advanced Biology*, *5*(5), 2000256. doi:10.1002/adbi.202000256 PMID:34028214

Boopathi, S. (2022). Performance Improvement of Eco-Friendly Near-Dry wire-Cut Electrical Discharge Machining Process Using Coconut Oil-Mist Dielectric Fluid. *World Scientific: Journal of Advanced Manufacturing Systems*.

Boopathi, S. (2023a). *Deep Learning Techniques Applied for Automatic Sentence Generation*. doi:10.4018/978-1-6684-3632-5.ch016

Boopathi, S. (2023b). Internet of Things-Integrated Remote Patient Monitoring System: Healthcare Application. In *Dynamics of Swarm Intelligence Health Analysis for the Next Generation* (pp. 137–161). IGI Global. doi:10.4018/978-1-6684-6894-4.ch008

Boopathi, S. (2023c). Securing Healthcare Systems Integrated With IoT: Fundamentals, Applications, and Future Trends. In Dynamics of Swarm Intelligence Health Analysis for the Next Generation (pp. 186–209). IGI Global.

Boopathi, S., Alqahtani, A. S., Mubarakali, A., & Panchatcharam, P. (2023). Sustainable developments in near-dry electrical discharge machining process using sunflower oil-mist dielectric fluid. *Environmental Science and Pollution Research International*, 1–20. doi:10.100711356-023-27494-0 PMID:37199846

Boopathi, S., & Davim, J. P. (2023a). Applications of Nanoparticles in Various Manufacturing Processes. In *Sustainable Utilization of Nanoparticles and Nanofluids in Engineering Applications* (pp. 1–31). IGI Global. doi:10.4018/978-1-6684-9135-5.ch001

Boopathi, S., & Davim, J. P. (2023b). *Sustainable Utilization of Nanoparticles and Nanofluids in Engineering Applications*. IGI Global. doi:10.4018/978-1-6684-9135-5

Boopathi, S., Khare, R., KG, J. C., Muni, T. V., & Khare, S. (2023). Additive Manufacturing Developments in the Medical Engineering Field. In Development, Properties, and Industrial Applications of 3D Printed Polymer Composites (pp. 86–106). IGI Global.

Boopathi, S., Khare, R., Jaya Christiyan, K. G., Muni, T. V., & Khare, S. (2023). Additive manufacturing developments in the medical engineering field. In Development, Properties, and Industrial Applications of 3D Printed Polymer Composites (pp. 86–106). IGI Global. doi:10.4018/978-1-6684-6009-2.ch006

Boopathi, S., Thillaivanan, A., Mohammed, A. A., Shanmugam, P., & VR, P. (2022). Experimental investigation on Abrasive Water Jet Machining of Neem Wood Plastic Composite. *IOP: Functional Composites and Structures, 4*, 025001.

Boopathi, S., Umareddy, M., & Elangovan, M. (2023). Applications of Nano-Cutting Fluids in Advanced Machining Processes. In *Sustainable Utilization of Nanoparticles and Nanofluids in Engineering Applications* (pp. 211–234). IGI Global. doi:10.4018/978-1-6684-9135-5.ch009

Carbonell, P., Radivojevic, T., & García Martín, H. (2019). Opportunities at the Intersection of Synthetic Biology, Machine Learning, and Automation. In ACS Synthetic Biology (Vol. 8, Issue 7, pp. 1474–1477). ACS Publications. doi:10.1021/acssynbio.8b00540

Deheyab, A. O. A., Alwan, M. H., Rezzaqe, I. K. A., Mahmood, O. A., Hammadi, Y. I., Kareem, A. N., & Ibrahim, M. (2022). An overview of challenges in medical image processing. *ACM International Conference Proceeding Series*, 511–516. 10.1145/3584202.3584278

Domakonda, V. K., Farooq, S., Chinthamreddy, S., Puviarasi, R., Sudhakar, M., & Boopathi, S. (2022). Sustainable Developments of Hybrid Floating Solar Power Plants: Photovoltaic System. In Human Agro-Energy Optimization for Business and Industry (pp. 148–167). IGI Global.

French, K. E. (2019). Harnessing synthetic biology for sustainable development. *Nature Sustainability*, *2*(4), 250–252. doi:10.103841893-019-0270-x

Gowri, N. V., Dwivedi, J. N., Krishnaveni, K., Boopathi, S., Palaniappan, M., & Medikondu, N. R. (2023). Experimental investigation and multi-objective optimization of eco-friendly near-dry electrical discharge machining of shape memory alloy using Cu/SiC/Gr composite electrode. *Environmental Science and Pollution Research International*, 1–19. doi:10.100711356-023-26983-6 PMID:37126160

Gunasekaran, K., Boopathi, S., & Sureshkumar, M. (2022). Analysis of a Cryogenically Cooled Near-Dry WEDM Process using Different Dielectrics. *Materials Technology*, *56*(2), 179–186.

Hanumanthakari, S., Gift, M. D. M., Kanimozhi, K. V., Bhavani, M. D., Bamane, K. D., & Boopathi, S. (2023a). Biomining Method to Extract Metal Components Using Computer-Printed Circuit Board E-Waste. In *Handbook of Research on Safe Disposal Methods of Municipal Solid Wastes for a Sustainable Environment* (pp. 123–141). IGI Global. doi:10.4018/978-1-6684-8117-2.ch010

Hanumanthakari, S., Gift, M. M., Kanimozhi, K., Bhavani, M. D., Bamane, K. D., & Boopathi, S. (2023b). Biomining Method to Extract Metal Components Using Computer-Printed Circuit Board E-Waste. In *Handbook of Research on Safe Disposal Methods of Municipal Solid Wastes for a Sustainable Environment* (pp. 123–141). IGI Global. doi:10.4018/978-1-6684-8117-2.ch010

Hema, N., Krishnamoorthy, N., Chavan, S. M., Kumar, N. M. G., Sabarimuthu, M., & Boopathi, S. (2023). A Study on an Internet of Things (IoT)-Enabled Smart Solar Grid System. In *Handbook of Research on Deep Learning Techniques for Cloud-Based Industrial IoT* (pp. 290–308). IGI Global. doi:10.4018/978-1-6684-8098-4.ch017

Javaid, M., & Haleem, A. (n.d.). *4D printing applications in medical field: A brief review*. Academic Press.

Jeevanantham, Y. A., Saravanan, A., Vanitha, V., Boopathi, S., & Kumar, D. P. (2022). Implementation of Internet-of Things (IoT) in Soil Irrigation System. *IEEE Explore*, 1–5.

Jiang, L., Zhao, J., Lian, J., & Xu, Z. (2018). Cell-free protein synthesis enabled rapid prototyping for metabolic engineering and synthetic biology. *Synthetic and Systems Biotechnology*, *3*(2), 90–96. doi:10.1016/j.synbio.2018.02.003 PMID:29900421

Kagadis, G. C., Kloukinas, C., Moore, K., Philbin, J., Papadimitroulas, P., Alexakos, C., Nagy, P. G., Visvikis, D., & Hendee, W. R. (2013). Cloud computing in medical imaging. *Medical Physics*, *40*(7), 70901. doi:10.1118/1.4811272 PMID:23822402

Kannan, E., Trabelsi, Y., Boopathi, S., & Alagesan, S. (2022). Influences of cryogenically treated work material on near-dry wire-cut electrical discharge machining process. *Surface Topography : Metrology and Properties*, *10*(1), 015027. doi:10.1088/2051-672X/ac53e1

Klangsin, P., & Harding, A. K. (1998). Medical waste treatment and disposal methods used by hospitals in Oregon, Washington, and Idaho. *Journal of the Air & Waste Management Association, 48*(6), 516–526.

Koshariya, A. K., Kalaiyarasi, D., Jovith, A. A., Sivakami, T., Hasan, D. S., & Boopathi, S. (2023). AI-Enabled IoT and WSN-Integrated Smart Agriculture System. In *Artificial Intelligence Tools and Technologies for Smart Farming and Agriculture Practices* (pp. 200–218). IGI Global. doi:10.4018/978-1-6684-8516-3.ch011

Koshariya, A. K., Khatoon, S., Marathe, A. M., Suba, G. M., Baral, D., & Boopathi, S. (2023). Agricultural Waste Management Systems Using Artificial Intelligence Techniques. In *AI-Enabled Social Robotics in Human Care Services* (pp. 236–258). IGI Global. doi:10.4018/978-1-6684-8171-4.ch009

Li, F., Wang, L., Liu, C., Wu, D., & Song, H. (2018). Engineering exoelectrogens by synthetic biology strategies. *Current Opinion in Electrochemistry*, *10*, 37–45. doi:10.1016/j.coelec.2018.03.030

Li, Z., Zhang, X., Müller, H., & Zhang, S. (2018). Large-scale retrieval for medical image analytics: A comprehensive review. *Medical Image Analysis*, *43*, 66–84. doi:10.1016/j.media.2017.09.007 PMID:29031831

Maguluri, L. P., Ananth, J., Hariram, S., Geetha, C., Bhaskar, A., & Boopathi, S. (2023). Smart Vehicle-Emissions Monitoring System Using Internet of Things (IoT). In Handbook of Research on Safe Disposal Methods of Municipal Solid Wastes for a Sustainable Environment (pp. 191–211). IGI Global.

Maheswari, B. U., Imambi, S. S., Hasan, D., Meenakshi, S., Pratheep, V., & Boopathi, S. (2023). Internet of Things and Machine Learning-Integrated Smart Robotics. In Global Perspectives on Robotics and Autonomous Systems: Development and Applications (pp. 240–258). IGI Global. doi:10.4018/978-1-6684-7791-5.ch010

Maier, A., Syben, C., Lasser, T., & Riess, C. (2019). A gentle introduction to deep learning in medical image processing. *Zeitschrift für Medizinische Physik*, *29*(2), 86–101. doi:10.1016/j.zemedi.2018.12.003 PMID:30686613

Mohanty, A., Jothi, B., Jeyasudha, J., Ranjit, P., Isaac, J. S., & Boopathi, S. (2023). Additive Manufacturing Using Robotic Programming. In *AI-Enabled Social Robotics in Human Care Services* (pp. 259–282). IGI Global. doi:10.4018/978-1-6684-8171-4.ch010

Mohanty, A., Venkateswaran, N., Ranjit, P. S., Tripathi, M. A., & Boopathi, S. (2023). Innovative Strategy for Profitable Automobile Industries: Working Capital Management. In Handbook of Research on Designing Sustainable Supply Chains to Achieve a Circular Economy (pp. 412–428). IGI Global.

Nguyen, P. Q., Huang, X., Collins, D. S., Collins, J. J., & Lu, T. (2023). Harnessing synthetic biology to enhance ocean health. *Trends in Biotechnology*, *41*(7), 860–874. Advance online publication. doi:10.1016/j.tibtech.2022.12.015 PMID:36669947

Paul, G., Roy, S., Sarkar, S., Hanumaiah, N., & Mitra, S. (2013). Investigations on influence of process variables on crater dimensions in micro-EDM of γ-titanium aluminide alloy in dry and oil dielectric media. *International Journal of Advanced Manufacturing Technology*, *65*(5–8), 1009–1017. doi:10.100700170-012-4235-8

Pramila, P., Amudha, S., Saravanan, T., Sankar, S. R., Poongothai, E., & Boopathi, S. (2023). Design and Development of Robots for Medical Assistance: An Architectural Approach. In Contemporary Applications of Data Fusion for Advanced Healthcare Informatics (pp. 260–282). IGI Global.

Rahamathunnisa, U., Sudhakar, K., Murugan, T. K., Thivaharan, S., Rajkumar, M., & Boopathi, S. (2023). *Cloud Computing Principles for Optimizing Robot Task Offloading Processes*. doi:10.4018/978-1-6684-8171-4.ch007

Rahamathunnisa, U., Sudhakar, K., Murugan, T. K., Thivaharan, S., Rajkumar, M., & Boopathi, S. (2023). Cloud Computing Principles for Optimizing Robot Task Offloading Processes. In *AI-Enabled Social Robotics in Human Care Services* (pp. 188–211). IGI Global. doi:10.4018/978-1-6684-8171-4.ch007

Ramudu, K., Mohan, V. M., Jyothirmai, D., Prasad, D., Agrawal, R., & Boopathi, S. (2023). Machine Learning and Artificial Intelligence in Disease Prediction: Applications, Challenges, Limitations, Case Studies, and Future Directions. In Contemporary Applications of Data Fusion for Advanced Healthcare Informatics (pp. 297–318). IGI Global.

Reddy, M. A., Gaurav, A., Ushasukhanya, S., Rao, V. C. S., Bhattacharya, S., & Boopathi, S. (2023). Bio-Medical Wastes Handling Strategies During the COVID-19 Pandemic. In Multidisciplinary Approaches to Organizational Governance During Health Crises (pp. 90–111). IGI Global. doi:10.4018/978-1-7998-9213-7.ch006

Saravanan, M., Vasanth, M., Boopathi, S., Sureshkumar, M., & Haribalaji, V. (2022). Optimization of Quench Polish Quench (QPQ) Coating Process Using Taguchi Method. *Key Engineering Materials*, *935*, 83–91. doi:10.4028/p-z569vy

Sathish, T., Sunagar, P., Singh, V., Boopathi, S., Sathyamurthy, R., Al-Enizi, A. M., Pandit, B., Gupta, M., & Sehgal, S. S. (2023). Characteristics estimation of natural fibre reinforced plastic composites using deep multi-layer perceptron (MLP) technique. *Chemosphere*, *337*(June), 139346. doi:10.1016/j.chemosphere.2023.139346 PMID:37379988

Sathiyaseelan, G., Srinath, P., Melwyn, J. G., Bhagyanathan, C., Socrates, S., & Boopathi, S. (2023). Tribology and hardness measurement of biosilica dispersed AA7075 metal matrix composites. *Materials Today: Proceedings*. Advance online publication. doi:10.1016/j.matpr.2023.02.440

Schuergers, N., Werlang, C., Ajo-Franklin, C. M., & Boghossian, A. A. (2017a). A synthetic biology approach to engineering living photovoltaics. *Energy & Environmental Science*, *10*(5), 1102–1115. doi:10.1039/C7EE00282C PMID:28694844

Schuergers, N., Werlang, C., Ajo-Franklin, C. M., & Boghossian, A. A. (2017b). A synthetic biology approach to engineering living photovoltaics. *Energy & Environmental Science*, *10*(5), 1102–1115. doi:10.1039/C7EE00282C PMID:28694844

Seifert, R., Weber, M., Kocakavuk, E., Rischpler, C., & Kersting, D. (2021). Artificial Intelligence and Machine Learning in Nuclear Medicine: Future Perspectives. *Seminars in Nuclear Medicine*, *51*(2), 170–177. doi:10.1053/j.semnuclmed.2020.08.003 PMID:33509373

Selvakumar, S., Adithe, S., Isaac, J. S., Pradhan, R., Venkatesh, V., & Sampath, B. (2023). A Study of the Printed Circuit Board (PCB) E-Waste Recycling Process. In Sustainable Approaches and Strategies for E-Waste Management and Utilization (pp. 159–184). IGI Global.

Selvakumar, S., Shankar, R., Ranjit, P., Bhattacharya, S., Gupta, A. S. G., & Boopathi, S. (2023). E-Waste Recovery and Utilization Processes for Mobile Phone Waste. In *Handbook of Research on Safe Disposal Methods of Municipal Solid Wastes for a Sustainable Environment* (pp. 222–240). IGI Global. doi:10.4018/978-1-6684-8117-2.ch016

Sengeni, D., Padmapriya, G., Imambi, S. S., Suganthi, D., Suri, A., & Boopathi, S. (2023). Biomedical Waste Handling Method Using Artificial Intelligence Techniques. In *Handbook of Research on Safe Disposal Methods of Municipal Solid Wastes for a Sustainable Environment* (pp. 306–323). IGI Global. doi:10.4018/978-1-6684-8117-2.ch022

Senthil, T., Puviyarasan, M., Babu, S. R., Surakasi, R., Sampath, B., & Associates. (2023). Industrial Robot-Integrated Fused Deposition Modelling for the 3D Printing Process. In Development, Properties, and Industrial Applications of 3D Printed Polymer Composites (pp. 188–210). IGI Global.

Senthil, T. S. R. Ohmsakthi vel, Puviyarasan, M., Babu, S. R., Surakasi, R., & Sampath, B. (2023). Industrial Robot-Integrated Fused Deposition Modelling for the 3D Printing Process. In Development, Properties, and Industrial Applications of 3D Printed Polymer Composites (pp. 188–210). IGI Global. doi:10.4018/978-1-6684-6009-2.ch011

Singh, V. (2014). Recent advancements in synthetic biology: Current status and challenges. *Gene*, *535*(1), 1–11. doi:10.1016/j.gene.2013.11.025 PMID:24269673

Subha, S., Inbamalar, T., Komala, C., Suresh, L. R., Boopathi, S., & Alaskar, K. (2023). A Remote Health Care Monitoring system using internet of medical things (IoMT). *IEEE Explore*, 1–6.

Sun, Z. Z., Hayes, C. A., Shin, J., Caschera, F., Murray, R. M., & Noireaux, V. (2013). Protocols for implementing an Escherichia coli based TX-TL cell-free expression system for synthetic biology. *Journal of Visualized Experiments*, *79*(79), e50762. doi:10.3791/50762 PMID:24084388

Tinafar, A., Jaenes, K., & Pardee, K. (2019). Synthetic Biology Goes Cell-Free. *BMC Biology*, *17*(1), 1–14. doi:10.118612915-019-0685-x PMID:31395057

Wisniewski, A., Zimmerman, M., Crews, T. Jr, Haulbrook, A., Fitzgerald, D. C., & Sistino, J. J. (2020). Reducing the impact of perfusion medical waste on the environment. *The Journal of Extra-Corporeal Technology*, *52*(2), 135–141. doi:10.1051/ject/202052135 PMID:32669740

Chapter 4
Deep Learning Techniques Revolutionizing Biomedical Applications:
Arrhythmia Detection, Cardiac Sensed Signals, and Cell-Free Synthetic Biology

Pranjit Das
Department of Computer Science and Engineering, Koneru Lakshmaiah Education Foundation, India

D. Vetrithangam
Department of Computer Science and Engineering, Chandigarh University, India

G. Hari Krishnan
Sree Vidyanikethan Engineering College, Mohan Babu University, India

Ashok Kumar
Department of Computer Science, Banasthali Vidyapith, India

R. Vijaya Kumar Reddy
Department of Computer Science and Engineering, Koneru Lakshmaiah Education Foundation, India

ABSTRACT

Deep learning techniques have revolutionized biomedical applications such as arrhythmia detection, analysis of cardiac sensed signals, and cell-free synthetic biology. This chapter provides an introduction to each topic, discusses the role of deep learning techniques, highlights recent advancements, and explores the challenges and future directions in these areas. It also provides an introduction to cardiac sensed signals and their relevance in cardiology and explores the role of deep learning in analyzing and interpreting cardiac signals. Cell-free synthetic biology is also discussed, including applications in gene circuit design and metabolic engineering. Deep learning techniques are increasingly being used in biomedical applications and are important for personalized medicine, explainability and interpretability, real-time monitoring and intervention, adaptive optimization, and integration with other technologies.

DOI: 10.4018/978-1-6684-6577-6.ch004

INTRODUCTION

In recent years, deep learning has emerged as a transformative technology that has revolutionized numerous fields, including biomedical research and healthcare. Deep learning techniques, fueled by advancements in computational power and the availability of large-scale datasets, have shown remarkable potential in addressing complex challenges and improving outcomes in various biomedical applications (Yatika et al., 2023; Yupapin et al., 2023; Zekrifa et al., 2023). This chapter aims to provide a research-oriented introduction to the use of deep learning techniques in key biomedical domains, namely arrhythmia detection, bioremediation, cardiac sensed signals, cell-free synthetic biology, climate change, and cognitive disorders. From a research standpoint, deep learning offers several advantages over traditional approaches in biomedical research (Sathiyaseelan et al., 2023; Sengeni et al., 2023a; Yatika et al., 2023). It excels at automatically extracting intricate patterns and features from large datasets, enabling the development of highly accurate and efficient models. Moreover, deep learning models are capable of learning hierarchical representations, which can capture complex relationships and nuances in the data. These capabilities have opened up new possibilities for solving critical challenges and advancing knowledge in various biomedical fields (Ching et al., 2018; Kang et al., 2022).

The importance of deep learning techniques in biomedical research cannot be understated. In the context of arrhythmia detection, for example, accurate and timely identification of irregular heart rhythms can significantly improve patient outcomes. Traditional approaches to arrhythmia detection often rely on manual feature extraction and rule-based algorithms, which can be limited in their ability to handle diverse and complex patterns (Hanumanthakari et al., 2023; Reddy et al., 2023). Deep learning models, on the other hand, can automatically learn relevant features directly from raw electrocardiogram (ECG) signals, leading to more accurate and efficient arrhythmia detection (Subha et al., 2023). Bioremediation is the process of using living organisms or their byproducts to remove pollutants from the environment. Deep learning techniques can leverage large-scale genomic, environmental, and chemical datasets to identify novel bioremediation candidates and optimize remediation processes. This research-driven approach can lead to more targeted and effective solutions for environmental cleanup (Carbonell et al., 2019; Nazir et al., 2023).

Deep learning techniques offer significant advantages for analyzing and interpreting complex physiological data. Traditional signal processing techniques require handcrafted features and domain-specific knowledge, but deep learning models can learn directly from raw data, capturing intricate patterns and relationships that may be difficult to extract manually. This enables enhanced cardiac arrhythmia diagnosis and monitoring, providing clinicians with valuable insights for patient care and treatment planning (Boopathi, 2023b; Chandrika et al., 2023; Hema et al., 2023; Syamala et al., 2023; Venkateswaran, Vidhya, Naik, et al., 2023b). Cell-free synthetic biology is an emerging field that focuses on harnessing biological components outside of living cells. Deep learning techniques can be applied to design and optimize cell-free systems, such as gene circuits and metabolic pathways, for various applications. Climate change is a pressing global concern that requires robust data analysis and predictive modeling. Deep learning techniques have demonstrated promise in climate change research by analyzing large-scale climate datasets, forecasting future climate patterns, and aiding in the understanding of complex climate dynamics (Parvaneh et al., 2019a). These research-oriented applications of deep learning can contribute to our understanding of climate change impacts, enabling more effective mitigation and adaptation strategies.

Deep learning techniques can assist in early detection, prediction, and intervention strategies for cognitive disorders. By leveraging neuroimaging data and other clinical information, deep learning models can uncover subtle biomarkers and patterns indicative of cognitive decline. This research-driven approach has the potential to improve diagnostic accuracy, personalize treatments, and contribute to advancements in cognitive disorder research. Deep learning is a subfield of machine learning that focuses on training artificial neural networks with multiple layers to learn and extract representations from complex data (Sengeni et al., 2023b). It has gained tremendous popularity and success in various domains, including computer vision, natural language processing, robotics, and healthcare. Deep learning has been used to improve computer vision tasks such as image classification, object detection, and image segmentation (Anitha et al., 2023). Convolutional neural networks (CNNs) have been used to analyze and understand visual data, while recurrent neural networks (RNNs) and transformer models have made significant strides in natural language processing tasks such as machine translation, sentiment analysis, and speech recognition. These models can learn complex linguistic patterns and generate coherent text, enabling advancements in chatbots, virtual assistants, and language generation systems (Carbonell et al., 2019).

Deep learning techniques are essential for biomedical applications, as biomedical data is complex, high-dimensional, and often heterogeneous. Traditional approaches to analyzing and interpreting this data often require handcrafted features and rule-based algorithms, which can be time-consuming and limited in their ability to capture intricate patterns. Deep learning, on the other hand, excels at automatically learning representations and extracting meaningful features directly from raw data, which can discover complex relationships and patterns that may not be apparent to human experts (T. S. Senthil et al., 2023). This capability is especially valuable in biomedical research, where identifying subtle biomarkers, predicting disease progression, and making accurate diagnoses are essential. Deep learning techniques have shown great promise in a wide range of biomedical applications, such as medical imaging, genetic data analysis, drug discovery, clinical decision support systems, and personalized medicine. It has the potential to improve diagnostic accuracy, provide more personalized treatment plans, accelerate drug discovery, and enhance patient outcomes. By leveraging the power of deep learning, researchers and healthcare professionals can gain deeper insights into complex biomedical data, leading to improved understanding, diagnosis, and treatment of diseases (Hammad et al., 2021).

Deep learning techniques have the ability to handle large-scale datasets, allowing for more robust model training and generalization. This allows for more reliable and effective solutions, and their ability to automatically learn and discover patterns has the potential to revolutionize medical research, diagnosis, and treatment, improving healthcare outcomes for individuals and populations.

DEEP LEARNING TECHNIQUES FOR ARRHYTHMIA DETECTION

Arrhythmia refers to an abnormality in the heart's rhythm, characterized by irregular or abnormal electrical activity. It is a significant medical condition that can have severe consequences, including palpitations, dizziness, chest pain, and even life-threatening events like stroke or sudden cardiac arrest. Arrhythmias are prevalent and can affect individuals of all ages, from infants to the elderly. The heart's electrical system controls the timing and coordination of its contractions, ensuring efficient blood circulation throughout the body. Disruptions in this electrical system can lead to various types of arrhythmias, such as atrial fibrillation, ventricular tachycardia, or bradycardia (Boopathi et al., 2021; Sampath et al., 2022). These conditions can significantly impact a person's quality of life and increase the risk of cardiovascular complications.

The significance of arrhythmia in healthcare stems from the need for accurate detection, diagnosis, and treatment. Timely identification of arrhythmias is crucial for appropriate management and intervention, as it allows healthcare professionals to implement targeted therapies, lifestyle modifications, or implantable devices to regulate the heart's rhythm. Arrhythmia detection plays a critical role in risk stratification, prognosis assessment, and guiding clinical decision-making (Venkateswaran, Vidhya, Naik, et al., 2023a). Traditionally, arrhythmia detection relied on manual interpretation of electrocardiogram (ECG) signals by clinicians. This process involved visual inspection of ECG waveforms, identification of abnormal patterns, and manual measurements of intervals and durations. However, manual analysis is time-consuming, subjective, and prone to human error, especially when dealing with large volumes of ECG data (Parvaneh et al., 2019b).

Deep learning techniques have emerged as a powerful approach for arrhythmia detection, offering advantages over traditional methods. By leveraging neural networks with multiple layers, deep learning models can automatically learn discriminative features from ECG signals, reducing the reliance on manual feature extraction. This allows for more efficient and accurate identification of arrhythmias, aiding in early detection and prompt intervention. Deep learning models for arrhythmia detection can be trained on large annotated datasets, enabling them to learn complex patterns and relationships (Mohanty, Jothi, et al., 2023). These models can capture subtle abnormalities in ECG waveforms that may be challenging for human observers to discern. Moreover, deep learning techniques facilitate the integration of diverse clinical data, such as patient demographics, medical history, and additional diagnostic tests, to enhance the accuracy and reliability of arrhythmia detection algorithms. The application of deep learning techniques in arrhythmia detection has the potential to improve patient outcomes, streamline clinical workflows, and provide valuable decision support for healthcare professionals. It can contribute to early detection of arrhythmias, enabling timely interventions and reducing the risk of adverse events. Furthermore, deep learning models can assist in remote monitoring of patients, facilitating personalized care and reducing the burden on healthcare systems (Hammad et al., 2021).

Deep learning techniques offer a promising approach to address the challenges associated with arrhythmia detection by leveraging advanced algorithms and large-scale datasets. By improving the efficiency and accuracy of arrhythmia detection, deep learning has the potential to enhance patient care and outcomes in cardiology.

Role of Deep Learning in Arrhythmia Detection

Deep learning techniques are essential for arrhythmia detection by leveraging their capabilities to learn complex patterns and representations from ECG data (Murat et al., 2020). Feature Learning: Deep learning models can automatically learn relevant features directly from ECG signals, reducing the need for manual feature extraction. This is particularly advantageous in arrhythmia detection, as ECG waveforms contain subtle temporal and spatial patterns that may be challenging for traditional algorithms to capture (Boopathi & Sivakumar, 2016; Ramudu et al., 2023; Sampath et al., 2022). Deep learning models can discover discriminative features at different levels of abstraction, allowing for more accurate and comprehensive representation of arrhythmias.

Deep learning models excel at classification tasks, and arrhythmia detection heavily relies on accurate classification of ECG signals into different rhythm classes. By leveraging deep neural networks, such as convolutional neural networks (CNNs) and recurrent neural networks (RNNs), these models can effectively learn the complex relationships between ECG waveforms and arrhythmia types. Deep learning models can handle diverse arrhythmia patterns and adapt to variations across different patients, leading to improved classification performance (Palaniappan et al., 2023a; Venkateswaran, Vidhya, Naik, et al., 2023a).

ECG signals can be affected by various sources of noise, such as muscle artifacts, baseline drift, and electrode movement. Traditional arrhythmia detection algorithms may struggle to handle such noise and variations. Deep learning models have the ability to learn robust representations from noisy and varied data. They can capture both global and local dependencies in the ECG signals, making them more resilient to noise and enhancing the overall performance of arrhythmia detection algorithms. Scalability and Generalization: Deep learning models can handle large-scale datasets, which is crucial for arrhythmia detection due to the availability of extensive ECG databases. These models can be trained on diverse datasets, incorporating a wide range of arrhythmia cases and patient demographics. This scalability enables the development of highly generalizable models that can accurately detect arrhythmias in unseen patients or in real-time monitoring scenarios.

Deep learning models can be optimized for real-time arrhythmia detection, allowing for immediate and continuous monitoring of patients. This is particularly valuable in critical care settings, where prompt detection and intervention are essential. Deep learning algorithms can process ECG signals in real-time, providing rapid feedback and alerts to healthcare professionals, enabling timely clinical decision-making (Babu et al., 2023; Jeevanantham et al., 2023).

Transfer learning, a technique in deep learning, allows models to leverage knowledge learned from one task or dataset to improve performance on another task or dataset. Pre-trained deep learning models, trained on large-scale datasets from related domains, can be fine-tuned for arrhythmia detection tasks, even with limited labeled data. This transfer learning approach reduces the need for large annotated datasets specific to arrhythmia detection, making it more feasible for practical implementation.

Recent Advancements and State-of-the-Art Models for Arrhythmia Detection

Deep learning techniques have improved accuracy and performance in arrhythmia detection, with state-of-the-art models demonstrating exceptional capabilities in detecting (Figure 1) and classifying various arrhythmia types (Izci et al., 2019).

1. **Convolutional Neural Networks (CNNs):**
    - One-dimensional CNNs have been widely used for arrhythmia detection from ECG signals. These models can capture temporal dependencies and extract informative features from ECG waveforms (Agrawal et al., 2023; Arunprasad & Boopathi, 2019; Boopathi, 2023c, 2023a; Chandrika et al., 2023).
    - Recent advancements include the development of deeper CNN architectures with residual connections, such as DenseNet, to improve feature extraction and classification performance.
2. **Recurrent Neural Networks (RNNs):**
    - RNNs, particularly Long Short-Term Memory (LSTM) and Gated Recurrent Unit (GRU) networks, have been successfully employed for arrhythmia detection (Dhanya et al., 2023; Domakonda et al., 2022; Durairaj et al., 2023; Gnanaprakasam et al., 2023).
    - These models are capable of capturing sequential dependencies in ECG signals, making them effective in detecting arrhythmias with long-term patterns.

Figure 1. Developments on arrhythmia detection using deep learning techniques

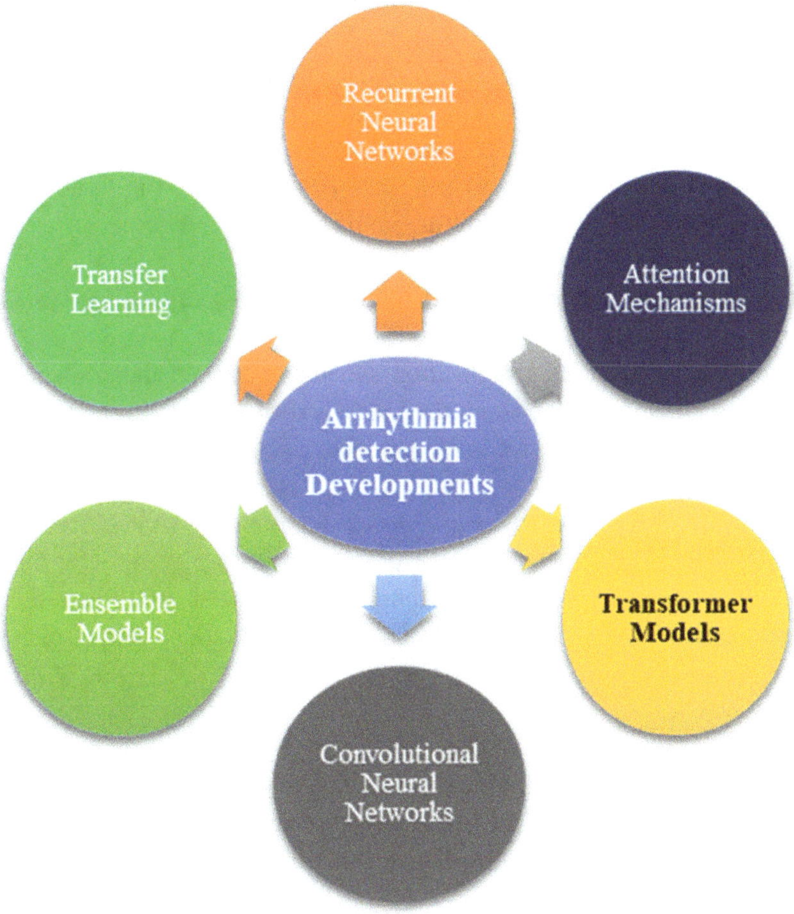

3. **Attention Mechanisms:**
   - Attention mechanisms have been introduced to enhance the performance of deep learning models for arrhythmia detection.
   - These mechanisms allow the model to focus on relevant segments of the ECG signal, amplifying informative features and suppressing noise, leading to improved accuracy.
4. **Transformer Models:**
   - Transformer models, originally developed for natural language processing tasks, have shown promise in arrhythmia detection.
   - These models leverage self-attention mechanisms to capture both local and global dependencies in the ECG signal, enabling accurate classification of arrhythmias.
5. **Ensemble Models:**
   - Ensemble models, which combine multiple deep learning models or multiple runs of the same model, have been utilized to improve the robustness and generalization of arrhythmia detection systems.

   - By leveraging the diversity of multiple models, ensemble approaches can achieve better performance and handle uncertainty in arrhythmia classification.
6. **Transfer Learning:**
   - Transfer learning has emerged as an effective technique for arrhythmia detection, where pre-trained models from related domains, such as natural images or general healthcare data, are fine-tuned for arrhythmia classification.
   - Transfer learning enables leveraging the knowledge captured in large-scale datasets, even with limited labeled ECG data, resulting in improved performance.
7. **Explainable AI:**
   - Recent advancements in explainable AI for arrhythmia detection aim to provide interpretable and transparent models.
   - Techniques such as attention visualization, saliency mapping, and gradient-based attribution methods help clinicians understand the model's decision-making process, enhancing trust and facilitating integration into clinical practice.

Deep learning techniques have been used to enhance accuracy, generalization, and real-world applicability in arrhythmia detection. These advances provide promising avenues for further research and development, paving the way for improved arrhythmia diagnosis, risk stratification, and personalized treatment approaches. Further advancements and clinical adoption are needed (Izci et al., 2019; Murat et al., 2020).

- *Data Variability and Generalization*: ECG signals can vary significantly across individuals due to factors such as age, gender, comorbidities, and device placement. Deep learning models need to be trained on diverse and representative datasets to ensure robustness and generalization. Future directions involve collecting large-scale, diverse, and annotated ECG datasets to improve model performance and adaptability.
- *Data Quality and Noise*: ECG signals can be affected by noise, artifacts, and recording errors, which can impact the accuracy of arrhythmia detection models. Handling noisy data and developing techniques to reduce noise interference remains a challenge. Future directions involve exploring data augmentation techniques, denoising algorithms, and developing robust models that can handle noisy ECG signals effectively.
- *Interpretable and Explainable Models*: Deep learning models often act as black boxes, making it challenging to interpret their decisions and provide explanations to clinicians. Interpretable and explainable AI techniques need to be developed to enhance trust, understand model outputs, and facilitate integration into clinical practice. Future directions involve exploring methods to visualize and explain the features learned by deep learning models for arrhythmia detection.
- *Limited Annotated Data*: Annotated ECG data for training deep learning models are often scarce and time-consuming to obtain, especially for rare arrhythmias or specific patient populations. Addressing the challenge of limited annotated data requires collaborative efforts to create comprehensive and standardized annotated datasets. Transfer learning and semi-supervised learning approaches can also be explored to leverage knowledge from related domains and utilize unlabeled data.

- *Real-time Monitoring and Deployment*: Arrhythmia detection models need to operate in real-time, enabling continuous monitoring and timely intervention. Future directions involve optimizing deep learning models for real-time performance, developing efficient hardware implementations, and integrating them with wearable devices or clinical monitoring systems.
- *Clinical Validation and Integration*: For successful clinical adoption, deep learning models for arrhythmia detection need to undergo rigorous validation studies to demonstrate their clinical utility, accuracy, and reliability. Future directions involve conducting large-scale clinical trials and collaborations with healthcare professionals to validate the performance of these models in real-world clinical settings.
- *Ethical and Regulatory Considerations*: Deep learning models raise ethical and regulatory considerations, including patient privacy, data security, and algorithm biases. Future directions involve developing guidelines and frameworks to address these concerns and ensure responsible development, deployment, and use of deep learning models in arrhythmia detection.
- *Personalized Medicine*: The future of arrhythmia detection lies in personalized medicine, where deep learning models can capture individual variations and assist in tailored treatment plans. Future directions involve integrating patient-specific data, such as genetic information, medical history, and lifestyle factors, to develop personalized arrhythmia detection models and enable precision medicine approaches.

DEEP LEARNING TECHNIQUES FOR BIOREMEDIATION

Bioremediation is a process that uses biological organisms or their by-products to remove, degrade, or transform pollutants from contaminated environments. It is cost-effective, environmentally friendly, and capable of treating a wide range of pollutants. Traditional bioremediation involves using indigenous microorganisms, plants, or their enzymatic activities to degrade or transform pollutants (Gunasekaran & Boopathi, 2023; Hanumanthakari et al., 2023; Harikaran et al., 2023; Hema et al., 2023; Janardhana et al., 2023).

- *Bio stimulation*: Bio stimulation aims to enhance the activity and growth of naturally occurring microorganisms in the contaminated site. This approach involves adding nutrients, oxygen, or electron acceptors to stimulate the metabolic processes of microorganisms, promoting the degradation of pollutants.
- *Bioaugmentation*: Bioaugmentation involves introducing or enhancing specific microorganisms or microbial consortia into the contaminated site to improve biodegradation capabilities. This approach is particularly useful when the indigenous microbial population is insufficient or lacks the necessary metabolic pathways to degrade certain pollutants.
- *Phytoremediation*: Phytoremediation utilizes plants to uptake, sequester, or degrade pollutants from the soil, water, or air. Different mechanisms, such as phytoextraction, phytodegradation, and Phyto stabilization, are employed depending on the contaminants and plant species involved.
- *Composting*: Composting is an effective technique for the treatment of organic pollutants, such as petroleum hydrocarbons and organic waste. It involves the controlled decomposition of organic materials by microorganisms under favourable conditions, resulting in the transformation of pollutants into more stable and less toxic forms.

- *Microembolisation*: Microembolisation utilizes fungi to degrade or transform various pollutants. Fungi possess unique enzymatic capabilities, allowing them to break down complex organic compounds, such as polycyclic aromatic hydrocarbons (PAHs) and certain pesticides.

Deep learning techniques can be used to advance bioremediation strategies, extract valuable insights from complex environmental data, optimize bioremediation processes, and design more efficient remediation strategies. It has the potential to improve pollutant degradation, accelerate clean-up times, and improve overall success rate of remediation projects.

Role of Deep Learning in Bioremediation Processes

Deep learning techniques can be used to optimize bioremediation strategies, providing benefits in bioremediation (Karthik et al., 2023; Koshariya, Kalaiyarasi, et al., 2023; Koshariya, Khatoon, et al., 2023; Kumar et al., 2023; Kumara et al., 2023), as shown in Figure 2.

Figure 2. Role of deep learning in bioremediation processes

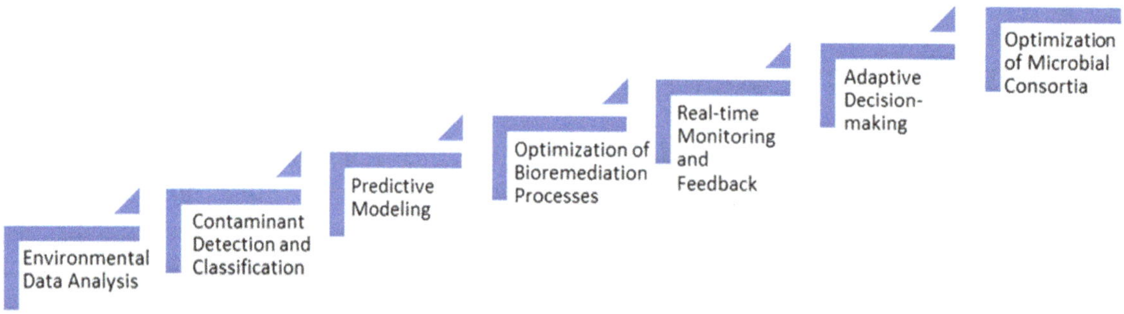

- *Environmental Data Analysis*: Deep learning models can process and analyze large volumes of environmental data, including chemical concentrations, soil characteristics, hydrological parameters, and microbial community information. By leveraging deep learning algorithms, patterns and correlations within the data can be identified, enabling a better understanding of the relationships between contaminants and environmental factors.
- *Contaminant Detection and Classification*: Deep learning models can be trained to detect and classify different types of contaminants based on data from sensors, spectroscopy, or chemical analyses. This can aid in the early identification of pollutants and their distribution in the environment, facilitating targeted remediation efforts.
- *Predictive Modeling*: Deep learning algorithms can generate predictive models that forecast the behavior and fate of contaminants in the environment. By incorporating historical data and various environmental parameters, these models can provide insights into the movement and transformation of pollutants over time, assisting in decision-making for remediation strategies.

- *Optimization of Bioremediation Processes*: Deep learning techniques can optimize bioremediation processes by determining the optimal conditions and parameters for enhanced pollutant degradation. By analyzing large datasets, deep learning models can identify key factors that influence bioremediation performance and recommend optimal strategies for maximizing pollutant removal or transformation.
- *Real-time Monitoring and Feedback*: Deep learning models can be integrated with real-time monitoring systems to provide continuous feedback on the progress and effectiveness of bioremediation processes. By analyzing data from sensors or remote sensing platforms, these models can detect changes in pollutant concentrations, microbial activity, or environmental conditions, allowing for timely adjustments and intervention.
- *Adaptive Decision-making*: Deep learning models can assist in adaptive decision-making during bioremediation projects. By continuously analyzing and learning from new data, these models can update and refine their predictions and recommendations, adapting to changing environmental conditions and optimizing remediation strategies accordingly.
- *Optimization of Microbial Consortia*: Deep learning can aid in the design and optimization of microbial consortia for bioremediation. By analyzing genomic and metagenomic data, deep learning models can identify potential synergistic interactions between microorganisms and optimize their composition to enhance pollutant degradation capabilities.

Applications of Deep Learning in Identifying and Optimizing Bioremediation Strategies

1. *Site Characterization*: Deep learning models can analyze geospatial and environmental data to identify suitable sites for bioremediation projects. By considering factors such as contaminant concentrations, soil properties, hydrological conditions, and microbial diversity, these models can assist in site selection for optimal remediation outcomes (Astuti et al., 2021; Naushin et al., 2022; Wahlang et al., 2020).
2. *Remediation Technique Selection*: Deep learning models can evaluate different bioremediation techniques based on site-specific data and recommend the most appropriate approach. By considering factors such as contaminant type, concentration, accessibility, and environmental conditions, these models can help in selecting the most effective and cost-efficient remediation technique.
3. *Bioremediation Optimization*: Deep learning models can optimize bioremediation parameters, such as nutrient dosages, aeration rates, pH levels, and temperature, to maximize pollutant degradation. By analyzing the interactions between environmental variables and bioremediation performance, these models can suggest optimal conditions for enhancing microbial activity and pollutant removal.
4. *Decision Support Systems*: Deep learning models can be integrated into decision support systems that provide real-time recommendations and insights during bioremediation projects. These systems can process and interpret data from monitoring systems, predict outcomes, and provide actionable guidance for remediation practitioners.

Deep learning can be used to refine, accelerate, and customize bioremediation strategies for more efficient and successful remediation outcomes.

DEEP LEARNING TECHNIQUES FOR CARDIAC SENSED SIGNALS

Cardiac sensed signals, such as ECG, provide valuable information about the electrical activity of the heart and are essential for diagnosing and monitoring cardiac conditions. These signals capture the depolarization and repolarization patterns of the heart, allowing clinicians to assess the heart's rhythm, identify abnormalities, and make informed decisions regarding patient care (Figure 3). Cardiac sensed signals play a significant role in cardiology, enabling the detection, diagnosis, and management of conditions such as arrhythmias, myocardial infarction, and heart failure. Traditional signal processing techniques are used to analyze and interpret cardiac signals (Mohanty, Venkateswaran, et al., 2023; Nishanth et al., 2023; Palaniappan et al., 2023b; Pramila et al., 2023; Rahamathunnisa et al., 2023; Ramudu et al., 2023).

Figure 3. Deep learning techniques for cardiac sensed signals

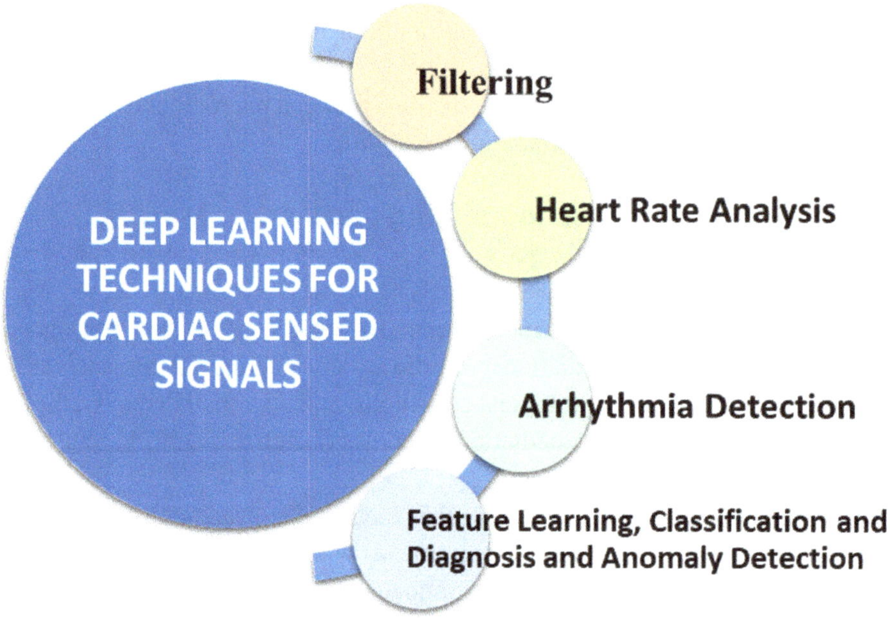

- *Filtering*: Filtering techniques, such as low-pass, high-pass, or band-pass filters, are used to remove noise and artifacts from cardiac signals, enhancing the quality of the signal for further analysis.
- *Feature Extraction*: Feature extraction methods are employed to identify relevant characteristics or landmarks in the cardiac signal. Commonly extracted features include R-peaks (QRS complex), P-waves, T-waves, and other wave morphology parameters.
- *Heart Rate Analysis*: Heart rate variability (HRV) analysis involves quantifying the variation in time intervals between successive heartbeats. HRV parameters provide insights into the autonomic regulation of the heart and can be used to assess the risk of cardiovascular diseases.

- *Arrhythmia Detection*: Traditional algorithms, such as rule-based algorithms or template matching, are used to detect and classify arrhythmias based on specific criteria defined by medical experts.

Role of Deep Learning in Analyzing and Interpreting Cardiac Signals

Deep learning techniques have been used to analyze and interpret cardiac sensed signals, offering advantages over traditional signal processing methods (Izci et al., 2019; Parvaneh et al., 2019b).

- *Feature Learning*: Deep learning models can automatically learn discriminative features directly from raw cardiac signals, eliminating the need for manual feature engineering. Convolutional neural networks (CNNs) can capture hierarchical representations of the signal, enabling more comprehensive and effective feature extraction.
- *Classification and Diagnosis*: Deep learning models, such as recurrent neural networks (RNNs) and deep neural networks (DNNs), can be trained to classify cardiac signals into different arrhythmia classes or identify specific cardiac conditions. These models can leverage the temporal dependencies and long-term dependencies in the signals for accurate classification and diagnosis.
- *Anomaly Detection*: Deep learning models can detect anomalies in cardiac signals, identifying patterns that deviate from normal physiological behavior. This capability is particularly useful in identifying rare or novel arrhythmias or detecting subtle changes in the cardiac signal that may indicate early signs of cardiac dysfunction.

Applications of Deep Learning in Cardiac Arrhythmia Diagnosis and Monitoring

Deep learning techniques have been used to diagnose and monitor cardiac arrhythmias (Kwon & Dong, 2022).

- *Arrhythmia Classification*: Deep learning models can classify different types of arrhythmias from ECG signals, enabling automated and accurate arrhythmia diagnosis. These models can differentiate between atrial fibrillation, ventricular tachycardia, bradycardia, and other arrhythmia classes, aiding in appropriate treatment decisions.
- *Risk Stratification*: Deep learning models can assess the risk of adverse cardiac events by analyzing cardiac signals and clinical data. By considering features such as HRV parameters, ECG morphology, and patient demographics, these models can predict the likelihood of future cardiac events, facilitating personalized treatment plans.
- *Continuous Monitoring*: Deep learning models can be deployed in wearable devices or remote monitoring systems to provide continuous monitoring of cardiac signals. Real-time analysis and interpretation of ECG data allow for early detection of arrhythmias or changes in cardiac activity, enabling timely intervention and reducing the risk of complications.

Case Studies and Recent Developments

Deep learning techniques are being used to diagnose and monitor cardiac arrhythmias, with examples of successful applications, comparisons to traditional methods, and potential impact on patient outcomes.

DEEP LEARNING TECHNIQUES FOR CELL-FREE SYNTHETIC BIOLOGY

Cell-free synthetic biology is an innovative approach that uses components of living cells to create functional biological systems. It offers applications in biotechnology, bioengineering, drug discovery, and biocomputing. By bypassing the complexities associated with living cells, it provides a versatile platform for designing and implementing artificial biological systems with enhanced control and programmability. Traditional approaches in cell-free synthetic biology involve using cell extracts to perform biochemical reactions outside of intact cells (Samikannu et al., 2022; Sampath et al., 2023; Sankar et al., 2023; Selvakumar, Adithe, et al., 2023; Selvakumar, Shankar, et al., 2023). These approaches require careful selection and preparation of the cellular components, optimization of reaction conditions, and control of reaction kinetics. Examples include in vitro transcription-translation systems, enzyme cascades, and metabolic pathway engineering using purified enzymes (Carbonell et al., 2019; Tinafar et al., 2019).

Role of Deep Learning in Designing and Optimizing Cell-Free Systems

Deep learning techniques are essential for designing and optimizing cell-free systems, offering advantages over traditional approaches (Jiang et al., 2018; Sun et al., 2013), as shown in Figure 4.

Figure 4. Role of deep learning in designing and optimizing cell-free systems

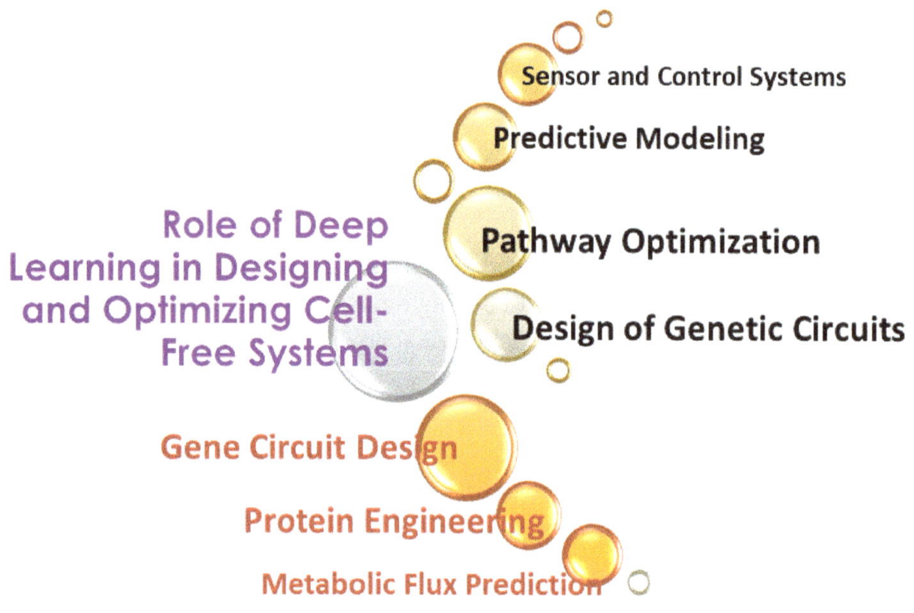

- *Design of Genetic Circuits*: Deep learning models can aid in the design and optimization of genetic circuits for cell-free systems. These models can analyze large datasets of genetic sequences, regulatory elements, and desired circuit behaviour to learn patterns and predict the functionality of synthetic genetic constructs. Deep learning models can assist in selecting suitable genetic components, optimizing circuit topology, and predicting the desired output of the designed circuit.

- *Pathway Optimization*: Deep learning algorithms can optimize metabolic pathways in cell-free systems by considering various parameters, such as enzyme kinetics, substrate availability, and product yield. By analysing large-scale metabolic data and utilizing optimization algorithms, deep learning models can identify optimal pathway configurations, enzyme combinations, and reaction conditions for enhanced productivity and efficiency.
- *Predictive Modelling*: Deep learning models can generate predictive models that simulate the behaviour of cell-free systems. By training on experimental data, these models can predict the outcomes of biochemical reactions, enzyme kinetics, and metabolic fluxes in a cell-free environment. This enables researchers to simulate and optimize system behavior before experimental implementation, reducing the time and cost associated with trial-and-error experiments.
- *Sensor and Control Systems*: Deep learning techniques can be applied to develop sensor and control systems for cell-free synthetic biology. By integrating machine learning algorithms with real-time monitoring data, deep learning models can analyze and interpret sensor outputs to provide feedback control, enabling dynamic regulation of biochemical reactions and optimization of system performance.

Deep learning techniques provide a data-driven and systematic approach to designing and optimizing cell-free synthetic biology systems, accelerating the development of functional and efficient systems for biotechnology and bioengineering applications. They also offer novel solutions and improved performance in gene circuit design and metabolic engineering (Perez et al., 2016).

- *Gene Circuit Design*: Deep learning models can assist in the design and optimization of gene circuits for synthetic biology applications. These models can analyze large datasets of genetic sequences, regulatory elements, and circuit behaviors to learn patterns and predict the functionality of synthetic genetic constructs. Deep learning algorithms can aid in selecting suitable genetic components, optimizing circuit topology, and predicting the desired output of the designed gene circuit.
- *Pathway Design and Optimization*: Deep learning algorithms can optimize metabolic pathways in metabolic engineering projects. By considering various parameters such as enzyme kinetics, substrate availability, and product yield, deep learning models can analyze large-scale metabolic data and identify optimal pathway configurations, enzyme combinations, and reaction conditions. This enables researchers to design and engineer efficient metabolic pathways for the production of desired compounds.
- *Protein Engineering*: Deep learning techniques can be employed in protein engineering to predict protein structures, properties, and functions. By training on large protein sequence and structure databases, deep learning models can generate accurate predictions of protein structures, identify critical amino acids for specific functions, and optimize protein properties for desired applications. This can facilitate the design of novel enzymes with enhanced catalytic activity or stability.
- *Metabolic Flux Prediction*: Deep learning models can predict metabolic fluxes in cellular systems, aiding in metabolic engineering projects. By training on experimental data and considering various environmental and genetic factors, these models can simulate and predict metabolic flux distributions in different conditions. This information is crucial for understanding cellular metabolism, identifying bottlenecks, and guiding genetic modifications for improved metabolic performance.

Case Study

Deep learning techniques are being used to design gene circuits and metabolic engineering, such as gene circuit design (Sun et al., 2013).

- *Designing Optimal Gene Circuits*: Case studies showcasing the use of deep learning in designing gene circuits for specific applications, such as biosensors, gene therapies, or metabolic pathway control. These studies can highlight the improvement in circuit performance achieved through deep learning-based design approaches.
- *Predictive Models for Metabolic Engineering*: Case studies demonstrating the development of predictive models using deep learning techniques to guide metabolic engineering projects. These studies can show how deep learning models accurately predict metabolic flux distributions, guide genetic modifications, and optimize pathway designs for improved production of valuable compounds (T. Senthil et al., 2023; Syamala et al., 2023; Venkateswaran, Vidhya, Ayyannan, et al., 2023; Venkateswaran, Vidhya, Naik, et al., 2023b; Vennila et al., 2022; Zekrifa et al., 2023).
- *Protein Engineering*: Case studies focusing on deep learning-based protein engineering approaches. These studies can demonstrate how deep learning models aid in protein structure prediction, enzyme design, and optimization of protein properties for various applications, such as biofuels production or pharmaceutical development.
- *Integration of Deep Learning with Experimental Data*: Case studies showcasing the integration of deep learning models with experimental data, such as omics data or high-throughput screening results. These studies can demonstrate how deep learning algorithms analyze and interpret complex datasets to uncover patterns, guide experimental design, and accelerate the discovery and optimization of gene circuits and metabolic pathways.

Deep learning techniques have had a significant impact on gene circuit design and metabolic engineering, enabling the development of more efficient and tailored biological systems.

FUTURE DIRECTIONS AND DEVELOPMENTS IN DEEP LEARNING TECHNIQUES

Arrhythmia Detection

1. *Multimodal Data Integration*: Future developments may involve integrating multiple data sources, such as cardiac signals, patient demographics, and medical imaging data, into deep learning models for improved arrhythmia detection. This multimodal data integration can provide a more comprehensive view of cardiac health and enhance the accuracy of arrhythmia diagnosis (Parvaneh et al., 2019a).
2. *Explainable AI for Clinical Decision Support*: Enhancing the explainability and interpretability of deep learning models in arrhythmia detection is crucial for their adoption in clinical practice. Future developments may focus on developing explainable AI techniques that can provide transparent and interpretable insights to clinicians, enabling them to make informed decisions based on the deep learning model's predictions.

3. *Real-Time Risk Assessment*: Deep learning models can be further developed to provide real-time risk assessment of arrhythmias. By continuously analyzing cardiac signals and patient data, these models can assess the risk of developing arrhythmias, predict the likelihood of adverse events, and provide timely warnings or interventions for at-risk patients.

Cardiac Sensed Signals

1. *Personalized Medicine*: Future developments in deep learning for cardiac sensed signals may involve advancing towards personalized medicine approaches. Deep learning models can be trained on large-scale patient data, including clinical information, genetics, and cardiac signals, to predict individual patient outcomes, guide treatment decisions, and tailor interventions for improved patient care (Parvaneh et al., 2019a).
2. *Explain ability and Interpretability*: Deep learning models are often considered black boxes, making it challenging to interpret the reasoning behind their predictions. Future developments may focus on enhancing the explain ability and interpretability of deep learning models for cardiac arrhythmia detection. This involves developing techniques to visualize and interpret the learned representations, providing clinicians with insights into why a particular prediction or diagnosis was made.
3. *Real-Time Monitoring and Intervention*: Deep learning models can be deployed in real-time monitoring systems to continuously analyse cardiac signals and provide instant feedback or intervention when abnormal patterns or arrhythmias are detected. Future developments may involve integrating deep learning algorithms into wearable devices or implantable sensors to enable real-time monitoring and early intervention, thus preventing adverse cardiac events.

Bioremediation

1. *Intelligent Environmental Monitoring*: Future developments in deep learning for bioremediation may involve the integration of deep learning algorithms with environmental monitoring systems. These systems can analyze large volumes of environmental data, including pollutants, soil conditions, and microbial activities, to provide real-time insights for adaptive and targeted bioremediation strategies (Paul et al., 2021).
2. *Adaptive Optimization*: Deep learning models can be trained to adapt and optimize bioremediation strategies based on changing environmental conditions. This includes dynamically adjusting nutrient supplementation, microbial inoculation, or process parameters to maximize the efficiency and effectiveness of bioremediation processes.
3. *Integration with Robotics and Automation*: Future developments may focus on integrating deep learning techniques with robotics and automation in bioremediation processes. This includes the development of autonomous robotic systems equipped with deep learning algorithms for environmental sensing, decision-making, and the implementation of bioremediation strategies in complex and hazardous environments.

Deep learning techniques are used to improve accuracy, interpretability, and real-time capabilities in cell-free synthetic biology, cardiac sensed signals, bioremediation, and arrhythmia detection.

Cell-Free Synthetic Biology

- *Enhanced Design Automation*: Future developments in deep learning for cell-free synthetic biology may focus on improving the automation and efficiency of the design process. This includes developing advanced deep learning models that can rapidly generate optimal genetic circuit designs, predict circuit behavior accurately, and optimize pathway configurations for desired outputs (Perez et al., 2016).
- *Multi-objective Optimization*: Deep learning techniques can be further advanced to support multi-objective optimization in cell-free synthetic biology. This involves simultaneously optimizing multiple parameters, such as production yield, resource utilization, and pathway robustness, to achieve desired system performance. Multi-objective deep learning algorithms can aid in balancing trade-offs and designing more versatile and adaptable cell-free systems.
- *Feedback Control Systems*: Integrating feedback control systems with deep learning models can enable real-time monitoring and regulation of cell-free systems. This allows dynamic adjustments of reaction conditions, resource allocation, or enzyme expression levels based on sensor inputs and desired system performance. Future developments may focus on developing intelligent feedback control systems that can adapt and optimize cell-free processes in real-time.

CONCLUSION

Deep learning techniques have shown promise in the areas of arrhythmia detection, analysis of cardiac sensed signals, and cell-free synthetic biology. In arrhythmia detection, deep learning models such as convolutional neural networks and recurrent neural networks have demonstrated high performance in classifying arrhythmias accurately. However, challenges remain in addressing interpretability and explainability, so future research should focus on developing techniques that provide insights into the reasoning behind the models' predictions.

Deep learning approaches have shown promise in detecting abnormal patterns and classifying cardiac arrhythmias. They allow for the extraction of complex features and the identification of subtle variations in the signals, improving diagnostic accuracy. Real-time monitoring systems leveraging deep learning can enable early detection and intervention, potentially preventing adverse cardiac events. Future developments should focus on integrating these models into wearable devices and implantable sensors, enabling continuous monitoring and personalized healthcare.

Deep learning techniques offer opportunities for designing and optimizing genetic circuits and metabolic pathways in cell-free synthetic biology. By optimizing resource allocation, pathway configurations, and reaction conditions, deep learning can enhance the production of desired outputs. Future directions should explore multi-objective optimization, feedback control systems, and integration with robotics and automation to further advance the field.

Deep learning techniques for biomedical applications should focus on addressing challenges and exploring new directions. These include improving interpretability and explainability of deep learning models, enabling personalized medicine approaches, integrating multimodal data sources, and improving real-time monitoring and intervention capabilities. By pushing the boundaries of deep learning in biomedical applications, we can unlock its full potential and improve healthcare outcomes.

ABBREVIATIONS

CNNs - Convolutional neural networks
DNNs - Deep Neural Networks
ECG - Electrocardiogram
GRU - Gated Recurrent Unit
HRV - Heart Rate Variability
LSTM - Long Short-Term Memory
RNNs - Recurrent Neural Networks

REFERENCES

Agrawal, A. V., Pitchai, R., Senthamaraikannan, C., Balaji, N. A., Sajithra, S., & Boopathi, S. (2023). Digital Education System During the COVID-19 Pandemic. In Using Assistive Technology for Inclusive Learning in K-12 Classrooms (pp. 104–126). IGI Global. doi:10.4018/978-1-6684-6424-3.ch005

Anitha, C., R, K. C., Vivekanand, C. V., Lalitha, S. D., Boopathi, S., & R, Revathi. (2023, February). Artificial Intelligence driven security model for Internet of Medical Things (IoMT). *IEEE Explore*. doi:10.1109/ICIPTM57143.2023.10117713

Arunprasad, R., & Boopathi, S. (2019). Chapter-4 Alternate Refrigerants for Minimization Environmental Impacts: A Review. In Advances in Engineering Technology (p. 75). AkiNik Publications.

Astuti, Y., Ahda, Y., & Putri, D. H. (2021). Development of Bioremediation Learning Materials based on Contextual Teaching and Learning (CTL). *Journal of Physics: Conference Series*, *1940*(1), 12126. doi:10.1088/1742-6596/1940/1/012126

Babu, B. S., Kamalakannan, J., Meenatchi, N., M, S. K. S., S, K., & Boopathi, S. (2023). Economic impacts and reliability evaluation of battery by adopting Electric Vehicle. *IEEE Explore*, 1–6. doi:10.1109/ICPECTS56089.2022.10046786

Boopathi, S. (2023a). An Investigation on Friction Stir Processing of Aluminum Alloy-Boron Carbide Surface Composite. In *Advances in Processing of Lightweight Metal Alloys and Composites* (pp. 249–257). Springer. doi:10.1007/978-981-19-7146-4_14

Boopathi, S. (2023b). Deep Learning Techniques Applied for Automatic Sentence Generation. In Promoting Diversity, Equity, and Inclusion in Language Learning Environments (pp. 255–273). IGI Global. doi:10.4018/978-1-6684-3632-5.ch016

Boopathi, S. (2023c). Securing Healthcare Systems Integrated With IoT: Fundamentals, Applications, and Future Trends. In Dynamics of Swarm Intelligence Health Analysis for the Next Generation (pp. 186–209). IGI Global.

Boopathi, S., Gavaskar, T., Dogga, A. D., Mahendran, R. K., Kumar, A., Kathiresan, G., N., V., Ganesan, M., Ishwarya, K. R., & Ramana, G. V. (2021). *Emergency medicine delivery transportation using unmanned aerial vehicle (Patent Grant)*. Academic Press.

Boopathi, S., & Sivakumar, K. (2016). Optimal parameter prediction of oxygen-mist near-dry wire-cut EDM. *Inderscience: International Journal of Manufacturing Technology and Management, 30*(3–4), 164–178. doi:10.1504/IJMTM.2016.077812

Carbonell, P., Radivojevic, T., & García Martín, H. (2019). Opportunities at the Intersection of Synthetic Biology, Machine Learning, and Automation. In ACS Synthetic Biology (Vol. 8, Issue 7, pp. 1474–1477). ACS Publications. doi:10.1021/acssynbio.8b00540

Chandrika, V., Sivakumar, A., Krishnan, T. S., Pradeep, J., Manikandan, S., & Boopathi, S. (2023). Theoretical Study on Power Distribution Systems for Electric Vehicles. In *Intelligent Engineering Applications and Applied Sciences for Sustainability* (pp. 1–19). IGI Global. doi:10.4018/979-8-3693-0044-2.ch001

Ching, T., Himmelstein, D. S., Beaulieu-Jones, B. K., Kalinin, A. A., Do, B. T., Way, G. P., Ferrero, E., Agapow, P. M., Zietz, M., Hoffman, M. M., Xie, W., Rosen, G. L., Lengerich, B. J., Israeli, J., Lanchantin, J., Woloszynek, S., Carpenter, A. E., Shrikumar, A., Xu, J., ... Greene, C. S. (2018). Opportunities and obstacles for deep learning in biology and medicine. *Journal of the Royal Society, Interface, 15*(141), 20170387. doi:10.1098/rsif.2017.0387 PMID:29618526

Dhanya, D., Kumar, S. S., Thilagavathy, A., Prasad, D., & Boopathi, S. (2023). Data Analytics and Artificial Intelligence in the Circular Economy: Case Studies. In Intelligent Engineering Applications and Applied Sciences for Sustainability (pp. 40–58). IGI Global.

Domakonda, V. K., Farooq, S., Chinthamreddy, S., Puviarasi, R., Sudhakar, M., & Boopathi, S. (2022). Sustainable Developments of Hybrid Floating Solar Power Plants: Photovoltaic System. In Human Agro-Energy Optimization for Business and Industry (pp. 148–167). IGI Global.

Durairaj, M., Jayakumar, S., Karpagavalli, V., Maheswari, B. U., Boopathi, S., & ... (2023). Utilization of Digital Tools in the Indian Higher Education System During Health Crises. In *Multidisciplinary Approaches to Organizational Governance During Health Crises* (pp. 1–21). IGI Global. doi:10.4018/978-1-7998-9213-7.ch001

Gnanaprakasam, C., Vankara, J., Sastry, A. S., Prajval, V., Gireesh, N., & Boopathi, S. (2023). Long-Range and Low-Power Automated Soil Irrigation System Using Internet of Things: An Experimental Study. In Contemporary Developments in Agricultural Cyber-Physical Systems (pp. 87–104). IGI Global.

Gunasekaran, K., & Boopathi, S. (2023). Artificial Intelligence in Water Treatments and Water Resource Assessments. In *Artificial Intelligence Applications in Water Treatment and Water Resource Management* (pp. 71–98). IGI Global. doi:10.4018/978-1-6684-6791-6.ch004

Hammad, M., Iliyasu, A. M., Subasi, A., Ho, E. S. L., & El-Latif, A. A. A. (2021). A Multitier Deep Learning Model for Arrhythmia Detection. *IEEE Transactions on Instrumentation and Measurement, 70*, 1–9. doi:10.1109/TIM.2020.3033072

Hanumanthakari, S., Gift, M. M., Kanimozhi, K., Bhavani, M. D., Bamane, K. D., & Boopathi, S. (2023). Biomining Method to Extract Metal Components Using Computer-Printed Circuit Board E-Waste. In *Handbook of Research on Safe Disposal Methods of Municipal Solid Wastes for a Sustainable Environment* (pp. 123–141). IGI Global. doi:10.4018/978-1-6684-8117-2.ch010

Harikaran, M., Boopathi, S., Gokulakannan, S., & Poonguzhali, M. (2023). Study on the Source of E-Waste Management and Disposal Methods. In *Sustainable Approaches and Strategies for E-Waste Management and Utilization* (pp. 39–60). IGI Global. doi:10.4018/978-1-6684-7573-7.ch003

Hema, N., Krishnamoorthy, N., Chavan, S. M., Kumar, N., Sabarimuthu, M., & Boopathi, S. (2023). A Study on an Internet of Things (IoT)-Enabled Smart Solar Grid System. In *Handbook of Research on Deep Learning Techniques for Cloud-Based Industrial IoT* (pp. 290–308). IGI Global. doi:10.4018/978-1-6684-8098-4.ch017

Izci, E., Ozdemir, M. A., Degirmenci, M., & Akan, A. (2019). Cardiac arrhythmia detection from 2d ecg images by using deep learning technique. *TIPTEKNO 2019 - Tip Teknolojileri Kongresi*, 1–4. doi:10.1109/TIPTEKNO.2019.8895011

Janardhana, K., Singh, V., Singh, S. N., Babu, T. R., Bano, S., & Boopathi, S. (2023). Utilization Process for Electronic Waste in Eco-Friendly Concrete: Experimental Study. In Sustainable Approaches and Strategies for E-Waste Management and Utilization (pp. 204–223). IGI Global.

Jeevanantham, Y. A., A, S., V, V., J, S. Isaac., Boopathi, S., & Kumar, D. P. (2023). Implementation of Internet-of Things (IoT) in Soil Irrigation System. *IEEE Explore*, 1–5. doi:10.1109/ICPECTS56089.2022.10047185

Jiang, L., Zhao, J., Lian, J., & Xu, Z. (2018). Cell-free protein synthesis enabled rapid prototyping for metabolic engineering and synthetic biology. *Synthetic and Systems Biotechnology*, *3*(2), 90–96. doi:10.1016/j.synbio.2018.02.003 PMID:29900421

Kang, M., Ko, E., & Mersha, T. B. (2022). A roadmap for multi-omics data integration using deep learning. *Briefings in Bioinformatics*, *23*(1), bbab454. Advance online publication. doi:10.1093/bib/bbab454 PMID:34791014

Karthik, S., Hemalatha, R., Aruna, R., Deivakani, M., Reddy, R. V. K., & Boopathi, S. (2023). Study on Healthcare Security System-Integrated Internet of Things (IoT). In Perspectives and Considerations on the Evolution of Smart Systems (pp. 342–362). IGI Global.

Koshariya, A. K., Kalaiyarasi, D., Jovith, A. A., Sivakami, T., Hasan, D. S., & Boopathi, S. (2023). AI-Enabled IoT and WSN-Integrated Smart Agriculture System. In *Artificial Intelligence Tools and Technologies for Smart Farming and Agriculture Practices* (pp. 200–218). IGI Global. doi:10.4018/978-1-6684-8516-3.ch011

Koshariya, A. K., Khatoon, S., Marathe, A. M., Suba, G. M., Baral, D., & Boopathi, S. (2023). Agricultural Waste Management Systems Using Artificial Intelligence Techniques. In *AI-Enabled Social Robotics in Human Care Services* (pp. 236–258). IGI Global. doi:10.4018/978-1-6684-8171-4.ch009

Kumar, P., Sampath, B., Kumar, S., Babu, B. H., & Ahalya, N. (2023). Hydroponics, Aeroponics, and Aquaponics Technologies in Modern Agricultural Cultivation. In Trends, Paradigms, and Advances in Mechatronics Engineering (pp. 223–241). IGI Global.

Kumara, V., Mohanaprakash, T., Fairooz, S., Jamal, K., Babu, T., & Sampath, B. (2023). Experimental Study on a Reliable Smart Hydroponics System. In *Human Agro-Energy Optimization for Business and Industry* (pp. 27–45). IGI Global. doi:10.4018/978-1-6684-4118-3.ch002

Kwon, S. H., & Dong, L. (2022). Flexible sensors and machine learning for heart monitoring. *Nano Energy*, *102*, 107632. doi:10.1016/j.nanoen.2022.107632

Mohanty, A., Jothi, B., Jeyasudha, J., Ranjit, P. S., Isaac, J. S., & Boopathi, S. (2023). Additive Manufacturing Using Robotic Programming. In *AI-Enabled Social Robotics in Human Care Services* (pp. 259–282). IGI Global. doi:10.4018/978-1-6684-8171-4.ch010

Mohanty, A., Venkateswaran, N., Ranjit, P., Tripathi, M. A., & Boopathi, S. (2023). Innovative Strategy for Profitable Automobile Industries: Working Capital Management. In Handbook of Research on Designing Sustainable Supply Chains to Achieve a Circular Economy (pp. 412–428). IGI Global.

Murat, F., Yildirim, O., Talo, M., Baloglu, U. B., Demir, Y., & Acharya, U. R. (2020). Application of deep learning techniques for heartbeats detection using ECG signals-analysis and review. *Computers in Biology and Medicine*, *120*, 103726. doi:10.1016/j.compbiomed.2020.103726 PMID:32421643

Naushin, M., Saraswat, A., & Abhishek, K. (2022). Deep Learning Approach to Deal with E-Waste. In *Lecture Notes in Electrical Engineering* (Vol. 858, pp. 123–136). Springer. doi:10.1007/978-981-19-0840-8_9

Nazir, S., Dickson, D. M., & Akram, M. U. (2023). Survey of explainable artificial intelligence techniques for biomedical imaging with deep neural networks. *Computers in Biology and Medicine*, *156*, 106668. doi:10.1016/j.compbiomed.2023.106668 PMID:36863192

Nishanth, J., Deshmukh, M. A., Kushwah, R., Kushwaha, K. K., Balaji, S., & Sampath, B. (2023). Particle Swarm Optimization of Hybrid Renewable Energy Systems. In *Intelligent Engineering Applications and Applied Sciences for Sustainability* (pp. 291–308). IGI Global. doi:10.4018/979-8-3693-0044-2.ch016

Palaniappan, M., Tirlangi, S., Mohamed, M. J. S., Moorthy, R. M. S., Valeti, S. V., & Boopathi, S. (2023a). Fused deposition modelling of polylactic acid (PLA)-based polymer composites: A case study. In Development, Properties, and Industrial Applications of 3D Printed Polymer Composites (pp. 66–85). IGI Global. doi:10.4018/978-1-6684-6009-2.ch005

Palaniappan, M., Tirlangi, S., Mohamed, M. J. S., Moorthy, R. S., Valeti, S. V., & Boopathi, S. (2023b). Fused Deposition Modelling of Polylactic Acid (PLA)-Based Polymer Composites: A Case Study. In Development, Properties, and Industrial Applications of 3D Printed Polymer Composites (pp. 66–85). IGI Global.

Parvaneh, S., Rubin, J., Babaeizadeh, S., & Xu-Wilson, M. (2019). Cardiac arrhythmia detection using deep learning: A review. *Journal of Electrocardiology*, *57*, S70–S74. doi:10.1016/j.jelectrocard.2019.08.004 PMID:31416598

Paul, O., Jasu, A., Lahiri, D., Nag, M., & Ray, R. R. (2021). in Situ and Ex Situ Bioremediation of Heavy Metals: The Present Scenario. *Journal of Environmental Engineering and Landscape Management*, *29*(4), 454–469. doi:10.3846/jeelm.2021.15447

Perez, J. G., Stark, J. C., & Jewett, M. C. (2016). Cell-free synthetic biology: Engineering beyond the cell. *Cold Spring Harbor Perspectives in Biology*, *8*(12), a023853. doi:10.1101/cshperspect.a023853 PMID:27742731

Pramila, P., Amudha, S., Saravanan, T., Sankar, S. R., Poongothai, E., & Boopathi, S. (2023). Design and Development of Robots for Medical Assistance: An Architectural Approach. In Contemporary Applications of Data Fusion for Advanced Healthcare Informatics (pp. 260–282). IGI Global.

Rahamathunnisa, U., Sudhakar, K., Murugan, T. K., Thivaharan, S., Rajkumar, M., & Boopathi, S. (2023). Cloud Computing Principles for Optimizing Robot Task Offloading Processes. In *AI-Enabled Social Robotics in Human Care Services* (pp. 188–211). IGI Global. doi:10.4018/978-1-6684-8171-4.ch007

Ramudu, K., Mohan, V. M., Jyothirmai, D., Prasad, D., Agrawal, R., & Boopathi, S. (2023). Machine Learning and Artificial Intelligence in Disease Prediction: Applications, Challenges, Limitations, Case Studies, and Future Directions. In Contemporary Applications of Data Fusion for Advanced Healthcare Informatics (pp. 297–318). IGI Global.

Reddy, M. A., Gaurav, A., Ushasukhanya, S., Rao, V. C. S., Bhattacharya, S., & Boopathi, S. (2023). Bio-Medical Wastes Handling Strategies During the COVID-19 Pandemic. In Multidisciplinary Approaches to Organizational Governance During Health Crises (pp. 90–111). IGI Global. doi:10.4018/978-1-7998-9213-7.ch006

Samikannu, R., Koshariya, A. K., Poornima, E., Ramesh, S., Kumar, A., & Boopathi, S. (2022). Sustainable Development in Modern Aquaponics Cultivation Systems Using IoT Technologies. In *Human Agro-Energy Optimization for Business and Industry* (pp. 105–127). IGI Global.

Sampath, B., Pandian, M., Deepa, D., & Subbiah, R. (2022). Operating parameters prediction of liquefied petroleum gas refrigerator using simulated annealing algorithm. *AIP Conference Proceedings*, *2460*(1), 070003. doi:10.1063/5.0095601

Sampath, B., Sasikumar, C., & Myilsamy, S. (2023). Application of TOPSIS Optimization Technique in the Micro-Machining Process. In Trends, Paradigms, and Advances in Mechatronics Engineering (pp. 162–187). IGI Global.

Sankar, K. M., Booba, B., & Boopathi, S. (2023). Smart Agriculture Irrigation Monitoring System Using Internet of Things. In *Contemporary Developments in Agricultural Cyber-Physical Systems* (pp. 105–121). IGI Global. doi:10.4018/978-1-6684-7879-0.ch006

Sathiyaseelan, G., Srinath, P., Melwyn, J. G., Bhagyanathan, C., Socrates, S., & Boopathi, S. (2023). Tribology and hardness measurement of biosilica dispersed AA7075 metal matrix composites. *Materials Today: Proceedings*. Advance online publication. doi:10.1016/j.matpr.2023.02.440

Selvakumar, S., Adithe, S., Isaac, J. S., Pradhan, R., Venkatesh, V., & Sampath, B. (2023). A Study of the Printed Circuit Board (PCB) E-Waste Recycling Process. In Sustainable Approaches and Strategies for E-Waste Management and Utilization (pp. 159–184). IGI Global.

Selvakumar, S., Shankar, R., Ranjit, P., Bhattacharya, S., Gupta, A. S. G., & Boopathi, S. (2023). E-Waste Recovery and Utilization Processes for Mobile Phone Waste. In *Handbook of Research on Safe Disposal Methods of Municipal Solid Wastes for a Sustainable Environment* (pp. 222–240). IGI Global. doi:10.4018/978-1-6684-8117-2.ch016

Sengeni, D., Padmapriya, G., Imambi, S. S., Suganthi, D., Suri, A., & Boopathi, S. (2023). Biomedical Waste Handling Method Using Artificial Intelligence Techniques. In *Handbook of Research on Safe Disposal Methods of Municipal Solid Wastes for a Sustainable Environment* (pp. 306–323). IGI Global. doi:10.4018/978-1-6684-8117-2.ch022

Senthil, T., Puviyarasan, M., Babu, S. R., Surakasi, R., Sampath, B., & Associates. (2023). Industrial Robot-Integrated Fused Deposition Modelling for the 3D Printing Process. In Development, Properties, and Industrial Applications of 3D Printed Polymer Composites (pp. 188–210). IGI Global.

Senthil, T. S. R., Ohmsakthi vel, Puviyarasan, M., Babu, S. R., Surakasi, R., & Sampath, B. (2023). Industrial Robot-Integrated Fused Deposition Modelling for the 3D Printing Process. In Development, Properties, and Industrial Applications of 3D Printed Polymer Composites (pp. 188–210). IGI Global. doi:10.4018/978-1-6684-6009-2.ch011

Subha, S., Inbamalar, T. M., R, K. C., Suresh, L. R., Boopathi, S., & Alaskar, K. (2023, February). A Remote Health Care Monitoring system using internet of medical things (IoMT). *IEEE Explore*. doi:10.1109/ICIPTM57143.2023.10118103

Sun, Z. Z., Hayes, C. A., Shin, J., Caschera, F., Murray, R. M., & Noireaux, V. (2013). Protocols for implementing an Escherichia coli based TX-TL cell-free expression system for synthetic biology. *Journal of Visualized Experiments*, *79*(79), e50762. doi:10.3791/50762 PMID:24084388

Syamala, M., Komala, C., Pramila, P., Dash, S., Meenakshi, S., & Boopathi, S. (2023). Machine Learning-Integrated IoT-Based Smart Home Energy Management System. In *Handbook of Research on Deep Learning Techniques for Cloud-Based Industrial IoT* (pp. 219–235). IGI Global. doi:10.4018/978-1-6684-8098-4.ch013

Tinafar, A., Jaenes, K., & Pardee, K. (2019). Synthetic Biology Goes Cell-Free. *BMC Biology*, *17*(1), 1–14. doi:10.118612915-019-0685-x PMID:31395057

Venkateswaran, N., Vidhya, K., Ayyannan, M., Chavan, S. M., Sekar, K., & Boopathi, S. (2023). A Study on Smart Energy Management Framework Using Cloud Computing. In 5G, Artificial Intelligence, and Next Generation Internet of Things: Digital Innovation for Green and Sustainable Economies (pp. 189–212). IGI Global. doi:10.4018/978-1-6684-8634-4.ch009

Venkateswaran, N., Vidhya, R., Naik, D. A., Raj, T. F. M., Munjal, N., & Boopathi, S. (2023). Study on Sentence and Question Formation Using Deep Learning Techniques. In *Digital Natives as a Disruptive Force in Asian Businesses and Societies* (pp. 252–273). IGI Global. doi:10.4018/978-1-6684-6782-4.ch015

Vennila, T., Karuna, M., Srivastava, B. K., Venugopal, J., Surakasi, R., & Sampath, B. (2022). New Strategies in Treatment and Enzymatic Processes: Ethanol Production From Sugarcane Bagasse. In Human Agro-Energy Optimization for Business and Industry (pp. 219–240). IGI Global.

Wahlang, I., Sharma, P., Sanyal, S., Saha, G., & Maji, A. K. (2020). Deep learning techniques for classification of brain MRI. *International Journal of Intelligent Systems Technologies and Applications*, *19*(6), 571–588. doi:10.1504/IJISTA.2020.112441

Yatika, G., Kumar, S. R., Patel, P. B., Singh, D. P., Rajkamal, M., & Boopathi, S. (2023). Experimental investigation of RHA biochar and tamarind fibre epoxy composite. *Materials Today: Proceedings*. Advance online publication. doi:10.1016/j.matpr.2023.02.439

Yupapin, P., Trabelsi, Y., Nattappan, A., & Boopathi, S. (2023). Performance improvement of wire-cut electrical discharge machining process using cryogenically treated super-conductive state of Monel-K500 alloy. *Iranian Journal of Science and Technology. Transaction of Mechanical Engineering*, *47*(1), 267–283. doi:10.100740997-022-00513-0

Zekrifa, D. M. S., Kulkarni, M., Bhagyalakshmi, A., Devireddy, N., Gupta, S., & Boopathi, S. (2023). Integrating Machine Learning and AI for Improved Hydrological Modeling and Water Resource Management. In *Artificial Intelligence Applications in Water Treatment and Water Resource Management* (pp. 46–70). IGI Global. doi:10.4018/978-1-6684-6791-6.ch003

Chapter 5
Arrhythmia Detection and Classification on Cardiac Sensed Signals Using Deep Learning Techniques

Jay Prakash Maurya
https://orcid.org/0000-0002-5574-5822
Samrat Ashok Technological Institute, India

Manish Manoria
Bansal Institute of Research Technology and Science, India

Sunil Joshi
Samrat Ashok Technological Institute, India

ABSTRACT

Arrhythmia is a general type of cardiac disease in persons between 30-40 years of age. Cardiac system in human body generates electrical pulses that can be captured and plotted through electrical system called ECG. Computer-aided diagnosis system (CADS) is a good approach to help the healthcare field for early, regular, and accurate diagnosis and treatment plan during critical care conditions. Deep learning-based CADS can helps in critical condition for more quick diagnosis and treatment in countries where doctor ratio is comparatively low. With the help of machine learning (ML) algorithm, intervariable relationships may be used for prediction. However, machine learning algorithms are also limited due to its datasets availability, established framework, and clinician unfamiliarity. This chapter aims to provide an idea of arrhythmia and CADS approach using cascaded deep learning model of CNN, LSTM, GRU, and RNN. The chapter focuses on techniques used in past years, comparative studies, and direction of research as future improvements in respective fields.

DOI: 10.4018/978-1-6684-6577-6.ch005

INTRODUCTION

Cardiologists use electrocardiography (ECG) as a core diagnostic tool because it provides invaluable information about the electrical activity of the heart. The ECG records the heart's rhythmic patterns and offers crucial data for determining cardiac health and making a variety of diagnoses. Analysis of ECG plays a critical role in the assessment of cardiovascular health in contemporary medical diagnosis. Through the use of an ECG, the electrical activity of the heart may be observed, and this information can be used to diagnose, treat, and prevent numerous cardiac diseases early on. Arrhythmia is a type of disease where the abnormal rate and rhythm of ECG signals are seen. Arrhythmia detection and classification are particularly important because of their potential to signal life-threatening circumstances and direct necessary actions. The significance of arrhythmia detection through ECG analysis is aligned with key contexts like Early Diagnosis and Risk Stratification, Guiding Treatment Decisions, Monitoring Treatment, Preventing Sudden Cardiac Death, Personalized Medicine, and Long-term Monitoring. Medical treatment in a critical situation needs immediate response from medical experts and guidance for saving lives. As the availability of medical experts in the population is also very low, in this situation computer-aided diagnosis system can help society. CADS is a combine system that may include hardware and software systems at the end. In the advanced software era, machine learning and deep learning help systems to be intelligent enough to make a decision as per previous learnings. Deep learning can support the medical field in early diagnosis, detection recommendations, etc. in this field.

This section contains information about heart disease and its attributes through which heart disease can be diagnosed from ECG signals. ECG signals sensed through an electrical conduction system have a morphology using which the status of the heart can be observed. Automation in the area of the healthcare sector inspires machine learning / deep-learning models to work for the identification of heart-related diseases.

Cardiac Conduction

In the central heating system analogy, the pump, pipes, and radiators are useless without being connected to a power source. To function, the pump requires electricity. Electricity is also used by the human heart, which has a comparable need for a power supply. Thankfully, we don't need to plug ourselves into the wall because the heart can generate its electrical impulses and regulate the path the impulses go using a specialized conduction pathway. The cardiac conduction system has 5 elements shown in Figure 1.

- Sino-atrial (SA) node: Natural Peacemaker
- Atrio-ventricular (AV) node: This periodically discharges electrical stimulation; the frequency is determined by the body's need.
- Bundle of His:
- Left and right bundle branches:
- Purkinje fibers:

Figure 1. Cardiac conduction system

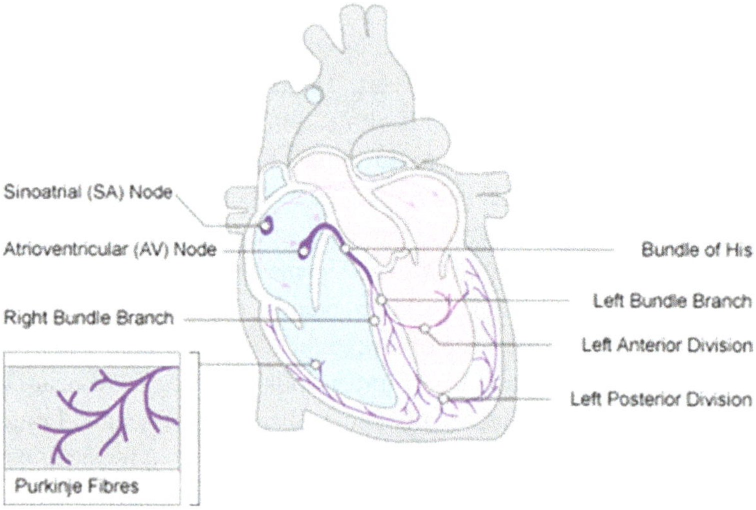

There are three heart stages in a single heartbeat

- Atrial depolarization
- Ventricular depolarization
- Atrial and ventricular repolarization.

The electrical system is attached to the human body at the time of ECG capturing, as shown in Figure 2. ECG signal plotted on the graph has two types of information i) measured time interval of heart conduction system ii) measured the strength of the electrical wave. Time interval helps to find whether the heart activity is regular or irregular, it is slow or fast and strength gives an idea about the workload on the heart through contraction of the Arial muscles. The structure of the normal ECG signal and the meaningful segment is shown in Figure 3. Three important segments exist in ECG signal structure atrial depolarization (P- wave), ventral depolarization (QRS complex wave), and repolarization (T-wave). Any kind of disorder generated from heart conduction affects the ECG, known as Arrhythmia. Different types of same are mentioned below.

Types of Arrhythmias

ECG signal structure shows the status of the heart mentioning, two primary information. The first is to check the time interval and the second is to measure the strength of an electric signal. The span of time & amplitude it takes for an electrical wave to move through the heart's electrical conduction system can be checked by a cardiologist. The pattern in ECG signal, regular or irregular and too fast or slow helps cardiologists to decide and predict the disease. Figure 4 explains different types of irregularity based on three important waves, atrial depolarization (P- wave), ventral depolarization (QRS complex wave), and repolarization (T-wave). Types of arrhythmia disorder can be explained using given waves & below a sequence of terms.

Figure 2. Heart electric conduction and signal leads

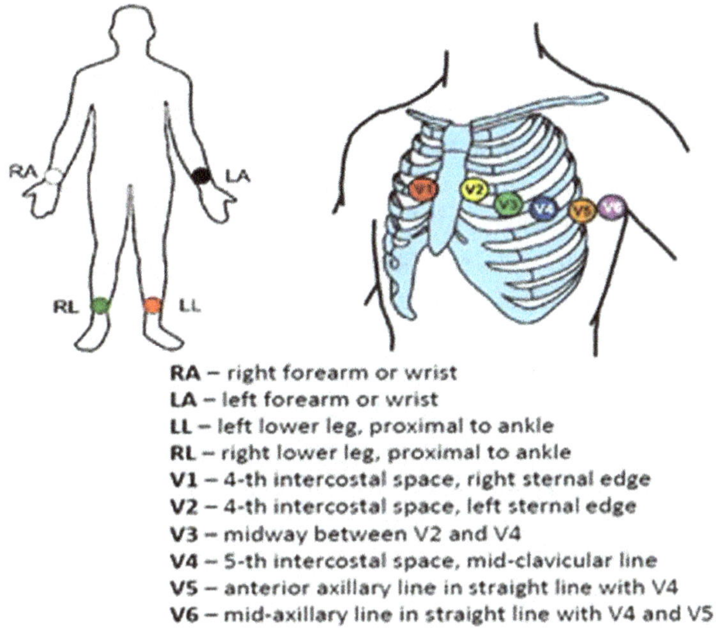

Figure 3. ECG waveform (three segments)

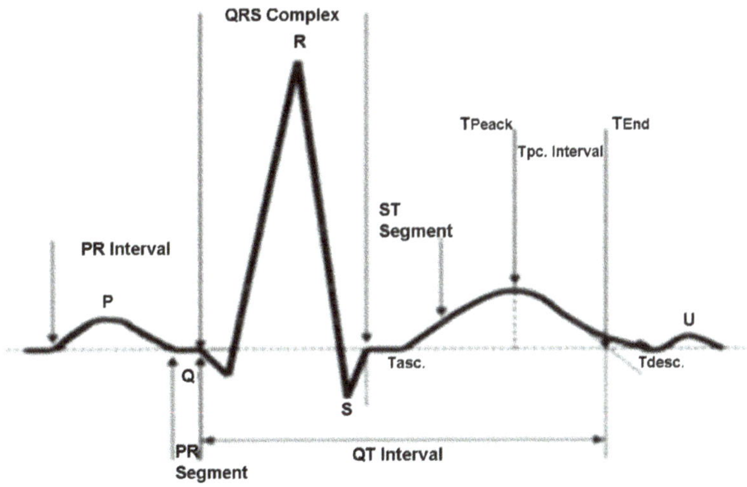

- AF (Atrial Fibrillation): is a disorder of atria having an instantaneous and irregular beating. The situation is rapid atrial rate approx. 400-600, where the p wave is not appeared having very low amplitude.
- Right Bundle Branch Block (RBBB) and Left Bundle Branch Block (LBBB): Bundle Branch Block is an aberrant QRS morphology caused by a disruption in the usual conduction system. The right bundle usually depolarizes the right ventricle (RV). The right bundle does not activate in an RBBB. Instead, the impulse from the left bundle is diffused via the Left Ventricle (LV) and

subsequently to the RV, depolarizing the right ventricle. An abnormal QRS morphology results from this pattern of electric spread. The LV is normally depolarized by the left bundle. The left bundle isn't activated in an LBBB. Instead, impulses from the right bundle travel to the RV and then to the LV, depolarizing the LV.

- Premature Atrial Contraction (PAC) and Premature Ventricular Contraction (PVC): if the regular rhythm the of heartbeat is broken in by a premature sound of beat generated from atria, named PAC. If premature sound looks raised from ventricles called PVC shown in Figure 5 (f).
- Ectopic beats: When a place outside of the sinus node, but within the atria, generates action potentials quicker than the sinus node, atrial tachycardia and other ectopic atrial rhythms arise. This ectopic focus takes over as the heart's primary pacemaker. Atrial tachycardia is a rhythm in which the atrial rate exceeds 100 beats per minute. The phrase "ectopic atrial rhythm" is used when the heart rate is fewer than 100 beats per minute.
- Myocardial Infarction (MI): A heart attack (Myocardial Infarction) occurs when the blood supply to your heart is suddenly decreased. Your heart muscle can't get the nutrition and oxygen it needs to function if there isn't enough blood flow in the heart muscle or arteries and causing damage. In ECG signal MI can be identified by following two patterns.
- ST segment depression or T-wave inversion
- ST-segment elevation or new RBBB/LBBB
- Fusion beat: When electrical impulses from various sources act on the same region of the heart at the same time, this is known as a fusion beat. If it acts on the ventricles, it's called a Ventricular Fusion Beat (VFB), but it's called an Atrial Fusion Beat (AFB) when it operates on the atrial chambers.
- Sinus bradycardia: A sinus rhythm with a lower-than-normal rate is known as sinus bradycardia (fewer than 60 beats per minute).

Figure 4. Cardiac disorder classification attribute

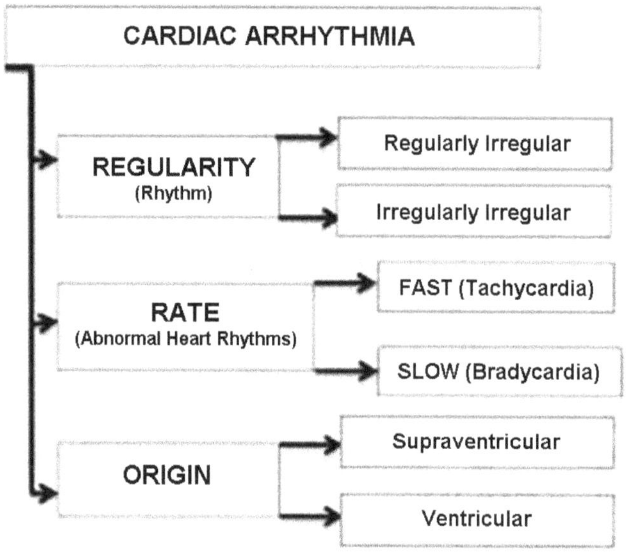

- Tachycardia: It occurs in a condition where the heart rate exceeds the resting rate. In adults 100 beats/ min is considered Tachycardia. Following patterns in ECG are shown in Figure 5(j,k,l). Types of tachycardia are mentioned below.
- Supraventricular Tachycardia (SVT): In the upper part of the heart the abnormal electric activity is caused due to fast heart rhythm.
- Sinus Tachycardia: When the heart generates an electrical signal faster than normal, causes an increase in heart rate.
- Ventricular Tachycardia (VT): In the lower part of the heart the abnormal electric activity is caused due to fast heart rhythm. It has more than three abnormally long consecutive QRS complex heart rhythms with a length of more than 120 milliseconds and an ST-T vector pointing in the opposite direction of the QRS deflection.
- Atrial Flutter (AFL): This is a common abnormal heart rhythm, initiated in the atrial chamber of the heart. The first time, it is usually associated with a fast heart rate and is classified as a type of supraventricular tachycardia.
- Ventricular Flutter (VF): a tachycardia affecting the ventricles with a pace of over 150–300 beats per minute is an unstable arrhythmia. VF is a stage that can occur between VT and fibrillation, and it can result in abrupt cardiac death. It is described by a sinusoidal waveform that lacks a distinct delineation of the T-waves and QRS.
- Ventricular fibrillation: is a cardiac arrhythmia in which the heart quivers rather than pumps due to disordered electrical activity in the ventricles, which is characterized by irregular unformed QRS complexes and no discernible P-waves. In the absence of therapy, VFib leads to cardiac arrest, loss of consciousness, and death
- Idioventricular rhythm: The ventricular rate is less than 60 beats per minute in an idioventricular rhythm, which is very close to VT. As a result, sluggish ventricular tachycardia is the name given to the idioventricular rhythm.
- Ventricular bigeminy: It is a type of irregular cardiac rhythm in which each sinus beat is followed by an ectopic beat and pauses frequently.
- Pacemaker rhythm: The repercussions of pacemaker implantation, regardless of pacing mode, due to poor atrioventricular synchronization or desynchrony are referred to as pacemaker clinical syndrome. It's an iatrogenic illness caused by medical treatment. Pacemaker syndrome is more likely in people who have a low heart rate before getting a pacemaker. Patients with pacemaker syndrome may need to have their pacemakers adjusted or a new lead implanted to properly coordinate the timing of atrial and ventricular contractions.

Arrhythmia Detection and Classification

Arrhythmia also named cardiac arrhythmia or heart arrhythmia, is a collection of disorders characterized by an irregular, rapid, or slow heartbeat. A fast heart rate (over 100 beats per minute in adults) is referred to as tachycardia, whereas a slow heart rate (below 60 beats per minute) is referred to as bradycardia. The Challenges faced to identify Arrhythmia are as follows:

1. It is very typical to judge the symptoms of Arrhythmia during the period of ECG signal capturing. (Ceylan & Özbay, 2007).

2. Period and amplitude of ECG signal might vary from person to person and also depends on other parameters like age, physical condition, lifestyle as well gender (Ceylan & Özbay, 2007; Joshi, Chandran, Jayaraman, & Kulkarni, 2009).

Figure 5. Different waveforms for types of arrhythmias

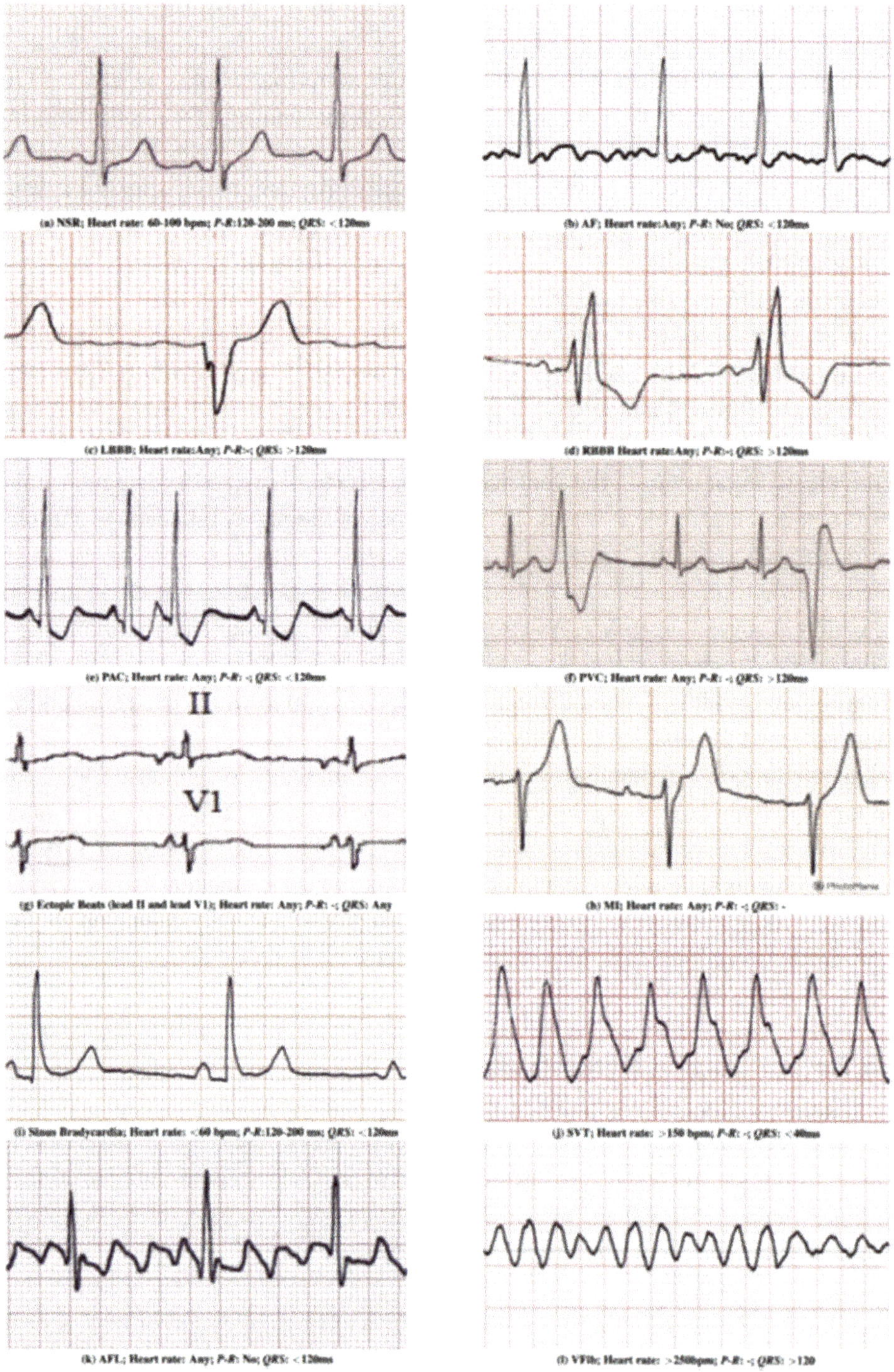

3. Physical state (sleeping or running or walking) of the person can also affect the morphology of the ECG signal.
4. Resources that capture ECG may involve some discrepancies due to noise, interference, and artifacts also lead to morphological change.
5. The volume of ECG signals may lead to a higher chance of a false diagnosis of Arrhythmia at the time of ECG analysis (Adams & Choi, 2012; Dinakarrao, Jantsch, & Shafique, 2019).

Deep Learning Methods

As machine learning is a very popular area of research work in recent years lot of applications has been benefited by incorporating machine learning like image classification, video recommendation, text mining, social network analysis, and many more. The rapid growth of size in available data for machine learning algorithms, and change in hardware used leads to change in an important step that can outperform in this situation. Deep Learning is an area of machine learning that works on feature engineering, representation learning, and transformation. The journey of deep learning has been started in 1943 when a simple McCulloch-Pitts (MCP) was a network as a prototype for artificial neurons. After MCP lot of other new models have been introduced for implementation, and a new method was del inspired through MCP.

Multilayer Perceptron

MLP is the most widely used supervised neural network, and it is effective in learning complex systems. The MLP architecture varies, but it typically consists of several layers of neurons connected in a feed-forward fashion. Each neuron is made up of the weighted total of its inputs after they have been processed by a non-linear function. (Goodfellow et al., 2016).

Figure 6. Multilayer perceptron

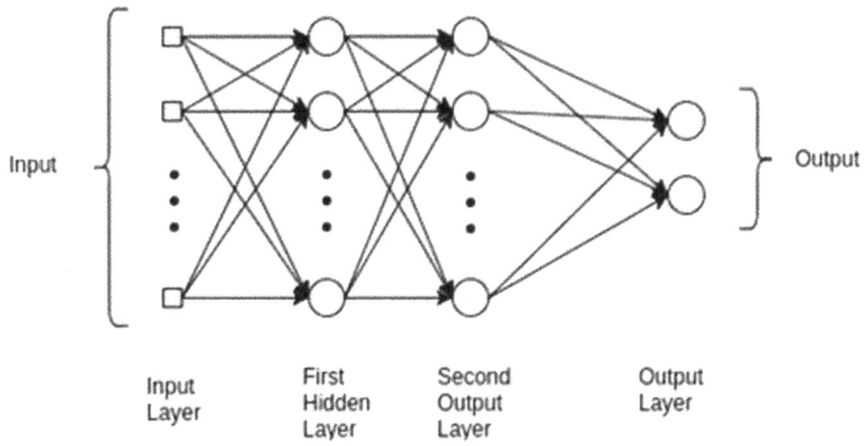

Convolutional Neural Network

CNN is a network having multiple connected layers using feed-forward connections. The important layers involved in CNN are convolutional, normalization, pooling, and fully connected layers. The first three are responsible for feature engineering and the last is responsible for classification. Figure 3 shows a simple architecture of CNN that can be used for the classification task. (Ciresan, Meier, & Schmidhuber, 2012). In deep neural networks, CNN is a simple and popular architecture in which is gradient-based approach is used for optimization (LeCun, Bottou, Bengio, &Haffner, 1998).

Figure 7. CNN architecture

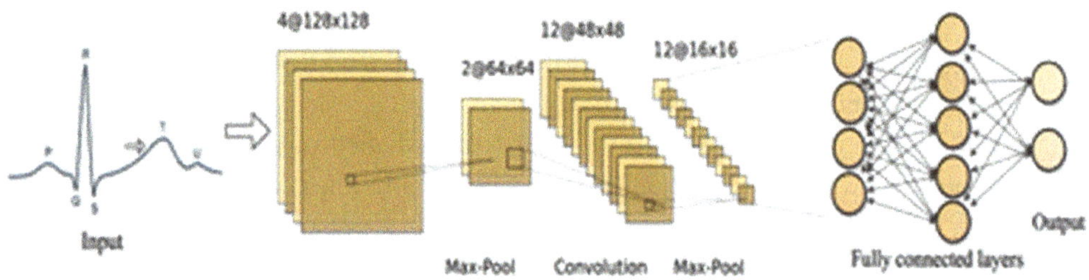

Deep Belief Network

In 2006, Hinton proposed DBNs are made up of multiple unsupervised networks with Restricted Boltzmann Machine (RBM) layers. Each sub-invisible network's layer becomes the visible layer of the next. The hidden or invisible layers are conditionally independent and are not related to each other. RBM was initially proposed in the year 1986, for modeling binary random variables (Keyvanrad & Homayounpour, 2015) into binary valued data. Each RBM has a joint configuration network over both visible and hidden layers determined by its energy in comparison to that of all other joint configuration networks.

Recurrent Neural Network

This is an adjunct of artificial neural network (ANN). RNN is an important model for handling sequential input data and time series data. Information can be stored in loops within adjacent levels using RNN. RNN is a popular network that trains upcoming events by reasoning from past experiences. Every time stamp, RNN takes input, updates it on the hidden layer, and predicts the value. For training the weight, RNN also uses gradient descent methods and it has a highly dynamic behavior due to the activation function used on taken on hidden layer shown in Figure 9.

Figure 8. Deep belief network (DBN) architecture

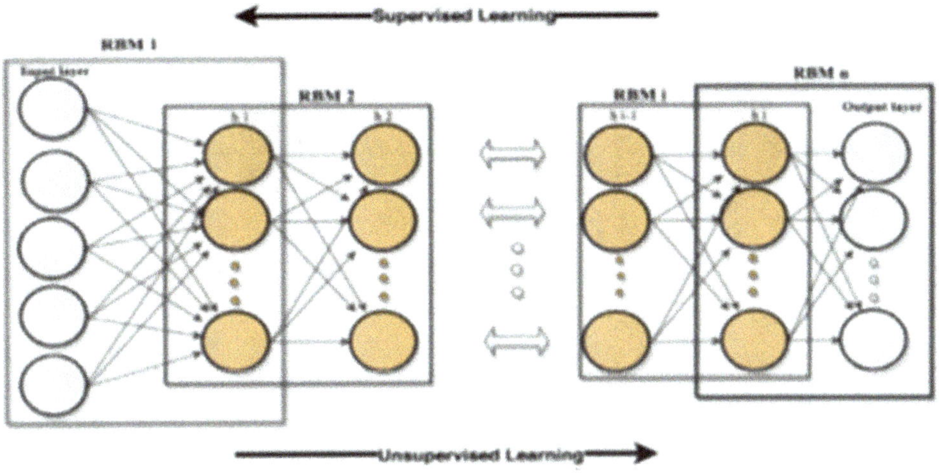

Figure 9. Architecture of RNN

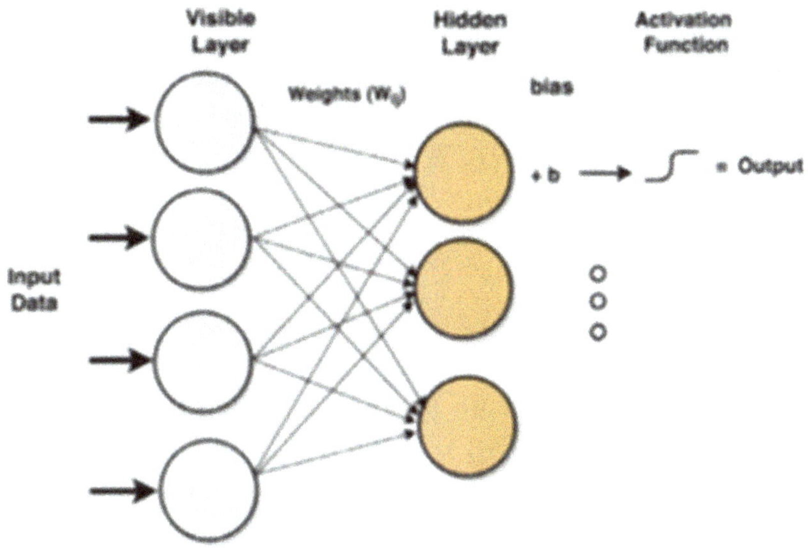

Bidirectional Recurrent Neural Network

BRNN uses two bidirectional RNNs simultaneously, one for getting information from the past and the second for future states connecting two hidden layers attached in different directions (Schuster & Paliwal, 1997). BRNNs may be introduced to improve the amount of input for the network architecture use of BRNNs has been shown in Figure 10.

Figure 10. Bidirectional recurrent neural network (BRNN) architecture

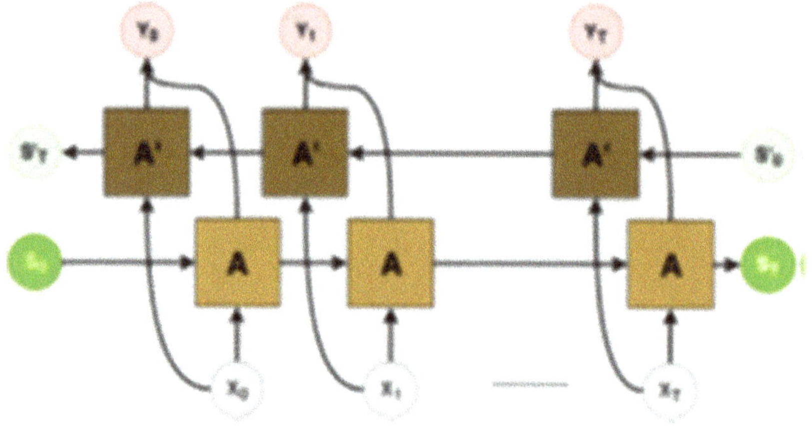

Long Short-Term Memory

LSTM is a special case of RNN in which temporal sequence and long-range dependency are being avoided. LSTM has a chain structure in repeating modules that has a memory block of one or more memory cells Figure 11. This memory cell uses pair of multiple gates as input and output. The information placed over the memory block is being updated over time steps through input and output gates. These gates control the flow of information from input/output to a memory cell.

Figure 11. (a) Structure of long short-term memory (LSTM) architecture, (b) functional structure of calls (LSTM)

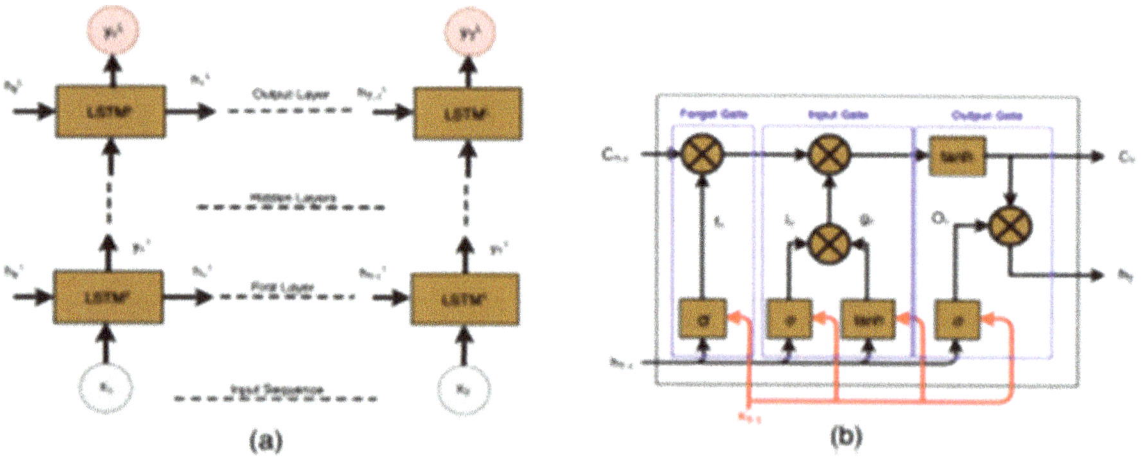

Gated Recurrent Unit (GRU)

This is a simple LSTM having computational complexity less than LSTM (Chung et al., 2014). GRU is improved in terms of less time needed in the training process. GRU makes the balance of the interior flow of unit's information using gates, two gates input gate and forgets gate create a new gate named updated gate. This updated gate is the core section responsible for balancing previous activation and candidate activation. The architecture of basic GRU is shown in Figure 12.

Figure 12. Architecture of gated recurrent unit (GRU)

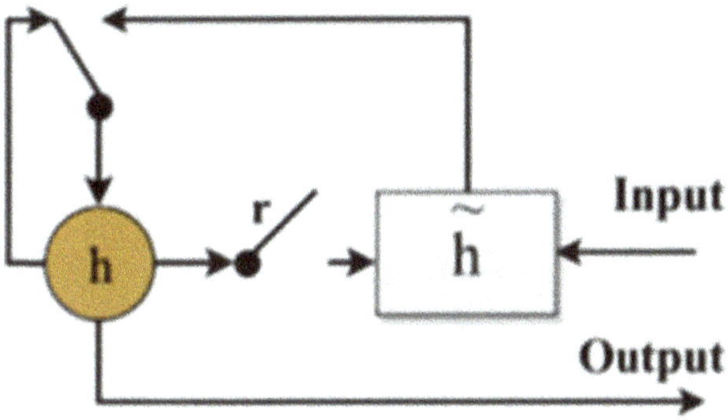

RESEARCH METHODOLOGY

The contemporary look over of past articles (period 2017-2020) for this review paper has been taken in given steps on different websites PubMed, Google Scholar, Sci-Hub, etc. Based on certain keywords approximately 97 articles have been collected for record [Identification]. In the screen duplicate articles have been removed and after these 73 articles have been screened. In the review process, 38 articles have been fully accessed among them 30 were taken for systematic review. The other was reported concerning of terminology and inclusion of a dedicated field of research domain and application.

The flow process adopted for the dedicated for review work taken for this paper is shown in Figure 13. Keywords used for collecting paper searching are Electrocardiogram (ECG), Deep Learning (DL), Deep Neural Network (DNN), Convolutional Neural Network (CNN), Deep Belief Network (DBN, Recurrent Neural Network (RNN), Long Short-Term Memory (LSTM), Gated Recurrent Unit (GRU), (used in individual or in combination). Seeking review flow (given above), the literature was reviewed (important are shown below). based on a review of different papers databases used in previous years, different performance parameters, and the contribution of different deep learning methods in this area discussed in this section.

According to Oh et al. (2018), CNN-based algorithms have been shown to be helpful in the classification of arrhythmias. Dynamic classification algorithms capable of learning both short- and long-term contents of the signal in an efficient manner are being used for such applications, according to a recent trend in research in this field. CNN has performed admirably in classifying various forms of arrhythmia. One of the most effective learning tools for this application would be this amazing approach. Atrial Fibrillation, Supraventricular Ectopic Beats, and Ventricular Ectopic Beats were correctly classified using the GRU/LSTM, CNN, and LSTM, respectively. It is also stated that using the correct type of Deep Learning approach can significantly increase classification performance for the relevant application.

According to Ali Haider Khan (n.d.), the automated systems anticipate good accuracy outcomes for arrhythmia diagnosis, but they are still not widely adopted by healthcare professionals because recent research used time-series data, which is inflexible in varied application scenarios. Furthermore, time-series data from an ECG with signal leads is insufficient for steady baseline drifts, muscle contraction, and power line interface. In general, a cardiologist's preferred strategy for screening cardiac patients is to use ECG scans with 12 leads. The successful result of the developed arrhythmia detection systems is hampered by (i) manual feature selection, (ii) methodologies for feature extraction, and (iii) algorithms for classification, the most important of which is the use of imbalanced data for classification.

Figure 13. Review steps

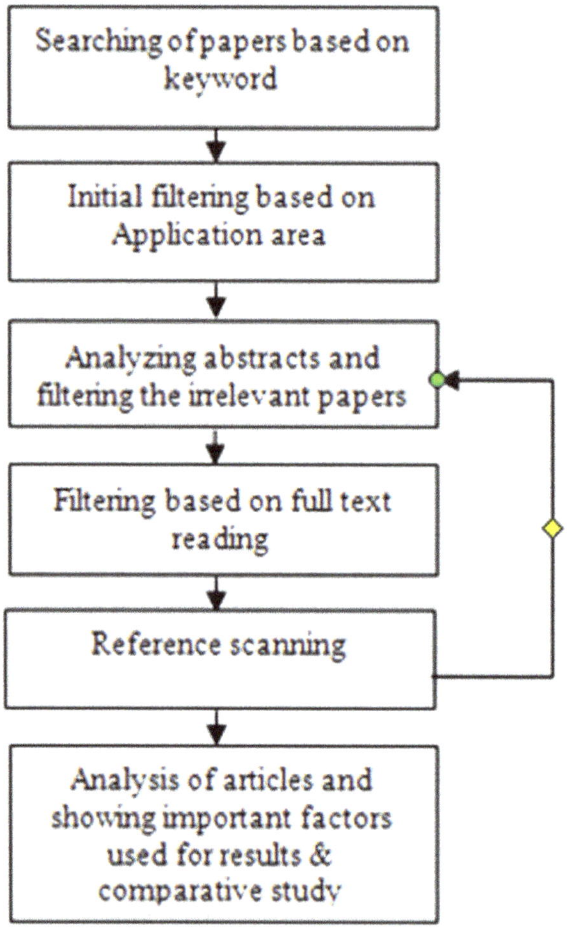

Singh et al. (2018) worked with a cascaded model of RNN, LSTM, and GRU, on the MIT-BIH database. The experimental model and its configuration worked well and proved the results better than a single observation using the given model. When the number of iterations is 5, the hidden layers are 3, and there are 64, 256, and 100 neurons per hidden layer, the RNN, and LSTM has an accuracy of 88.1 percent, which is better than RNN and GRU. RNN has an accuracy of 85.4 percent and GRU has an accuracy of 82.5 percent, both of which are less than LSTM accuracy. The model is implemented without any pre-processing, directly using signals from the MIT BIH database. As a result, the constructed model is far less complex than typical machine learning algorithms for the detection and classification of arrhythmia.

Maknickas and Maknickas (2017) proposed automatic diagnosis model for arrhythmia using combination of two models CNN and LSTM on MIT BIH dataset. The classification is target on eight ECG signals, including normal sinus rhythm. For experimental work 192x128 gray scale image was used generated after 1-Dimenesional to 2- Dimensional conversion. For cropping ECG images R peak wave peak marker is used and 92 data points before and after signal has been taken. It was placed before the model's last completely linked layer in dropout regularization trials. The dropout rate was 0.5. This method shows accuracy 99.01 percent, specificity 99.57 percent, and sensitivity 97.67 percent as a result means shows better detection through the proposed method.

Can et al. (2020) shows the study and experiment on the dynamics and trainability of RNN using a mathematical concept of a random matrix, and mean-field theories. The conclusion of the paper was additional gates in memory-based networks like LSTM, GRU, or RNN do not necessarily improve performance, the influence of gates on dynamics and trainability must be clear to develop a novel architecture.

Madan et al. (2022) proposed a model for automation of detection and classification, the author proposed a hybrid deep learning model using CNN and LSTM. The paper contributes two-fold, first 1D signal is transformed into a 2D scalogram for noise filtering and feature extraction, and after that 2D CNN and LSTM is applied for experimentation. The experiment results in 99% accuracy in the detection of arrhythmia. These 2D scalogram images provide robustness to the proposed system.

Shumai (n.d.). experimented in A spatial information fusion model based on ResNet and a temporal information fusion model based on BiLSTM is proposed in a work on the identification of arrhythmias in long-term ECG. The generated ECG feature map's local feature extraction section allows the model to efficiently combine the location data of nearby neighbors, and the global feature extraction section uses the BiLSTM network to obtain the correlation of the global features through autonomous learning across multiple spaces. Results are 99.4% accurate in terms of diagnosis, and the algorithm has strong clinical value and recognition performance.

Database

Most of the publications employ existing ECG recordings available as internet databases to adhere to ethical considerations. To train CADS systems, usually ECG databases, like, MITDB, PhysioNet, PTB, and others, have been categorized as normal and pathological sets of ECG. Table-1 lists commonly used ECG databases in the community.

Table 1. Some available database related to ECG

Name of Database	Disease	Sampling Information
Computing in Cardiology / PhysioNet	NSR AF	Real-time data collected on 44.1 KHz with resolution of 24 bit
PhysioNet, The ECG-ID Database	NSR AF	20 sec (time period) collected on 500 KHz with a resolution of 12 bit with a variation of ± 10 mV on normal
Long Term ST Database (LTSTDB).	NSR SVEB VEB	250 samples / second on 12-bit resolution, range of ± 10 mv
UCI cardiac arrhythmia	NSR Old (MI Sinus Bradycardia)	------
CinC Challenge 2000 Datasets	Sleep Apnea NSR	Signals sampled at 100 samples / second, 16 bit /sample with LSB priority in each pair,
MIT-BIH	Normal Sinus Rhythm (NSR)	Digitized at 250 KHz
Fantasia Database-PhysioBank	NSR (Activity: Watching movie)	------
Physikalisch- Technische Bundesanstalt (PTB)	MI	Signals sampled at 1000 samples / second resolution:16-bit range of ± 16.348 mv
PAF Prediction Challenge Database	PAF	16 bits/ sample, 128 samples/ signal in a second

Performance Parameters in Review

Common statistical methods to judge the performance of the classifier. Two classes have been decided as per the result of the learning method i.e positive class and negative class. The result of the prediction may be either correct or incorrect. In this way, classification can be summarized using four possible factors: True positive (TP): correct prediction of positive class, True Negative (TN): correct prediction of negative class, False positive (FP): incorrect prediction of positive class, and False Negative (FN): incorrect prediction of negative class.

$$Accuracy = \frac{TP + TN}{TP + TN + FP + FN}$$

$$Specificity = \frac{TN}{TN + FP}$$

$$Sensitivity = \frac{TP}{TP + FN}$$

$$Precision = \frac{TP}{TP + FP}$$

$$Recall = \frac{TP}{TP + FN}$$

$$PPV = \frac{TP}{TP + FP}$$

$$NPV = \frac{TN}{TN + FN}$$

$$F1 = \left(\frac{recall^{-1} + Precesion^{-1}}{2}\right)^{-1} - 2.\frac{precesion.recall}{precesion + recall}$$

Review Results

The outcomes of review are being taken on 2 bases, first comparison of performance of paper and second which technique used for feature extraction or classification. This exploratory review try to identify which techniques and databases highest standing gap in research where emerging research is going on. It has been identified that CNN-based techniques can effectively work in the field of arrhythmia classification. Both long-term and short-term content of signal can be used for learning in classification applications and can improve the results. The Results have been presented in table 2, CNN proves excellent capability in this field. The goal of this review was to look into the most potent deep learning method for various types of arrhythmias. The GRU/LSTM, CNN, and LSTM properly recognized Atrial Fibrillation, Supraventricular Ectopic Beats, and Ventricular Ectopic Beats, respectively. It's also been discovered that employing the appropriate Deep Learning methods can considerably improve categorization and related application performance.

In this review, the contribution of different variant techniques in percentage has been shown in the graph, and complexity over different datasets has been shown in Figure 14.

Figure 14. Contribution of different techniques in studied papers

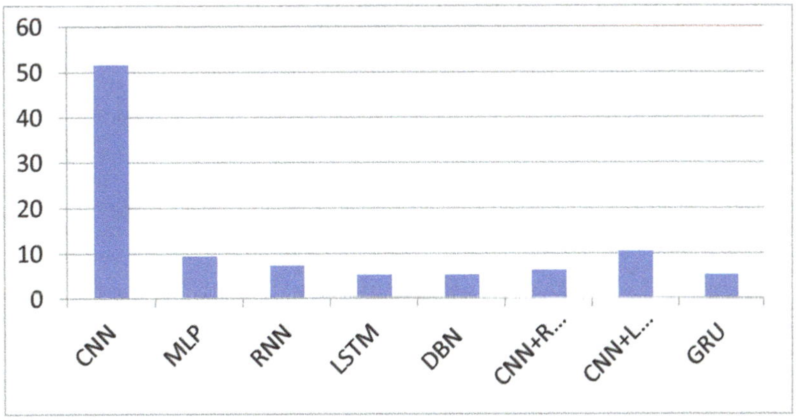

Table 2. Performance results from reviewed papers

Technique	Worked		Database used	Performance	Application	Ref. Taken
	Feature Extraction	Classification				
GRU-based	Time slice last heartbeat, wavelet energy	GRU + LSTM	PhysioNet Challenge 2017	Classwise F1(NSR)-0.90, F1(AF)-0.79, F1 (Av)-0.79	Atrial Fibrillation	Can, Tankut, et al et al. (2020)
Bi-GRU-based	Raw & wavelet	GRU	MIT-BIH DB	RAW F1 -score: 0.91 Wavelet: F1-Score:0.90	AF NSR	Minxiang Ye et at (2020)
LSTM	DWT	Bidirectional LSTM	MIT-DB	Accuracy: 0.9939	NSR PVC Paced Beat (PB) LBBB RBBB	Yildirim (2018)
LSTM	LSTM (Bidirectional)		MIT-DB	Sensitivity: 98.32 Accuracy: 98.51 Specificity: 98.67 Positive Predictive Accuracy: 98.39 (In %)	Atrial Fibrillation	Faust et al. (2018)
RNN-based	Recurrent Neural Network Configuration		MITDB	Accuracy: 95 (In %) F-score: 0.941 Precision: 1.00 Recall: 0.889	NSR AF	Sujadevi et al. (2018)
RNN-based	LSTM Layers -2 FCN-2		MIT-DB: [11 records 24 records 44 records]	Accuracy: 99.4, Specificity: 99.7, Sensitivity: 97.6, Positive Predictivity: 97.6 Accuracy: 99.6, Specificity: 99.8, Sensitivity: 97.5, Positive Predictivity: 97.9 Accuracy: 99.7, Specificity: 99.9, Sensitivity: 97.1, Positive Predictivity: 98.1	VEB	Zhang et al. (2019)
RNN-based	All possible Intervals (QRS and P, T)		PhysioNet challenge 2017	0.78- (F1- Score)	Normal sinus rhythm Atrial Fibrillation	Maknickas (2018)
RNN-based	Structure Premature-or-Escape- Flag	GRNN	MITDB SVDB LTSTDB-I	Accuracy: 97.4, Sensitivity:85.7, Specificity: 98.3 SVDB: Accuracy: 97.2, Sensitivity:77.2, Specificity:99.2	SVEB VEB	Wang et al. (2019)
DBN-based	DBN + RBM		MIT-DB AFDB	Precision: 25.5, Recall: 29.3, Accuracy: 58.7, Specificity: 70.5 After Gating Precision: 65.00, Recall: 68.1, Accuracy: 81.00, Specificity: 85.00	AF	Taji et al. (2018)
CNN based	CNN	Fully Convolution Network	Custom	Precision:94.71 Recall: 94.68, Accuracy: 94.7	Fatal Heart Monitoring	Tang et al. (2018)
CNN based	Transformation Technique (Wavelet)	CNN	MIT-DB	Accuracy: 81.07, Sensitivity: 74.96	AF	Xu et al. (2018)

continues on following page

Table 2. Continued

Technique	Worked		Database used	Performance	Application	Ref. Taken
	Feature Extraction	Classification				
CNN based	Convolutional		MITDB	Sensitivity: 97.91, Specificity: 99.61 Accuracy: 99.11, [Augmented] Sensitivity: 97.26, Specificity: 99.67 Accuracy: 98.92, [native]	AF	Ullah, Amin, et al (2020)
CNN based	Convolutional	Linear Classifier	Customized Database	Sensitivity: 95.2, Specificity: 99.0,	AF	Poh et al. (2018)
CNN based	1-D Convolutional	1-D Convolutional	MIT-DB	Error Rate:0.32 Positive Predictivity: 99.91 Sensitivity: 99.77	QRS Detection	Xiang et al. (2018)
CNN based	CNN	Fully convolutional Network	PhysioNet Challenge 2017	F 1 -score: Training Data: 99 Test Data: 89 (Average)	AF (NSR)	Sodmann et al. (2018)
CNN based	Segment and Epi-Endo)	Fully convolutional Network	9 patients 90 PVC beats	Accuracy: 78 (Segment) Accuracy: 90 (Epi-Endo)	PVC	Yang et al. (2018)
CNN based +MLP		convolutional	MIT-DB	f 1 -score: 66.09 Accuracy: 96.27	AAMI	Tao Wang et, al (2020)
CNN-BiLSTM	CNN	LSTM	PhysioNet 2017 MITDB	Sensitivity 74.3%, Specificity: 97.5%, F1 Score: 80% (Mean) F1- Score: 82.6 (best case)	NSR AF (Other rhythms)	Liang, Yongbo, et al. (2020)
CNN based	convolutional	Fully convolutional Network	Physio-Bank	Accuracy: 88 (Arrythymia) Accuracy: 87 (NSR)	NSR VT AF Ventricular	Savalia and Emamian (2018)
CNN based	convolutional	Fully convolutional Network	CUDB MIT-BIH (VFDB)	Accuracy: 99.26 Specificity: 99.44 Sensitivity:97.07	VFib VT	Nguyen et al. (2018)
CNN based	CNN + Fully convolutional Network		PhysioNet 2017- Challenge	Accuracy:96.99 (5 Sec Rec) Accuracy:98.13 (20 Sec Rec)	AF + NSR Abnormal Rhythms	Fan et al. (2018)
MLP-based ECG arrhythmia	RR interval (Pre,Post,Local average Global average)	Multilayer Perceptron	MIT-DB	Accuracy: 99.68	NSR SVEB VEB NSR Heartbeats (Cannot be Classified)	Sannino al.(2018)
MLP-based ECG arrhythmia	Multilayer Perceptron	Support Vector Machine	2000 Physio-Net Challenge	Sensitivity:100 Accuracy:100,	OSA Detection	Li et al. (2018a)

Results of all the techniques used for respective research focus on the following points for which consideration and its accumulated behavior must be considered for progress in this direction.

- Variant of different techniques like CNN/ DBN/ RNN/ LSTM/ GRU
- Complexity consideration with used resources and available data.
- Performance parameter and application area of results

Working with feature selection, extraction of essential features, the pattern of QRS, V & T, and other ECG signal taxonomies is vital and beneficial for early diagnosis and prevention of cardiac disorders. As a future effort, the effect of multiple lead ECG signals on improving the experimental cases in the automobile framework of the optimized model is being investigated.

THEORETICAL FRAMEWORK

Because of the inter-patient variability in ECG data, traditional approaches may not produce effective findings. Additionally, the increasing bulk of data may have a negative impact on the efficiency and accuracy of existing methods. The strategies given in the literature have been applied to smaller datasets; nevertheless, the performance needs be tested on bigger datasets for generalization. There have been methods published that use 2-D ECG signals, but there are no clear details on how the 1-D ECG signal is converted to 2-D pictures for use with 2-D CNN models to our knowledge. The majority of approaches has only been tested on a few forms of arrhythmia and must be reviewed on all major types. It should be mentioned that the performance of 1-D ECG signal techniques can be improved further. We proposed work focus towards cascaded model using CNN and bidirectional LSTM to on 12 lead ECG signal data. Experimental setup and also include

- Proposed cascade model configured for effective arrhythmia classification with single and multiple lead data.
- Spectrograms (2-D pictures) proposing to use, which are created using STFT from a 1-D ECG signal.
- Data augmentation applied to depict ECG signals in a 2-D picture.
- Experimental performance comparison with existed learning architectures existing models.

Figure 3 shows a quick idea about proposed work specifying the work to be carried against found gap. Proposed model aimed to configure such a cascade system that work quick responsive, optimized model for Arrhythmia classification effective for society in medical health care field.

In the above proposed model (Figure-15) diagram preprocessing and data augmentation are very important for the optimal training of deep learning machines. Some of the key process methods are noise filtering, signal normalization, data augmentation those helps in whole preprocessing part of proposed model. Discreate wavelet transform (DWT) applied for dropping baseline wander noise while notch filter was applied for powerline interference. After applying noise removal techniques, R-peak detection and segmentation can be made easy. For fair comparisons among data common scales of signals helps out though signal normalization. There are various normalization techniques available like min-max scaling, z- score, peak amplitude, percentage change, integral, waveform alignment, and

Figure 15. Proposed model

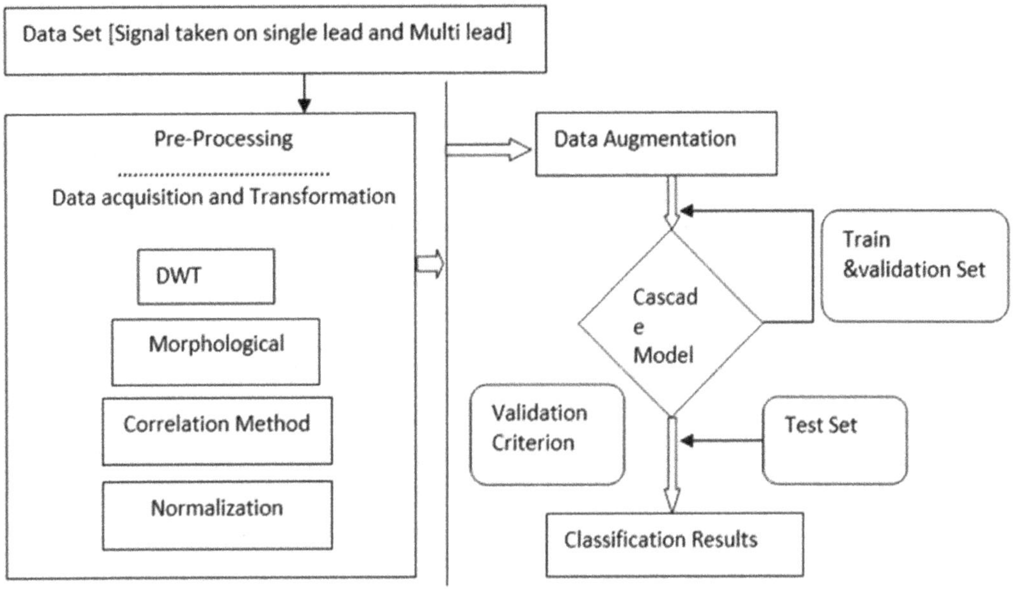

segment based. Proposed model uses Z-score normalization which focus on transforming signal into a distribution with mean 0 and standard deviation 1. Percentage change normalization is also an effective method to work. Data augmentation method uses variations in amplitude, scaling, shifting and rotation method applying on ECG signals. Combination of these augmentation techniques allows the model to learn high-level features that are more representative of the underlying physiological characteristics of the heart's electrical activity, rather than being overly sensitive to specific recording conditions or noise that means improves model generalizations. Functional equation is given below for noise and augmentation techniques applied in proposed work.

ECG_filtered(t) = ECG_original(t) - BaselineWander(t)
ECG_cleaned(t) = ECG_filtered(t) - PowerlineInterference(t)
ECG_normalized(t) = (ECG(t) - mean(ECG)) / std(ECG)
ECG_normalized(t) = ((ECG(t) - reference_point) / reference_point) * 100

SOLUTION AND RECOMMENDATION

Working towards an experiment from the above-discussed theoretical framework following algorithm steps are used for the solution of discussed problems.

Algorithm
Step 1. **Reading Input ECG data:** ECG Data is collected from MIT Database.
Step 2. **Data Cleaning:** Removing irrelevant structures not appropriate for Work.

Step 3. **Segmenting and Feature Extraction:** Extracting exact area and information present in ECG waveform for signal peak.
Step 4. **Feature Selection:** Selecting appropriate features information
Step 5. **Data Augmentation:** Creating Synthetic data for the increasing amount of learning.
Step 6. **Creating Training & Test Sample:** Partitioning the full dataset
Step 7. **Create Cascaded model (RNN + LSTM):** Creating a cascade model for learning and tested on sample
Step 8. **To calculate and compare the performance:** Performance is measured and compared with previous models.

The MIT-BIH dataset has 47 records of 30 min each and in which 40% of records are cardiac patients. This dataset contained different signal based on leads like V1, V2, V4, V5, and ML2. To begin, the data was divided 70-30. Signals from ML2 were chosen, and each record's signal was divided into 720-720 length chunks. Because the sampling rate was 360, 360 segments before the R-Peak value and 360 segments after the R-Peak value insured that almost three beats would be present in one chunk, and the core beat would be classified against 15 different arrhythmic labels. The accuracy, specificity, and sensitivity of the classification were the three main measurements used to evaluate the performance of CNN models. A confusion matrix is used to characterize these measurements. True Positive exists when there is an arrhythmia that was accurately recognized, hence the proportion of True Positive should be large when compared to other values for the algorithm to perform well and for accuracy. The total number of beats correctly classified as a Normal Beat or an Arrhythmia is referred to as classification accuracy.

Table 3. Confusion matrix

	True (N)	**False (A)**
True (N)	15820	1298
False (A)	1835	9188

As per the model implementation is concerned with proposed CNN and LSTM + RNN model is tested and shows the results on the basis of accuracy, sensitivity, and Specificity. Confusion matrix and classification results have been displayed in Table 3 and Table 4 respectively.

Table 4. Classification results

Model With Augmentation	**Accuracy**	**Sensitivity**	**Specificity**
CNN	88.87	89.61	87.62
LSTM	89.06	89.15	88.14
RNN	85.1	80.13	82.25
LSTM +RNN	89.89	93.46	84.36

The aforementioned findings were obtained using a three-hidden-layer CNN with five iterations, 64, 256 neurons per hidden layer, and two pooling layers. In all of the aforementioned models that are applied with augmentation, the LSTM model performs better. In order to achieve results that are superior to those of another existing model, we are proposing to implement CNN and LSTM cascade models in future work. Our suggested approach does the same and leaves potential for additional research in the field of binary classification (Arrhythmia detection), where there hasn't been any significant work. Even higher categorization accuracy can be achieved by increasing the number of epochs. The study shows that long short-term memory yields the best outcomes in the binary classification of ECG arrhythmia, and that convolution neural networks on the MIT BIH dataset can be used in classification work in the future. The hidden layer's epoch and neuron counts may be increased even more during the classification process.

Figure 16. Result and comparison

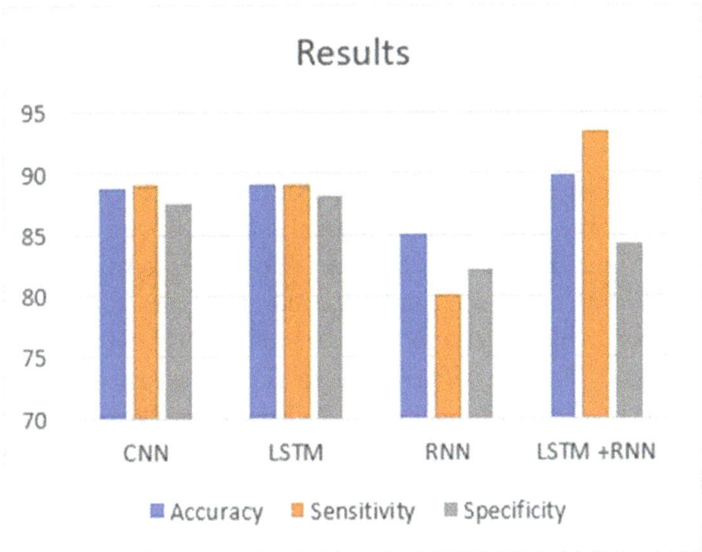

The outcomes of the effort are displayed in Figure 16. The accuracy, specificity, and sensitivity of the prior research in this area are compared. Results from experiments show that the suggested strategy works better. In comparison to earlier efforts, the proposed technique yields more sensitivity, accuracy, and less specificity. The aforementioned findings were derived from a 5-iteration, 3-hidden-layer CNN with 64–256 neurons per hidden layer and 2 pooling layers. In the complete above model applied, the LSTM model performs better since these models are enhanced. When compared to the solo model, the RNN and LSTM cascaded model performs better. LSTM performs well when applied to an ECG time series data sequence, whereas CNN performs better when applied directly to picture data. The work may be extended to customized models and create more patterns of cascaded models based on the classification problem, dataset properties, and application of work. The deeper network may result well but overfitting is possible. This work can be emphasized noise reduction at the early preprocessing step to get the most accurate dataset for work.

FUTURE RESEARCH DIRECTION

The future research direction in this felid of research is how to handle data variability of arrhythmia signals so that accuracy can be increased. In order to produce superior results in comparison to another current model, as shown in Figure 5 and Table 2, we propose implementing CNN and LSTM cascade models along with reconfiguration of the attributes. The following stage of the future study will involve multiclass classification on multiple datasets. Increasing the number of epochs or rearranging the structure may produce results with greater precision. The experiment's findings demonstrate that long-term memory and RNN work well for binary classification of ECG arrhythmias and that future classification work on the MIT BIH dataset can be done with Faster R-CNN. The hidden layer's epoch and neuron counts may be increased even more during the classification process.

CONCLUSION

This chapter focus on review and experimental work using deep learning methods, applicable in arrhythmia classification model. From the results and previous review article it can be concluded that the results if arrhythmia classification using deep learning techniques are totally based on the dataset used and how effective dataset is learned [know about properties] and which DL technique can work well in that scenario. Vast amount of data collected from hospital, critical patients in relative filed, ECG collection method (single lead or multi-lead) always gives new challenges to work with real time application. Careful experimental work with customized and cascaded model may work well till certain situation. Moreover, performance improvement always increases resource cost in this area, so a balance model must find for applying in real world. Working with ECG images dataset is also a recent experimental domain to understand, classify and predict arrhythmia. Transfer learning and recent optimization method may be used on this image dataset. Arrhythmia classification work is being carried from last 20 years but a reliable and sustainable system is needed for helping healthcare professional and society to save lives.

ACKNOWLEDGMENT

A special thanks to Dr. Manish Manoria, Dr. Sunil Joshi for supporting every time. I am also very thankful to Dr. Pankaj Manoria, (Cardiologist), who provide me a direct support to align deep learning area with health care domain and better understanding of ECG equipment and signal analysis for arrhythmia.

REFERENCES

Acharya, U. R., Fujita, H., Lih, O. S., Hagiwara, Y., Tan, J. H., & Adam, M. (2017). Automated detection of arrhythmias using different intervals of tachycardia ECG segments with convolutional neural network. *Information Sciences*, *405*, 81–90. doi:10.1016/j.ins.2017.04.012

Acharya, U. R., Fujita, H., Oh, S. L., Hagiwara, Y., Tan, J. H., Adam, M., & San Tan, R. (2019). Deep convolutional neural network for the automated diagnosis of congestive heart failure using ecg signals. *Applied Intelligence*, *49*(1), 16–27. doi:10.100710489-018-1179-1

Acharya, U. R., Fujita, H., Oh, S. L., Raghavendra, U., Tan, J. H., Adam, M., Gertych, A., & Hagiwara, Y. (2018). Automated identification of shockable and non-shockable life-threatening ventricular arrhythmias using convolutional neural network. *Future Generation Computer Systems*, *79*, 952–959. doi:10.1016/j.future.2017.08.039

Acharya, U. R., Oh, S. L., Hagiwara, Y., Tan, J. H., Adam, M., Gertych, A., & San Tan, R. (2017c). A deep convolutional neural network model to classify heart- beats. *Computers in Biology and Medicine*, *89*, 389–396. doi:10.1016/j.compbiomed.2017.08.022 PMID:28869899

Al Rahhal, M. M., Bazi, Y., Al Zuair, M., Othman, E., & BenJdira, B. (2018). Convo- lutional neural networks for electrocardiogram classification. *Journal of Medical and Biological Engineering*, *38*(6), 1014–1025. doi:10.100740846-018-0389-7

Andersen, R. S., Peimankar, A., & Puthusserypady, S. (2019). A deep learning ap- proach for real-time detection of atrial fibrillation. *Expert Systems with Applications*, *115*, 465–473. doi:10.1016/j.eswa.2018.08.011

Andreotti, F., Carr, O., Pimentel, M. A., Mahdi, A., & De Vos, M. (2017). Comparing feature-based classifiers and convolutional neural networks to detect ar- rhythmia from short segments of ecg. In 2017 computing in cardiology (CINC) (pp. 1–4). IEEE.

Babic, A., & Sepehri, A. A. (2019a). Extraction of diagnostic infor- mation from phonocardiographic signal using time-growing neural network. In L. Lhotska, L. Sukupova, I. Lackovi ´c, & G. S. Ibbott (Eds.), *World congress on med- ical physics and biomedical engineering 2018* (pp. 849–853). Singapore: Springer Singapore.

Benjamin, E. J., Virani, S. S., Callaway, C. W., Chamberlain, A. M., Chang, A. R., Cheng, S., & Deo, R. (2018). Heart disease and stroke statistics-2018 up- date: A report from the american heart association. *Circulation*, *137*(12), e67. doi:10.1161/CIR.0000000000000558 PMID:29386200

Bizopoulos, P., & Koutsouris, D. (2018). Deep learning in cardiology. *IEEE Reviews in Biomedical Engineering, 12*, 168–193.

Can, T. (2020). Gating Creates Slow Modes And Controls Phase-space Complexity In GRUs And LSTMs. *PMLR Proceedings*. http://proceedings.mlr.press/v107/can20a.

Chamatidis, I., Katsika, A., & Spathoulas, G. (2017). Using deep learning neural net- works for ECG based authentication. In *2017 International Carnahan Conference on Security Technology (ICCST)* (pp. 1–6). IEEE.

Chandra, B., Sastry, C. S., Jana, S., & Patidar, S. (2017). Atrial fibrillation detection using convolutional neural networks. In 2017 Computing in cardiology (CINC) (pp. 1–4). IEEE.

Chen, M., Wang, G., Xie, P., Sang, Z., Lv, T., Zhang, P., & Yang, H. (2018). Region aggregation network: Improving convolutional neural network for ECG characteristic detection. In *2018 40th Annual international conference of the ieee engineering in medicine and biology society (EMBC)* (pp. 2559–2562). IEEE.

Dewangan, N. K., & Shukla, S. (2015). A survey on ECG signal feature extraction and analysis techniques. *International Journal of Innovative Research in Electrical, Electronics, Instrumentation and Control Engineering, 3*(6).

Dinakarrao, S. M. P., Jantsch, A., & Shafique, M. (2019). Computer-aided arrhythmia diagnosis with bio-signal processing: A survey of trends and techniques. *ACM Computing Surveys*, *52*(2), 23.

Dua, D., & Graff, C. (2017). *UCI machine learning repository*. http://archive.ics.uci. edu/ml

Fan, X., Yao, Q., Cai, Y., Miao, F., Sun, F., & Li, Y. (2018). Multiscaled fusion of deep convolutional neural networks for screening atrial fibrillation from single lead short ECG recordings. *IEEE Journal of Biomedical and Health Informatics*, *22*(6), 1744–1753. doi:10.1109/JBHI.2018.2858789 PMID:30106699

Faust, O., Shenfield, A., Kareem, M., San, T. R., Fujita, H., & Acharya, U. R. (2018). Au- tomated detection of atrial fibrillation using long short-term memory network with RR interval signals. *Computers in Biology and Medicine*, *102*, 327–335. doi:10.1016/j.compbiomed.2018.07.001 PMID:30031535

Gharehbaghi, A., & Babic, A. (2018). Structural risk evaluation of a deep neural net- work and a Markov model in extracting medical information from phonocardio- graphy. *Studies in Health Technology and Informatics*, *251*, 157–160. PMID:29968626

Gharehbaghi, A., Lindén, M., & Babic, A. (2019b). An artificial intelligent-based model for detecting systolic pathological patterns of phonocardiogram based on time-growing neural network. *Applied Soft Computing*, *83*, 105615. doi:10.1016/j.asoc.2019.105615

Gharehbaghi, A., & Lindn, M. (2018). A deep machine learning method for classifying cyclic time series of biological signals using time-growing neural network. *IEEE Transactions on Neural Networks and Learning Systems*, *29*(9), 4102–4115. doi:10.1109/TNNLS.2017.2754294 PMID:29035230

Ghiasi, S., Abdollahpur, M., Madani, N., Kiani, K., & Ghaffari, A. (2017). Atrial fibrillation detection using feature based algorithm and deep convolutional neural network. In 2017 Computing in cardiology (CINC) (pp. 1–4). IEEE.

H., Ma, S., Cui, J., Xiao, W., & Liu, L. (2022, August 11). *Deep Learning-Based Data Augmentation and Model Fusion for Automatic Arrhythmia Identification and Classification Algorithms.* doi:10.1155/2022/1577778

He, K., Zhang, X., Ren, S., & Sun, J. (2015). Spatial pyramid pooling in deep convolu- tional networks for visual recognition. *IEEE Transactions on Pattern Analysis and Machine Intelligence*, *37*(9), 1904–1916. Advance online publication. doi:10.1109/TPAMI.2015.2389824 PMID:26353135

Hinton, G. E., & Sejnowski, T. J. (1986). Learning and relearning in Boltzmann machines. *Parallel Distributed Processing: Explorations in the Microstructure of Cognition, 1*, 282-317.

Huff, J. (2006). *ECG workout: Exercises in arrhythmia interpretation*. Lippincott Williams & Wilkins.

Isin, A., & Ozdalili, S. (2017). Cardiac arrhythmia detection using deep learning. *Procedia Computer Science*, *120*, 268–275. doi:10.1016/j.procs.2017.11.238

Ji, J., Chen, X., Luo, C., & Li, P. (2018). A deep multi-task learning approach for ECG data analysis. In *2018 IEEE EMBS International conference on biomedical & health informatics (BHI)* (pp. 124–127). IEEE 10.1109/BHI.2018.8333385

Joshi, A. J., Chandran, S., Jayaraman, V. K., & Kulkarni, B. D. (2009). *Hybrid SVM for multiclass arrhythmia classification. In 2009 IEEE international conference on bioinformatics and biomedicine.* IEEE.

JunT. J.NguyenH. M.KangD.KimD.KimD.KimY.-H. (2018). *Ecg arrhythmia classification using a 2-d convolutional neural network.* arXiv:1804.06812.

Kamaleswaran, R., Mahajan, R., & Akbilgic, O. (2018). A robust deep convolutional neural network for the classification of abnormal cardiac rhythm using single lead electrocardiograms of variable length. *Physiological Measurement*, *39*(3), 035006. doi:10.1088/1361-6579/aaaa9d PMID:29369044

Kasper, D. L., Fauci, A. S., Hauser, S. L., Longo, D. L., Jameson, J. L., & Loscalzo, J. (2018). Harrison's principles of internal medicine. Mc- Graw Hill Professional.

Keyvanrad, M. A., & Homayounpour, M. M. (2015). Deep belief network training im- provement using elite samples minimizing free energy. *International Journal of Pattern Recognition and Artificial Intelligence*, *29*(05), 1551006. doi:10.1142/S0218001415510064

Khan, A. H. (n.d.). *Arrhythmia Classification Techniques Using Deep Neural Network.* Www.hindawi.com

Krizhevsky, A., Sutskever, I., & Hinton, G. E. (2012). ImageNet classification with deep convolutional neural networks. *Advances in Neural Information Processing Systems.*

Labati, R. D., Muñoz, E., Piuri, V., Sassi, R., & Scotti, F. (2018). Deep-ECG: Convolu- tional neural networks for ECG biometric recognition. *Pattern Recognition Letters.*

LeCun, Y., Bottou, L., Bengio, Y., & Haffner, P. (1998). Gradient-based learning applied to document recognition. *Proceedings of the IEEE.* https://doi.org/. 72679110.1109/5

Li, K., Pan, W., Li, Y., Jiang, Q., & Liu, G. (2018a). A method to detect sleep apnea based on deep neural network and hidden Markov model using single-lead ECG signal. *Neurocomputing*, *294*, 94–101. doi:10.1016/j.neucom.2018.03.011

Li, Y., Pang, Y., Wang, J., & Li, X. (2018b). Patient-specific ECG classification by deeper CNN from generic to dedicated. *Neurocomputing*, *314*, 336–346. doi:10.1016/j.neucom.2018.06.068

Liang, Y. (2020, January). Deep Learning Algorithm Classifies Heartbeat Events Based On Electrocardiogram Signals. *Frontiers*, *1*. Www.frontiersin.org PMID:33117191

Limam, M., & Precioso, F. (2017). Atrial fibrillation detection and ECG classification based on convolutional recurrent neural network. In 2017 computing in cardiology (CINC) (pp. 1–4). IEEE.

Liu, M., & Kim, Y. (2018). Classification of heart diseases based on ECG signals using long short-term memory. In *2018 40th Annual international conference of the IEEE engineering in medicine and biology society (EMBC)* (pp. 2707–2710). IEEE. 10.1109/EMBC.2018.8512761

Liu, W., Huang, Q., Chang, S., Wang, H., & He, J. (2018). Multiple-feature-branch conolutional neural network for myocardial infarction diagnosis using electrocardiogram. *Biomedical Signal Processing and Control*, *45*, 22–32. doi:10.1016/j.bspc.2018.05.013

Loni, M., Sinaei, S., Zoljodi, A., Daneshtalab, M., & Sjödin, M. (2020). Deepmaker: A multi-objective optimization framework for deep neural networks in embedded systems. *Microprocessors and Microsystems*, *73*, 102989. doi:10.1016/j.micpro.2020.102989

Loni, M., Zoljodi, A., Sinaei, S., Daneshtalab, M., & Sjödin, M. (2019). Neu- ropower: Designing energy efficient convolutional neural network architecture for embedded systems. In *International conference on artificial neural networks* (pp. 208–222). Springer.

Lui, H. W., & Chow, K. L. (2018). Multiclass classification of myocardial infarction with convolutional and recurrent neural networks for portable ECG devices. *Informatics in Medicine Unlocked*, *13*, 26–33. doi:10.1016/j.imu.2018.08.002

Luo, K., Li, J., Wang, Z., & Cuschieri, A. (2017). Patient-specific deep architectural model for ECGclassification. *Journal of Healthcare Engineering*.

Madan, P., Singh, V., Singh, D. P., Diwakar, M., Pant, B., & Kishor, A. (2022, April 2). A *Hybrid Deep Learning Approachfor ECG-BasedArrhythmia Classification*. MDPI. doi:10.3390/bioengineering9040152

Majumdar, A., & Ward, R. (2017). Robust greedy deep dictionary learning for ECGar- rhythmia classification. In *2017 International joint conference on neural networks (IJCNN)* (pp. 4 400–4 407). IEEE.

Maknickas, V., & Maknickas, A. (2017). Atrial fibrillation classification using QRS complex features and LSTM. In 2017 computing in cardiology (CINC) (pp. 1–4). IEEE. doi:10.22489/CinC.2017.350-114

Mathews, S. M., Kambhamettu, C., & Barner, K. E. (2018). A novel application of deep learning for single-lead ECG classification. *Computers in Biology and Medicine*, *99*, 53–62. doi:10.1016/j.compbiomed.2018.05.013 PMID:29886261

Minxiang. (n.d.). Feature selection and extraction in sequence labeling for arrhythmia detection. *Fourth International Balkan Conference on Communications and Networking*.

Nguyen, M. T., Van Nguyen, B., & Kim, K. (2018). Deep feature learning for sudden cardiac arrest detection in automated external defibrillators. *Scientific Reports*, *8*(1), 1–12. doi:10.103841598-018-33424-9 PMID:30464177

Oh, S. L., Ng, E. Y., San Tan, R., & Acharya, U. R. (2018). Automated diagnosis of arrhythmia using combination of cnn and lstm techniques with variable length heart beats. *Computers in Biology and Medicine*, *102*, 278–287. doi:10.1016/j.compbiomed.2018.06.002 PMID:29903630

Paszke, A., Gross, S., Massa, F., Lerer, A., Bradbury, J., Chanan, G., . . . Antiga, L. (2019). Pytorch: An imperative style, high-performance deep learning library. In Advances in neural information processing systems (pp. 8024–8035). Academic Press.

Pater, C. (2005). Methodological considerations in the design of trials for safety as- sessment of new drugs and chemical entities. *Current Controlled Trials in Cardiovascular Medicine*, *6*(1), 1. doi:10.1186/1468-6708-6-1 PMID:15691384

Plesinger, F., Nejedly, P., Viscor, I., Halamek, J., & Jurak, P. (2017). Automatic detec- tion of atrial fibrillation and other arrhythmias in holter ECG recordings using rhythm features and neural networks. In 2017 Computing in cardiology (CINC) (pp. 1–4). IEEE.

Poh, M.-Z., Poh, Y. C., Chan, P.-H., Wong, C.-K., Pun, L., Leung, W. W.-C., Wong, Y.-F., Wong, M. M.-Y., Chu, D. W.-S., & Siu, C.-W. (2018). Diagnostic assessment of a deep learning system for detecting atrial fibrillation in pulse waveforms. *Heart (British Cardiac Society)*, *104*(23), 1921–1928. doi:10.1136/heartjnl-2018-313147 PMID:29853485

Pourbabaee, B., Roshtkhari, M. J., & Khorasani, K. (2017). Deep convolutional neu- ral networks and learning ECGfeatures for screening paroxysmal atrial fibrilla- tion patients. *IEEE Transactions on Systems, Man, and Cybernetics. Systems*, *48*(12), 2095–2104. doi:10.1109/TSMC.2017.2705582

Qiao, J., Wang, G., Li, W., & Li, X. (2018). A deep belief network with PLSR for non- linear system modeling. *Neural Networks*, *104*, 68–79. doi:10.1016/j.neunet.2017.10.006 PMID:29729561

Rocher Medical Center. (n.d.). *Intercity digital electrocardiogram Alliance' ideal database* (e-hol-03-0202-003). http://thew-project.org/Database/E-HOL-03-0202-003.html

Rubin, J., Parvaneh, S., Rahman, A., Conroy, B., & Babaeizadeh, S. (2018). Densely connected convolutional networks for detection of atrial fibrillation from short single-lead ECG recordings. *Journal of Electrocardiology*, *51*(6), S18–S21. doi:10.1016/j.jelectrocard.2018.08.008 PMID:30122456

Sadrawi, M., Lin, C.-H., Lin, Y.-T., Hsieh, Y., Kuo, C.-C., Chien, J. C., Haraikawa, K., Abbod, M., & Shieh, J.-S. (2017). Arrhythmia evaluation in wearable ECG devices. *Sensors (Basel)*, *17*(11), 2445. doi:10.339017112445 PMID:29068369

Sannino, G., & De Pietro, G. (2018). A deep learning approach for ECG-based heartbeat classification for arrhythmia detection. *Future Generation Computer Systems*, *86*, 446–455. doi:10.1016/j.future.2018.03.057

Savalia, S., & Emamian, V. (2018). Cardiac arrhythmia classification by multi-layer perceptron and convolution neural networks. *Bioengineering (Basel, Switzerland)*, *5*(2), 35. doi:10.3390/bioengineering5020035 PMID:29734666

Sawhney, N. S., Anousheh, R., Chen, W. C., & Feld, G. K. (2009). *Diagnosis and management of typical atrial flutter*. . doi:10.1016/j.ccl.2008.09.010

Sayantan, G., Kien, P., & Kadambari, K. (2018). Classification of ECG beats using deep belief network and active learning. *Medical & Biological Engineering & Computing*, *56*(10), 1887–1898. doi:10.100711517-018-1815-2 PMID:29651694

Scheba, P., Scebba, G. C., Zhang, J., Delai, M., & Karlen, W. (2017). Beat by beat: Classifying cardiac arrhythmias with recurrent neural networks. In 2017 computing in cardiology (CINC) (pp. 1–4). IEEE.

Schuster, M., & Paliwal, K. K. (1997). Bidirectional recurrent neural networks. *IEEE Transactions on Signal Processing*, *45*(11), 2673–2681. doi:10.1109/78.650093

Shashikumar, S. P., Shah, A. J., Clifford, G. D., & Nemati, S. (2018). Detection of paroxysmal atrial fibrillation using attention-based bidirectional recurrent neural networks. In *Proceedings of the 24th ACM SIGKDD international conference on knowledge discovery & data mining* (pp. 715–723). ACM. 10.1145/3219819.3219912

Singh, S., Pandey, S. K., Pawar, U., & Janghel, R. R. (2018). Classification of ECG Arrhythmia using Recurrent Neural Networks. *Procedia Computer Science*, *132*, 1290–1297. doi:10.1016/j.procs.2018.05.045

Zhang, J., Zhang, Q., & Wei, X. (2018). Classification of ECG signals based on 1d convolution neural network. In *2017 IEEE 19th international conference on e-health networking, applications and services (healthcom)* (pp. 1–6). IEEE.

KEY TERMS AND DEFINITIONS

CADS: Computer Aided Diagnosis system is a technology that involves the use of computer algorithms and artificial intelligence (AI) to assist healthcare professionals in the process of diagnosing medical conditions.

DL: Deep Learning is a subset of machine learning that involves training artificial neural networks with multiple layers, also known as deep neural networks, to learn and make predictions from data.

DWT: DWT stands for Discrete Wavelet Transform. It is another widely used signal processing technique, primarily used for time-frequency analysis and feature extraction

ECG: Stands for Electrocardiogram. It is a medical test that measures the electrical activity of the heart over a period of time on signal plot.

ML: Machine Learning (ML) is a subfield of artificial intelligence (AI) that focuses on the development of algorithms and statistical models that allow computers to learn and improve their performance on a specific task without being explicitly programmed.

STFT: STFT stands for Short-Time Fourier Transform. It is a widely used signal processing technique in the field of audio and image processing, specifically in time-frequency analysis.

Chapter 6
Modeling the Microbial Cells for Biotechnological Applications

Fatma Ece Altinisik Kaya
Marmara University, Turkey & Eberhard Karl University of Tübingen, Germany

Fatma Gizem Avci
Üsküdar University, Turkey

ABSTRACT

Synthetic biology is an engineering approach that designs biological systems to have new functions for useful purposes. Rational cell design is crucial for synthetic biology and can be considered a strategy that includes increasing the production efficiency of a cell or developing a new product by changing or adding specific biochemical reactions and regulations using modern metabolic engineering techniques. Metabolic flux and control analyses with models created in silico following cellular genome mapping have also contributed significantly to rational cell design. Models make the prediction of behaviors possible without performing a vast number of experiments when biological systems are modified. The modeling of microbial cells for the biosynthesis of high-value products is gaining increasing attention in biotechnology. In this chapter, the main focus will be on the concept of the genome scale metabolic modeling of microbial cells in biotechnological applications with an emphasis on biofuels and pharmaceuticals.

Recent advancements in synthetic biology have opened up remarkable opportunities for enhancing the design and performance of microbial cells. By harnessing synthetic biology tools, scientists can modify the metabolic functions of microorganisms for the efficient industrial manufacturing of diverse and valuable products (Chandran et al., 2008; Singh, 2018). There are two main approaches in synthetic biology, bottom-up and top-down. In the bottom-up approach, a singular discrete genetic element is utilized to modify a module. This alteration's influence is then extended throughout the entire biological network. This process can lead to a transformation of the original biological system's inherent function, potentially giving rise to novel functions and even what can be termed 'artificial life'. The top-down approach operates without a deterministic understanding of the genotype-phenotype relationship. Instead, top-down techniques leverage random or semi-random whole-genome engineering to enhance the available genetic

DOI: 10.4018/978-1-6684-6577-6.ch006

diversity. By doing so, a diverse collection of genetic variants is generated, creating a pool from which superior phenotypes can be identified and isolated (Dahabieh et al., 2020).

In silico models have contributed significantly to rational cell design (Kulkarni, 2016; Stephanopoulos et al., 1998; Torres et al., 2002). With the models, numerous designs can be examined using computer simulations without performing experiments (Baldwin et al., 2015; Zheng et al., 2010). Modeling plays a crucial role in synthetic biology approaches in many ways. Models can be used to predict the behavior of the metabolic pathways and modifications to enhance production, to select the optimal pathways for the synthesis of desired product, to identify the target genes and reactions, to predict the effect of genetic modifications on the overall cellular metabolism, and to explore various substrate utilization strategies for a given organism in the bottom-up approach while they are useful for understanding cellular behavior, predicting phenotypic responses, bioprocess optimization, and strain improvement in top-down approach (Ding et al., 2022; Liang et al., 2011; Roberts et al., 2013; San León et al., 2022).

The rapid development in the availability of multi-omics data (genes, genomes, metagenomes, etc.) and automated model construction and curation tools have accelerated modeling studies (Gyori et al., 2021). The modeling of microbial cells for the biosynthesis of high-value products like biofuels, pharmaceuticals, and natural compounds is gaining increasing attention in biotechnology (Camacho-Zaragoza et al., 2016; Chen et al., 2020; Krämer et al., 2003). Models for several industrial workhorses including *Escherichia coli* (Alper et al., 2005; Shah et al., 2021), *Saccharomyces cerevisiae* (Asadollahi et al., 2009; Ishchuka et al., 2022), and *Corynebacterium glutamicum* (Becker et al., 2011; Cheng et al., 2019) have been constructed.

In this chapter, the main focus will be on the genome-scale metabolic models of microbial cells in biotechnological applications with an emphasis on the production of biofuels and pharmaceuticals.

COMPUTATIONAL ASPECTS IN METABOLIC ENGINEERING

Synthetic biology and metabolic engineering are interdependent disciplines that collaborate to thrive and attain their distinct objectives. While synthetic biology provides the components and insights about diverse biological processes, metabolic engineering employs this knowledge to optimize the path of biological synthesis for a specific desired compound (García-Granados et al., 2019).

Metabolic engineering aims to maximize cellular processes that belong to a particular organism in order to synthesize the molecule of interest from a substrate that is preferably inexpensive and simple. Genetic engineering tools are used in metabolic engineering studies to remove genes from the organism that hinder synthesis as well as to insert genes of partial or whole metabolic pathways for the production of a certain substance. Additionally, these tools enable the introduction of point mutations that alter the metabolic flux by lowering the activity or expression of native proteins. But to make all of these modifications, specific knowledge about the enzymes involved in the reactions, the metabolic pathways that include these reactions, and details about the organism where the modifications will be made are required; for this reason, organisms' genomes have been sequenced, proteins have been characterized, and metabolic studies have provided a wealth of extremely useful tools and information (Caspi et al., 2013). In recent years, especially with the introduction of the CRISPR/Cas9 techniques to the community, genetic manipulations and interventions are quite simple. Technological advances in sequencing technologies have enabled more information about the genomes of both known and newly discovered organisms. Additionally, the catalog of enzymes and biochemical reactions has grown, and specific biological phenomena (such as viruses or phage infections) have been characterized (Goodwin et al., 2016; Liu et al., 2015). New technologies in computational techniques have been developed rapidly. All these achievements gain new perspectives on metabolic engineering.

Computational techniques in metabolic engineering have a promising key role to improve and design new strains and increase the productivity of organisms. Available information in different databases which are related to the biological materials of the organisms is used to evaluate and characterize the metabolic models of the organisms and then these metabolic models are improved or redesigned to have strains with higher production yields (Cameron et al., 2014; Woolston et al., 2013). These models help to understand the behavior of the system, the prediction of the product yield, and the choice of process conditions easier. Several uses of mathematical models and computational simulations are provided, including, but not limited to, analyzing gene and reaction knockouts, proposing possible genetic interventions for an overproducer of a metabolite, comparing inputs and outputs, iterative improvement, gap-filling strategies, discovering regulatory interactions, quantifying predictable phenotypes from optimality principles, and designing environmental and genetic perturbations (Greenberg et al., 2011). The use of mathematical models can be effective for comprehending biological systems, and computational simulations can provide a clearer understanding of the processes, changes, and predictions. With all this progress and knowledge, scientists will be able to develop microorganisms as little factories that enable the production of chemical compounds at greater yields/rates (Jouhten, 2012).

Among different mathematical models, genome-scale metabolic models have proven to be an effective tool in systems biology during the past few decades with their applications in several fields. Genome-scale metabolic models, for instance, can be used to design drug targets for significant pathogens, predict metabolic interactions in microbial communities, or predict cellular behavior under various biological conditions. They have also been used to study evolutionary processes or provide a justification for laboratory experiments (Scossa et al., 2020).

GENOME-SCALE METABOLIC RECONSTRUCTION: HOW TO CREATE A METABOLIC MODEL

Understanding the systems biology of the metabolic pathways for an organism can be achieved by a highly mathematical, organized platform that is provided by a metabolic reconstruction. Genome-scale metabolic models (GEMs) are a set of stoichiometry-based, mass-balanced metabolic reactions in an organism built using gene-protein-reaction (GPR) associations that are formulated based on genome annotation data and experimentally obtained information. Basically, by using genomic data of an organism, metabolic pathways are created and the summation of all these pathways constructs the metabolic model of the organism. The availability of information regarding the physiology, biochemistry, and genetics of the target organism increases the predictive capability of the constructed models. Adding these constraints to the model from different areas enhances the quality of the metabolic model and the results of the simulations would reflect the reality better (Thiele et al., 2010).

Many tools have been developed for GEMs construction and these tools benefit from several databases. These databases and tools will be detailed in the following section.

Databases and Tools

Commonly utilized databases are listed below in the process of construction or reconstruction of metabolic models.

Kyoto Encyclopedia of Genes and Genomes (KEGG): It is a database containing information on genes, proteins, reactions, and pathways. Annotated genes of specific organisms are shown on the metabolic pathway maps as green on KEGG. Currently, the database contains 1,071,815 pathways in total and 9,219 (932 eukaryotes, 7876 bacteria, 411 archaea) genomes, and 48,712,402 genes information (data from 12/08/2023) (Kanehisa et al., 2023; Kanehisa, 2019; ; Kanehisa and Goto, 2000).

UniProt: It is a database for annotated proteins. Protein sequences and functional information of proteins are found in the database. 569,793 sequences are available (data from 12/08/2023) (Bateman et al., 2021).

BioCyc: BioCyc has 20,040 pathways/genomes available in the database to search (data from 12/08/2023) (Karp et al., 2019).

MetaCyc: Primary and secondary metabolic pathways, as well as related metabolites, processes, enzymes, and genes, are all included in the MetaCyc database (Caspi et al., 2014).

EcoCyc: A database that is specific to *E. coli* pathways and genomes (Keseler et al., 2011).

BRENDA: It is a comprehensive enzyme repository. Molecular and biochemical information about enzymes is found in the database (Chang et al., 2021).

ModelSEED and *RAST:* RAST is a toolbox to annotate raw genome data to meaningful proteins/enzymes. RAST results which contain annotated genomes directly are sent to the MODELSEED to construct a draft metabolic model (Aziz et al., 2008; Seaver et al., 2021).

STRING: A database to show known and predicted protein interactions (Szklarczyk et al., 2021).

After proteins have been annotated from the genomic data, databases like those mentioned above (e.g. KEGG and UniProt) can be used to get more information about the structure of the enzymes, the reactions they catalyze, the substrates and products they produce, and the mechanisms they use. Additional data on the reversibility and preferred physiological direction of enzyme-catalyzed reactions are needed for accurate metabolic reconstructions, and this data can be found in databases like BRENDA or MetaCyc.

Data gathered from the databases mentioned above is used to create a reconstruction (Figure 1). To locate all the metabolic genes in the target organism, databases like KEGG and BioCyc can be utilized in cooperation with one another. The majority of reconstructions have been created manually because there are currently quite a few automatic/semi-automatic tools that enable the building of these reconstructions. The draft reconstruction of the relevant organism is created using homologous genes and reactions obtained from previous reconstructions. A network of metabolic and non-metabolic pathways can be created using tools like RAVEN (Reconstruction, Analysis, and Visualization of Metabolic Networks) and ModelSEED by assembling data into pathways. RAVEN is a widely used MATLAB toolbox for GEMs reconstruction, refinement, and restriction-based modeling and simulation (Wang et al., 2018). RAVEN toolbox has an annotated protein 'fasta' file as input. After introducing the fasta file to the RAVEN and the determination of the automation parameters, GEMs can be constructed in a short time. It allows for major improvements such as the convergence of restructurings from various sources and improved performance with Constraint-Based Reconstruction and Analysis (COBRA) (Heirendt et al., 2019). RAVEN has better import/export support with different file types. It also has improved compatibility with COBRA, especially as it provides a better bidirectional model conversion function, by resolving previously conflicting function names. As a MATLAB toolbox, RAVEN provides a single unified environment for both model reconstruction and simulation analysis and allows scripting for more flexible operations (Mendoza et al., 2019; Wang et al., 2018).

Despite the advantages of automated tools like saving time and decreasing the workload at the beginning, the reconstructed models still need some additions like biomass reaction or upper/lower bound adjustments (Xavier et al., 2017). These tools enable the creation of a rough draft quickly while yet allowing for the fine adjustments needed whenever new experimental data is discovered including substrate utilization rate, oxygen consumption rate, maximum specific growth rate, etc. The simulation process makes sure that all of the reactions present in the reconstruction are suitably balanced. In the end, a completely accurate reconstruction can provide more knowledge about how the target organism functions (Mendoza et al., 2019).

Figure 1. A visual representation of the metabolic network construction process

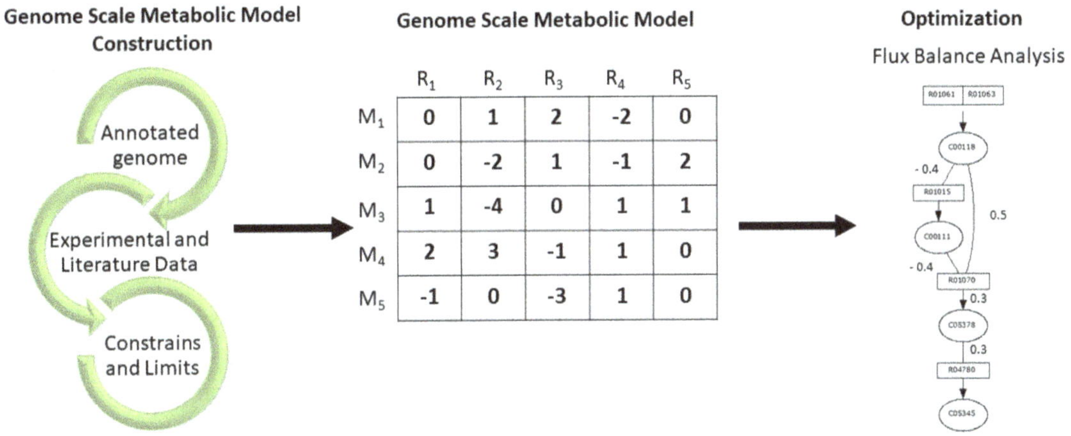

Metabolic Model Refinement

Due to the huge variability and diversity of organisms, an initial metabolic reconstruction of a genome is often far from representing reality. There could be potential mistakes in the metabolic pathway databases like including some enzymatic activities catalyzed by the proteins that are not encoded by the genome. In semi-automatically produced reconstructions, it is also possible that some routes are incorrectly predicted and do not take place as expected. As a result, a methodical verification is performed to ensure that there are no discrepancies and that each entry listed is accurate. Additionally, any information gleaned from one of the numerous metabolic reaction and genome databases can be supported by a literature search. This gives the reconstruction more confidence that the enzyme and the reaction it catalyzes genuinely take place in that organism(Gawand et al., 2013; Medlock et al., 2020).

The reconstruction should be considered incorporating any newly identified reactions that are not currently listed in the databases. The experimental phase and the coding phase alternate during this iterative procedure. The model will be modified to anticipate the metabolic and phenotypical output of the cell as new details about the target organism are discovered (Feist et al., 2008).

Flux Balance Analysis With COBRA

Cellular metabolism is a large and complex network. Therefore, *in silico* modeling and simulations assist the investigation of metabolic networks. Metabolic networks can be derived from whole genome sequences by identifying which enzymes are present and linking them to formal chemical reactions (Waller et al., 2020).

One of the most widely used tools for systems biology analyses using GEMs is the COBRA toolbox (Heirendt et al., 2019). COBRA toolbox is a MATLAB software package for the quantitative prediction of cellular and multicellular biochemical networks with constraint-based modeling. The toolkit enables the use of a wide range of constraint-based analytical techniques for investigating the metabolic properties of biosystems with complexity ranging from single cells to microbial communities, and ultimately multicellular animals (Campbell et al., 2017; Schellenberger et al., 2011). Depending on the user's preferred programming language, the toolbox offers a variety of alternative variations that can be used. It applies a comprehensive range of fundamental and advanced modeling methods, including reconstructing and model building, as well as biased and unbiased model-driven analysis methods. In addition, COBRA is widely used to model, analyze, and predict various metabolic phenotypes using genome-scale biochemical networks (Stalidzans et al., 2018).

In order to constrain intracellular fluxes, COBRA algorithms leverage biological knowledge and data, and they have recently been expanded to take a wide range of new omics techniques into account (O'Brien et al., 2015).

When compared to the time scales involved with substrate uptake and cell division, flux balance analysis (FBA) implies that metabolite concentrations in the cell approach a pseudo-steady-state. By making this supposition, mass balance equations that are created using databases of biochemical process stoichiometry can be used to constrain fluxes. Another constraint with the mass balance equations is the objective function. When simulating an FBA, an objective function is required like maximizing/minimizing biomass or substrate utilization/metabolite production. The accuracy of these predicted flux values during the simulation is highly sensitive to the objective function. GEMs of microbial metabolism have been developed in large numbers because these models can be efficiently simulated and rely mostly on carefully curated databases of metabolic processes. These presumptions help understand metabolic functionality and forecast the outcomes of gene modification, but they fall short of fully incorporating the phenotypic findings from omics analysis (King et al., 2016).

COBRA algorithms and different omics data have a very close relationship. Flux values during the simulation can be validated by using omic data like metabolomics (metabolite concentration), proteomics (protein concentrations), fluxomics, etc. To integrate omics data of various forms, COBRA approaches constitute an extensible and computationally effective framework. They have been effectively applied to understand omics data and enhance strain designs in several research studies. Due to the minimal number of parameters that must be fitted from experimental data, COBRA approaches have the benefit of frequently suggesting strain designs without requiring a lot of experimental evidence. These techniques are particularly effective at identifying metabolic alterations that link product synthesis to cell proliferation (Long et al., 2015; Wisselink et al., 2010).

Metabolic Model Repositories

After the construction/reconstruction of a metabolic model of a target organism, the model should be uploaded to one of the metabolic model databases. Also without the construction of a metabolic model, already available models can be downloaded from these databases and then can be investigated for further analysis. The most common ones are given below:

BiGG: A database to search metabolic models, reaction metabolites, or genes. The database has many metabolic models of various types of organisms and these models can be downloaded as '.json' of .xml file type (King et al., 2016).

Metabolic Atlas: It is an open source collection for GEMs. The database enables both the download and upload of metabolic models (Robinson et al., 2020).

BioModels: BioModels is an open-access repository of mathematical models of biological and biomedical systems. The database contains 1,066 manually curated, 1,467 non-curated, and 833 auto-generated metabolic models (data from 12/08/2023) (Glont et al., 2018; Malik-Sheriff et al., 2020).

DESIGN-BUILD-TEST-LEARN (DBTL) CYCLE

Modern omics analyses can successfully define the genetic, regulatory, and metabolic phenotypes of engineered microorganisms, but it is still difficult to design genetic interventions to produce a particular phenotype. Timelines for creating and testing strain designs have been drastically shortened by recent advancements in genetic engineering techniques, making it possible to efficiently iterate between experiment and analysis in a closed loop for the first time. However, manual biological reasoning about the mechanisms causing phenotypic changes is made more challenging by the size and complexity of multi-omics information (Nielsen et al., 2016).

Consequently, Design-Build-Test-Learn (DBTL) cycle has a significant role in metabolic engineering. These data sets' dimensionality can be reduced by using conventional statistical methods, which can also be used to find patterns shared by strains with high performance. These techniques have been used successfully in several research studies, however, they do not fully exploit the relationships between genes, proteins, and metabolic networks. As a result, model-aided design, which integrates experimental data and produces useful but unconventional design predictions, is gaining popularity. It is done by using modeling frameworks from systems biology(Liu et al., 2015).

In the DBTL cycle, while the *Design* component defines the problem and selects the desired path and main method; the *Build* component selects, synthesizes, and assembles parts for inclusion in the method; the *Test* component validates engineered strains for target molecule production, transcripts, proteins, and metabolites; the *Learn* component analyzes the *Test* data and reports the next iterations of the loop. The *Learn* and *Design* phases' objective is to create better strain designs by using the characterization of previously produced strains. The *Learn* stage can be carried out in the simplest form by looking at biological characteristics associated with better strain performance and overexpressing those presumably implicated in the pathway of interest (Ibrahim et al., 2021).

Figure 2 illustrates the coordinated processes necessary to initiate the manufacture of a metabolite within a non-native host. The creation of GEMs along with FBA for a thorough investigation of putative metabolic pathways precipitates the *Design* phase. The validation and meticulous quality assessment of the GEMs, which are crucial prerequisites to accurate simulations, are central to this phase. After confirming the validity of GEMs, the focus of the investigation shifts to the evaluation of gene editing techniques that are enhanced to promote metabolite biosynthesis (Kitano et al., 2023; Opgenorth et al., 2019).

These prescriptive genetic treatments, which are implemented through skilled metabolic engineering methods, take on the mantle of the *Build* phase. With the necessary genetic foundations to clarify metabolite production, novel microbial strains are created in the *in vitro* experimentation crucible. The cultivation of emerging microbial progeny is now set to begin as the *Test* phase approaches. These cellular envoys, which have been grown and collected, are proof that the desired metabolite yield was achieved. The evaluation of these samples acts as a trial run and success indicator for the production of metabolites. The *Learn* phase starts after taking note of the lessons learned from each iteration. It embraces data synthesis and analysis while thinking about the future course, making it a crucial decision nexus. The culmination of experimental observations gives rise to refined procedures and tactics, supporting succeeding DBTL cycles or laying out a clear path toward successful microbial refinement.

The coupling of GEMs and systems biology approaches resonates powerfully inside the industrial canvas. These paradigms offer a proactive shield against inefficiency as an effective remedy for the waste of time and money. Notably, systems biology opens up a different vision while the development of a bioproduct workhorse can be punctuated only by empirical methods. Although dependent on the model's faithfulness to real-world settings, it serves as a powerful ally in efficiently condensing data vastness and producing yield projections (Carbonell et al., 2018).

DBTL cycle technique is an elegant framework that fosters systematic steps and seamlessly blends the realms of in silico and in vitro methodologies. In metabolic modeling studies, by following the next step with such a guide, the experimental design can be done much more easily and it plays an important role in quickly analyzing and solving whether the result obtained in each option is suitable for the next step (Ibrahim et al., 2021).

Figure 2. A diagram of DBTL cycle with example procedures

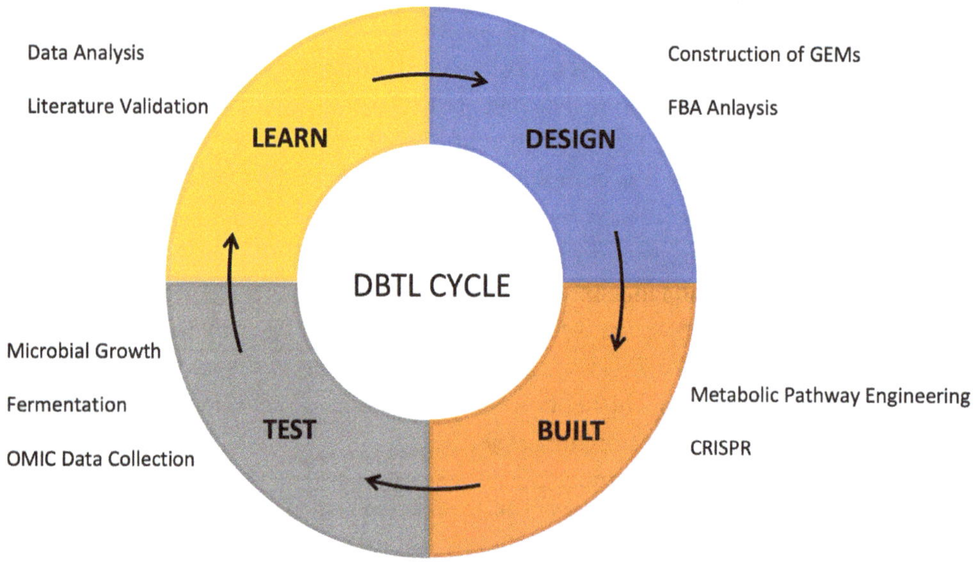

APPLICATIONS OF GEMS

By using the microbial fermentation approach, it is possible to obtain novel compounds that have not been produced via conventional processes. During this fermentation route, high yield, high productivity, and process robustness are the major criteria for the evaluation of process efficiency, and these are dependent on the host microorganism. Different approaches have been applied to increase the production capabilities of microorganisms including natural screening, mutagenesis, genetic engineering, and metabolic engineering. Although improved production has been achieved in many processes, there are still some problems to solve like toxic intermediate accumulation and metabolic stress. The primary reason for these issues is a limited comprehension of microbial physiology (Liu et al., 2010; Park et al., 2008).

Metabolic engineering approaches that generally use mutagenesis and screening for the identification of strains with improved production capabilities are needed to have high-yield production strains. But it is not very possible to identify all the strains because of the huge number of possible permutations. Systems biology tools make the targeted strain improvements possible to meet the demand for robust and high-yield production strains. GEMs have become increasingly significant for the investigation of metabolic networks with the advent of next-generation sequencing technologies. Over 6000 GEMs of several microorganisms and multicellular organisms have been constructed (Gu et al., 2019).

GEMs have been used in several applications including strain development for bio-based production, modeling the interactions among cells or organisms, drug targeting in pathogens, evaluation of human diseases, etc. (Gu et al., 2019). They are important tools for the design of microbial cell factories in terms of optimizing the whole biological network instead of the central metabolism and target biosynthetic pathways (Cho et al., 2022).

With the knowledge of biochemical reactions, GEMs enable the quantification of metabolic capacity. They can be used for analyzing and simulating the operation of metabolism in response to different stimuli (Lopes et al., 2017). Additionally, GEMs are widely used approaches to increase the rate and yield of microbial production. Although different algorithms with GEMs basis have been developed to estimate the effects of different genetic modifications on the production capacity, only a few of them have been validated experimentally for several products which have resulted in an increased rate and yield (King et al., 2016; Landon et al., 2019).

Biotechnological synthesis pathways offer the advantages of avoiding organic solvents, heavy metal catalysts, and strong acid/base solutions, which are commonly required in synthetic chemistry-based approaches (Jeandet et al., 2013). Thus, they provide environmentally friendly production conditions by avoiding environmental pollutants. Also, when a microbial cell is used, synthesis can generally be carried out under moderate (easily available) temperatures and atmospheric pressure (Rollié et al., 2012). Hence, the modeling of microbial cells for the biosynthesis of high-value products is gaining increasing attention in biotechnology (Camacho-Zaragoza et al., 2016; Chen et al., 2020; Krämer et al., 2003). Here, the main focus will be on the GEMs of microbial cells in biotechnological applications with an emphasis on the production of biofuels and pharmaceuticals. Common algorithms used in these applications are given in Table 1.

Table 1. Common algorithms in strain design and improvement

Algorithm	Function	Reference
MOMA	Gene deletion	(Mahadevan and Schilling, 2003)
OptKnock	Gene deletion	(Burgard et al., 2003)
OptGene	Gene deletion	(Patil et al., 2005)
OptORF	Gene deletion	(Kim et al., 2010)
OptGpSampler	Sampling	(Megchelenbrink et al., 2014)
OptForce	Gene deletion, expression regulation	(Ranganathan et al., 2010)
ReacKnock	Gene deletion	(Xu et al., 2013)
OptStrain	Pathway prediction	(Pharkya et al., 2004)

Biofuel Production

The increase in the world population has brought an increase in energy consumption. The total global energy consumption supplied by fossil resources is approximately 84% whereas renewable resources contribute 11%. This use is estimated to increase by 50% and 25% will account for renewable energy by 2050 (EIA, 2021). The use of fossil fuels to meet the energy demand causes greenhouse gas emissions that harm the environment. Biofuels (e.g., bioalcohols, biodiesel, biogas, and biohydrogen) derived from biological sources such as plants, microorganisms, and waste material are considered sustainable and environment-friendly alternatives to fossil fuels (Jagadevan et al., 2018).

GEMs can be applied for biofuel production attempts (especially for fourth-generation biofuels which are obtained from engineered microorganisms) to increase production yields and to get a systematic framework for optimizing the flux for the desired pathways while balancing important cofactor and energy metabolites (Lee et al., 2008). Several studies which focus on the application of GEMs for biofuel production using different organisms have been published. Some of them are summarized below.

The high carbohydrate and lipid content, rapid growth, and being resistant to changing environmental conditions make the microalgae attractive biomass sources (Chisti, 2007). The advancements in bioinformatic tools and genetics make the manipulation of microalgae possible for improvements in carbon flow, light utilization, and lipid engineering (Gimpel et al., 2013). GEMs have gained importance in the new strain design with the progress in metabolic engineering studies. The GEMs for the microalgal model organism *Chlamydomonas reinhardtii* have been reconstructed, some of the notable ones are *i*AM303 (Manichaikul et al., 2009), *i*Cre1355 (Imam et al., 2015), *i*RC1080 (Chang et al., 2011), AlgaGEM (Gomes de Oliveira Dal'Molin et al., 2011) and a GEM by Salguero *et al.* (Salguero et al., 2018). Apart from *C. reinhardtii*, GEMs have been constructed for *Nannochloropsis salina* (Loira et al., 2017) and *Nannochloropsis gaditana* (Shah et al., 2017) to study lipid synthesis for biofuel production.

AlgaGEM which covers the metabolism of a compartmentalized algae cell was constructed using the organism *C. reinhardtii* genome. AlgaGEM is an extensive literature-based GEM that has functions of 866 unique ORFs, 1,862 metabolites, 2,249 gene-enzyme reaction association entries, and 1,725 unique reactions. Using the information on compartmentalization for algae and *Arabidopsis thaliana*, the reconstruction was compartmentalized into the cytoplasm, mitochondrion, plastid, and microbody. The model was validated using literature data by simulating the growth and metabolic functions. It was concluded that AlgaGEM can be used to identify bottlenecks and new targets to engineer microalgae for the production of biofuels (Gomes de Oliveira Dal'Molin et al., 2011).

Another GEM (*i*RC108) for *C. reinhardtii* was reconstructed by integrating biological and optical data (Chang et al., 2011). A novel light-modeling approach that predicts growth for a given light source, resolving wavelength, and photon flux was proposed. The transcripts in the network were verified experimentally. The model function was validated physiologically through simulation. The model showed high confidence in network contents and predictive applications. The application of the *i*RC108 to design an efficient light source for growth will be a guide for metabolic engineering and bioreactor design studies for biofuel production (Chang et al., 2011).

Li *et al.* utilized the model for the green microalgae *Chlorella vulgaris* (*i*CZ946) (Zuñiga et al., 2018) to reconstruct the models under photoautotrophic and heterotrophic conditions for nutrient optimization. Model-driven cultivation strategy has resulted in a 50% increase in the total volumetric yield of biomass, a 61% increase in fatty acid methyl ester (FAME) yield, and enhanced lutein yield nearly 3-fold compared to nitrogen starvation. The study provided a control methodology that integrates omics data and GEMs to optimize nutrient supplies based on the metabolic state of algal cells in different nutrient environments. The approach was proposed as a strategy for controlling the bioprocesses at a systems-biology level to limit nutrient inputs, reduce processing costs, and optimize biomanufacturing for the next generation of desirable biotechnology products (Li et al., 2019).

Apart from microalgae, oleaginous yeasts are one of the best hosts for lipid and diesel-like fuel production with their characteristics including high lipogenesis, growing on many different carbon sources, and easy large-scale cultivation. *Yarrowia lipolytica* is the model oleaginous yeast and has been engineered using several approaches including increasing lipogenic pathway flux, blocking degradation pathways, and increasing biosynthetic precursors to improve lipid production (Adrio, 2017). The first functional GEM for *Y. lipolytica* (*i*NL895) was built by Loira *et al.* (Loira et al., 2012). *S. cerevisiae* model was used as a scaffold and *in silico* tools and manual curation were combined. The model which has 2,002 reactions, 895 genes, and 1,847 metabolites was validated using published experimental data. A fair degree of concordance was obtained between the model's growth predictions and the experimental results with 0.65 accuracy (Loira et al., 2012).

Another GEM (*i*Y619_PCP) was developed by Pan and Hua (Pan and Hua, 2012). The model accounts for 619 genes, 843 metabolites, and 1,142 reactions. It was successful in the prediction of growth in minimal media and on different substrates with an accuracy of 0.83. Additionally, FBA with single-gene knockouts was applied to predict gene essentiality. Flux variability analysis (FVA) was employed to design new mutant strains that redirect fluxes toward lipid production.

Kavscek *et al.* reconstructed the metabolic network of *Y. lipolytica* (*i*MK735) using the GEM of baker's yeast as the scaffold. The model has 1,336 reactions that use 1,111 metabolites and are encoded by 735 genes. They aimed to stimulate growth and lipid production phases in their study. The model was validated experimentally. A successful fed-batch strategy was designed which prevents citrate excretion during the lipid production phase. It was hypothesized that a reduced aeration rate might induce lipid accumulation. The lipid content of the biomass was increased by 80% after combining these two strategies (Kavšcek et al., 2015).

Kim *et al.* slightly modified the previous model *i*MK735 to show the metabolic flux distribution of *Y. lipolytica* in nitrogen-limited conditions by the environmental version of minimization of metabolic adjustment (eMOMA). In this method, several novel targets were predicted. Some of the novel knockout targets were experimentally tested and the deletion of YALI0F30745g in *Y. lipolytica* resulted in a 1.45-fold increase in lipid accumulation compared to the wild-type strain (Kim et al., 2019).

In a recent study, a high-quality GEM (*i*Yli21) of *Y. lipolytica* type strain W29 was constructed (Guo et al., 2022). The model has 2,285 reactions, 1,868 metabolites, and 1,058 genes. It is reported that the model showed 85.7% accuracy in predicting nutrient utilization. The model is proposed as a promising tool in assisting the metabolic engineering of this important yeast cell factory.

Lipomyces starkeyi is another promising oleaginous yeast featuring a wide substrate spectrum, adaption to fermentation inhibitors, excellent fatty acid composition for high-quality biodiesel, and negligible lipid remobilization. But low lipid yields obtained from experiments are a drawback for the industrial use of *L. starkeyi.* A small-scale metabolic model of *L. starkeyi* NRRL Y-11557 with 112 metabolites, 123 reactions, and 3 cell compartments including extracellular, cytoplasm, and mitochondria was constructed. The theoretical lipid yields of several low-cost substrates were calculated according to FBA. The use of different substrate combinations was proposed as a promising strategy to increase lipid accumulation (Zhou et al., 2021).

The oleaginous yeast *Papiliotrema laurentii* UFV-1 is able to produce a high amount of lipids in short periods. Additionally, its fatty acid composition meets the requirements of quality standards for biodiesel production. The first GEM of *Papiliotrema laurentii* (papla-GEM) was reconstructed by Ventorim *et al.* (Ventorim et al., 2022). The model includes 796 genes, 2,465 reactions, and 2,127 metabolites. The experimental validation step was performed to evaluate growth and lipid accumulation physiology. The model has been indicated to be suitable for the identification of metabolic engineering targets to improve lipid production by this organism.

Cutaneotrichosporon oleaginosus ATCC 20509 is a fast-growing oleaginous yeast. It can grow in a wide range of carbon sources including glycerol byproduct of biodiesel production. Lipids derived from *C. oleaginosus* grown on glycerol have high concentrations of monounsaturated fatty acids which are perfect biodiesel components (Liang et al., 2010). The first GEM for *C. oleaginosus* ATCC 20509, *i*NP636 *i*NP636_*Coleaginosus*_ATCC20509, was built to better understand this yeast and form a basis for its industrial use (Pham et al., 2021). The model contains 1,553 reactions and 1,373 metabolites in 11 compartments. The model gave qualitative predictions at different carbon sources consistent with experimental data. ATP-citrate lyase has been suggested as a target to further improve lipid production.

A GEM of *S. cerevisiae*, *i*Sc-AMRS-1, was constructed by updating the existing model *i*LL672 (Kuepfer et al., 2005) to obtain a yeast strain that produces the isoprenoid farnesene efficiently. Four non-native metabolic reactions were integrated into central carbon metabolism in *S. cerevisiae*. The resultant strain produced 25% more farnesene with 75% less oxygen requirement (Meadows et al., 2016). The conjunction of GEMs with 13C metabolic flux analysis was used to evaluate the bottlenecks and imbalances in *S. cerevisiae*. The suggested engineering strategies have resulted in an increase in fatty acid production by 70% (d'Espaux et al., 2017) and the achievement of a fatty alcohol with a 1.2 g/L titer (Ghosh et al., 2016).

Cyanobacteria have also gained attention as a potential renewable energy source. They are capable of fixing CO_2 and performing oxygenic photosynthesis. They grow significantly faster than other photosynthetic organisms like microalgae and plants and have the biochemical pathways to convert solar energy to biochemical energy to produce several energy-rich compounds (Pathania et al., 2022). The first GEM for photosynthetic organisms/cyanobacteria was developed by Fu (Fu, 2009) using *Synechocystis* sp. PCC 6803 as the model. The reconstructed model includes 633 genes, 704 metabolites, and 831 metabolic reactions. Heterotrophic, photoautotrophic, and mixotrophic growth conditions were simulated. The ethanol fermentation pathway was inserted for the *in silico* predictions. Cell growth and ethanol production simulation results showed a high correlation with actual metabolic measurements.

Shabestary *et al.* used constraint-based modeling and optimization algorithms on a GEM of *Synechocystis* sp. PCC6803 to increase the productivity of fermentative fatty-acid and terpene-derived fuels. The model *i*JN678_2016 was reconstructed by adding several reactions to the previously reconstructed *Synechocystis* sp. PCC6803 model, *i*JN678 (Nogales et al., 2012). OptGene and MOMA were used for knockout strategies to increase biofuel productivity. OptKnock was used to determine knockouts that couple photoautotrophic growth and biofuel production. The results have shown that the high productivity of the alcohols such as 1-butanol needs the elimination of NADH sinks while fatty acid-based fuels and terpenes require imbalances in ATP and NADPH production and consumption. It was indicated that the FBA predicted productivities of these biofuels are at least 10-fold higher than previously reported ones. The results of the study could be a guide to engineer cyanobacteria to obtain higher biofuel productivities (Shabestary et al., 2016).

A Java tool was developed by Mohammedi *et al.* (Mohammadi et al., 2016) to merge three existing reconstructed metabolic networks of *Synechocystis* sp. PCC 6803 (Erdrich et al., 2014; Knoop et al., 2010, 2013). The results of FBA from the obtained model showed an increased yield in biofuel production for ethanol, isobutanol, 3-Methyl-1-butanol, 2-Methyl-1-butanol, and propanol. It is also reported that coverage of the metabolic network in terms of the number of metabolites and reactions was increased in the newly obtained model.

In the study of Hendry *et al.*, FBA was carried out with GEM of *Synechococcus* sp. PCC 7002 which is a fast-growing cyanobacterium. The model was developed from the existing *i*syp702 (Xu et al., 2013) and the updated model includes 742 reactions and 728 genes. It was validated by available experimental data. MOMA was used to predict mutants with single and double reaction knockouts that overproduce ethanol and butanol. It was predicted that only 10% of fixed carbon could be diverted for ethanol and butanol production (Hendry et al., 2016).

Genus *Anabena* is filamentous cyanobacteria. The vegetative cells of this genus differentiate into heterocysts enabling nitrogen fixation in the absence of nitrate or ammonia. These microorganisms produce biohydrogen during nitrogen fixation. *Anabaena variabilis* ATCC29413 has one of the highest hydrogen production rates and has been a popular candidate for studying biohydrogen production. A curated GEM for *A. variabilis*, *i*AM957, was reconstructed to study biohydrogen production in the study of Shahkouhi *et al.* (Shahkouhi et al., 2020). A regulated two-cell model was developed by integrating the model and gene expression data for vegetative and heterocyst cells and then used for the hydrogen production improvement in heterocysts. As a result, the regulated model proposed the activation of some reactions to provide redox cofactors which are required for improving hydrogen production by up to 60% by bidirectional hydrogenase.

Zymomonas mobilis is a facultative anaerobic ethanologen that acquires special attention due to its higher ethanol yield and tolerance and a broad pH range for production. It can efficiently produce ethanol from several substrates via the Entner-Doudoroff pathway (Widiastuti et al., 2011; Yang et al., 2016). The GEM of *Z. mobilis* ZM4, ZmoMBEL601 with 579 metabolites and 601 metabolic reactions, was reconstructed by Lee *et al.* (Lee et al., 2010). The physiological changes of *Z. mobilis* as it shifts from anaerobic to aerobic environments were investigated. Then the intensities of flux-sum and the maximum *in silico* yields of ethanol for *E. coli* and *Z. mobilis* were compared and analyzed. Metabolic pathways were introduced to enable *Z. mobilis* to utilize pentose sugars. The constructed model has represented the metabolic characteristics of *Z. mobilis* under different conditions which will be very useful for designing metabolic engineering strategies for various biotechnological applications including ethanol production.

In another study, Widiastuti *et al.* reconstructed a GEM network of *Z. mobilis* ZM4 (*i*ZM363) composed of 704 metabolites and 747 reactions to investigate the physiological characteristics. The model predictions were highly consistent with the experimental observations. The comparisons with *E. coli* and *S. cerevisiae* and flux coupling analysis have shown the functional role of *pdc* and *adh* genes in the ethanologenic activity. The model could be used to determine cell engineering targets to enhance ethanol production in *Z. mobilis* (Widiastuti et al., 2011).

Clostridium thermocellum (ATCC 27405) is an anaerobic thermophile that produces several industrially important products like ethanol, acetic acid, hydrogen, and formic acid. It also can degrade lignocellulosic biomass efficiently. Hence, it is a promising bacterium for metabolic engineering studies in terms of increasing the production efficiency of desired products. Roberts *et al.* constructed a GEM of *C. thermocellum*, *i*SR432, that consists of 577 reactions involving 525 intracellular metabolites, 432 genes, and a proteomic-based representation of a cellulosome. The results obtained from the analysis of the model showed that the reduction-oxidation state of the cell affected ethanol production. The model has been suggested as a predictive tool for rational manipulation of *C. thermocellum* metabolism (Roberts et al., 2010).

Garcia *et al.* developed an updated GEM of *C. thermocellum*, *i*CBI655, with improved prediction accuracy. Firstly, a formulated multi-omic integration protocol was used to evaluate redox metabolism and potential bottlenecks in biofuel production. And then, they used the metabolic model in combination with the previously developed ModCell to design the chassis cells for the production of esters and alcohols efficiently. High consistency scores obtained from the model showed its quality and applicability for future studies (Garcia et al., 2020).

Adlakha *et al.* reconstructed the metabolic network model of *Paenibacillus polymyxa* ICGEB2008, a potential biofuel producer, with 133 metabolites and 158 reactions. FBA was applied to investigate the effect of redox balance. In addition, the pathways involved in 2,3-butanediol, ethanol, and hydrogen productions were studied using a combined modeling and experimental approach. It was shown that the redox state of the nitrogen source is critical for metabolite production. The presence of the Bifid Shunt in the bacterium has been elucidated as a bypass for the glycolytic pathway. This study can be utilized as a basis for further metabolic engineering efforts to improve the efficiency of biofuel production by this strain (Adlakha et al., 2015).

A GEM of isobutanol-producing *Bacillus subtilis* BSUL03 was developed by introducing the isobutanol biosynthetic reactions to the previously constructed model (Oh et al., 2007). Elementary mode analysis (EMA) was employed to improve the isobutanol production in *B. subtilis*. A 2.3-fold increase in isobutanol production was obtained in LDH and PDHC mutants designated as BSUL05 (Li et al., 2012).

Pharmaceutical Production

Several important molecules and natural compounds have been used as pharmaceuticals since ancient times. The extraction of these molecules (animal- or plant-derived) is really difficult and they have lower yields. Hence the production of these pharmaceuticals using microorganisms is gaining more and more attention each year. Several drugs including antibiotics are produced by both natural and engineered microorganisms. Because of being environmentally friendly, having lower production costs, and higher production rates, they are more advantageous compared with chemical synthesis or extraction from natural resources. But the development of the metabolic pathways for the synthesis of these compounds becomes a bottleneck. The use of GEMs-guided metabolic engineering strategies assists in the design of efficient microbial cells for the production of these molecules (Hans et al., 2022; C. Xu et al., 2013). With the improvement in their quality and quantity, it is possible to engineer microbial cells by rewiring the metabolisms toward the production of pharmaceutical molecules. A number of studies that use GEMs for the production of pharmaceuticals are summarized below.

Lycopene is a valuable pharmaceutical with beneficial properties like the prevention of cardiovascular disease and cancers of the prostate or gastrointestinal tract. Alper *et al.* investigated the putative genes that affect the network properties and cellular phenotype for lycopene biosynthesis. The *E. coli* *i*JE660a GEM model (Reed et al., 2003) served as the basis for this stoichiometric network. The model was updated for lycopene synthesis and the final model comprised 965 metabolic fluxes and 546 metabolites. After the determination of the targets, they were tested experimentally by constructing single, double, and triple knockouts. A triple knockout construct yielded a 40% increase in lycopene synthesis over the parental strain (Alper et al., 2005).

Pichia pastoris is a well-known methylotrophic yeast that can produce heterologous proteins efficiently with highly inducible expression systems. It is also capable of post-translational modifications which are very important to have safe and effective therapeutics for human use. Sohn *et al.* reported the reconstruction of the GEM of *P. pastoris*, PpaMBEL1254, which has 1,254 reactions and 1,147 metabolites with 8 different compartments. In addition, the reactions for the production of human serum albumin and human superoxide dismutase were incorporated. The final model was used to investigate the effect of oxygen limitation on protein production. An optimum state that maximizes protein production was determined which will guide the development of different fermentation strategies (Sohn et al., 2010).

Saitua *et al.* assembled a dynamic GEM for glucose-limited, aerobic cultivations of *P. pastoris*. The model was applied to improve the production of recombinant human serum albumin in fed-batch culture. The simulation of single knockouts showed that the carbon flux deviation towards cysteine and tryptophan formation increased human serum albumin production. The deletion of methylene tetrahydrofolate dehydrogenase could enhance human serum albumin productivity by 630% (Saitua et al., 2017).

Daptomycin is a cyclic lipopeptide antibiotic that is produced by *Streptomyces roseosporus*. Huang *et al.* reconstructed a partial metabolic network of *S. roseosporus* based on the GEM of *Streptomyces coelicolor*. Three potential targets, *dptJ*, *zwf2*, and *dptI*, were determined and selected for genetic modification after a pathway flux analysis. Up to approximately 43.2% higher production amounts were obtained from the engineered strain in the fed-batch culture (Huang et al., 2012).

Licona-Cassani *et al.* reconstructed the GEM for the erythromycin producer *Saccharopolyspora erythraea* NRRL23338. The manually curated model including the primary and secondary metabolism pathways consisted of 2,075 genes, 1,482 reactions, and 1,646 metabolites. The effects of amino acid supply on erythromycin production were investigated as part of the model validation. The erythromycin production increased with valine, isoleucine, threonine, and proline addition. Supplementation with 2.5% carbon mole of valine increased the growth rate by 20% and simultaneously the erythromycin yield on biomass by 50% in bioreactor experiments (Licona-Cassani et al., 2012).

Penicillium chrysogenum is a filamentous fungus that produces penicillin naturally. The production titers and productivity has been increased at least 3-fold using classical strain improvement (CSI) methods. Agren *et al.* reconstructed a GEM (*iAL1006*) of the CSI-developed penicillin over-producing strain, *P. chrysogenum* Wisconsin54-1255. The FBA combined with transcriptome analysis was used to figure out metabolic bottlenecks and the effect of co-factor availability on penicillin yields. Increment of the NADPH availability and the modification of different precursor supplying pathways were suggested as potential metabolic engineering strategies for increasing penicillin production (Agren et al., 2013).

Terpenoids are a large class of secondary metabolites that are used as pharmaceuticals with their different biological activities. The anticancer drug Taxol and the antimalarial drug Artemisinin are two of the most well-known terpene-based drugs. Chung *et al.* used GEMs of *E. coli* (*i*JO1366) (Orth et al., 2011) and *S. cerevisiae* (*i*MM904) (Mo et al., 2009) besides *Arabidopsis thaliana* to characterize the metabolic states during terpenoid biosynthesis. Constraints-based flux analysis was used to simulate terpenoid biosynthesis. The GEMs of *E. coli* and *S. cerevisiae* were modified by adding necessary metabolic reactions for terpenoid biosynthesis. Higher terpenoid yields were obtained through the methylerythritol phosphate (MEP) pathway compared to the mevalonate pathway (MVA) pathway as in *A. thaliana*. A novel cofactor modification analysis (CMA) was applied to identify potential targets, including glyceraldehyde-3-phosphate dehydrogenase and malate dehydrogenase, for the improvement of terpenoid yields (Chung et al., 2013).

FK506 (tacrolimus) is a 23-membered polyketide macrolide that is commonly used as an immunosuppressant to prevent the rejection of organ transplants and can be produced by *Streptomyces tsukubaensis*. Huang *et al.* constructed the GEM of *S. tsukubaensis* to improve the FK506 production in their study. A draft model including 865 reactions and 621 metabolites was obtained. Several potential gene targets (*gdhA* and *ppc* for knockout; *dahP*, *pntAB*, *accA2*, and *zwf2* for overexpression) were identified. Each target gene was manipulated in the parent strain D852 for model validation. Higher production of FK506 was obtained in all engineered strains. The combined effect of genetic modifications has resulted in an increase from 143.5 mg/L to 398.9 mg/L, compared with the parent strain. Additionally, the use of fed-batch fermentation increased the final titer from 158.7 mg/L to 435.9 mg/L. The results demonstrated the validity of *in silico* modeling tools for biopharmaceuticals overproduction (Huang et al., 2013).

Squalene is a triterpene hydrocarbon that can be used as a cardio-protective agent, anti-tumor agent, natural antioxidant, and vaccine adjuvant. Microbial production of squalene serves as an alternative to meet the increasing demand instead of obtaining shark liver oil. Huang *et al.* studied the squalene biosynthesis in *Y. lipolytica* focusing on the acetyl-CoA metabolism based on GEM analysis, *i*YL_2.0. The push of carbon fluxes from intracellular precursors to the formation of acetyl-CoA, block of further conversion of acetyl-CoA by competing pathways, and pull of acetyl-CoA consumption flux towards the target metabolite was performed. A 50-fold improvement over the starting strain was obtained with the genetically engineered one (Huang et al., 2018).

Paramasivan *et al.* used the GEM of *S. cerevisiae*, *i*MM904, to find the gene deletion targets by their computational tool FOCuS (Flower Pollination and Clonal Selection) to improve squalene biosynthesis. The deletion of one of the novel identified targets, *LYS1*, resulted in a 2-fold increase in squalene. A double mutant, *LYS1* and *ADK1*, has increased the squalene yield by 2.38-fold. In addition, the integration of *tHMG1* and *POS5* genes and the use of a fed-batch strategy has increased the biosynthesis of squalene several folds (Paramasivan et al., 2019).

N-acetylglucosamine is a glucosamine derivative that has several applications in pharmaceutical fields. It is widely used for medicines due to its excellent effect on articular cartilage regeneration. Having high yield and high efficiency and being environmentally friendly make microbial biosynthesis of N-acetylglucosamine more attractive. Lu *et al.* used a model-based dynamic regulation strategy to engineer *E. coli* ATCC 25947 (DE3) strain for the overproduction of N-acetylglucosamine. The GEM of *E. coli* *i*ML1515 was obtained from the BIGG database and necessary modifications were made for N-acetylglucosamine production. The COBRA toolbox was used for the analysis and simulation of the gene knockout and metabolic flow regulation, promoting the increase in titer and yield of N-acetylglucosamine. Engineered bacteria produced 143.8 g/L N-acetylglucosamine in a 30-L bioreactor with a yield reaching 0.539 g/g glucose. This work significantly enhanced the N-acetylglucosamine production of *E. coli* (Lu, Wu, et al., 2022).

CURRENT CHALLENGES AND FUTURE PROSPECTS

With the advancement of modeling techniques and the discovery of multi-level constraint mechanisms, GEMs have undergone updates and expansions to become crucial tools in cell factory design. The process of model construction has been standardized (Heirendt et al., 2019; Schellenberger et al., 2011), and several automated-construction tools have been introduced. Correspondingly, application and analysis algorithms have been continuously refined. The integration of various data levels, including protein 3D structure, protein secretion pathways, and cell replication into GEMs has become possible to create a more accurate and comprehensive whole-cell models.

However, the integration of large-scale data has brought complexity to the model. Due to limitations in computing power, comprehensive analysis and sophisticated calculations can be time-consuming. This necessitates the simplification of algorithms and optimization for parallel computing adaptability. To handle vast datasets, machine learning methods have emerged as a preferred tool for data simplification and the selection of crucial variables. (Lu, Bi, et al., 2022).

Machine learning has become an integral part of biological research, as it can infer relationships between specific categories of variables. Recent advancements in machine learning have introduced new approaches that eliminate the need for fixed constraint parameters. Instead, these machine learning methods rely on training sets to automatically modify parameters, leading to more accurate models. (Lu, Bi, et al., 2022; Ye et al., 2022).

Machine learning has shown tremendous potential in various domains, including speech recognition, visual object recognition, object detection, drug discovery, and genome analysis (Lawson et al., 2021; Lecun et al., 2015; Moen et al., 2019). As a result, there is a growing trend of integrating machine learning approaches with GEMs. This integration allows for the precise prediction of metabolic flux regulation and characterization of interactions under complex conditions (Czajka et al., 2021; Lewis et al., 2021; Pearcy et al., 2021; Saini et al., 2021; Schinn et al., 2021), significantly enhancing the capabilities and utility of GEMs in various research fields.

Additionally, While GEMs are powerful in predicting cellular metabolic behaviors, they do not consider the diffusion process of matter and energy, relying on steady-state assumptions. To address this challenge, bioreactor models can be utilized to describe the distribution of matter and energy in the fermentation broth, effectively guiding process parameter control and facilitating production scale-up from laboratory to industrial levels (Moreno-Paz et al., 2022) Moreover, extending fundamental knowledge of *in vivo* gene and protein functions will help to create more realistic GEMs (Straathof et al., 2019).

CONCLUSION

The contribution of the GEMs to the design of more efficient cell factories will continue with the development of high-throughput measurement technologies and the growing number of sequenced genomes. For the acceleration of GEMs constructions, improvement of the model simulations, and the use of GEM-based strategies in microbial enhancement, much more effort should be made. The deep integration of machine learning and artificial intelligence, along with automated training and calibration using real data, further enhances the accuracy and efficiency of phenotypic prediction. This powerful combination will open up new possibilities for more accurate and reliable models in biological research.

REFERENCES

Adlakha, N., Pfau, T., Ebenhöh, O., & Yazdani, S. S. (2015). Insight into metabolic pathways of the potential biofuel producer, *Paenibacillus polymyxa* ICGEB2008. *Biotechnology for Biofuels*, *8*(1), 1–10. doi:10.118613068-015-0338-4 PMID:26413158

Adrio, J. L. (2017). Oleaginous yeasts: Promising platforms for the production of oleochemicals and biofuels. *Biotechnology and Bioengineering*, *114*(9), 1915–1920. doi:10.1002/bit.26337 PMID:28498495

Agren, R., Liu, L., Shoaie, S., Vongsangnak, W., Nookaew, I., & Nielsen, J. (2013). The RAVEN Toolbox and Its Use for Generating a Genome-scale Metabolic Model for *Penicillium chrysogenum*. *PLoS Computational Biology*, *9*(3), e1002980. Advance online publication. doi:10.1371/journal.pcbi.1002980 PMID:23555215

Alper, H., Jin, Y. S., Moxley, J. F., & Stephanopoulos, G. (2005). Identifying gene targets for the metabolic engineering of lycopene biosynthesis in *Escherichia coli*. *Metabolic Engineering*, *7*(3), 155–164. doi:10.1016/j.ymben.2004.12.003 PMID:15885614

Asadollahi, M. A., Maury, J., Patil, K. R., Schalk, M., Clark, A., & Nielsen, J. (2009). Enhancing sesquiterpene production in *Saccharomyces cerevisiae* through in silico driven metabolic engineering. *Metabolic Engineering*, *11*(6), 328–334. doi:10.1016/j.ymben.2009.07.001 PMID:19619667

Aziz, R. K., Bartels, D., Best, A. A., Dejongh, M., Disz, T., Edwards, R. A., Formsma, K., Gerdes, S., Glass, E. M., Kubal, M., Meyer, F., Olsen, G. J., Olson, R., Osterman, A. L., Overbeek, R. A., Mcneil, L. K., Paarmann, D., Paczian, T., Parrello, B., ... Zagnitko, O. (2008). The RAST Server: Rapid Annotations using Subsystems Technology. *BMC Genomics*, *9*, 75.

Baldwin, G., Bayer, T., Dickinson, R., Ellis, T., Freemont, P. S., Kitney, R. I., Polizzi, K., & Stan, G.-B. (2015). *Modeling Synthetic Biology Systems. In Synthetic Biology — A Primer* (pp. 89–108). Imperial College Press. doi:10.1142/9781783268801_0006

Bateman, A., Martin, M. J., Orchard, S., Magrane, M., Agivetova, R., Ahmad, S., Alpi, E., Bowler-Barnett, E. H., Britto, R., Bursteinas, B., Bye-A-Jee, H., Coetzee, R., Cukura, A., da Silva, A., Denny, P., Dogan, T., Ebenezer, T. G., Fan, J., Castro, L. G., ... Teodoro, D. (2021). UniProt: The universal protein knowledgebase in 2021. *Nucleic Acids Research*, *49*(D1), D480–D489. doi:10.1093/nar/gkaa1100 PMID:33237286

Becker, J., Zelder, O., Häfner, S., Schröder, H., & Wittmann, C. (2011). From zero to hero—Design-based systems metabolic engineering of *Corynebacterium glutamicum* for l-lysine production. *Metabolic Engineering*, *13*(2), 159–168. doi:10.1016/j.ymben.2011.01.003 PMID:21241816

Burgard, A. P., Pharkya, P., & Maranas, C. D. (2003). Optknock: A bilevel programming framework for identifying gene knockout strategies for microbial strain optimization. *Biotechnology and Bioengineering*, *84*(6), 647–657. doi:10.1002/bit.10803 PMID:14595777

Camacho-Zaragoza, J. M., Hernández-Chávez, G., Moreno-Avitia, F., Ramírez-Iñiguez, R., Martínez, A., Bolívar, F., & Gosset, G. (2016). Engineering of a microbial coculture of *Escherichia coli* strains for the biosynthesis of resveratrol. *Microbial Cell Factories*, *15*(1), 163. doi:10.118612934-016-0562-z PMID:27680538

Cameron, D. E., Bashor, C. J., & Collins, J. J. (2014). A brief history of synthetic biology. *Nature Reviews. Microbiology*, *12*(5), 381–390. doi:10.1038/nrmicro3239

Campbell, K., Xia, J., & Nielsen, J. (2017). The Impact of Systems Biology on Bioprocessing. *Trends in Biotechnology*, *35*(12), 1156–1168. doi:10.1016/j.tibtech.2017.08.011 PMID:28987922

Carbonell, P., Jervis, A. J., Robinson, C. J., Yan, C., Dunstan, M., Swainston, N., Vinaixa, M., Hollywood, K. A., Currin, A., Rattray, N. J. W., Taylor, S., Spiess, R., Sung, R., Williams, A. R., Fellows, D., Stanford, N. J., Mulherin, P., Le Feuvre, R., Barran, P., ... Scrutton, N. S. (2018). An automated Design-Build-Test-Learn pipeline for enhanced microbial production of fine chemicals. *Communications Biology*, *1*(1), 66. doi:10.103842003-018-0076-9 PMID:30271948

Caspi, R., Altman, T., Billington, R., Dreher, K., Foerster, H., Fulcher, C. A., Holland, T. A., Keseler, I. M., Kothari, A., Kubo, A., Krummenacker, M., Latendresse, M., Mueller, L. A., Ong, Q., Paley, S., Subhraveti, P., Weaver, D. S., Weerasinghe, D., Zhang, P., & Karp, P. D. (2014). The MetaCyc database of metabolic pathways and enzymes and the BioCyc collection of Pathway/Genome Databases. *Nucleic Acids Research*, *42*(D1), D459–D471. doi:10.1093/nar/gkt1103 PMID:24225315

Caspi, R., Dreher, K., & Karp, P. D. (2013). The challenge of constructing, classifying and representing metabolic pathways. *FEMS Microbiology Letters*, *345*(2), 85–93. doi:10.1111/1574-6968.12194 PMID:23746312

Chandran, D., Copeland, W. B., Sleight, S. C., & Sauro, H. M. (2008). Mathematical modeling and synthetic biology. *Drug Discovery Today: Disease Models*, *5*(4), 299-309. doi: 10.1016/j.ddmod.2009.07.002 PMID: 27840651. doi:10.1016/j.ddmod.2009.07.002

Chang, A., Jeske, L., Ulbrich, S., Hofmann, J., Koblitz, J., Schomburg, I., Neumann-Schaal, M., Jahn, D., & Schomburg, D. (2021). BRENDA, the ELIXIR core data resource in 2021: New developments and updates. *Nucleic Acids Research*, *49*(D1), D498–D508. doi:10.1093/nar/gkaa1025 PMID:33211880

Chang, R. L., Ghamsari, L., Manichaikul, A., Hom, E. F. Y., Balaji, S., Fu, W., Shen, Y., Hao, T., Palsson, B., Salehi-Ashtiani, K., & Papin, J. A. (2011). Metabolic network reconstruction of *Chlamydomonas* offers insight into light-driven algal metabolism. *Molecular Systems Biology*, *7*(1), 518. doi:10.1038/msb.2011.52 PMID:21811229

Chen, F., Yuan, L., Ding, S., Tian, Y., & Hu, Q. N. (2020). Data-driven rational biosynthesis design: From molecules to cell factories. *Briefings in Bioinformatics*, *21*(4), 1238–1248. doi:10.1093/bib/bbz065 PMID:31243440

Cheng, F., Yu, H., & Stephanopoulos, G. (2019). Engineering *Corynebacterium glutamicum* for high-titer biosynthesis of hyaluronic acid. *Metabolic Engineering*, *55*, 276–289. doi:10.1016/j.ymben.2019.07.003 PMID:31301358

Chisti, Y. (2007). Biodiesel from microalgae. *Biotechnology Advances*, *25*(3), 294–306. doi:10.1016/j.biotechadv.2007.02.001 PMID:17350212

Chung, B. K. S., Lakshmanan, M., Klement, M., Mohanty, B., & Lee, D. Y. (2013). Genome-scale in silico modeling and analysis for designing synthetic terpenoid-producing microbial cell factories. *Chemical Engineering Science*, *103*, 100–108. doi:10.1016/j.ces.2012.09.006

Czajka, J. J., Oyetunde, T., & Tang, Y. J. (2021). Integrated knowledge mining, genome-scale modeling, and machine learning for predicting *Yarrowia lipolytica* bioproduction. *Metabolic Engineering*, *67*, 227–236. doi:10.1016/j.ymben.2021.07.003 PMID:34242777

d'Espaux, L., Ghosh, A., Runguphan, W., Wehrs, M., Xu, F., Konzock, O., Dev, I., Nhan, M., Gin, J., Reider Apel, A., Petzold, C. J., Singh, S., Simmons, B. A., Mukhopadhyay, A., García Martín, H., & Keasling, J. D. (2017). Engineering high-level production of fatty alcohols by *Saccharomyces cerevisiae* from lignocellulosic feedstocks. *Metabolic Engineering*, *42*, 115–125. doi:10.1016/j.ymben.2017.06.004 PMID:28606738

Dahabieh, M. S., Thevelein, J. M., & Gibson, B. (2020). Multimodal Microorganism Development: Integrating Top-Down Biological Engineering with Bottom-Up Rational Design. *Trends in Biotechnology*, *38*(3), 241–253. doi:10.1016/j.tibtech.2019.09.006 PMID:31653446

Ding, X., Zheng, Z., Zhao, G., Wang, L., Wang, H., Yang, Q., Zhang, M., Li, L., & Wang, P. (2022). Bottom-up synthetic biology approach for improving the efficiency of menaquinone-7 synthesis in *Bacillus subtilis*. *Microbial Cell Factories*, *21*(1), 1–14. doi:10.118612934-022-01823-3 PMID:35643569

Erdrich, P., Knoop, H., Steuer, R., & Klamt, S. (2014). Cyanobacterial biofuels: New insights and strain design strategies revealed by computational modeling. *Microbial Cell Factories*, *13*(1), 1–15. doi:10.118612934-014-0128-x PMID:25323065

Feist, A. M., Herrgård, M. J., Thiele, I., Reed, J. L., & Palsson, B. (2009). Reconstruction of biochemical networks in microorganisms. *Nature Reviews Microbiology*, *7*(2), 129–143. PMID: 19116616 doi:10.1038/nrmicro1949

Fu, P. (2009). Genome-scale modeling of *Synechocystis* sp. PCC 6803 and prediction of pathway insertion. *Journal of Chemical Technology and Biotechnology*, *84*(4), 473–483. doi:10.1002/jctb.2065

Garcia, S., Thompson, R. A., Giannone, R. J., Dash, S., Maranas, C. D., & Trinh, C. T. (2020). Development of a Genome-Scale Metabolic Model of Clostridium thermocellum and Its Applications for Integration of Multi-Omics Datasets and Computational Strain Design. *Frontiers in Bioengineering and Biotechnology*, *8*, 1–16. doi:10.3389/fbioe.2020.00772 PMID:32974289

García-Granados, R., Lerma-Escalera, J. A., & Morones-Ramírez, J. R. (2019). Metabolic engineering and synthetic biology: Synergies, future, and challenges. *Frontiers in Bioengineering and Biotechnology*, *7*, 439905. doi:10.3389/fbioe.2019.00036 PMID:30886847

Gawand, P., Yang, L., Cluett, W. R., & Mahadevan, R. (2013). Metabolic model refinement using phenotypic microarray data. *Methods in Molecular Biology (Clifton, N.J.)*, *985*, 47–59. doi:10.1007/978-1-62703-299-5_3 PMID:23417798

Ghosh, A., Ando, D., Gin, J., Runguphan, W., Denby, C., Wang, G., Baidoo, E. E. K., Shymansky, C., Keasling, J. D., & Martín, H. G. (2016). 13C Metabolic Flux Analysis for Systematic Metabolic Engineering of *S. cerevisiae* for Overproduction of Fatty Acids. *Frontiers in Bioengineering and Biotechnology*, *4*, 76. doi:10.3389/fbioe.2016.00076 PMID:27761435

Gimpel, J. A., & Mayfield, S. P. (2013). Analysis of heterologous regulatory and coding regions in algal chloroplasts. *Applied Microbiology and Biotechnology*, *97*(10), 4499–4510. doi:10.100700253-012-4580-4 PMID:23179624

Glont, M., Nguyen, T. V. N., Graesslin, M., Hälke, R., Ali, R., Schramm, J., Wimalaratne, S. M., Kothamachu, V. B., Rodriguez, N., Swat, M. J., Eils, J., Eils, R., Laibe, C., Malik-Sheriff, R. S., Chelliah, V., Le Novère, N., & Hermjakob, H. (2018). BioModels: Expanding horizons to include more modeling approaches and formats. *Nucleic Acids Research*, *46*(D1), D1248–D1253. doi:10.1093/nar/gkx1023 PMID:29106614

Gomes de Oliveira Dal'Molin, C., Quek, L.-E., Palfreyman, R. W., & Nielsen, L. K. (2011). AlgaGEM - a genome-scale metabolic reconstruction of algae based on the *Chlamydomonas reinhardtii* genome. *BMC Genomics*, *12*(Suppl 4), 1–10. PMID: 22369158. doi:10.1186/1471-2164-12-S4-S5

Goodwin, S., McPherson, J. D., & McCombie, W. R. (2016). Coming of age: Ten years of next-generation sequencing technologies. In *Nature Reviews Genetics* (Vol. 17, Issue 6, pp. 333–351). Nat Rev Genet. doi:10.1038/nrg.2016.49

Greenberg, A. J., Hackett, S. R., Harshman, L. G., & Clark, A. G. (2011). Environmental and genetic perturbations reveal different networks of metabolic regulation. *Molecular Systems Biology*, *7*(1), 563. doi:10.1038/msb.2011.96 PMID:22186737

Gu, C., Kim, G. B., Kim, W. J., Kim, H. U., & Lee, S. Y. (2019). Current status and applications of genome-scale metabolic models. *Genome Biology*, *20*(1), 1–18. doi:10.118613059-019-1730-3 PMID:31196170

Guo, Y., Su, L., Liu, Q., Zhu, Y., Dai, Z., & Wang, Q. (2022). Dissecting carbon metabolism of *Yarrowia lipolytica* type strain W29 using genome-scale metabolic modelling. *Computational and Structural Biotechnology Journal*, *20*, 2503–2511. doi:10.1016/j.csbj.2022.05.018 PMID:35664225

Gyori, B. M., & Bachman, J. A. (2021). From knowledge to models: Automated modeling in systems and synthetic biology. In *Current Opinion in Systems Biology* (Vol. 28, p. 100362). Elsevier. doi:10.1016/j.coisb.2021.100362

Hans, S., Kumar, N., Gohil, N., Khambhati, K., Bhattacharjee, G., Deb, S. S., Maurya, R., Kumar, V., Reshamwala, S. M. S., & Singh, V. (2022). Rebooting life: Engineering non-natural nucleic acids, proteins and metabolites in microorganisms. *Microbial Cell Factories*, *21*(1), 1–12. doi:10.118612934-022-01828-y PMID:35643549

Heirendt, L., Arreckx, S., Pfau, T., Mendoza, S. N., Richelle, A., Heinken, A., Haraldsdóttir, H. S., Wachowiak, J., Keating, S. M., Vlasov, V., Magnusdóttir, S., Ng, C. Y., Preciat, G., Žagare, A., Chan, S. H. J., Aurich, M. K., Clancy, C. M., Modamio, J., Sauls, J. T., ... Fleming, R. M. T. (2019). Creation and analysis of biochemical constraint-based models using the COBRA Toolbox v.3.0. *Nature Protocols*, *14*(3), 639–702. doi:10.103841596-018-0098-2 PMID:30787451

Hendry, J. I., Prasannan, C. B., Joshi, A., Dasgupta, S., & Wangikar, P. P. (2016). Metabolic model of *Synechococcus* sp. PCC 7002: Prediction of flux distribution and network modification for enhanced biofuel production. *Bioresource Technology*, *213*, 190–197. doi:10.1016/j.biortech.2016.02.128 PMID:27036328

Huang, D., Li, S., Xia, M., Wen, J., & Jia, X. (2013). Genome-scale metabolic network guided engineering of *Streptomyces tsukubaensis* for FK506 production improvement. *Microbial Cell Factories*, *12*(1), 1–18. doi:10.1186/1475-2859-12-52 PMID:23705993

Huang, D., Wen, J., Wang, G., Yu, G., Jia, X., & Chen, Y. (2012). In silico aided metabolic engineering of *Streptomyces roseosporus* for daptomycin yield improvement. *Applied Microbiology and Biotechnology*, *94*(3), 637–649. doi:10.100700253-011-3773-6 PMID:22406858

Huang, Y. Y., Jian, X. X., Lv, Y. B., Nian, K. Q., Gao, Q., Chen, J., Wei, L. J., & Hua, Q. (2018). Enhanced squalene biosynthesis in *Yarrowia lipolytica* based on metabolically engineered acetyl-CoA metabolism. *Journal of Biotechnology*, *281*, 106–114. doi:10.1016/j.jbiotec.2018.07.001 PMID:29986837

Ibrahim, M., Raajaraam, L., & Raman, K. (2021). Modeling microbial communities: Harnessing consortia for biotechnological applications. *Computational and Structural Biotechnology Journal*, *19*, 3892–3907. doi:10.1016/j.csbj.2021.06.048 PMID:34584635

Imam, S., Schäuble, S., Valenzuela, J., Lõpez García De Lomana, A., Carter, W., Price, N. D., & Baliga, N. S. (2015). A refined genome-scale reconstruction of *Chlamydomonas* metabolism provides a platform for systems-level analyses. *The Plant Journal*, *84*(6), 1239–1256. doi:10.1111/tpj.13059 PMID:26485611

Ishchuka, O. P., Domenzain, I., Sanchez, B. J., Muñiz-Paredes, F., Martínez, J. L., Nielsen, J., & Petranovic, D. (2022). Genome-scale modeling drives 70-fold improvement of intracellular heme production in *Saccharomyces cerevisiae*. *Proceedings of the National Academy of Sciences of the United States of America*, *119*(30), e2108245119. doi:10.1073/pnas.2108245119 PMID:35858410

Jagadevan, S., Banerjee, A., Banerjee, C., Guria, C., Tiwari, R., Baweja, M., & Shukla, P. (2018). Recent developments in synthetic biology and metabolic engineering in microalgae towards biofuel production. *Biotechnology for Biofuels*, *11*(1), 1–21. doi:10.118613068-018-1181-1 PMID:29988523

Jeandet, P., Vasserot, Y., Chastang, T., & Courot, E. (2013). Engineering microbial cells for the biosynthesis of natural compounds of pharmaceutical significance. *BioMed Research International*, *780145*, 1–13. Advance online publication. doi:10.1155/2013/780145 PMID:23710459

Jouhten, P. (2012). Metabolic modeling in the development of cell factories by synthetic biology. *Computational and Structural Biotechnology Journal*, *3*(4), e201210009. doi:10.5936/csbj.201210009 PMID:24688669

Kanehisa, M. (2019). Toward understanding the origin and evolution of cellular organisms. *Protein Science*, *28*(11), 1947–1951. doi:10.1002/pro.3715 PMID:31441146

Kanehisa, M., Furumichi, M., Sato, Y., Kawashima, M., & Ishiguro-Watanabe, M. (2023). KEGG for taxonomy-based analysis of pathways and genomes. *Nucleic Acids Research*, *51*(D1), D587–D592. doi:10.1093/nar/gkac963 PMID:36300620

Kanehisa, M., & Goto, S. (2000). KEGG: Kyoto Encyclopedia of Genes and Genomes. *Nucleic Acids Research*, *28*(1), 27–30. doi:10.1093/nar/28.1.27 PMID:10592173

Karp, P. D., Billington, R., Caspi, R., Fulcher, C. A., Latendresse, M., Kothari, A., Keseler, I. M., Krummenacker, M., Midford, P. E., Ong, Q., Ong, W. K., Paley, S. M., & Subhraveti, P. (2019). The BioCyc collection of microbial genomes and metabolic pathways. *Briefings in Bioinformatics*, *20*(4), 1085–1093. doi:10.1093/bib/bbx085 PMID:29447345

Kavšcek, M., Bhutada, G., Madl, T., & Natter, K. (2015). Optimization of lipid production with a genome-scale model of Yarrowia lipolytica. *BMC Systems Biology*, *9*(1), 1–13. doi:10.118612918-015-0217-4 PMID:26503450

Keseler, I. M., Collado-Vides, J., Santos-Zavaleta, A., Peralta-Gil, M., Gama-Castro, S., Muniz-Rascado, L., Bonavides-Martinez, C., Paley, S., Krummenacker, M., Altman, T., Kaipa, P., Spaulding, A., Pacheco, J., Latendresse, M., Fulcher, C., Sarker, M., Shearer, A. G., Mackie, A., Paulsen, I., ... Karp, P. D. (2011). EcoCyc: A comprehensive database of Escherichia coli biology. *Nucleic Acids Research*, *39*(Database), D583–D590. Advance online publication. doi:10.1093/nar/gkq1143 PMID:21097882

Kim, J., & Reed, J. L. (2010). OptORF: Optimal metabolic and regulatory perturbations for metabolic engineering of microbial strains. *BMC Systems Biology*, *4*(1), 53. doi:10.1186/1752-0509-4-53 PMID:20426856

Kim, M., Park, B. G., Kim, E. J., Kim, J., & Kim, B. G. (2019). In silico identification of metabolic engineering strategies for improved lipid production in *Yarrowia lipolytica* by genome-scale metabolic modeling. *Biotechnology for Biofuels*, *12*(1), 1–14. doi:10.118613068-019-1518-4 PMID:31367232

King, Z. A., Lu, J., Dräger, A., Miller, P., Federowicz, S., Lerman, J. A., Ebrahim, A., Palsson, B. O., & Lewis, N. E. (2016). BiGG Models: A platform for integrating, standardizing and sharing genome-scale models. *Nucleic Acids Research*, *44*(D1), D515–D522. doi:10.1093/nar/gkv1049 PMID:26476456

Kitano, S., Lin, C., Foo, J. L., & Chang, M. W. (2023). Synthetic biology: Learning the way toward high-precision biological design. *PLoS Biology*, *21*(4), e3002116. doi:10.1371/journal.pbio.3002116 PMID:37099620

Knoop, H., Gründel, M., Zilliges, Y., Lehmann, R., Hoffmann, S., Lockau, W., & Steuer, R. (2013). Flux Balance Analysis of Cyanobacterial Metabolism: The Metabolic Network of *Synechocystis* sp. PCC 6803. *PLoS Computational Biology*, *9*(6), e1003081. doi:10.1371/journal.pcbi.1003081 PMID:23843751

Knoop, H., Zilliges, Y., Lockau, W., & Steuer, R. (2010). The metabolic network of *Synechocystis* sp. PCC 6803: Systemic properties of autotrophic growth. *Plant Physiology*, *154*(1), 410–422. doi:10.1104/pp.110.157198 PMID:20616194

Krämer, M., Bongaerts, J., Bovenberg, R., Kremer, S., Müller, U., Orf, S., Wubbolts, M., & Raeven, L. (2003). Metabolic engineering for microbial production of shikimic acid. *Metabolic Engineering*, *5*(4), 277–283. doi:10.1016/j.ymben.2003.09.001 PMID:14642355

Kuepfer, L., Sauer, U., & Blank, L. M. (2005). Metabolic functions of duplicate genes in *Saccharomyces cerevisiae*. *Genome Research*, *15*(10), 1421–1430. doi:10.1101/gr.3992505 PMID:16204195

Kulkarni, R. (2016). Metabolic Engineering: Biological Art of Producing Useful Chemicals. *Resonance*, *21*(3), 233–237. doi:10.100712045-016-0318-4

Landon, S., Rees-Garbutt, J., Marucci, L., & Grierson, C. (2019). Genome-driven cell engineering review: In vivo and in silico metabolic and genome engineering. *Essays in Biochemistry*, *63*(2), 267–284. doi:10.1042/EBC20180045 PMID:31243142

Lawson, C. E., Martí, J. M., Radivojevic, T., Jonnalagadda, S. V. R., Gentz, R., Hillson, N. J., Peisert, S., Kim, J., Simmons, B. A., Petzold, C. J., Singer, S. W., Mukhopadhyay, A., Tanjore, D., Dunn, J. G., & Garcia Martin, H. (2021). Machine learning for metabolic engineering: A review. *Metabolic Engineering*, *63*, 34–60. doi:10.1016/j.ymben.2020.10.005 PMID:33221420

Lecun, Y., Bengio, Y., & Hinton, G. (2015). Deep learning. *Nature*, *521*(7553), 436–444. doi:10.1038/nature14539 PMID:26017442

Lee, K. Y., Park, J. M., Kim, T. Y., Yun, H., & Lee, S. Y. (2010). The genome-scale metabolic network analysis of *Zymomonas mobilis* ZM4 explains physiological features and suggests ethanol and succinic acid production strategies. *Microbial Cell Factories*, *9*(1), 94. doi:10.1186/1475-2859-9-94 PMID:21092328

Lee, S. K., Chou, H., Ham, T. S., Lee, T. S., & Keasling, J. D. (2008). Metabolic engineering of microorganisms for biofuels production: From bugs to synthetic biology to fuels. *Current Opinion in Biotechnology*, *19*(6), 556–563. doi:10.1016/j.copbio.2008.10.014 PMID:18996194

Lewis, J. E., & Kemp, M. L. (2021). Integration of machine learning and genome-scale metabolic modeling identifies multi-omics biomarkers for radiation resistance. *Nature Communications*, *12*(1), 1–14. doi:10.103841467-021-22989-1 PMID:33976213

Li, C. T., Yelsky, J., Chen, Y., Zuñiga, C., Eng, R., Jiang, L., Shapiro, A., Huang, K. W., Zengler, K., & Betenbaugh, M. J. (2019). Utilizing genome-scale models to optimize nutrient supply for sustained algal growth and lipid productivity. *NPJ Systems Biology and Applications*, *5*(1), 33. Advance online publication. doi:10.103841540-019-0110-7 PMID:31583115

Li, S., Huang, D., Li, Y., Wen, J., & Jia, X. (2012). Rational improvement of the engineered isobutanol-producing *Bacillus subtilis* by elementary mode analysis. *Microbial Cell Factories*, *11*(1), 1–12. doi:10.1186/1475-2859-11-101 PMID:22862776

Liang, J., Luo, Y., & Zhao, H. (2011). Synthetic biology: Putting synthesis into biology. *Wiley Interdisciplinary Reviews. Systems Biology and Medicine*, *3*(1), 7–20. doi:10.1002/wsbm.104 PMID:21064036

Liang, Y., Cui, Y., Trushenski, J., & Blackburn, J. W. (2010). Converting crude glycerol derived from yellow grease to lipids through yeast fermentation. *Bioresource Technology*, *101*(19), 7581–7586. doi:10.1016/j.biortech.2010.04.061 PMID:20478702

Licona-Cassani, C., Marcellin, E., Quek, L. E., Jacob, S., & Nielsen, L. K. (2012). Reconstruction of the Saccharopolyspora erythraea genome-scale model and its use for enhancing erythromycin production. *Antonie van Leeuwenhoek. Antonie van Leeuwenhoek, 102*(3), 493–502. doi:10.100710482-012-9783-2 PMID:22847261

Liu, L., Agren, R., Bordel, S., & Nielsen, J. (2010). Use of genome-scale metabolic models for understanding microbial physiology. *FEBS Letters, 584*(12), 2556–2564. doi:10.1016/j.febslet.2010.04.052 PMID:20420838

Liu, R., Bassalo, M. C., Zeitoun, R. I., & Gill, R. T. (2015). Genome scale engineering techniques for metabolic engineering. *Metabolic Engineering, 32*, 143–154. doi:10.1016/j.ymben.2015.09.013 PMID:26453944

Loira, N., Dulermo, T., Nicaud, J. M., & Sherman, D. J. (2012). A genome-scale metabolic model of the lipid-accumulating yeast *Yarrowia lipolytica*. *BMC Systems Biology, 6*(1), 1–9. doi:10.1186/1752-0509-6-35 PMID:22558935

Loira, N., Mendoza, S., Paz Cortés, M., Rojas, N., Travisany, D., Di Genova, A., Gajardo, N., Ehrenfeld, N., & Maass, A. (2017). Reconstruction of the microalga *Nannochloropsis salina* genome-scale metabolic model with applications to lipid production. *BMC Systems Biology, 11*(1), 1–17. doi:10.118612918-017-0441-1 PMID:28676050

Long, M. R., Ong, W. K., & Reed, J. L. (2015). Computational methods in metabolic engineering for strain design. *Current Opinion in Biotechnology, 34*, 135–141. doi:10.1016/j.copbio.2014.12.019 PMID:25576846

Lopes, H., & Rocha, I. (2017). Genome-scale modeling of yeast: Chronology, applications and critical perspectives. *FEMS Yeast Research, 17*(5), 1–14. doi:10.1093/femsyr/fox050 PMID:28899034

Lu, J., Bi, X., Liu, Y., Lv, X., Li, J., Du, G., & Liu, L. (2022). In silico cell factory design driven by comprehensive genome-scale metabolic models: Development and challenges. *Systems Microbiology and Biomanufacturing, 3*(2), 207–222. doi:10.100743393-022-00117-4

Lu, J., Wu, Y., Deng, C., Liu, Y., Lv, X., Li, J., Du, G., & Liu, L. (2022). Model-based dynamic engineering of *Escherichia coli* for N-acetylglucosamine overproduction. *Biotechnology Notes, 3*, 15–24. doi:10.1016/j.biotno.2022.02.001

Mahadevan, R., & Schilling, C. H. (2003). The effects of alternate optimal solutions in constraint-based genome-scale metabolic models. *Metabolic Engineering, 5*(4), 264–276. doi:10.1016/j.ymben.2003.09.002 PMID:14642354

Malik-Sheriff, R. S., Glont, M., Nguyen, T. V. N., Tiwari, K., Roberts, M. G., Xavier, A., Vu, M. T., Men, J., Maire, M., Kananathan, S., Fairbanks, E. L., Meyer, J. P., Arankalle, C., Varusai, T. M., Knight-Schrijver, V., Li, L., Dueñas-Roca, C., Dass, G., Keating, S. M., ... Hermjakob, H. (2020). BioModels-15 years of sharing computational models in life science. *Nucleic Acids Research, 48*(D1), D407–D415. doi:10.1093/nar/gkz1055 PMID:31701150

Manichaikul, A., Ghamsari, L., Hom, E. F. Y., Lin, C., Murray, R. R., Chang, R. L., Balaji, S., Hao, T., Shen, Y., Chavali, A. K., Thiele, I., Yang, X., Fan, C., Mello, E., Hill, D. E., Vidal, M., Salehi-Ashtiani, K., & Papin, J. A. (2009). Metabolic network analysis integrated with transcript verification for sequenced genomes. *Nature Methods, 6*(8), 589–592. doi:10.1038/nmeth.1348

Meadows, A. L., Hawkins, K. M., Tsegaye, Y., Antipov, E., Kim, Y., Raetz, L., Dahl, R. H., Tai, A., Mahatdejkul-Meadows, T., Xu, L., Zhao, L., Dasika, M. S., Murarka, A., Lenihan, J., Eng, D., Leng, J. S., Liu, C. L., Wenger, J. W., Jiang, H., ... Tsong, A. E. (2016). Rewriting yeast central carbon metabolism for industrial isoprenoid production. *Nature*, *537*(7622), 694–697. doi:10.1038/nature19769 PMID:27654918

Medlock, G. L., & Papin, J. A. (2020). Guiding the Refinement of Biochemical Knowledgebases with Ensembles of Metabolic Networks and Machine Learning. *Cell Systems*, *10*(1), 109–119.e3. doi:10.1016/j.cels.2019.11.006 PMID:31926940

Megchelenbrink, W., Huynen, M., & Marchiori, E. (2014). optGpSampler: An Improved Tool for Uniformly Sampling the Solution-Space of Genome-Scale Metabolic Networks. *PLoS One*, *9*(2), e86587. doi:10.1371/journal.pone.0086587 PMID:24551039

Mendoza, S. N., Olivier, B. G., Molenaar, D., & Teusink, B. (2019). A systematic assessment of current genome-scale metabolic reconstruction tools. *Genome Biology*, *20*(1), 1–20. doi:10.118613059-019-1769-1 PMID:31391098

Mo, M. L., Palsson, B., & Herrgård, M. J. (2009). Connecting extracellular metabolomic measurements to intracellular flux states in yeast. *BMC Systems Biology*, *3*(1), 1–17. doi:10.1186/1752-0509-3-37 PMID:19321003

Moen, E., Bannon, D., Kudo, T., Graf, W., Covert, M., & Van Valen, D. (2019). Deep learning for cellular image analysis. *Nature Methods*, *16*(12), 1233–1246. doi:10.103841592-019-0403-1 PMID:31133758

Mohammadi, R., Fallah-Mehrabadi, J., Bidkhori, G., Zahiri, J., Javad Niroomand, M., & Masoudi-Nejad, A. (2016). A systems biology approach to reconcile metabolic network models with application to: *Synechocystis* sp. PCC 6803 for biofuel production. *Molecular BioSystems*, *12*(8), 2552–2561. doi:10.1039/C6MB00119J PMID:27265370

Moreno-Paz, S., Schmitz, J., Martins dos Santos, V. A. P., & Suarez-Diez, M. (2022). Enzyme-constrained models predict the dynamics of Saccharomyces cerevisiae growth in continuous, batch and fed-batch bioreactors. *Microbial Biotechnology*, *15*(5), 1434–1445. doi:10.1111/1751-7915.13995 PMID:35048533

Nielsen, J., & Keasling, J. D. (2016). Engineering Cellular Metabolism. *Cell, 164*(6), 1185–1197. doi:10.1016/j.cell.2016.02.004 PMID:26967285

Nogales, J., Gudmundsson, S., Knight, E. M., Palsson, B. O., & Thiele, I. (2012). Detailing the optimality of photosynthesis in cyanobacteria through systems biology analysis. *Proceedings of the National Academy of Sciences of the United States of America*, *109*(7), 2678–2683. doi:10.1073/pnas.1117907109 PMID:22308420

O'Brien, E. J., Monk, J. M., & Palsson, B. O. (2015). Using Genome-scale Models to Predict Biological Capabilities. *Cell, 161*(5), 971–987. doi:10.1016/j.cell.2015.05.019 PMID:26000478

Oh, Y. K., Palsson, B. O., Park, S. M., Schilling, C. H., & Mahadevan, R. (2007). Genome-scale reconstruction of metabolic network in *Bacillus subtilis* based on high-throughput phenotyping and gene essentiality data. *The Journal of Biological Chemistry, 282*(39), 28791–28799. doi:10.1074/jbc.M703759200 PMID:17573341

Opgenorth, P., Costello, Z., Okada, T., Goyal, G., Chen, Y., Gin, J., Benites, V., De Raad, M., Northen, T. R., Deng, K., Deutsch, S., Baidoo, E. E. K., Petzold, C. J., Hillson, N. J., Garcia Martin, H., & Beller, H. R. (2019). Lessons from Two Design-Build-Test-Learn Cycles of Dodecanol Production in *Escherichia coli* Aided by Machine Learning. *ACS Synthetic Biology, 8*(6), 1337–1351. doi:10.1021/acssynbio.9b00020 PMID:31072100

Orth, J. D., Conrad, T. M., Na, J., Lerman, J. A., Nam, H., Feist, A. M., & Palsson, B. (2011). A comprehensive genome-scale reconstruction of *Escherichia coli* metabolism—2011. *Molecular Systems Biology, 7*(1), 535. Advance online publication. doi:10.1038/msb.2011.65 PMID:21988831

Pan, P., & Hua, Q. (2012). Reconstruction and In Silico Analysis of Metabolic Network for an Oleaginous Yeast, *Yarrowia lipolytica*. *PLoS One, 7*(12), e51535. doi:10.1371/journal.pone.0051535 PMID:23236514

Paramasivan, K., Kumar, H. N. P., & Mutturi, S. (2019). Systems-based *Saccharomyces cerevisiae* strain design for improved squalene synthesis. *Biochemical Engineering Journal, 148*(April), 37–45. doi:10.1016/j.bej.2019.04.025

Park, J. H., & Lee, S. Y. (2008). Towards systems metabolic engineering of microorganisms for amino acid production. *Current Opinion in Biotechnology, 19*(5), 454–460. doi:10.1016/j.copbio.2008.08.007 PMID:18760356

Pathania, R., Srivastava, A., Srivastava, S., & Shukla, P. (2022). Metabolic systems biology and multi-omics of cyanobacteria: Perspectives and future directions. *Bioresource Technology, 343*, 126007. doi:10.1016/j.biortech.2021.126007

Patil, K. R., Rocha, I., Förster, J., & Nielsen, J. (2005). Evolutionary programming as a platform for in silico metabolic engineering. *BMC Bioinformatics, 6*(1), 1–12. doi:10.1186/1471-2105-6-308 PMID:16375763

Pearcy, N., Hu, Y., Baker, M., Maciel-Guerra, A., Xue, N., Wang, W., Kaler, J., Peng, Z., Li, F., & Dottorini, T. (2021). Genome-Scale Metabolic Models and Machine Learning Reveal Genetic Determinants of Antibiotic Resistance in *Escherichia coli* and Unravel the Underlying Metabolic Adaptation Mechanisms. *mSystems, 6*(4), e00913-20. Advance online publication. doi:10.1128/mSystems.00913-20 PMID:34342537

Pham, N., Reijnders, M., Suarez-Diez, M., Nijsse, B., Springer, J., Eggink, G., & Schaap, P. J. (2021). Genome-scale metabolic modeling underscores the potential of *Cutaneotrichosporon oleaginosus* ATCC 20509 as a cell factory for biofuel production. *Biotechnology for Biofuels, 14*(1), 1–17. doi:10.118613068-020-01838-1 PMID:33407779

Pharkya, P., Burgard, A. P., & Maranas, C. D. (2004). OptStrain: A computational framework for redesign of microbial production systems. *Genome Research, 14*(11), 2367–2376. doi:10.1101/gr.2872004 PMID:15520298

Ranganathan, S., Suthers, P. F., & Maranas, C. D. (2010). OptForce: An Optimization Procedure for Identifying All Genetic Manipulations Leading to Targeted Overproductions. *PLoS Computational Biology*, *6*(4), e1000744. doi:10.1371/journal.pcbi.1000744 PMID:20419153

Reed, J. L., Vo, T. D., Schilling, C. H., & Palsson, B. O. (2003). An expanded genome-scale model of Escherichia coli K-12 (iJR904 GSM/GPR). *Genome Biology*, *4*(9), 1–12. doi:10.1186/gb-2003-4-9-r54 PMID:12952533

Roberts, M. A. J., Cranenburgh, R. M., Stevens, M. P., & Oyston, P. C. F. (2013). Synthetic biology: Biology by design. *Microbiology*, *159*(Pt 7), 1219–1220. doi:10.1099/mic.0.069724-0 PMID:23744901

Roberts, S. B., Gowen, C. M., Brooks, J. P., & Fong, S. S. (2010). Genome-scale metabolic analysis of Clostridium thermocellum for bioethanol production. *BMC Systems Biology*, *4*(1), 31. Advance online publication. doi:10.1186/1752-0509-4-31 PMID:20307315

Robinson, J. L., Kocabaş, P., Wang, H., Cholley, P. E., Cook, D., Nilsson, A., Anton, M., Ferreira, R., Domenzain, I., Billa, V., Limeta, A., Hedin, A., Gustafsson, J., Kerkhoven, E. J., Svensson, L. T., Palsson, B. O., Mardinoglu, A., Hansson, L., Uhlén, M., & Nielsen, J. (2020). An atlas of human metabolism. *Science Signaling*, *13*(624), eaaz1482. Advance online publication. doi:10.1126cisignal.aaz1482 PMID:32209698

Rollié, S., Mangold, M., & Sundmacher, K. (2012). Designing biological systems: Systems Engineering meets Synthetic Biology. *Chemical Engineering Science*, *69*(1), 1–29. doi:10.1016/j.ces.2011.10.068

Saini, D. K., Rai, A., Devi, A., Pabbi, S., Chhabra, D., Chang, J. S., & Shukla, P. (2021). A multi-objective hybrid machine learning approach-based optimization for enhanced biomass and bioactive phycobiliproteins production in *Nostoc* sp. CCC-403. *Bioresource Technology*, *329*, 124908. doi:10.1016/j.biortech.2021.124908 PMID:33690058

Saitua, F., Torres, P., Pérez-Correa, J. R., & Agosin, E. (2017). Dynamic genome-scale metabolic modeling of the yeast *Pichia pastoris*. *BMC Systems Biology*, *11*(1), 1–21. doi:10.118612918-017-0408-2 PMID:28222737

Salguero, D. A. M., Fernández-Niño, M., Serrano-Bermúdez, L. M., Melo, D. O. P., Winck, F. V., Caldana, C., & Barrios, A. F. G. (2018). Development of a *Chlamydomonas reinhardtii* metabolic network dynamic model to describe distinct phenotypes occurring at different CO2 levels. *PeerJ*, *6*(9), e5528. Advance online publication. doi:10.7717/peerj.5528 PMID:30202653

San León, D., & Nogales, J. (2022). Toward merging bottom–up and top–down model-based designing of synthetic microbial communities. *Current Opinion in Microbiology*, *69*, 102169. doi:10.1016/j.mib.2022.102169 PMID:35763963

Schellenberger, J., Que, R., Fleming, R. M. T., Thiele, I., Orth, J. D., Feist, A. M., Zielinski, D. C., Bordbar, A., Lewis, N. E., Rahmanian, S., Kang, J., Hyduke, D. R., & Palsson, B. (2011). Quantitative prediction of cellular metabolism with constraint-based models: The COBRA Toolbox v2.0. *Nature Protocols*, *6*(9), 1290–1307. doi:10.1038/nprot.2011.308 PMID:21886097

Schinn, S. M., Morrison, C., Wei, W., Zhang, L., & Lewis, N. E. (2021). A genome-scale metabolic network model and machine learning predict amino acid concentrations in Chinese Hamster Ovary cell cultures. *Biotechnology and Bioengineering*, *118*(5), 2118–2123. doi:10.1002/bit.27714 PMID:33580712

Scossa, F., & Fernie, A. R. (2020). The evolution of metabolism: How to test evolutionary hypotheses at the genomic level. *Computational and Structural Biotechnology Journal*, *18*, 482–500. doi:10.1016/j.csbj.2020.02.009 PMID:32180906

Seaver, S. M. D., Liu, F., Zhang, Q., Jeffryes, J., Faria, J. P., Edirisinghe, J. N., Mundy, M., Chia, N., Noor, E., Beber, M. E., Best, A. A., DeJongh, M., Kimbrel, J. A., D'haeseleer, P., McCorkle, S. R., Bolton, J. R., Pearson, E., Canon, S., Wood-Charlson, E. M., ... Henry, C. S. (2021). The ModelSEED Biochemistry Database for the integration of metabolic annotations and the reconstruction, comparison and analysis of metabolic models for plants, fungi and microbes. *Nucleic Acids Research*, *49*(D1), D575–D588. doi:10.1093/nar/gkaa746 PMID:32986834

Shabestary, K., & Hudson, E. P. (2016). Computational metabolic engineering strategies for growth-coupled biofuel production by *Synechocystis*. *Metabolic Engineering Communications*, *3*, 216–226. doi:10.1016/j.meteno.2016.07.003 PMID:29468126

Shah, A. R., Ahmad, A., Srivastava, S., & Jaffar Ali, B. M. (2017). Reconstruction and analysis of a genome-scale metabolic model of *Nannochloropsis gaditana*. *Algal Research*, *26*, 354–364. doi:10.1016/j.algal.2017.08.014

Shah, M. V., Nazem-Bokaee, H., Antoney, J., Kang, S. W., Jackson, C. J., & Scott, C. (2021). Improved production of the non-native cofactor F420 in *Escherichia coli*. *Scientific Reports*, *11*(1), 1–16. doi:10.103841598-021-01224-3 PMID:34741069

Shahkouhi, A. M., & Motamedian, E. (2020). Reconstruction of a regulated two-cell metabolic model to study biohydrogen production in a diazotrophic cyanobacterium *Anabaena variabilis* ATCC 29413. *PLoS One*, *15*(1), 1–18. doi:10.1371/journal.pone.0227977 PMID:31978122

Singh, S. (2018). Synthetic biology: Omics tools and their applications. In *Synthetic Biology*. Omics Tools and Their Applications., doi:10.1007/978-981-10-8693-9

Sohn, S. B., Graf, A. B., Kim, T. Y., Gasser, B., Maurer, M., Ferrer, P., Mattanovich, D., & Lee, S. Y. (2010). Genome-scale metabolic model of methylotrophic yeast *Pichia pastoris* and its use for in silico analysis of heterologous protein production. *Biotechnology Journal*, *5*(7), 705–715. doi:10.1002/biot.201000078 PMID:20503221

Stalidzans, E., Seiman, A., Peebo, K., Komasilovs, V., & Pentjuss, A. (2018). Model-based metabolism design: Constraints for kinetic and stoichiometric models. *Biochemical Society Transactions*, *46*(2), 261–267. doi:10.1042/BST20170263 PMID:29472367

Stephanopoulos, G. N., Aristidou, A. A., & Nielsen, J. (1998). Metabolic Engineering: Principles and Methodologies. In *Metabolic Engineering* (Vol. 54). Academic Press.

Straathof, A. J. J., Wahl, S. A., Benjamin, K. R., Takors, R., Wierckx, N., & Noorman, H. J. (2019). Grand Research Challenges for Sustainable Industrial Biotechnology. *Trends in Biotechnology*, *37*(10), 1042–1050. doi:10.1016/j.tibtech.2019.04.002 PMID:31054854

Szklarczyk, D., Gable, A. L., Nastou, K. C., Lyon, D., Kirsch, R., Pyysalo, S., Doncheva, N. T., Legeay, M., Fang, T., Bork, P., Jensen, L. J., & von Mering, C. (2021). The STRING database in 2021: Customizable protein-protein networks, and functional characterization of user-uploaded gene/measurement sets. *Nucleic Acids Research*, *49*(D1), D605–D612. doi:10.1093/nar/gkaa1074 PMID:33237311

Thiele, I., & Palsson, B. (2010). A protocol for generating a high-quality genome-scale metabolic reconstruction. *Nature Protocols*, *5*(1), 93–121. doi:10.1038/nprot.2009.203 PMID:20057383

Torres, N. V., & Voit, E. O. (2002). Pathway Analysis and Optimization in Metabolic Engineering. In *Pathway Analysis and Optimization in Metabolic Engineering*. Cambridge University Press., doi:10.1017/CBO9780511546334

US Energy Information Administration (EIA). (2021). *International energy outlook 2021: With projections to 2050*. Author.

Ventorim, R. Z., Ferreira, M. A. de M., de Almeida, E. L. M., Kerkhoven, E. J., & da Silveira, W. B. (2022). Genome-scale metabolic model of oleaginous yeast *Papiliotrema laurentii*. *Biochemical Engineering Journal*, *180*, 108353. doi:10.1016/j.bej.2022.108353

Waller, T. C., Berg, J. A., Lex, A., Chapman, B. E., & Rutter, J. (2020). Compartment and hub definitions tune metabolic networks for metabolomic interpretations. *GigaScience*, *9*(1), 1–18. doi:10.1093/gigascience/giz137 PMID:31972021

Wang, H., Marcišauskas, S., Sánchez, B. J., Domenzain, I., Hermansson, D., Agren, R., Nielsen, J., & Kerkhoven, E. J. (2018). RAVEN 2.0: A versatile toolbox for metabolic network reconstruction and a case study on *Streptomyces coelicolor*. *PLoS Computational Biology*, *14*(10), e1006541. doi:10.1371/journal.pcbi.1006541 PMID:30335785

Widiastuti, H., Kim, J. Y., Selvarasu, S., Karimi, I. A., Kim, H., Seo, J. S., & Lee, D. Y. (2011). Genome-scale modeling and in silico analysis of ethanologenic bacteria *Zymomonas mobilis*. *Biotechnology and Bioengineering*, *108*(3), 655–665. doi:10.1002/bit.22965 PMID:20967753

Wisselink, H. W., Cipollina, C., Oud, B., Crimi, B., Heijnen, J. J., Pronk, J. T., & van Maris, A. J. A. (2010). Metabolome, transcriptome and metabolic flux analysis of arabinose fermentation by engineered Saccharomyces cerevisiae. *Metabolic Engineering*, *12*(6), 537–551. doi:10.1016/j.ymben.2010.08.003 PMID:20816840

Woolston, B. M., Edgar, S., & Stephanopoulos, G. (2013). Metabolic engineering: Past and future. *Annual Review of Chemical and Biomolecular Engineering*, *4*(1), 259–288. doi:10.1146/annurev-chembioeng-061312-103312 PMID:23540289

Xavier, J. C., Patil, K. R., & Rocha, I. (2017). Integration of Biomass Formulations of Genome-Scale Metabolic Models with Experimental Data Reveals Universally Essential Cofactors in *Prokaryotes*. *Metabolic Engineering*, *39*, 200–208. doi:10.1016/j.ymben.2016.12.002 PMID:27939572

Xu, C., Liu, L., Zhang, Z., Jin, D., Qiu, J., & Chen, M. (2013). Genome-scale metabolic model in guiding metabolic engineering of microbial improvement. *Applied Microbiology and Biotechnology*, *97*(2), 519–539. doi:10.100700253-012-4543-9 PMID:23188456

Xu, Z., Zheng, P., Sun, J., & Ma, Y. (2013). ReacKnock: Identifying Reaction Deletion Strategies for Microbial Strain Optimization Based on Genome-Scale Metabolic Network. *PLoS One*, *8*(12), e72150. doi:10.1371/journal.pone.0072150 PMID:24348984

Yang, S., Fei, Q., Zhang, Y., Contreras, L. M., Utturkar, S. M., Brown, S. D., Himmel, M. E., & Zhang, M. (2016). Zymomonas mobilis as a model system for production of biofuels and biochemicals. *Microbial Biotechnology*, *9*(6), 699–717. doi:10.1111/1751-7915.12408 PMID:27629544

Ye, C., Wei, X., Shi, T., Sun, X., Xu, N., Gao, C., & Zou, W. (2022). Genome-scale metabolic network models: From first-generation to next-generation. *Applied Microbiology and Biotechnology*, *106*(13), 4907–4920. doi:10.100700253-022-12066-y PMID:35829788

Zheng, Y., & Sriram, G. (2010). Mathematical modeling: Bridging the gap between concept and realization in synthetic biology. *BioMed Research International*, *541609*, 1–16. Advance online publication. doi:10.1155/2010/541609 PMID:20589069

Zhou, W., Wang, Y., Zhang, J., Zhao, M., Tang, M., Zhou, W., & Gong, Z. (2021). A metabolic model of Lipomyces starkeyi for predicting lipogenesis potential from diverse low-cost substrates. *Biotechnology for Biofuels*, *14*(1), 1–17. doi:10.118613068-021-01997-9 PMID:34210354

Zuñiga, C., Levering, J., Antoniewicz, M. R., Guarnieri, M. T., Betenbaugh, M. J., & Zengler, K. (2018). Predicting Dynamic Metabolic Demands in the Photosynthetic Eukaryote *Chlorella vulgaris*. *Plant Physiology*, *176*(1), 450–462. doi:10.1104/pp.17.00605 PMID:28951490

KEY TERMS AND DEFINITIONS

COBRA: COBRA (COnstraint-Based Reconstruction and Analysis) is a MATLAB software suite that is widely used for modeling, analyzing, and predicting a variety of metabolic phenotypes using genome-scale biochemical networks.

Flux Balance Analysis: Flux balance analysis (FBA) is a mathematical method to analyze the flow of metabolites through a metabolic network.

Genome-Scale Metabolic Model: It is a mathematical representation of an organism's metabolism that uses two matrices to connect genes, reactions, and metabolites. The S matrix links metabolites to reactions, and the rxnGeneMat matrix links reactions to the corresponding genes and enzymes.

Metabolic Engineering: It is the field of study that combines systematic analysis of metabolic pathways with genetic engineering methods to modify and improve cellular properties.

Rational Cell Design: It is the strategy of creating new cells with certain properties, based upon the ability to predict how the modifications will affect its behavior through models.

RAVEN: RAVEN (Reconstruction, Analysis, and Visualization of Metabolic Networks) is a MATLAB toolbox that is commonly used for genome-scale metabolic model reconstruction, curation, and simulation.

Systems Biology: It is a comprehensive quantitive analysis of the behavior and interaction of biological components including molecules, cells, organs, and cells.

Chapter 7
Molecular Pharming

R. Kelmani Chandrakanth
Gulbarga University, India

Sahanapriya Gubbewad
Gulbarga University, India

Ajay Kumar Oli
https://orcid.org/0000-0002-9634-2819
Shri Dharmasthala Manjunatheshwara University, Dharwad, India

ABSTRACT

Molecular pharming is a method that combines the use of plants and biotechnological techniques for the production of proteins and bioactive metabolites which are important to support human health and industries. Using genetic engineering techniques, transgenic plants are developed. These plants have applications in the field of agriculture, biofuel, chemical industries, and pharmaceutical industries. These transgenic crop plants with limited facilities can synthesize a large amount of biomass at a low cost, through which a novel source of molecular medicines like plasma proteins, antibodies, vaccines, and growth factors can be isolated. Plant molecular pharming can be used as an alternative to mammalian cells, yeast, or bacterial culture systems for the production of pharmaceutical proteins. This also helps in studying and understanding the mechanism of expression of genes in plants. This chapter deals with the current strategies and approaches used in plant molecular pharming including the strategies, applications, limitations, and biosafety considerations.

HISTORY

Molecular Pharming combines the use of tissue culture, medicinal plants, and the techniques of biotechnology for the production of pharmaceutical proteins from decades ago. The first recombinant therapeutic protein for human growth hormone in tobacco plants was developed (Andrea et.al.,1986). Before that, transgenic mice were also used in the production of proteins (J W Gordon et.al., 1981). A few years later, transgenic tobacco and potato plants were developed for human serum albumin production (Peter et.al.,1990). The concept of developing therapeutics, recombinant proteins, enzymes, and vaccines was

DOI: 10.4018/978-1-6684-6577-6.ch007

developed in the 2000s. Leading biotech firms gradually achieved results for a variety of clinical conditions, including cystic fibrosis and lymphoma. The progress toward commercialization of plant-based products takes much effort, and most of them are still in clinical trials.

INTRODUCTION

Pharmaceuticals are producing different biochemicals, vaccines, therapeutics, and monoclonal antibodies are being produced in bioreactors, but due to development in molecular biology, genetically modified organisms are being used in the production. Genetically modified organisms(GMOs) are those organisms whose genes are modified or altered by genetic engineering techniques using desired genes. These GMOs may be microbes, transgenic animals, or transgenic plants. The use of microbial cells, animal sources, or mammalian cells for biopharmaceutical production has both advantages and disadvantages. The drawbacks include the frequency of contamination of the cells to meet the needs for their growth, and it was expensive.

The best alternative to overcome these conditions was the use of plants that can produce large amounts of biomass at a low cost. Even plants have a low risk of contamination with toxic compounds and human pathogens (Michael et.al., 2022). Plants are used for the production of food supplements, antibiotics, industrially important enzymes, vaccines, and proteins (Shalini et.al.,2021).

Definition: Molecular pharming is a process in which genetically modified organisms are produced by the integration of the desired gene into bioreactor systems like plants, animals, and microbial cells are used to form a substance that has a biological activity (Shalini et.al.,2021). In simple terms, it can be defined as one of the applied fields of molecular biology.

Most people understand how to grow plants. If these recombinant protein genes are introduced into plants, it may be a feasible method to meet the vast production needs.

METHODS AND STRATEGIES USED IN MOLECULAR PHARMACOLOGY

Plant transformation is a method of inserting the DNA from an organism with the desired character into the plant genome. It provides an essential tool for much basic research, for the study of gene functions and interactions, protein-protein interaction, and development process as well as the development of plant bioreactors to produce vaccines, recombinant proteins, antibiotics, and so on. Methods of transformation in plants can be by three means.

Biological
Physical
Chemical

BIOLOGICAL

The vectors, such as *Agrobacterium tumefaciens* and *Agrobacterium rhizogens,* are gram-negative soil bacteria that cause crown gall disease. The T-DNA is used for transformations that consist of Ti plasmids (tumor-inducing) in *A.tumefaciens.Agrobacterium rhizogens* have the Ri plasmid (hairy root-inducing). Both the plasmid has a T-region which is transferred into the plant cell for integration in their genome and expression of the transgene. This method was successfully used for transformed a different variety of monocot plants.

The modified agrobacteria will not have the tumor-promoting and opine synthesis genes (disarmed agrobacteria) the desired foreign gene or selectable marker is replaced in that region, making the vectors useful for incorporating the foreign gene in the plant genome.

STEPS INVOLVED IN AGROBACTERIUM-MEDIATED GENE TRANSFER

Recognition of the chemical signals produced by the host organism.

The Vir gene in agro-bacterium gets activated because of which it attaches to plant cells.

Transportation and activation of the virulence proteins.

Production and transfer of T-DNA out of agrobacteria.

T-DNA transfers in to plant nucleus and integrates into the genome (this method is an exception for stable transformation).

The plant viruses will not integrate into the host genome but the vectors based on the T-DNA of *A.*tumefaciens will integrate. The host plant is infected with the plant virus with the desired protein gene, when the host plant is grown to a certain stage then the modified virus is inoculated with it, and as the virus replicates and spreads, within the plant the copies of desired DNA is produced from those plants the high amount of protein production is achieved in a short time (as shown in image). Normally this method is used to infect the cells of adult plants. The viruses such as the Caulimo virus, Geminivirus, and TMV are employed.

Magnifection is an expression platform developed by Icon Genetics, a German Plant Biotechnology Company. This technique uses the deconstructed virus that relays on the agrobacterium to deliver the viral vector into the plant (Y. Gleba et.al.,2005).

PHYSICAL

The gene gun, or biolistic, is used to introduce nucleic acids into plants, The gene containing DNA is coated with metal gold and tungsten particles are loaded into the gene gun and then at high speed transferred to the plant cells using the gene gun. These plant cells uptakes the DNA and are permitted to grow in new plants and later are cloned to obtain a large number of similar plants (as shown in the image). This technique is more often suitable to deliver DNA into cells of the plant (Paulina et.al., 2020). Transgenic plants obtained from particle bombardment typically contain a high average copy number of the expression cassette.

Microinjection involves the direct mechanical transfer of DNA into the nucleus or cytoplasm using a fine glass micropipette. while working under the microscope the protoplasts are immobilized in low melting agar, using suction force DNA is injected and then cells are cultured in vitro and regenerated into plants.

CHEMICAL

The polymer ethylene glycol (PEG), and polyvinyl chloride in the presence of divalent cations (calcium) are used to destabilize the plasma membrane of the plant protoplast and allow the DNA to enter the cell.

STRATEGIES

There are mainly two strategies for using plants in pharmaceuticals.

Transient expression system
Stable expression systems

TRANSIENT EXPRESSION SYSTEM

The transgene by using agrobacteria through agro-infiltration is introduced into the plant cells, but the transgene will not integrate into the genome of the plant. Apart from this, viral vectors or biolistic methods are used to transfer the foreign gene. This type of system is used in pilot experiments before proceeding to costly and time-consuming stable transformations.

SUSPENSION CULTURE

This technique is used as an alternative to transient systems, which involve the removal of cell walls and gene transfer, which are allowed to form the undifferentiated callus and used as bioreactors (Sajeewani et.al.,2022). The advantage of this technique is that purification and downstream processing is cheaper and easier.

STABLE EXPRESSION SYSTEMS

In this method, the foreign gene is inserted into the plant genome. It is achieved in dicotyledonous plants by an agrobacterium and in monocot plants through a biolistic method. A recombinant protein is produced over a long period. They are further classified as follows:

Stable nuclear transformation.
Stable plastid transformation

STABLE NUCLEAR TRANSFORMATION

The foreign gene is transferred into the host nuclear genome by the agrobacterium or gene gun method. The advantage of this method when used in crop plants is the recombinant protein accumulated in seeds, which can be harvested in dry forms. (Sajeewani et.al.,2022)

STABLE PLASTID TRANSFORMATION

This system was used for the first time in a tobacco plant in 1990. The exogenous DNA is inserted into the chloroplast genome at a specific site by homologous recombination. It has numerous advantages, like plastids showing maternal inheritance, allowing multiple gene expressions in a single transformation, and high levels of recombinant proteins being produced.

PHARMACEUTICAL PRODUCTS

Recombinant Proteins

A recombinant protein is a protein formed by the transgene which is expressed in a heterologous host. These are the manipulated forms of proteins generated by various methods to produce large quantities of proteins. These proteins are a powerful tool for combating many diseases that were previously hard to treat. Plants are now being used as the expression system for recombinant glucocerebrosidase, which was manufactured in carrot cells by Protalix biotherapeutics and benefited from the lack of salicylic acid residues on the glycans are used as the therapeutics for Gaucher disease (Ari et.al.,2018). Later, transgenic *Arabidopsis* was used to produce the human epidermal growth factor (HEGF) (Weidong et.al.,2020). Agrobacterium-mediated transformation was also used to produce HEGF in barley plants (Michael et.al., 2022).

Monoclonal Antibodies

These are man-made antibodies that are similar to human antibodies. Antibodies are a crucial part of the immune system which can be produced by transforming plants with genes coding for the antibodies from mammals. How the mammalian antibodies will function the same way these plant-derived plantibodies will (Poonam et.al.,2018). Previously monoclonal antibodies are typically produced in yeast cells, insect cells, and mammalian cells. Plant systems have recently been used to produce MAbs and their derivatives. The leaves, stems, roots, and mature plants can all be used for production. The transgenic tobacco plant was used to produce the antibodies (C.Rühl et.al., 2018). The antibody genes were incorporated in *Nicotiana benthamiana* by using agrobacteria, and proteins extracted from the leaves were purified and examined. The purification is done by filtration chromatography and polymer fusion, the purified protein can be examined or evaluated by ELISA, and Western Blot methods (Ewelina et.al., 2016). These plant-produced antibodies B38 and H4 can bind the receptor binding domain protein and show neutralization activity against SARS-CoV-2(Balamurugan et.al., 2020).

Table 1. Method of transformation used to produce recombinant proteins and antibodies

SI.NO	Recombinant Protein/ Antibodies	Plant Expression System	Method of Transformation	Reference No.
1.	Recombinant Glucocerebrosidase	Carrot cells	Enzyme replacement therapy	(Ari et.al.,2018)
2.	Human Epidermal Growth Factor (HEGF)	*Arabidopsis thaliana*	Agrobacterium-mediated transformation	(Weidong et.al.,2020)
3.	HEGF	Barley *(Hordeum vulgare* L).	Agrobacterium-mediated transformation	(Michael et.al., 2022)
4.	B38 & H4 antibody for SARS-CoV-2	*Nicotiana benthamiana*	Agro-infiltration	(Balamurugan et.al., 2020)

Edible Vaccines

Is a plant-based protein that acts as a vaccine against a particular disease when consumed it stimulates the immune system. The edible vaccine is a means of transferring the antigen into the body. The protein or fragments of protein act as an antigen and produces the antibodies which remain in the immune system, and are ready to attack the virus or bacteria that enter the body, (C. Rühl et.al., 2018). There are two ways used to express the protein or the peptide in the host plant and are developed based on plant viruses. Epitope presentation systems, and polypeptide expression systems. Epitope presentation systems use "short antigenic peptides bind to the coat protein that is displayed on the surface of assembled viral particles" (Paulina et.al., 2020), in polypeptide expression systems the complete unfused recombinant protein is collected within the plant. The plant-based vaccines are having the benefit of high functionality and are highly stable. Major safety issues are not found with that heat-killed vaccine or the whole-cell antigen.

Plants have been modified by using agrobacterium-mediated transient systems to produce the vaccine in lettuce by using Hepatitis B virus antigen as HBV is capable of causing chronic liver diseases. The tobacco plant *N. benthamiana* was used to create the influenza vaccine (John et.al., 2020), as well as the human papillomavirus vaccine by using HPV virus-like particles (Paulina et.al., 2020). The HB-sAg (hepatitis -b virus antigen) expressed in the transgenic plant was been used against the Hepatitis-B induced hepatocellular carcinoma. Many approaches for plant-based vaccines against the disease was been performed and manufactured by using lettuce, tomato, maize, and rice plants (Kathleen, 2017). The Hepatitis-C virus (HCV) core protein expressed in the tobacco plant and *Brassica napus* seeds were used against HCV-induced hepatocellular carcinoma (Kathleen, 2017). The Medicago USA by using the plant-based influenza virus-like particles, an influenza vaccine has been developed which has obtained the positive result for the phase I clinical trials (Johanna et.al., 2016).

Table 2. List of vaccines under clinical trial

SI.NO	Part of the Virus Used for Vaccine	Pathogen/Disease	Plant Expression System	Method of Transformation	Reference No
1.	Hepatitis-B virus antigen	HBV- Chronic liver disease	Lettuce	Agrobacterium-mediated transient system	(Mihaela et.al., 2018)
2.	Influenza	Influenza virus	*Nicotiana benthamiana*	Agrobacterium-mediated transformation	(John et.al., 2020)
3.	HPV-Virus like particles	Human Papilloma Virus	*Nicotiana benthamiana*	Agrobacterium-mediated transient system	(Paulina et.al., 2020)
4.	HBs antigen	Hepatitis -b induced hepatocellular carcinoma	Lettuce, Tobacco, Maize, and Rice	Agrobacterium-mediated transient system	(Kathleen, 2017)
5.	Hepatitis -C virus core protein	HCV induced hepatocellular	Tobacco plant and Brassica napus	Agrobacterium-mediated transient system	(Kathleen, 2017)
6.	Influenza viruses like particle	Influenza virus	*Nicotiana benthamiana*	Magnifection	(Johanna et.al., 2016)

THE BENEFITS OF MOLECULAR PHARMING

Culturing mammalian cells or microbes for the production of pharmaceutical molecules requires a large investment and running cost. Plants are used to overcome these limits.

They reduce the upstream cost of biopharmaceuticals (Pere et.al.,2019).
They reduce the risk of contamination from animals or human pathogens (Pere et.al.,2019).
They provide rapid and high-yield production.
They are environmentally sustainable.
They help in the long-term storage of biopharmaceuticals with no quality loss.
In the case of edible vaccines, the downstream process is easier.

Applications of Molecular Pharming

Producing modern medicines from plants is an extremely attractive method that has gained consumer acceptance because of animal-free protein. (Stefan et.al.,2019)

It is not only relevant for pharmaceutical molecules but can be used in the production of enzymes in industries.

It can be used in agriculture to improve crop yield or to form resistant plants.

Limitations

Labour is quite expensive in crop plants with stable transformations. During the stationary phase, large amounts of proteins cannot be isolated from transiently transformed plants or suspension cultures.

Biosafety Issues

in healthcare systems like hospitals, diagnostic laboratories, biological laboratories, and so on. Precaution can be taken by using plants as a promising tool for producing efficient and inexpensive commercial products (Vinaya 2022). Biosafety is a protocol used to handle genetically modified organisms (GMOs) that are capable of causing harm to animals, plants, humans, and the environment. Biosafety rules and regulations help to understand and implement risk management related to GMOs.

Ecological Concerns

The introduction of GMOs into biodiversity may result in the elimination of wild varieties among the genetic population. This alters the diversity of organisms in the ecosystem and also leads to genetic contamination. GMOs will increase selection pressure on the target and non-target organisms.

Public Attitude

Consumers will react differently depending on their perception of the risks and benefits of genetically modified products. As these GMOs are produced to get benefits for mankind, researchers, scientists, media, and NGOs have to educate the people regarding their benefits.

Environmental Concern

The development of GMOs may increase the net yield or the amount of desired products. These may have a negative impact on the ecosystem by developing resistance after prolonged use, may lead to secondary pest problems, or will be a reason for the loss of traditional variety.

REFERENCES

Andrea, B., Karin, S., Diana, T., Klaus, H., Marjori, A. M., & Antionius, J. M. M. (1986). The expression of a nopaline synthase - human growth hormone chimeric gene in transformed tobacco and sunflower callus tissue. *Plant Molecular Biology*.

Ari, Z., Michael, W., Betina, H., & Gregory, M. P. (2018). Taligiucerase alfa: Safety and efficacy across clinical studies in adults and children with Gaucher disease. *Orphanet Journal of Rare Diseases*, *13*(1), 36. Advance online publication. doi:10.118613023-018-0776-8

Balamurugan, S., Kaewta, R., Suwimon, M., Arunee, T., & Waranyoo, P. (2020, November). Monoclonal Antibodies B38 and H4 Produced in Nicotiana benthamiana Neutralize SARS-CoV-2 in vitro. *Frontiers in Plant Science*. PMID:33329653

Ewelina, L., Tomasz, K., Szymon, O., & Tomasz, S. (2016). Extraction and purification methods in downstream processing of plant-based recombinant proteins. *Protein Expression and Purification*. PMID:26742898

Gleba, Y., Klimyuk, V., & Marillonnet, S. (2005). Magnifection- A new platform for expressing recombinant vaccines in plants. *Vaccine*, *23*(17-18), 23. doi:10.1016/j.vaccine.2005.01.006 PMID:15755568

Gordon & Ruddle. (1981). *Integration and stable germ line transmission of genes injected into mouse pronuclei*. doi:10.1126/science.6272397

Johanna, M., & George, P. L. (2016). Molecular pharming — VLPs made in plants. *Current Opinion in Biotechnology*, *37*, 201–206. doi:10.1016/j.copbio.2015.12.007 PMID:26773389

John, S. T. (2020). *First human efficacy study of a plant-derived influenza vaccine*. doi:10.1016/S0140-6736(20)32010-9Lancet

Kathleen H. (2017). Reconceptualizing cancer immunotherapy based on plant production systems. *Future Sci*.

Michael, P., Inger, B. O., Jon, M. B., & Henrik, B. P. (2022). *Modulation of Barely (Hordeum vulgare L.)* Grain Protein Sink-Source Relations Towards Human Epidermal Growth Factor Instead of B-hordein Storage Protein. Springer Nature.

Mihaela-O, D., Catalin, L., Lisa, P., Andre, V. E., Jihong, L. C., Catalin, T., Luliana, C., Sonya, C., Adrian, O., Vlad, T., Alexandru, B., Stephan, U., Crina, S., & Norica, B.-N. (2018). Oral administration of a chimeric Hepatitis B Virus S/preS1 antigen produced in lettuce triggers infection-neutralizing antibodies in mice. *Vaccine*, *36*(38), 5789–5795. Advance online publication. doi:10.1016/j.vaccine.2018.07.072 PMID:30082163

Paulina, N. N., & Albertha, R. (2020, December 6). Immunogenicity of Plant-Produced Human Papillomavirus (HPV) Virus-Like Particles (VLPs). *Vaccines*, *8*(4), 740. Advance online publication. doi:10.3390/vaccines8040740

Pere, M. A., Richard,, M. T., Derry,, A., & Pedro,, C. B., Merce, B., Paul, C., & Teresa, C. (2019). A simplified techno-economic model for the molecular pharming of antibodies. *Biotechnology and Bioengineering*. PMID:31209856

Peter, C. S., Ben, M. M. D., Barbara, S., Theo, C. V., Peter, J. M. V. D. E., & André, H. (1990). *Production of Correctly Processed Human Serum Albumin in Transgenic Plants*. Academic Press.

Poonam, S., Varsha, S., Anju, N., Joycee, J., Vandana, G., Ajay, R., & Smita, B. (2018). Plantibodies as biopharmaceuticals: A review. *Journal of Pharmacognosy and Phytochemistry*.

Sajeewani, R. K., & Imalka, T. (2022). Molecular Farming: Implication for Future Pharmaceutical products Sri Lankan. *Journal of Health Sciences*.

Shalini, D., Prodipta, R., & Upasana, B. (2021). Molecular pharming for the betterment of crop yielding. *International Research Journal of Plant Science*.

Stefan, S., Nicole, R., Holger, S., Stefan, R., & Matthias, B. (2019). *Critical Analysis of the Commercial Potential of Plants for the Production of Recombinant Proteins hypothesis and theory*. Academic Press.

Vinaya, R. (2022). *Metabolic engineering of plants – Molecular Pharming. Science – Policy Brief for the Multistakeholder Forum on Science, Technology and Innovation for the SDGs*. Academic Press.

Weidong, Q., Tingting, G., Xinxin, L., Jinnan, G., Muhammad, N., Yaying, L., & Yongxin, G. (2020, August 19). Molecular Pharming of the Recombinant Protein hEGF-hEGF Concatenated with Oleosin Using Transgenic *Arabidopsis*. *Genes*, *11*(9), 959. Advance online publication. doi:10.3390/genes11090959

Chapter 8
New Trends in Developing Therapeutic Proteins by Synthetic Cells on Diseased Tissues

Chittipolu Ajaykumar
https://orcid.org/0000-0001-8243-3405
Vision College of Pharmaceutical Sciences, Jawaharlal Nehru Technological University, India

Ravinder Doneti
https://orcid.org/0000-0003-1262-186X
Osmania University, India

Pasha Akbar
Osmania University, India

Venkata Satya Sai Nagasri Abhishek Chembuli
Northeastern University, USA

ABSTRACT

A number of disorders can be treated promisingly with therapeutic proteins. Pharmaceutical corporations are focused on protein-based research in their pursuit for cutting-edge, efficient treatments. The manufacturing and application of therapeutic proteins have recently advanced significantly as a consequence of the enormous attention given by researchers. Biotechnological developments have accelerated and streamlined the manufacturing of therapeutically important proteins to fight a variety of terminal illnesses. Although protein-based treatments have assumed a prominent role in drug research and development and improved human safety, a number of difficulties still exist, including problems with safety, protein stability, and immunogenicity and degradation. The authors made an effort to address all potential contributing aspects to the creation of therapeutic proteins as well as any obstacles that researchers may be facing at the moment.

DOI: 10.4018/978-1-6684-6577-6.ch008

INTRODUCTION

Until recently small molecules like antibiotics, hormones, analgesics and pharmaceuticals which synthesized and produced from microbes but due to technological advancements in recent times planning for biotech drugs. Capable of treating diseases in precise manner, proteins can be engineered in laboratories including their bonds, sequence of amino acids and there dimensional arrangements (Chenthamara et al., 2019). Since with knowledge on proteins are capable of preventing and curing diseases and disorders. The therapeutic proteins can be relieved from many conditions such as cancer, strokes, heart attack, anaemia, haemophilia and cancer. It is well acknowledged that proteins and peptides offer excellent therapeutic potential for treating a variety of illnesses and disorders (Falanga et al., 2013; Haisma & de Hon, 2006). Additionally, the market value and availability of protein- and peptide-based medicines have expanded as a result of recent advancements in pharmaceutical biotechnology. Numerous recombinant proteins have received FDA approval as biotechnology drugs, vaccines, enzymes, natural and recombinant cytokines, and interferon's. Because these recombinant therapeutic proteins serve a variety of functions, including illness diagnostics, prevention, disease management, and/or treatment (Mitragotri et al., 2014; Usmani et al., 2017).

Furthermore, it is known that one-third of the entities introduced as novel medications under the category of biologics are known as protein therapies. For the treatment of chronic disorders, recombinant technology has created a variety of protein medications that are now freely accessible as therapies at affordable prices. Because they have been successful in treating many potentially deadly diseases including diabetes, heart conditions, and cancer, therapeutic proteins are becoming more and more popular. It's interesting to note that proteins have now been shown to be useful as vaccinations that aid in triggering the immune system's natural defences (Akash et al., 2015; Mitragotri et al., 2014; Prado et al., 2017). These therapeutic proteins have demonstrated their effectiveness in the pharmaceutical sector through the cloning and expression of cDNA that encodes heterologous proteins. But in order to distribute them to the intended areas safely and effectively, a crucial formulation is needed. To do this, it is crucial to comprehend all of the fundamental specifications for the creation and administration of protein-based medicines. Immune responses and other undesirable effects have reportedly been brought on by various protein-based therapies (Akash et al., 2015; Hu et al., 2006; Uribe-Echeverry & Lopez-Gartner, 2017).

We provide in-depth insights into a number of fundamental elements in the current book chapter that may be crucial in making proteins therapeutic and may affect their synthesis, formulation, or delivery as medications. We will jointly explore the fundamentals, synthesis, method of administration, and possible effects of both proteins and peptides as medicines because they have both demonstrated successful therapeutic outcomes against a variety of medical situations.

TYPES OF THERAPEUTIC PROTEINS

The proteins and peptides have known for several properties they differ in certain features such as length of polymers, the count of amino acids were <50 in case of peptides but the minimum chain length were found to be >100 amino acids.

As shown in Figure 1, they divided them into groups according to whether they had been FDA-approved. Proteins in Group Ib, which include erythropoietin for the treatment of chronic anaemia and hepatitis C, are known to induce a variety of haematological and body immunological responses. Group

Ic proteins mainly utilized therapeutically for changing disease pathophysiology; notable examples are Botulinum toxin for dystonia and lepirudin for thrombocytopenia. Group IIa are known to directly attach to chemicals or organisms and either activate a signalling pathway or inhibit it. One such is cetuximab, a human monoclonal antibody that has been used to treat head neck and colorectal related malignancies. Anakinra used to active rheumatoid arthritis, also been considered Group IIa drug. The compounds like denileukin diftitox and Ibritumomab tiuxetan can be used to treat persistent cutaneous T-cell lymphoma, non-lymphoma Hodgkin's respectively. These compounds have been listed under Group IIb. Group III employed as vaccines include hepatitis B surface antigen and anti-rhesus (rh) immunoglobulin G vaccination, used in regular postpartum prevention of Rh(D) immunisation in Rh(D)-negative women (Deutsch & Tomer, 2013; Hu et al., 2006; Huang et al., 2010; Jäger et al., 2020; Tayal & Kalra, 2008).

Last but not least, proteins that belong to Group IV are employed as diagnostic tools, such as imaging agents for cancer detection and diagnostics for infectious diseases. In conclusion, every protein is unique place among biomedical goods with the potential to be useful in every area linked to biologics, but notably recombinant human proteins that have the benefit of being FDA-approved. A variety of prokaryotic and eukaryotic based therapeutic proteins developed described above. For instance, the most often employed therapeutic proteins in the treatment of various diseases are L-glutaminase (an antileukemia enzyme), L-asparaginase (a chemotherapeutic enzyme), methionine gamma-lyase (perhaps an anticancer agent) (Deutsch & Tomer, 2013; Hu et al., 2006; Huang et al., 2010; Jäger et al., 2020; Tayal & Kalra, 2008).

Production and Purification of Therapeutic Proteins from Synthetic Cells

Significant improvements have been achieved in the production, purification, and sourcing procedures for therapeutic proteins. Low-molecular-weight proteins are often synthesised artificially, whereas living cells produce proteins with a high level of amino acids. Utilizing recombinant DNA technology, a large number of proteins are created, for instance using mammalian Chinese hamster ovary cells (CHO). Some of the proteins made with CHO include interferons, factor VIII, erythropoietin (Burnett & Burnett, 2020; Mohammadinejad et al., 2019; O'Flaherty, Bergin, Flampouri, Mota, Obaidi, Quigley, Xie, & Butler, 2020).

Recombinant therapeutic proteins must be manufactured in high quality in order to be used in humans. Yeasts, insects, bacteria and mammalian cells are used as expression system to produce therapeutic proteins with a human-like glycosylation pattern. Recently, metabolic engineering has also made use of gene-editing tools to produce successful clones and goods which can able to synthesize the therapeutic proteins even in the tissues itself against the diseases. Technological advances in cell engineering such as CRISPR-Cas techniques, ribosome engineering RNAi are capable as efficient in synthesis of therapeutic proteins. Specifically a new generation CRISPR/Cas9, zinc finger nucleases (ZFNs), transcription activator-like effector nucleases (TALENs), and recombinase-mediated cassette exchange (RMCE) are all viable methods for gene editing tools and related advancements have been described in terms of genetic modification, with a focus on the use of CRISPR/Cas9 for the "multiplexing gene-editing approach" for genetic modification of yeast and CHO cells, which ultimately results in rapid product development with consistency, improved product yield, quality, and affordability (Burnett & Burnett, 2020; Frenzel et al., 2013; Gupta & Shukla, 2018).

Table 1. List of expression cells can produce a variety of therapeutic proteins

Expression Cells	Therapeutic Proteins	Therapeutic Application
CHO cell	Interferon beta Factor VIII	Sclerosis Haemophilia
E. Coli	Insulin Human growth factor Interferon alpha Interleukin-1 Interleukin-2 Interferon beta Interferon gamma Streptokinase	Diabetes Hypo pituitary dwarfism Leukaemia, cancers, hepatitis Auto-immune diseases Renal carcinoma Sclerosis Granulomatous disease Myocardial infraction
E.Coli, Bacillus sp., Ercinia sp. Saccharomyces cerevisiae Pseudomonas sp., & putida	L-asparaginase Hepatitis vaccine Methionine gamma-lyase	Mast cell tumour Hepatitis cancers

The E coli & CHO cells as microbial host cells for synthesis of therapeutic proteins. The most advanced system's E coli due to its inexpensive active growth, well known genetics & cellular functions & productivity scale, some of the drawback can be expected from the approach such as inclusion body formation, codon bias, endotoxin production & post-translational changes. Some of the new approaches such as cofactor supplementation, co-expression with molecular chaperones & fusion tag insertion into the cells can prevent formation of inclusion bodies soluble expression of proteins (Akash et al., 2015). Along with baceterial cells Yeast based therapeutic proteins are considered due to their quick development, ease of genetic modification, need for less expensive growth media, availability of gene sequences, and capacity to produce PTMs, yeasts make good candidates for use as expression hosts for therapeutic proteins. The expressed proteins shown extracellular expressions and codon bias. Recombinant proteins produced in this manner exhibit codon bias and extracellular expressions. The most popular expression host systems for the generation of recombinant biopharmaceuticals has a long history of usage in the manufacture of medicines for human use (Burnett & Burnett, 2020; Frenzel et al., 2013; Gupta & Shukla, 2018; Kim et al., 2015).

In order to increase the cellular expression of therapeutic proteins to gram level/litre changes in trafficking & secretory processes had changed which alter the endosome and Golgi trafficking. Chen et al. provided a detailed account of the creation of a synthetic biology toolbox and how it has been used in the fields of drug manufacture and screening (2018). Recombinant protein accumulates intracellular as a result of overexpression, which lowers product titters. When employed as a therapy, hyper mannosylation of proteins causes a quicker blood clearance (Wang et al., 2014).

TRANSGENIC ANIMALS AND PLANTS

Numerous human therapeutic proteins have been produced using animal cell cultures as well. In contrast to the microbial manufacturing approach, animal cell cultures are capable of posttranslational modifications and may produce proteins with physiological activity. Animal cell bioreactors are quite expensive since the process requires a rich growth media. However, the transgenic technology gaining much attention due to similar properties, post-translational modifications & can produce various therapeutic proteins. For instance, the transgene of interest is released in the animal's milk after being expressed in its mammary glands, which include those of other animals like sheep and goats (Corchero et al., 2013; Gomes et al., n.d.).

Transgenic animals have been used to express recombinant protein-based treatments as growth factors, enzymes, vaccines and monoclonal antibodies. Transgenic animals have a transgene inserted into their genome that codes for a recombinant protein, and they are able to pass this gene on to their progeny. Modern protein supply methods include using transgenic animal milk and transgenic chicken eggs. In this method, recombinant proteins naturally secrete themselves, producing the right PTMs. However, the ethicality of creating transgenic animals is under doubt. The protein products made from transgenic animals may include zoonotic pathogens (Desai et al., 2010; Khodarovich et al., 2013; Pogue et al., 2002).

In addition to microbial host and transgenic animals, several transgenic plants have been used to manufacture different types of therapeutic proteins. The DNA of the protein is put into the plant genome in order to create a sizable amount of the desired protein (Desai et al., 2010).

It is possible to increase the production of recombinant biopharmaceuticals using transgenic plants which are safe, low in cost, scale of production, stability of protein, neutral metabolites & can produce glycosylated protein. The kind of production can be a great asset in creating cancer immunotherapy drugs. To get good quality of therapeutic protein we must take in counts of nature of plant, sub-cellular localization of protein expression, expression cassettes, post translation modifications, extraction capability and purification methods (Desai et al., 2010; Donini & Marusic, 2019; Gerszberg & Hnatuszko-Konka, 2022; Owczarek et al., 2019).

In-depth research on glycoengineering has been done to lessen the immunogenicity of recombinant proteins produced in plants. Pesticides, herbicides, and product contamination with hazardous plant metabolites are drawbacks of this system. The rigorous purification procedure and the management of the transgenic expression level are two additional difficulties with this approach. Recombinant proteins are primarily produced via plant tissue-based methods, plant cell cultures, and transgenic plant building. The transgene is often inserted directly into the plant cells using direct methods like biolistic bombardment or the PEG-mediated technology, bacterial infection (agroinfection), viral infection, or both. Recombinant protein expression in the chosen cell compartment or plant organ is one of the main benefits of these expression techniques. Human therapeutic proteins made in plants frequently have a glycosylation pattern more akin to plants than humans (Bahera et al., 2021).

Insect Cells

Insect cells gaining ground rapidly as a platform in production of therapeutic proteins with several comparative advantages with mammalian call includes ease culture, osmolar tolerance, and level of expression when infected with gene expression systems. Various therapeutic proteins are also expressed using insect cell as host cells. The baculovirus expression system is one of the widely used expressions in insect cells. After reaching the required cell density, the insect cells are infected by baculovirus expression gene even though the glycosylation process is also shows similarity but distinct with mammalian system can overcome with mammalian glycosyltransformants in baculoviruses or by introducing desired gene with mammalian n-glycosyltransferases in virus or insect cells (Golfam et al., 2021; Sproles et al., 2021; Zhu, 2012).

Table 2. List of transgenic plants produces the therapeutic proteins along with their application

Host	Therapeutic Proteins	Application
Tobacco	Protein C Chloroplast Interferon's Serum albumin Epidermal growth factor Lactoferrin Angiotensin converting enzyme Glucocerebrosidase	Anticoagulant Growth hormone Hepatitis B&C Surgery, burns and cirrhosis Wound repair Antimicrobial Hypertension Gaucher's disease
Rice	Interferon alpha and beta	Hepatitis B&C
Maize	Aprotinin	Transplant surgery
Turnip	Interferon alpha & Beta	Hepatitis B&C
Canola	Hirudin	Antithrombotic
Arabidopsis	Enkephalins	Antihyperanalgesic

Table 3. List of transgenic animals produces the therapeutic proteins along with their applications

Host	Therapeutic Proteins	Applications
Sheep, pig	Plasminogen activator	Thrombosis
Sheep, cow and pig	Factor Vii, IX	Haemophilia
Sheep, cow	Fibrinogen	Wound healing
Sheep, cow	Human serum albumin	Maintain blood volume
Goat, cow and chicken	Monoclonal antibodies	Vaccine
Cow	Lactoferrin	GI infection
Goat	Antithrombin III	Thrombosis
Sheep	Anti-trypsin	Emphysema
Rabbit	Glucosidase	Pompe's disease
Goat	Protein C	Thrombosis
Goat	Pro 542	HIV

The cell lines like S2, Sf21, High Five TM cells and Sf9 cells are employed. The productivity of proteins can be achieved by vector development, transcription, translation strategies, proteolysis, prevention and biomass are effects the nature of the expressed proteins (Roldão et al., 2014). Recently gene silencing by cellular mechanisms & silencing the vector gene & cell based engineering gaining importance for expression of therapeutic protein in the insect cells (Cid & Bolívar, 2021; O'Flaherty, Bergin, Flampouri, Mota, Obaidi, Quigley, Xie, & Butler, 2020).

Table 4. List of expression cell systems produces the therapeutic proteins along with their details for production and purifications

Host	Cost for Production	Storage Cost	Glycosylation	Pathogenicity	Time Scale for Production
Mammalian cells	High	Expensive	Yes	High	Long
Insect cells	High	Expensive	Yes (Minor differ)	Medium	Long
Yeast	Medium	Expensive	Yes(minor differ)	Low	Medium
Bacteria	Low	Expensive	No	Medium	Short
Transgenic plants	Low-medium	Low	Yes(Minor differ)	Low	Long
Transgenic animals	High	Expensive	Yes	High	Very long

EFFECT OF THERAPEUTIC PROTEINS ON HEALTH CARE

Therapeutic proteins' impact on healthcare A number of potentially fatal diseases, including different types of cancer, diabetes mellitus, and auto-immune diseases, can be effectively treated with therapeutic proteins. This has been shown to have a significant positive impact on the expanding field of study focused on the results and effective distribution of these therapeutic proteins appears to offer the potential to produce ground-breaking treatments for a number of diseases. It appears that the future will see the eradication of many deadly diseases thanks to the employment of therapeutic proteins (Akash et al., 2015; Khuroo, 2020; Mukherjee et al., 2018).

CELL FREE EXPRESSION SYSTEMS

The activation of transcription and translation utilising unprocessed cellular extracts rather than whole cells is known as cell-free gene expression (CFE). Protein synthesis may be started in vitro utilising CFE by supplementing the extracts with exogenous resources such as amino acids, nucleotides, and a high-energy phosphorylated substrate since cellular extracts maintain the original cellular transcriptional and translational machinery. CFE allows for unprecedented control over the molecular setting for gene expression and metabolism by removing the restriction of maintaining life. This allows the experimenter to directly alter the system, such as by adding non-native substrates, purified proteins or RNAs, or recombinant DNA templates. Additionally, cell-free experiments have the unique ability to concentrate resource utilisation on a particular genetic network or the biosynthesis of a single product. They can also get around restrictions on molecular transport across the cell wall and circumvent mechanisms that have evolved to help species survive (Brooks & Alper, 2021; Campo, n.d.).

The cell wall, a highly selective barrier that isolates processes of interest from direct user input, poses a significant obstacle to the majority of biochemical studies of cellular activity. In contrast, CFE devices provide direct biological function probing by providing experimenters with unrestricted access to the reaction environment. Thus, an important application area for CFE systems has emerged: the capacity to prototype discrete genetic components or intricate genetic designs (such genetic response networks or metabolic pathways) in vitro before putting them into use in cells (Khuroo, 2020; Lamnabhi-Lagarrigue et al., 2017; Majumder, n.d.). As a biological equivalent of combinatorial synthetic chemistry, these studies may be carried out on smaller reaction sizes and automated to enable quick testing of huge combinations of genetic components. In this section, we go through how to utilise CFE to prototype ever more complicated biological functions, starting with libraries of basic genetic components and ending with the creation of synthetic minimum cells. Few well-known genetic components (including promoters, ribosome-binding sites, and transcriptional terminators) hinder the rational and predictable engineering of biological systems, especially for non-model or challenging-to-culture hosts. Even well-characterized components typically fail to perform as predicted in living cells due to pathway crosstalk, or unexpected action from endogenous regulators of gene expression. CFE systems have begun to address these problems by enabling rapid, high-throughput in vitro prototyping procedures. Early proof-of-concept studies showed that T7 bacteriophage transcription could be monitored for elongation and termination using cell-free reporter assays (Vojvoda et al., 2022). Cell-free prototyping of more complex regulatory elements, such as quorum-sensing promoters, light-sensing promoters, allosteric transcription factors, synthetic translational riboregulators, and riboswitches, became widely available following later demonstrations that native bacterial polymerases generated enough mRNA in vitro to activate cell-free protein synthesis.

FACTORS AFFECTING THE THERAPEUTIC DRUG DELIVERY

The current clinical applications of therapeutic proteins are limited due to attributes for absorption & distribution which impact on the pharmacokinetics and pharmacodynamics so it is much important to delivery of synthesized proteins in formulation, storage and delivery methods. Proteins which have less proteolytic activity in GIT can use protease inhibitors or by carrier systems for example the oral delivery of insulin can be achieved by formulating a hexyl-insulin- amphiphillic polymer for oral drug delivery (Sivadasan et al., 2021).

The nasal mucosa secretions are rich in vasculature form can rapidly absorb the proteins but due to their small route which made impossible to absorb form this route to overcome this administered through adeno-associated virus vector in efficient transfer of gene to lung cells produce efficient therapeutic proteins transfer to blood systems directly (Mukherjee et al., 2022; Sivadasan et al., 2021). Extensively all proteins can be efficient in cloning parenteral route with maximum absorption and escapes from enzymatic & penetration barriers but the limitation comes from the half-life of the drug. The conditions counter by creating sustainable formulation of polymers, liposomes, implants, nanoparticles and oil based injections (Dąbrowska & Abedon, 2019; Jain, 2020).

For instance drugs like leuprolide acetate made into polymer poly (D-L, lactide-co-glycolide) approved by US FDA for sustained release and effectively releasing therapeutic proteins. In recent advancements focus shifted for local & systemic delivery at same time for that the best suited delivery by pulmonary route increase the rate of absorption due to large surface are, exclude first pass metabolism (Butreddy et al., 2021). Therapeutic proteins must delivered in dry form for stability reason (dry inhalers) used for TB vaccine, IgG, insulin, peroxide, calcitonin, hormones like insulin, growth hormone, parathyroid hormone delivered by this route only (Sun et al., 2017).

CHALLENGES AND FUTURE DIRECTION

A significant number of therapeutic proteins have been developed and synthesised over the course of many decades thanks to advancements and improvements in DNA-based genetic engineering technologies, which are being used to treat a wide range of life-threatening illnesses and disorders. Pharmaceutical biotechnology and genetic engineering have had a considerable influence on the healthcare system and are expected to grow in importance in the future, but there are still many problems that need to be solved in the here and now and in the future. We go through these issues and how to make therapeutic proteins the optimal treatments in the next subsections. Problems with development Pharmaceutical scientists are now only allowed to concentrate on molecules that have the greatest potential for success in clinical trial programmes due to the drastically increased overall budget and timescales for the creation of therapeutic proteins.

About 10 years pass between the discovery of novel compounds and their introduction to the pharmaceutical market, although the probability of success (POS) is not a guarantee. Small compounds have a POS of 6–7%, but monoclonal antibodies and fusion proteins have a POS of over 17%. (Shih, 2012) Therefore, it is crucial to choose molecules with maximal POS problems with safety and immunogenicity there is fierce competition and it is challenging to distinguish the most effective therapeutic product among comparable goods since several therapeutic products have been produced to modify the same therapeutic target and/or pathway (Shih, 2012; Strohl, 2009). As an illustration, there are now five anti-TNF therapeutic proteins on the market: Enbrel, Cimzia, Remicade, Humira, and Simponi. An autoinjector is used to deliver SC doses of each of these medications on a monthly basis (Strohl, 2009). The safety and clinical efficacy of these therapeutic proteins—which may be attained by a variety of parameters such as disease condition, target biology, potency, safety margin, dose, and patient population selection—are their most crucial qualities. The most effective treatment agent may be distinguished from the others based on its safety and clinical efficacy (Presta, 2008; Strohl, 2009).

In terms of safety and immunogenicity, the following variables should be taken into account: Molecules that offer the greatest margin of safety are priority number one, followed by delivery methods, longer half-lives, tissue distribution, enzymatic stability, and molecular solubility underdevelopment (Ashford, England, & Akhtar, 2021; Posner et al., 2019; Williams, Layfield, Vallance, Patel, Bicknell, Trim, & Vaiyapuri, 2019). The treatment of diseases and the effectiveness of medications are now significantly hampered by immunogenicity. In a therapeutic protein formulation, even a trace quantity of particle material, such as protein aggregates, is thought to trigger immunogenicity. Nearly all therapeutic proteins, including those with just human sequences, can cause some patients to develop neutralising or nonneutralizing antibodies. Increased neutralising antibodies to particular therapeutic proteins and peptides are being employed as a kind of therapy for these individuals (Williams, Layfield, Vallance, Patel, Bicknell, Trim, & Vaiyapuri, 2019). However, a significant problem with these antibodies is a decrease in the therapeutic effectiveness of the injected protein. Due to the creation of neutralising antibodies, the quick clearance of the protein therapy, and/or its altered tissue distribution, it can happen at any time. Life-threatening situations are the clinical effects of immunogenicity in individuals receiving therapeutic proteins. Therapeutic proteins administered to patients neutralise endogenous proteins, resulting in long-term harmful toxicities. Inflammation, hypersensitivity, and moderate skin reactions to severe anaphylaxis are other immunological responses. Some of these immunological responses can be effectively controlled by changing the dose schedule or by co-administering corticosteroids to reduce inflammation. Patients' immunogenicity is influenced by a number of variables, including: 1) structural characteristics (glycosylation and sequence variation); 2) impurities present during preparation; 3) storage conditions (denaturation, or aggregation brought on by oxidation); 4) dosage and duration of treatment; 5) administration route; 6) formulation type; and 7) patients' genetic characteristics (Schellekens, 2005; Williams, Layfield, Vallance, Patel, Bicknell, Trim, & Vaiyapuri, 2019). Recent developments in pharmaceutical biotechnology and genetic engineering have directly improved the clinical safety of protein therapies. For instance, the development of entirely human antibody therapies and the humanization of rodent antibodies have now substantially bypassed immunogenic responses brought on by the introduction of nonhuman antibodies. Pharmaceutical researchers are always looking for T-cell epitopes in protein therapies that can be used to increase immunological tolerance by activating Treg cells (Lei, Zhang, Fang, Akash, Shi, Sun, Xu, & Chen, 2013; Lei, Zhang, Fang, Akash, Shi, Sun, Xu, & Chen, 2013). Although it is possible for these two to reduce unfavourable immunogenic responses, it is yet unknown if they will assist reduce the likelihood of immunogenicity seen with antibody-based therapies. Immunogenicity related to protein treatments cannot be properly predicted using recognised analytical techniques. To pick a molecule with a low immunogenic profile as a clinical treatment candidate, pharmaceutical experts continually make an effort to evaluate the potential for immunogenicity during the discovery phase (Jiskoot et al., 2012; Lei, Zhang, Fang, Akash, Shi, Sun, Xu, & Chen, 2013). Protein stability is the most rapidly expanding category of pharmaceutical active ingredients now being employed in a variety of clinical contexts is therapeutic proteins. But their stability prevents them from being the best clinical medicines. When kept in the high concentrations needed for their use, therapeutic proteins have a tendency to agglomerate. The overall activity of therapeutic proteins tends to decline with accumulation, and they can cause immune responses. In order to engineer antibodies for stability, one can execute target mutations of aggregation-prone areas by determining the size and position of those sites in protein formulations. The temperature may also have an impact on the stability of therapeutic proteins. Extracted proteins are often kept in storage for a long time to preserve their functionality and original structural integrity. Typically, proteins should be kept at or below 4 °C. Therapeutic proteins frequently degrade when stored at room temperature. Another aspect of protein instability is the creation of protein particles during in vivo sustained release, which can be caused by protein instability during sustained delivery (Akash, Rehman, & Chen, 2014; Akash, Rehman, Sun et al, 2014; Jiskoot et al., 2012). In contrast to the immunogenicity of traditional protein therapies, this may trigger an immune response in individuals receiving treatment with prolonged release formulations of protein therapeutics. The stability of therapeutic proteins is being increased using a variety of techniques. The first strategy is to change the protein's amino acid sequences. The second is to improve how therapeutic proteins are made. Therapeutic proteins' stability has been found to be improved by thermosensitive polymers. A range of medicinal compounds have been successfully delivered using biodegradable polymers, which have undergone extensive evaluation. Nontoxic nanostructured materials can also improve protein stability. Degradation of proteins is another difficulty that arises at various times during the development and distribution of the medicine to its target areas is protein breakdown. Aggregation is a significant route for protein breakdown during the production and storage of therapeutic proteins. When proteins are not in their natural settings, they frequently become unstable (Jiskoot et al., 2012; Lei, Zhang, Fang, Akash, Shi, Sun, Xu, & Chen, 2013). Extracellular fluids and cell compartments have a wide range of conditions. Therapeutic proteins may not work as intended if specific parameters are not upheld during storage. Proteolysis, aggregation, and poor buffer conditions all have the potential to reduce the therapeutic action of proteins. Understanding and measuring different chemical degradation pathways such oxidation, deamidation, chemical cross-linking, disulfide changes, and fragmentation is crucial. To stop therapeutic proteins from degrading, many tactics are employed. The most popular method involves encasing therapeutic proteins in inert, biocompatible polymers. Enzyme inhibitors can be given along with therapeutic proteins to limit protein breakdown. Protein-excipient interactions, section is altering the amino acid sequences in the protein and/or improving the formulation of therapeutic proteins using various types of polymers, it is possible to halt the immunogenicity and instability of therapeutic proteins. Therapeutic proteins must be incorporated into polymers that are inert, biocompatible, and preferably biodegradable in nature. Protein-excipient interactions must be thoroughly assessed throughout the formulation development of therapeutic proteins in order to improve the overall stability of protein therapies and reduce the likelihood of any sort of immunogenicity. Pharmaceutical researchers are assessing the interactions between proteins and excipients using a variety of methodologies (Akash, Rehman, Sun et al, 2014; Jiskoot et al., 2012). Pharmaceutical researchers are always battling to stop therapeutic proteins from being quickly eliminated and being broken down by the hepatic first-pass metabolism. Numerous attempts have been made to get around these obstacles thus far. These initiatives include the non-invasive administration of therapeutic proteins using methods (other than the oral route) that avoid the delivered therapeutic proteins' hepatic metabolism (Kamerzell et al., 2011). Additionally, invasive therapeutic protein administration is another method that may be used to stop hepatic first-pass metabolism. Additionally, the biological half-lives of the majority of therapeutic proteins are brief. This issue may be resolved by applying fusion protein technology to increase the half-life of therapeutic proteins, as well as by encapsulating and/or conjugating desirable therapeutic proteins with biocompatible polymers. It has recently been successful to encapsulate IL-1Ra, a naturally occurring anti-inflammatory antagonist of pro-inflammatory cytokines, in PF127. Studies conducted in both the lab and in animals have revealed that PF127 retains the stability, bioactivity, and therapeutic potentials of IL-1Ra while also greatly extending its sustained release (Akash, Rehman, & Chen, 2014; Ashford, England, & Akhtar, 2021; Jiskoot et al., 2012; Williams, Layfield, Vallance, Patel, Bicknell, Trim, & Vaiyapuri, 2019).

CONCLUSION

Therapeutic proteins have become more important as a kind of treatment for a variety of disorders during the past few of decades. A rising number of therapeutic proteins' therapeutic properties are accompanied by techniques for their manufacture and purification that have grown in significance, making these beneficial agents available on the pharmaceutical market. There are many various types of synthetic cells being used, including bacterial cells, insect cells, plant cells, and animal cells, all of which have the capacity to produce therapeutic proteins. With advances in biotechnology and genome-wide analysis, protein production has become simpler. To improve the quality, safety, and effectiveness of these agents, several parameters still need to be carefully taken into account during the synthesis, purification, and formulation of therapeutic proteins. This indicates that a suitable and logical formulation of protein-based treatments is needed in order to utilise these therapeutic proteins for the treatment of diverse ailments. Additionally, a deeper comprehension of the pharmacokinetic characteristics of therapeutic proteins and how these characteristics relate to the pharmacodynamic effects of these agents would promote the creation and effective administration of therapeutic proteins and peptides. An accurate prediction of the biodisposition of these drugs in both clinical and nonclinical situations requires a thorough examination of the characteristics and variables that might impact these parameters. Biotechnological developments have accelerated and streamlined the manufacturing of therapeutically important proteins to fight a variety of potentially deadly illnesses. To take advantage of these potential, a wide range of new technologies for the creation and improvement of protein therapies are predicted. Future developments in protein treatments might lead to significant improvements in synthesis, effectiveness, safety, immunogenicity, and delivery.

ABBREVIATIONS

FDA - Food and Drug Administration
cDNA - complementary DNA
CHO Cells - Chinese hamster ovary cells
ZFNs - Zinc finger nucleases
TALENs - Transcription activator-like effector nucleases
CRISPR/Cas9 - Clustered regularly interspaced short palindromic repeats and CRISPR- associated protein 9.
RNAi - RNA interference
PTMs - Posttranslational modification
CFE - Cell-free gene expression
DNA - Deoxyribonucleic acid
GIT - Gastrointestinal tract
Anti-TNF - Anti-Tumour necrosis factor
IL-1Ra - Interleukin-1Ra
PF - Pluronic F-127

REFERENCES

Akash, M. S., Rehman, K., & Chen, S. (2014, July 1). Effects of coffee on type 2 diabetes mellitus. *Nutrition (Burbank, Los Angeles County, Calif.), 30*(7-8), 755–763. doi:10.1016/j.nut.2013.11.020 PMID:24984989

Akash, M. S., Rehman, K., Sun, H., & Chen, S. (2014, May 1). Assessment of release kinetics, stability and polymer interaction of poloxamer 407-based thermosensitive gel of interleukin-1 receptor antagonist. *Pharmaceutical Development and Technology, 19*(3), 278–284. doi:10.3109/10837450.2013.775158 PMID:23506246

Akash, M. S., Rehman, K., Tariq, M., & Chen, S. (2015). Development of therapeutic proteins: Advances and challenges. *Turkish Journal of Biology, 39*(3), 343–358. doi:10.3906/biy-1411-8

Ashford, M. B., England, R. M., & Akhtar, N. (2021, May). Highway to success—Developing advanced polymer therapeutics. *Advanced Therapeutics, 4*(5), 2000285. doi:10.1002/adtp.202000285

Bahera, B. K., Prasad, R., & Behera, S. (2021). *Therapeutic Proteins: Production and Delivery. InLife Sciences Industry*. Springer.

Brooks, S. M., & Alper, H. S. (2021, March 2). Applications, challenges, and needs for employing synthetic biology beyond the lab. *Nature Communications*, *12*(1), 1–6. doi:10.103841467-021-21740-0 PMID:33654085

Burnett, M. J., & Burnett, A. C. (2020, March). Therapeutic recombinant protein production in plants: Challenges and opportunities. *Plants, People, Planet*, *2*(2), 121–132. doi:10.1002/ppp3.10073

Butreddy, A., Gaddam, R. P., Kommineni, N., Dudhipala, N., & Voshavar, C. (2021, August 18). PLGA/PLA-Based long-acting injectable depot microspheres in clinical use: Production and characterization overview for protein/peptide delivery. *International Journal of Molecular Sciences*, *22*(16), 8884. doi:10.3390/ijms22168884 PMID:34445587

Campo, A. M. (n.d.). *NMR Metabolomics for Optimizing Cell-Free Protein Synthesis* [Doctoral dissertation]. Wright State University.

Chenthamara, D., Subramaniam, S., Ramakrishnan, S. G., Krishnaswamy, S., Essa, M. M., Lin, F. H., & Qoronfleh, M. W. (2019, December). Therapeutic efficacy of nanoparticles and routes of administration. *Biomaterials Research*, *23*(1), 1–29. doi:10.118640824-019-0166-x PMID:31832232

Cid, R., & Bolívar, J. (2021, July 21). Platforms for production of protein-based vaccines: From classical to next-generation strategies. *Biomolecules*, *11*(8), 1072. doi:10.3390/biom11081072 PMID:34439738

Corchero, J. L., Gasser, B., Resina, D., Smith, W., Parrilli, E., Vázquez, F., Abasolo, I., Giuliani, M., Jäntti, J., Ferrer, P., Saloheimo, M., Mattanovich, D., Schwartz, S. Jr, Tutino, M. L., & Villaverde, A. (2013, March 1). Unconventional microbial systems for the cost-efficient production of high-quality protein therapeutics. *Biotechnology Advances*, *31*(2), 140–153. doi:10.1016/j.biotechadv.2012.09.001 PMID:22985698

Dąbrowska, K., & Abedon, S. T. (2019, October 30). Pharmacologically aware phage therapy: Pharmacodynamic and pharmacokinetic obstacles to phage antibacterial action in animal and human bodies. *Microbiology and Molecular Biology Reviews*, *83*(4), e00012–e00019. doi:10.1128/MMBR.00012-19 PMID:31666296

Desai, P. N., Shrivastava, N., & Padh, H. (2010, July 1). Production of heterologous proteins in plants: Strategies for optimal expression. *Biotechnology Advances*, *28*(4), 427–435. doi:10.1016/j.biotechadv.2010.01.005 PMID:20152894

Deutsch, V. R., & Tomer, A. (2013, June). Advances in megakaryocytopoiesis and thrombopoiesis: From bench to bedside. *British Journal of Haematology*, *161*(6), 778–793. doi:10.1111/bjh.12328 PMID:23594368

Donini, M., & Marusic, C. (2019, March). Current state-of-the-art in plant-based antibody production systems. *Biotechnology Letters*, *41*(3), 335–346. doi:10.100710529-019-02651-z PMID:30684155

Falanga, A., Marchetti, M., & Vignoli, A. (2013, February). Coagulation and cancer: Biological and clinical aspects. *Journal of Thrombosis and Haemostasis*, *11*(2), 223–233. doi:10.1111/jth.12075 PMID:23279708

Frenzel, A., Hust, M., & Schirrmann, T. (2013, July 29). Expression of recombinant antibodies. *Frontiers in Immunology, 4*, 217. doi:10.3389/fimmu.2013.00217 PMID:23908655

Gerszberg, A., & Hnatuszko-Konka, K. (2022, March 17). Compendium on food crop plants as a platform for pharmaceutical protein production. *International Journal of Molecular Sciences, 23*(6), 3236. doi:10.3390/ijms23063236 PMID:35328657

Golfam, R., Kiarostami, K., Lohrasebi, T., Esfahani, K., & Ghazvini, H. (2021, June 1). Application of molecular farming based on barley plant (Hordeum vulgare L.) platform: A Review. *Research on Crops, 22*(2).

Gomes, A.R., Byregowda, S.M., Veeregowda, B.M., & Balamurugan, V. (n.d.). *An overview of heterologous expression host systems for the production of recombinant proteins*. Academic Press.

Gupta, S. K., & Shukla, P. (2018, December). Glycosylation control technologies for recombinant therapeutic proteins. *Applied Microbiology and Biotechnology, 102*(24), 10457–10468. doi:10.100700253-018-9430-6 PMID:30334089

Haisma, H. J., & de Hon, O. (2006, May). Gene doping. *International Journal of Sports Medicine, 27*(04), 257–266. doi:10.1055-2006-923986 PMID:16572366

Hu, W., Wang, Y., & Zhu, Z. (2006, January). A perspective on fish gonad manipulation for biotechnical applications. *Chinese Science Bulletin, 51*(1), 1–6. doi:10.100711434-005-1055-3

Huang, C. Jr, Lowe, A. J., & Batt, C. A. (2010, June). Recombinant immunotherapeutics: Current state and perspectives regarding the feasibility and market. *Applied Microbiology and Biotechnology, 87*(2), 401–410. doi:10.100700253-010-2590-7 PMID:20422181

Jäger, U., Barcellini, W., Broome, C. M., Gertz, M. A., Hill, A., Hill, Q. A., Jilma, B., Kuter, D. J., Michel, M., Montillo, M., Röth, A., Zeerleder, S. S., & Berentsen, S. (2020, May 1). Diagnosis and treatment of autoimmune hemolytic anemia in adults: Recommendations from the First International Consensus Meeting. *Blood Reviews, 41*, 100648. doi:10.1016/j.blre.2019.100648 PMID:31839434

Jain, K. K. (2020). An overview of drug delivery systems. *Drug Delivery System*, 1–54. PMID:31435914

Jiskoot, W., Randolph, T. W., Volkin, D. B., Middaugh, C. R., Schöneich, C., Winter, G., Friess, W., Crommelin, D. J., & Carpenter, J. F. (2012, March 1). Protein instability and immunogenicity: Roadblocks to clinical application of injectable protein delivery systems for sustained release. *Journal of Pharmaceutical Sciences, 101*(3), 946–954. doi:10.1002/jps.23018 PMID:22170395

Kamerzell, T. J., Esfandiary, R., Joshi, S. B., Middaugh, C. R., & Volkin, D. B. (2011, October 1). Protein–excipient interactions: Mechanisms and biophysical characterization applied to protein formulation development. *Advanced Drug Delivery Reviews, 63*(13), 1118–1159. doi:10.1016/j.addr.2011.07.006 PMID:21855584

Khodarovich, Y. M., Goldman, I. L., Sadchikova, E. R., & Georgiev, P. G. (2013, December). Expression of eukaryotic recombinant proteins and deriving them from the milk of transgenic animals. *Applied Biochemistry and Microbiology, 49*(9), 711–722. doi:10.1134/S0003683813090020

Khuroo, M. S. (2020, September 1). Chloroquine and hydroxychloroquine in coronavirus disease 2019 (COVID-19). Facts, fiction and the hype: A critical appraisal. *International Journal of Antimicrobial Agents*, *56*(3), 106101. doi:10.1016/j.ijantimicag.2020.106101 PMID:32687949

Kim, H., Yoo, S. J., & Kang, H. A. (2015, February 1). Yeast synthetic biology for the production of recombinant therapeutic proteins. *FEMS Yeast Research*, *15*(1), 1–6. PMID:25130199

Lamnabhi-Lagarrigue, F., Annaswamy, A., Engell, S., Isaksson, A., Khargonekar, P., Murray, R. M., Nijmeijer, H., Samad, T., Tilbury, D., & Van den Hof, P. (2017, January 1). Systems & control for the future of humanity, research agenda: Current and future roles, impact and grand challenges. *Annual Reviews in Control*, *43*, 1–64. doi:10.1016/j.arcontrol.2017.04.001

Lei, Y., Zhang, S., Fang, L., Akash, M. S., Shi, W., Sun, K., Xu, Y., & Chen, S. (2013). A sensitive and specific enzyme immunoassay for detecting tartrazine in human urinary samples. *Analytical Methods*, *5*(4), 925–930. doi:10.1039/c2ay26121a

Majumder, S. (n.d.). *Engineering Platforms for Mimicking Cellular Processes using Cell-Free Expression Systems* [Doctoral dissertation].

Mitragotri, S., Burke, P. A., & Langer, R. (2014, September). Overcoming the challenges in administering biopharmaceuticals: Formulation and delivery strategies. *Nature Reviews. Drug Discovery*, *13*(9), 655–672. doi:10.1038/nrd4363 PMID:25103255

Mohammadinejad, R., Shavandi, A., Raie, D. S., Sangeetha, J., Soleimani, M., Hajibehzad, S. S., Thangadurai, D., Hospet, R., Popoola, J. O., Arzani, A., & Gómez-Lim, M. A. (2019). Plant molecular farming: Production of metallic nanoparticles and therapeutic proteins using green factories. *Green Chemistry*, *21*(8), 1845–1865. doi:10.1039/C9GC00335E

Mukherjee, A., Bisht, B., Dutta, S., & Paul, M. K. (2022, April 4). Current advances in the use of exosomes, liposomes, and bioengineered hybrid nanovesicles in cancer detection and therapy. *Acta Pharmacologica Sinica*, *43*(11), 1–8. doi:10.103841401-022-00902-w PMID:35379933

Mukherjee, S., Joardar, N., Sengupta, S., & Babu, S. P. (2018, November 1). Gut microbes as future therapeutics in treating inflammatory and infectious diseases: Lessons from recent findings. *The Journal of Nutritional Biochemistry*, *61*, 111–128. doi:10.1016/j.jnutbio.2018.07.010 PMID:30196243

O'Flaherty, R., Bergin, A., Flampouri, E., Mota, L. M., Obaidi, I., Quigley, A., Xie, Y., & Butler, M. (2020, November 1). Mammalian cell culture for production of recombinant proteins: A review of the critical steps in their biomanufacturing. *Biotechnology Advances*, *43*, 107552. doi:10.1016/j.biotechadv.2020.107552 PMID:32416132

Owczarek, B., Gerszberg, A., & Hnatuszko-Konka, K. (2019, January 8). A brief reminder of systems of production and chromatography-based recovery of recombinant protein biopharmaceuticals. *BioMed Research International*, *2019*, 2019. doi:10.1155/2019/4216060 PMID:30729123

Pogue, G. P., Lindbo, J. A., Garger, S. J., & Fitzmaurice, W. P. (2002, September). Making an ally from an enemy: Plant virology and the new agriculture. *Annual Review of Phytopathology*, *40*(1), 45–74. doi:10.1146/annurev.phyto.40.021102.150133 PMID:12147754

Posner, J., Barrington, P., Brier, T., & Datta-Mannan, A. (2019). Monoclonal antibodies: past, present and future. *Concepts and Principles of Pharmacology*, 81-141.

Prado, M. S., Bendtzen, K., & Andrade, L. E. (2017, September 2). Biological anti-TNF drugs: Immunogenicity underlying treatment failure and adverse events. *Expert Opinion on Drug Metabolism & Toxicology*, *13*(9), 985–995. doi:10.1080/17425255.2017.1360280 PMID:28772079

Presta, L. G. (2008, August 1). Molecular engineering and design of therapeutic antibodies. *Current Opinion in Immunology*, *20*(4), 460–470. doi:10.1016/j.coi.2008.06.012 PMID:18656541

Roldão, A., Cox, M., Alves, P., Carrondo, M., & Vicente, T. (2014). *Industrial large scale of suspension culture of insect cells*. Academic Press.

Schellekens, H. (2005, September 1). Immunologic mechanisms of EPO-associated pure red cell aplasia. *Best Practice & Research. Clinical Haematology*, *18*(3), 473–480. doi:10.1016/j.beha.2005.01.016 PMID:15792922

Shih, H. H. (2012). *Discovery process for antibody-based therapeutics. InDevelopment of Antibody-Based Therapeutics*. Springer.

Sivadasan, D., Sultan, M. H., Madkhali, O., Almoshari, Y., & Thangavel, N. (2021, August 18). Polymeric Lipid Hybrid Nanoparticles (PLNs) as Emerging Drug Delivery Platform—A Comprehensive Review of Their Properties, Preparation Methods, and Therapeutic Applications. *Pharmaceutics*, *13*(8), 1291. doi:10.3390/pharmaceutics13081291 PMID:34452251

Sproles, A. E., Fields, F. J., Smalley, T. N., Le, C. H., Badary, A., & Mayfield, S. P. (2021, March 1). Recent advancements in the genetic engineering of microalgae. *Algal Research*, *53*, 102158. doi:10.1016/j.algal.2020.102158

Strohl, W.R. (2009). *Therapeutic monoclonal antibodies: Past, present, and future*. Academic Press.

Sun, X., Xu, C., Wu, G., Ye, Q., & Wang, C. (2017, June 1). Poly (lactic-co-glycolic acid): Applications and future prospects for periodontal tissue regeneration. *Polymers*, *9*(6), 189. doi:10.3390/polym9060189 PMID:30970881

Tayal, V., & Kalra, B. S. (2008, January 28). Cytokines and anti-cytokines as therapeutics—An update. *European Journal of Pharmacology*, *579*(1-3), 1–2. doi:10.1016/j.ejphar.2007.10.049 PMID:18021769

Uribe-Echeverry, P. T., & Lopez-Gartner, G. A. (2017, June 1). Fungal immunomodulatory proteins in the context of biomedicine. *Frontiers in Bioscience (Elite Edition)*, *9*(2), 286–306. doi:10.2741/e803 PMID:28410152

Usmani, S. S., Bedi, G., Samuel, J. S., Singh, S., Kalra, S., Kumar, P., Ahuja, A. A., Sharma, M., Gautam, A., & Raghava, G. P. (2017, July 31). THPdb: Database of FDA-approved peptide and protein therapeutics. *PLoS One*, *12*(7), e0181748. doi:10.1371/journal.pone.0181748 PMID:28759605

Vojvoda, E.J., Burrington, L.R., & Oza, J.P. (2022). Development of next-generation diagnostic tools using synthetic biology. In *New Frontiers and Applications of Synthetic Biology* (pp. 287-330). Academic Press.

Wang, H., Chen, X., Huang, Z., Zhou, B., Jia, G., Liu, G., & Zhao, H. (2014). Expression and purification of porcine PID1 gene in Escherichia coli. *Turkish Journal of Biology*, *38*(4), 523–527. doi:10.3906/biy-1403-38

Williams, H. F., Layfield, H. J., Vallance, T., Patel, K., Bicknell, A. B., Trim, S. A., & Vaiyapuri, S. (2019, June 20). The urgent need to develop novel strategies for the diagnosis and treatment of snakebites. *Toxins*, *11*(6), 363. doi:10.3390/toxins11060363 PMID:31226842

Zhu, J. (2012, September 1). Mammalian cell protein expression for biopharmaceutical production. *Biotechnology Advances*, *30*(5), 1158–1170. doi:10.1016/j.biotechadv.2011.08.022 PMID:21968146

Chapter 9
Pathomechanisms of Proteins in Polycystic Ovary Syndrome (PCOS)

Apoorva Jain
Research Institute for Biomedical Sciences, Dharwad, India

Ajay Kumar Oli
https://orcid.org/0000-0002-9634-2819
Research Institute for Biomedical Sciences, Dharwad, India

ABSTRACT

Polycystic ovary syndrome (PCOS) is the most common endocrine disorder in females of reproductive age. It is one of the main causes of infertility in females with a prevalence of 5-10%. PCOS is characterized by Oligo or anovulation, hyperandrogenism, and polycystic ovaries. PCOS is also associated with obesity, type 2 diabetes, hyperinsulinemia, insulin resistance, lipid abnormalities, and cardiovascular diseases. The early diagnosis of PCOS is complicated and complex due to the overlapping symptoms of PCOS. Genetic variance, epigenetic changes, and disturbed lifestyles lead to the development of pathophysiological disturbances and disturbed lifestyles in women. At the molecular level, different proteins and molecular and signaling pathways are involved in disease progression. Proteomics is advantageous in linking both genotype and phenotype and contributing to biomarker discovery. Investigating the molecular mechanism and proteins underlying PCOS is imperative to gain insight into pathophysiology of PCOS and formulate novel diagnostic and treatment strategies.

INTRODUCTION

Polycystic ovary syndrome is a serious medical condition in females of reproductive age. It is estimated to be affecting 5-20% of women. However, among them, 70% of the females are unrecognized (Gnanadass et.al.,2021). Primarily PCOS is characterized by complex psychological, metabolic, and reproductive condition associated with excess androgen production, anovulation/oligovulation, and insulin resistance

DOI: 10.4018/978-1-6684-6577-6.ch009

(Witchel et.al., 2019). It is observed from the study that 65-95% of all women with PCOS are also found to be suffering with insulin resistance (Manu et.al.,2021). There are different criteria for diagnosis of PCOS, Rotterdam criteria, American Society for Reproductive Medicine (ASRM), Androgen Excess Society (AES), of these three criteria the Rotterdam criteria is widely accepted clinically. According to 2003 Rotterdam criteria diagnosis the women with PCOS must present at least two of the following three characteristics: hyperandrogenism (clinical/biochemical), ovulatory dysfunction, and Polycystic ovaries with >12 follicles and volume of >10ml on ultrasonography (Eiras et.al.,2022).

The Human Genome Project has sequenced complete Human DNA and about 30000 genes have been identified, the function of these genes would be understood by proteomics which aims to correlate the structural and functional diversity of proteins with biological processes which could be the better diagnostic and prognostic test for various diseases. Advancements in proteomics allow an analysis of various low molecular weight proteins that patterns of diseases which could be useful for early detection and prognosis of those diseases (Kavallaris et.al.,2005). Even though the correlation between gene and protein is useful in many instances the posttranscriptional and posttranslational mechanisms alter this correlation by altering the abundance, function, and network of proteins. These mechanisms make proteins more complex and diverse when compared to genes (Oliveira et.al.,2021).

Recent development in the field of advanced proteomic technology provides an opportunity to identify the current challenges associated with the diagnosis and treatment of PCOS. As proteins are involved in almost all biological processes the total protein or peptide complement from samples could of the potential biomarker of PCOS (Atiomo et.al.,2009). This proteomic approach has been used in the study of other diseases such as breast cancer, lung cancer, ovarian cancer and, chronic kidney diseases, Alzheimer's disease (Lu,2014; Joanna,2021; Tsuji,2002). Thus, the cumulative effect of modified proteins the product of genes which are mutated along with this the various other factors like genetic inheritance and environment leads to large complications in case of PCOS.

THE PROTEINS INVOLVED IN POLYCYSTIC OVARY SYNDROME

As PCOS is associated with causing metabolic disorders and change in Endocrine hormone release there are number of pathways or mechanisms involved in causing PCOS in one or the other ways. These metabolic disorder mechanisms involve large number of proteins synthesized via specific gene or chromosome any change or the mutation in these would lead to production of defective proteins would lead to physiological symptoms of puberty. In this chapter we have highlighted some of the major proteins and enzymes whose mutation would change the Pathomechanisms of Polycystic Ovary Syndrome.

21 Hydroxylase Enzyme

21 hydroxylase enzyme function is to produce the hormone called cortisol aldosterone. Cortisol assists in the maintenance of blood sugar levels and protect the body from stress. Aldosterone helps in salt retention and regulates the amount of salt in the kidneys. (National library of medicine) PCOS women are characterized by hyperandrogenism which contributes to the inappropriate secretion of gonadotropin. The androgen has a negative feedback mechanism on gonadotropin release which results in a surge of LH hormone. The mechanism involved here is that the increased 17 hydroxylase activity enhances the conversion of progesterone to 17 hydroxyprogesterone (Jaffery Chang R & Ralph, R.K., 2014). During

steroidogenesis 17 hydroxyprogesterone is converted into 11 deoxycortisol catalyzed by the enzyme 21 hydroxylase which is encoded by gene CYP21. The deficiency of this enzyme is responsible for the increased serum levels of 17 hydroxyprogesterone in PCOS women. 17 hydroxyprogesterone stimulates GnRH more than in normal women. The level of this hormone acts as the biomarker for PCOS (medlineplus,2010; Chang et.al.,2014). The studies have shown that children with premature pubarche and adolescent girls with hyperandrogenism were heterozygous for mutations in CYP21 (Morreale et.al.,1994; Azziz et.al.,1995).

P450c17 Enzyme

The enzyme P450c17 converts pregnenolone and progesterone into 17-hydroxypregnenolone and 17-hydroxyprogesterone respectively. Later the conversion of these steroids into dehydroepiandrosterone (DHEA) and 4-androstenedione is done by the P450c17 enzyme. This enzyme is encoded by the CYP17 gene located at 10q24.3.(Witchel et.al., 2000). It was reported that the increased expression of P450c17 enzymatic activity and increased transactivation of CYP17 promoter is found in ovarian theca cells of women with PCOS. However, the dysregulation of CYP17 expression is also identified at the level of mRNA stability in theca cells of PCOS (Witchel et.al.,1997). Although the CYP17 gene does not seem to play a significant role in the pathophysiology of PCOS the post-transcriptional regulation of this gene product might play role in the pathophysiology of PCOS (Leonard et.al.,1987).

P450scc Enzyme

The enzyme P450scc is also referred to as the cholesterol side chain cleavage enzyme. Steroid hormone biosynthesis is catalyzed by the P450scc enzyme. This enzyme is encoded by the gene CYP11 which is located at 15q24 (Wickenheisser et.al.,2005). Deletion of the CYP11 gene in studies conducted on rabbits shows that there is an elimination of steroid synthesis indicating that this enzyme is responsible for steroidogenesis (Unluturk et.al., 2007). The upregulation or downregulation of the expression of CYP11 gene polymorphisms results in an increase or decrease production of androgen. The studies carried out on the CYP11 gene associated with PCOS reported the association of microsatellite polymorphism (TTTTA) in the CYP11A1 promoter region (Frank;2002). The association of CYP11A1 (TTTTA)n repeat polymorphism with PCOS was confirmed in India (Miller,1988).

Follistatin

Follistatin is a glycoprotein that functions as a high-affinity binding protein for activin. Activin is a dimeric glycoprotein that belongs to the TGF beta superfamily responsible for inducing FSH, insulin secretion, and ovarian follicular maturation and inhibits LH-stimulated ovarian androgen production. The increased activin due to elevated follistatin reduces FSH concentration, arrests follicular maturation, augments androgen production, and impairs insulin release. As these are the characteristics of females with PCOS it could be concluded that follistatin cause PCOS (Gharani et.al.,1997).

Capn10

CAPN10 is also known as Caplain10. It is calcium-dependent cysteine proteases present on chromosome2q37.3. The protein released by this gene is heterodimer which is mainly associated with type 2 diabetes. Chromosomal locus having CAPN10 encodes cysteine proteases calpain 10. This Calpain 10 plays a vital role in insulin secretion. Any defect or abnormality in this CAPN10 leads to PCOS as it is associated with insulin resistance and type 2 diabetes (Reddy et.al.,2014).

Aromatase

The aromatase enzyme is encoded by the gene CYP19. In women with PCOS hormonal disruption such as an increase in androgens and luteinizing hormone (LH), and a reduction in follicle-stimulating hormone (FSH) is accompanied by an abnormality in ovarian morphology. FSH gonadotropin is essential for follicle maturation and ovarian steroidogenesis which activates FSH receptors (FSHRs) in granulosa cells of antral follicles leading to the transcription of aromatase enzyme catalyzing the conversion of androgen to estrogen which is the main component for follicle maturation. In healthy women normally follicle development in the granulosa cells starts expression of the aromatase mRNA when follicle growth reaches approximately 6-7 mm in diameter. In the case of PCOS patients the follicle development arrest at 5-8 mm in diameter. Thus, the granulosa cells express minimal aromatase at the larger diameter and produce little estrogen (Prapas et.al.,2009; Ajmal et.al.,2019).

5 Alpha Reductases

5 alpha reductases are an enzyme responsible for the metabolic processing of androgens that catalyze the irreversible conversion of cortisol to dihydrocortisol in the liver and testosterone to 5 alpha DHT. In girl's increase in 5 alpha-reductase activity indicates the risk of developing PCOS in early childhood. Thus, a change in the 5 alpha-reductase activity is helpful in the pathogenesis of PCOS causing abnormal androgen levels which is the critical characteristic of PCOS (Yang et.al.,2015; Magoffin et.al.,2006; Remer et.al.,2008).

HUMAN FOLLICULAR FLUID PROTEINS

Follicular fluid is the major reproductive fluid of PCOS patients used for proteomic analysis which affects folliculogenesis (Graupp et.al.,2011). There are proteins functioning at various processes of follicular development like amphiregulin, heparan sulfate proteoglycan 2, tumor necrosis factor, alpha-induced protein 6, plasminogen, and lymphatic vessel endothelial hyaluronan receptor 1 which were found to be deregulated in PCOS (Torchen et.al.,2016). In another study, it is identified that the abnormal expression of HFF proteins involved in lipid metabolism in PCOS patients experienced the inability to ovulate normally (Ambekar et.al.,2015). An upregulation of alpha 1 antitrypsin in HFF of PCOS women failed in the maturation of oocytes. There are about nine differentially expressed protein biomarkers were identified in women with endometrial cancer and PCOS which include transgelin, pyruvate kinase M1/M2, Gelsolin-like capping protein (macrophage capping protein), glutathione S transferase P, leucine aminopeptidase (cytosol aminopeptidase), peptidyl-prolyl cis-trans isomerase, cyclophilin A, manganese superoxide dismutase (Dai et.al.,2012).

FERRITIN AND HAPTOGLOBIN

Ferritin is an iron storage protein it is critical in iron homeostasis and metabolism. The concentration of serum ferritin act as a biomarker to estimate the level of iron storage in the body. It is suggested that the increased level of iron stored is associated with both insulin resistance and metabolic syndrome which are important criteria of PCOS (Swain et.al.,2016). Haptoglobin is an acute-phase circulating protein synthesized in the liver under influence of the cytokines like TNF alpha and interleukin 6 which enhances iron loss. This haptoglobin is identified to be inhibiting the synthesis of prostaglandins and haptoglobin concentration in the serum correlates with BMI and is considered a marker for adiposity it could be the potential biomarker for PCOS (Galazis et.al.,2013).

APOLIPOPROTEIN

Apolipoprotein is required for cholesterol transport in the ovary for the biosynthesis of steroid hormones in the granulosa cells. Reduction in the level of apolipoprotein is been observed in the serum of PCOS women. This reduced levels of Apo-AI in the granulosa cells of PCOS women affects cholesteryl ester uptake which in turn affects the normal steroidogenesis in the ovary of women. It is investigated from the studies that the knockdown of Apo-AI reduced the steroidogenic enzymes transcripts in the human ovarian granulosa cell line (KGN), in contrast to this the overexpression of Apo-AI increased the expression level of steroidogenic enzymes like CYP11A, 17-hydroxysteroid dehydrogenase (HSD), 3-HSD, and CYP19 in the granulose cells. Particularly CYP19 expression level reduction due to apolipoprotein hampers the conversion of testosterone to estradiol, a higher level of testosterone is the critical symptom of PCOS (Tiongco et.al.,2019; Alvarez et.al.,2009; Choi et.al.,2010; Li et.al.,2017; Liu et.al.,2019).

FUTURE PROSPECTIVE

Proteomics is greatly benefiting the clinical research with the potential to develop a better diagnostic and prognostic approach to identify potential new therapeutic targets and to move in the direction of identifying the individual patient therapy. Although proteomics has proved to be promising in the discovery of biomarkers, further advancements are still required to enhance the performance and reproducibility of established proteomics tools before they could be used routinely in the clinical laboratory. Researchers have investigated the use of blood or urine proteomes as these samples are readily available. Another preventing factor for the widespread use of proteomics in the clinical laboratory is the cost. Most proteomics technologies use complex instrumentation, critical computing power, and expensive consumables. Along with this expertise for operating and carrying out the research is required especially for high throughput analysis of MS data. It has been suggested that this technology could be the most reliable and cost-effective if bulk samples are being analyzed.

CONCLUSION

PCOS is one of the most common diseases affecting women of reproductive age, which is characterized by hyperandrogenism, insulin resistance and abnormal gonadotropin secretions. PCOS disturbs both reproductive and metabolic functions. High insulin resistance, bad dietary choices, weekend immune system are the main factors that promote the PCOS. And it remains difficult to manage due to a lack of knowledge regarding the pathophysiological mechanism of PCOS. Although the biochemical manifestations and clinical characteristics of PCOS are typical. PCOS is still a complicated pathophysiology condition with no gold standard for diagnosis. In general, the serum Anti Mullerian Hormone (AMH) levels reflect the size of the ovarian follicle. In clinical practice the use of serum AMH for the assessment of ovarian reserve has been used for diagnosis of PCOS. The numerous proteins involved in various biological functions differentially expressed in normal and PCOS women provide important insight for a better understanding of Proteomic identification and would be a valid tool for predicting and tracking the progression of comorbidities in women with PCOS. The major challenges that remained in this are the technical issues like detecting low abundance proteins and maintaining of sample stability and correlating the changes in protein with the disease process.

REFERENCES

Ajmal, N., Khan, S., & Shaikh, R. (2019). Polycystic ovary syndrome (PCOS) and genetic predisposition: A review article. *European Journal of Obstetrics, Gynecology, and Reproductive Biology*, *3*, 1–6. PMID:31403134

Alvarez Blasco, F., Martinez Garcia, M. A., Luque Ramirez, M., Parraza, N., & Millan, J. (2009). Role of Haptoglobin in Polycystic ovary syndrome (PCOS), Obesity and Disorders of glucose tolerance in premenopausal women. *PLoS One*, *4*(5), 1–7. doi:10.1371/journal.pone.0005606 PMID:19440331

Ambekar, A. S., Kelkar, D. S., Pinto, S. M., Sharma, R., Hinduja, I., Zaveri, K., Pandey, A., Prasad, K., Gowda, H., & Mukherjee, S. (2015). Proteomics of follicular fluid from women with polycystic ovary syndrome suggests molecular defects in follicular development. *The Journal of Clinical Endocrinology and Metabolism*, *100*(2), 744–753. doi:10.1210/jc.2014-2086 PMID:25393639

Atiomo, W., Khalid, S., Parameshweran,S., Houda,M., Layfield,R. (2009). Proteomic Biomarkers for the diagnosis and risk stratification of polycystic ovary syndrome: a systematic review. *BJOG: An International Journal of Obstetrics and Gynecology, 116*(2), 137-143.

Azziz, R., Bradley, E. L., Potter, H. D., & Boots, L. R. (1995). Adrenal androgen excess in women: Lack of a role for 17-hydroxylase and 17,20-lyase dysregulation. *The Journal of Clinical Endocrinology and Metabolism*, *80*(2), 400–405. PMID:7852496

Chang, R.J., & Kazer, R.R. (2014). Polycystic ovary syndrome. *The global library of women's medicine.*

Choi, D. H., Lee, W. S., Won, M., Park, M., Park, H. O., Kim, E., Lee, K. H., & Bae, J. (2010). The Apolipoprotein A-I level in downregulated in the grnualosa cells of patients with polycystic ovary syndrome and affects steroidogesis. *Journal of Proteome Research*, *9*(9), 4329–4336. doi:10.1021/pr100008e PMID:20426491

CYP21A2 gene. (2010). https://medlineplus.gov/genetics/gene/cyp21a2/

Dai, G., & Lu, G. (2012). Different protein expression patterns associated with polycystic ovary syndrome in human follicular fluid during controlled ovarian hyperstimulation. *Reproduction, Fertility, and Development*, *24*(7), 893–904. doi:10.1071/RD11201 PMID:22935150

Eiras, M. C., Pinheiro, D. P., Mubarac Romcy, K. A., Ferriani, R. A., Reis, R. M., & Furtado, C. L. (2022). Polycystic ovary syndrome: The Epigenetic Behind the Disease. *Reproductive Sciences (Thousand Oaks, Calif.)*, *29*(3), 1–15. doi:10.100743032-021-00516-3 PMID:33826098

Franks, S. (2002). Adult polycystic ovary syndrome begins in childhood. *Best Practice & Research. Clinical Endocrinology & Metabolism*, *16*(2), 263–272. doi:10.1053/beem.2002.0203 PMID:12064892

Galazis, N., Pang, Y. L., Galazi, M., Haoula, Z., Layfield, R., & Atiomo, W. (2013). Proteomic biomarkers of endometrial cancer risk in women with polycystic ovary syndrome: A systematic review and biomarker database integration. *Gynecological Endocrinology*, *29*(7), 638–644. doi:10.3109/09513590.2013.777416 PMID:23527552

Gharani, N., Waterworth, D. M., Batty, S., White, D., Gilling-Smith, C., Conway, G. S., McCarthy, M., Franks, S., & Williamsom, R. (1997). Association of the steroid synthesis gene CYP11a with polycystic ovary syndrome and hyperandrogenism. *Human Molecular Genetics*, *6*(3), 397–402. doi:10.1093/hmg/6.3.397 PMID:9147642

Gnanadass, S.A., Prabhu, Y.D., & Gopalakrishnan, A.V. (2021). Association of metabolic and inflammatory markers with polycystic ovarian syndrome (PCOS): An update. *Archives of Gynecology and obstestrics, 303*(3), 631-643.

Graupp, M., Wehr, E., Schweighofer, N., Pieber, T. R., & Obermayer, B. (2011). Association of genetic variants in the two isoforms of 5alpha-reductase, SRD5A1 and SRD5A2, in lean patients with polycystic ovary syndrome. *European Journal of Obstetrics, Gynecology, and Reproductive Biology*, *157*(2), 175–179. doi:10.1016/j.ejogrb.2011.03.026 PMID:21530059

Joanna, T., & Luczak, M. (2021). Applying proteomics and integrative omics strategies to decipher the chronic kidney disease-Related Atherosclerosis. *International Journal of Molecular Sciences*, *22*(14), 1–19. PMID:34299112

Kavallaris, M., & Marshall, G. M. (2005). Proteomics and disease: Opportunities and challenges. *The Medical Journal of Australia*, *182*(11), 575–579. doi:10.5694/j.1326-5377.2005.tb06817.x PMID:15938685

Leonard, P. J., & Miller, W. L. (1987). Cloning and sequence of the human gene for P450c17 (steroid 17α-hydroxylase/ 17,20lyase): Similarity with the gene for P450. *DNA (Mary Ann Liebert, Inc.)*, *6*(5), 439–448. doi:10.1089/dna.1987.6.439 PMID:3500022

Li, L., Feng, Q., Ye, M., He, Y., Yao, A., & Shi, K. (2017). Metabolic effect of obesity on polycystic ovary syndrome in adolescents: A meta-analysis. *Journal of Obstetrics & Gynaecology*, *37*(8), 1036–1047. doi:10.1080/01443615.2017.1318840 PMID:28657375

Liu, Q., Xie, Y. J., Qu, L. H., Zhang, M. X., & Mo, Z. C. (2019). Dyslipidemia involvement in the development of polycystic ovary syndrome. *Taiwanese Journal of Obstetrics & Gynecology*, *58*(4), 447–453. doi:10.1016/j.tjog.2019.05.003 PMID:31307731

Luo, L., Dong, L., Yan, Q., Cao,S., Wen, X., Huang, Y., Huang, X., Wu, R., & Ping Ma, X. (2014). *Research progress in applying proteomics technology to explore early diagnosis biomarkers of breast cancer, lung cancer and ovarian cancer*. Academic Press.

Magoffin, D. A. (2006). Ovarian enzyme activities in women with polycystic ovary syndrome. *Fertility and Sterility*, *86*(1), 9–11. doi:10.1016/j.fertnstert.2006.03.015 PMID:16798289

Manu, M., Soni, T., & Prabhakar, P.K. (2021). Pathophysiology of polycystic ovarian Syndrome. *Intech Open*.

Miller, W. L. (1988). Molecular biology of steroid hormone synthesis. *Endocrine Reviews*, *9*(3), 295–318. doi:10.1210/edrv-9-3-295 PMID:3061784

Morreale, H. E., Pazos, F., Potau, N., Garcia-Robles, R., Sancho, J. M., & Varela, C. (1994). Ovarian suppression with triptorelin and adrenal stimulation with adrenocorticotropin in functional hyperadrogenism: Role of adrenal and ovarian cytochrome P450c17 alpha. *Fertility and Sterility*, *62*(3), 521–530. doi:10.1016/S0015-0282(16)56940-4 PMID:8062947

Oliveira, A. G., & Fiorotto, R. (2021). Novel approach to liver disease diagnosis and modelling. *Translational Gastroenterology and Hepatology*, *6*(19), 1–14. PMID:33824923

Prapas, N., Karkanaki, A., Prapas, I., Kalogiannidis, I., Katsikis, I., & Panidis, D. (2009). Genetics of Polycystic ovary syndrome. *Hippokratia*, *13*(4), 216–223. PMID:20011085

Reddy, K. R., Deepika, M. L. N., Supriya, K., Latha, K. P., Lakshmana Rao, S. S., Usha Rani, V., & Jahan, P. (2014). CYP11A1 microsatellite (tttta)n polymorphism in PCOS women from South India. *Journal of Assisted Reproduction and Genetics*, *31*(7), 857–863. doi:10.100710815-014-0236-x PMID:24793009

Remer, T., Maser-Gluth, C., & Wudy, S. A. (2008). Glucocorticoid measurements in health and disease-metabolic implications and the potential of 24-h urine analyses. *Mini-Reviews in Medicinal Chemistry*, *8*(2), 153–170. doi:10.2174/138955708783498096 PMID:18289099

Swain, N., Cirenza, C., & Samanta, L. (2016). Role of proteomics in female infertility. *Proteomics in Human Reproduction*, 45-63.

Tiongco, R. E., Rivera, N., Clemente, B., Dizon, D., Salita, C., & Pineda-Cortel, M. R. (2019). Serum Ferritin as a candidate diagnostic biomarker of polycystic ovarian syndrome: A meta-analysis. *Biomarkers*, *24*(5), 484–491. doi:10.1080/1354750X.2019.1620335 PMID:31096807

Torchen, L. C., Idkowiak, J., Fogel, N. R., O'Neil, D. M., Shackleton, C. L., Wiebke, A., & Dunaif, A. (2016). Evidence for Increased 5alpha-Reductase Activity During Early Childhood in Daughters of Women with Polycystic Ovary Syndrome. *The Journal of Clinical Endocrinology and Metabolism*, *101*(5), 2069–2075. doi:10.1210/jc.2015-3926 PMID:26990942

Tsuji, T., Shiozaki, A., Kohno, R., Yoshizato, K., & Shimohama, S. (2002) Article. *Neurochemical Research*, *27*(10), 1245–1253. doi:10.1023/A:1020941929414 PMID:12462422

Unluturk, U., Harmanci, A., Kocaefe, C., & Yildiz, B. O. (2007). The genetic basis of the polycystic ovary syndrome: A literature review including discussion of PPAR-gamma. *PPAR Research, 2007*, 1–23. doi:10.1155/2007/49109 PMID:17389770

Wickenheisser, J. K., Nelson-DeGrave, V. L., & McAllister, J. M. (2005). Dysregulation of cytochrome P450 17α-hydroxylase messenger ribonucleic acid stability in theca cells isolated from women with polycystic ovary syndrome. *The Journal of Clinical Endocrinology and Metabolism, 90*(3), 1720–1727. doi:10.1210/jc.2004-1860 PMID:15598676

Witchel, S. F., & Aston, C. E. (2000). The role of heterozygosity for CYP21 in the polycystic ovary syndrome. *Journal of Pediatric Endocrinology & Metabolism, 13*(5), 1315–1317. PMID:11117678

Witchel, S. F., Lee, P. A., Suda-Hartman, M., & Hoffman, E. P. (1997). Hyperandrogenism and manifesting heterozygotes for 21-hydroxylase deficiency. *Biochemical and Molecular Medicine, 62*(2), 151–158. doi:10.1006/bmme.1997.2632 PMID:9441866

Witchel, S. F., Oberfield, S. E., & Pena, A. S. (2019). Polycystic ovary syndrome: Pathophysiology presentation and treatment with emphasis on adolescent girls. *Journal of the Endocrine Society, 3*(8), 1545–1573. doi:10.1210/js.2019-00078 PMID:31384717

Yang, F., Ruan, Y. C., Yang, Y., Wang, K., Liang, S., Han, Y., Teng, X.-M., & Yang, J.-Z. (2015). Follicular hyperandrogenism downregulates aromatase in luteinizing granulosa cells in polycystic ovary syndrome women. *Reproduction (Cambridge, England), 150*(4), 289–296. doi:10.1530/REP-15-0044 PMID:26199450

Chapter 10
Smart Drugs in Cognitive Disorders

Jatin Phogat
https://orcid.org/0000-0003-4586-7719
Panjab University, Chandigarh, India

Alka Bali
Panjab University, Chandigarh, India

Nishika Kapoor
Panjab University, Chandigarh, India

ABSTRACT

More than 65% of the elderly population globally suffers from dementia. The term "smart drug" refers to any drug substance that positively impacts mental skills. Nootropics are agents that are used for boosting the performance of the brain and to increase one's concentration. These are also called cognitive enhancers as these can enhance memory, creativity, awareness, and attention. Natural sources, as well as synthetic approaches, both, are being employed by researchers worldwide for developing smart drugs. Smart drugs are intended to enhance academic, social, as well as career performance. These drugs work by boosting the oxygen supply and nutrition to the brain and are useful for treating various neurodegenerative disorders. Several mechanistic pharmacological approaches have been explored for the development of smart drugs and enhancement of brain activity, the most prominent ones being glutamatergic, dopaminergic, serotonergic, and cholinergic pathways. This chapter reviews the various approaches for developing synthetic smart drugs and their applications and limitations.

INTRODUCTION

Every individual aspires to be smarter, learn more in less time, think and act more quickly, and command a great memory. Generally, cognitive performance can be improved with positive lifestyle changes including sufficient and good-quality sleep, stress reduction, meditation, etc. Besides, these, several substances in

DOI: 10.4018/978-1-6684-6577-6.ch010

our traditional systems of medicine have been utilized since times immemorial to boost brain performance, enhance concentration and cognition and to ameliorate mental alertness. "Smart drugs", also termed as "nootropics" (Giurgea & Salama, 1977), refer to this class of medications. These "cognitive enhancers" are the drugs or supplements that enhance cognitive performance. Smart drugs exhibit their actions by enhancing specific mental functions like memory, creativeness, awareness, consciousness and attention. In recent times, several researchers are working in the area of developing new potential smart drugs from synthetic as well as natural sources. Cognition means the capacity of processing existing information and applying knowledge to solve problems. This virtually involves all brain processes including memory, attention, language, executive functions, perception, and psychomotor functions. Cognition-enhancing drugs have the ability to increase intelligence and memory, and hence, these have a great impact on the health of individuals. These have applications for enhancement of cognitive performance of brain areas, particularly, in the event of some injury or degeneration. Majority of these medications require several doses to be administered before any discernible improvement is noted in the subject. Very few amongst these can work at the outset, after a single delivery. These are generally well tolerated and are utilized as supportive therapies for individuals with CNS disorders such as Alzheimer's disease, schizophrenia, hyperkinetic disorder, and senile dementia. Further, these have applications for use in acute, sub-acute, and chronic states of memory, consciousness, and learning difficulties. Besides improving the mental health of individuals, smart drugs are also intended to enhance academic or professional performance, and this way, these command a sizeable market in the current competitive scenario. There are several smart drugs that are commercially available, however, the drugs with acute and long-term effects in children and adolescents whose brains are under development are still elusive. People frequently encounter obstacles in resource-constrained areas, including financial limitations, a lack of understanding, and insufficient healthcare infrastructure, which together prevent their access to smart pharmaceuticals. These differences highlight the requirement for focused actions to provide equal access to cognitive enhancers and stop the escalation of already-existing socioeconomic inequities. A thorough methodology is required to implement a smart drug program in resource-constrained places. This entails developing cost-effective formulations, educating the public, integrating services into the current healthcare system, creating ethical standards, investigating local choices, working with organizations, and continual monitoring. This comprehensive approach attempts to improve the availability, cost, and appropriate use of cognitive-enhancing drugs in various areas.

The aim of this chapter is to investigate the idea of smart drugs, their mechanisms of action, possible benefits in increasing cognitive abilities, and the ethical concerns associated with their use. This chapter seeks to offer an in-depth understanding of smart drugs and their effects by exploring the scientific, ethical, and sociological aspects. In this chapter, we have explored the intricate mechanisms of action of various nootropics, encompassing the dopaminergic, cholinergic, glutamatergic, and serotonergic pathways. This comprehensive analysis offers profound insights into the realm of cognitive enhancement and their potential applications..

This chapter is important because it gives readers knowledge about the workings and possible advantages of smart drugs. It increases moral consciousness regarding cognitive enhancement and its effects on people and society. The chapter supports educated decision-making on the use of smart medicines for cognitive improvement.

BACKGROUND

Nootropics are a diverse class of substances, which are also referred to as "smart drugs" in English language journals (Schifano et al., 2022; Malik et al., 2007). The name is made up of the Greek words "tropein", which means to lead, and "nöos", which denotes thinking (Giurgea, 1973). Cornelius E. Giurgea coined the word "nootropic" in 1972-73 to refer to drugs that primarily activate cognitive processes like memory and learning, particularly, in conditions where these processes are impeded (Giurgea, 1973). Nootropics, or "cognition-enhancing substances," are the substances having the capability to enhance cognitive abilities including memory and attention without causing significant adverse events. One of the primary mechanisms of these drugs is the modification of various neurotransmitter systems involved in cognition and memory, particularly, cholinergic pathways in the brain. These agents can also work by increasing blood flow, or by promoting neural plasticity.

These essentially modulate the metabolism of the central nervous system's (CNS) neuronal cells (Zhao et al., 2006, 2001; Suliman et al., 2016). There are referred to collectively as cognitive enhancers, and have been broadly classified based on their source of origin as natural nootropics and synthetic nootropics. (Froestl et al., 2012). Another method of classification of nootropics is based on their mechanistic basis i.e., the neurotransmitter systems mimicked or targeted by them in the brain, e.g., dopamine, acetylcholine, serotonin, glutamate, etc.

MODE OF ACTION

Smart drugs generally work by increasing the supply of oxygen and nutrients to the brain. The brain comprises of 3% of the total body weight and receives 15% of body's total oxygen supply. One of the major mechanisms of smart drugs is of the arteries and veins supplying blood to the brain. Since the brain can only utilize glucose for generation of energy, therefore, the functioning of neurons is critically dependent upon the supply of nutrients and oxygen, which, in turn, can be enhanced by the use of smart drugs. Smart drugs constitute a major approach for treating cognitive impairment in patients suffering from neurodegenerative disorders like Alzheimer's disease and several other diseases including schizophrenia, epilepsy, attention deficit hyperactivity disorder (ADHD), aging and stroke. Smart drugs affect brain activity through several diverse mechanisms or pathways involving a multitude of receptor systems (Figure. 1). The major accepted mechanisms include positive modulation of the dopaminergic cholinergic and the glutamatergic pathways.

Dopaminergic pathway: Dopamine is a major neurotransmitter with role and implication in several CNS disorders. The ventral tegmental area (VTA) in the brain, which is responsible for the production of dopamine, plays a significant role in motivation, thinking and emotions (Figure. 2). Dopamine is responsible for the reward pathway in this brain area. The release of dopamine produces a feeling of reward which stimulates mood and hence, enhances cognition. Smart drugs up-regulate the dopaminergic pathway and cause the release of dopamine in the ventral tegmental area, thus, contributing towards an enhanced memory, mood and cognition. For instance, the wakefulness-promoting drug modafinil also raises dopamine levels in an indirect manner. It improves cognitive function and alertness while treating narcolepsy and shift work sleep disorder.

Figure 1. Smart drugs diagram

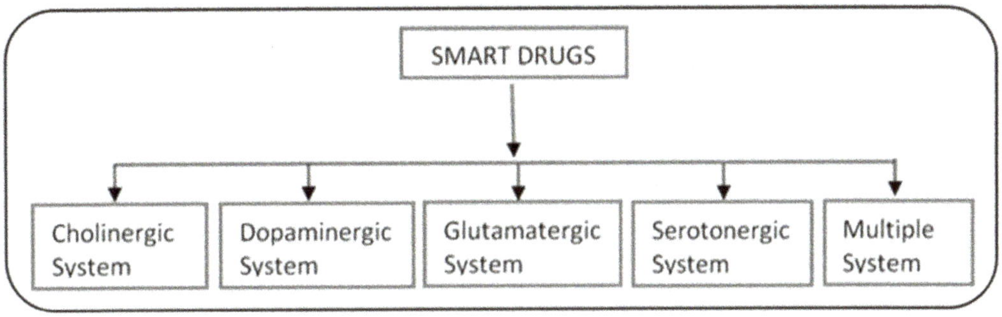

Figure 2.

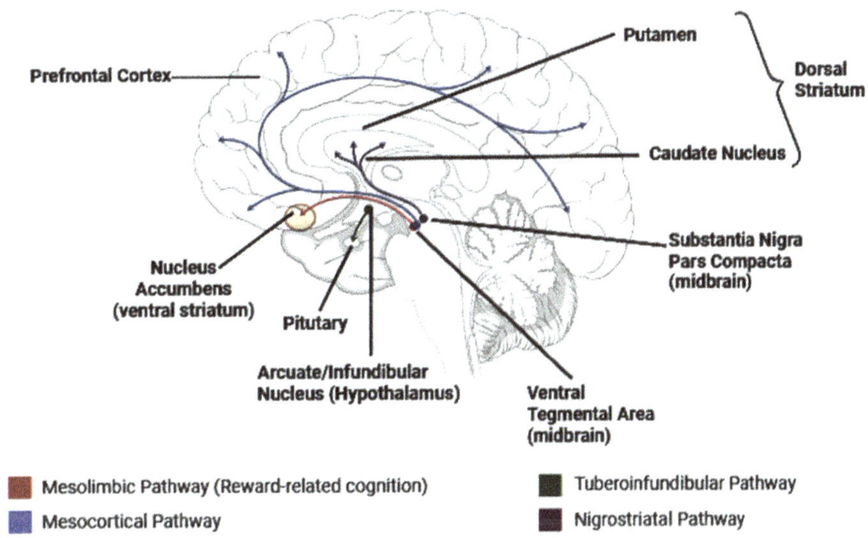

Cholinergic pathway: The neurotransmitter most closely linked to memory and overall cognitive acuity is acetylcholine (ACh). The cholinergic pathway is associated with this neurotransmitter, which is also up-regulated by several smart drugs to modulate cognitive functions. The neuronal nicotinic acetylcholine receptors (nAChRs), located on the presynaptic neurons (Figure. 3), have an effect on hippocampal synaptic transmission, as these stimulate the release of glutamate (Nishizaki et al., 1999). Activation of these nicotinic ACh receptors is stimulated by the activation of protein kinase C (PKC), which subsequently maintains the phosphorylation of the receptor and sustains the release of glutamate. In several clinical cases, cognitive dysfunction is related to a decrease in cholinergic function, and this can be treated by the use of smart drugs which cause stimulation of the central cholinergic activity. The major geriatric neurodegenerative disorder, Alzheimer's disease, is characterized by a loss of neuronal cholinergic activity in the hippocampal region of the brain. Hence, smart drugs having mechanistic basis in cholinergic system, find use in the symptomatic treatment of the Alzheimer's disease (Müller et al., 1997). Acetylcholinesterase is inhibited by the drug donepezil, which is used to treat Alzheimer's disease. It improves cognitive function in people with cognitive impairment by halting acetylcholine breakdown.

Glutamatergic pathway: Glutamate is another essential neurotransmitter involved in cognitive processes. Glutamatergic transmission represents a major neuronal pathway in the brain responsible for excitatory responses and synaptic plasticity (Figure. 4).

Figure 3.

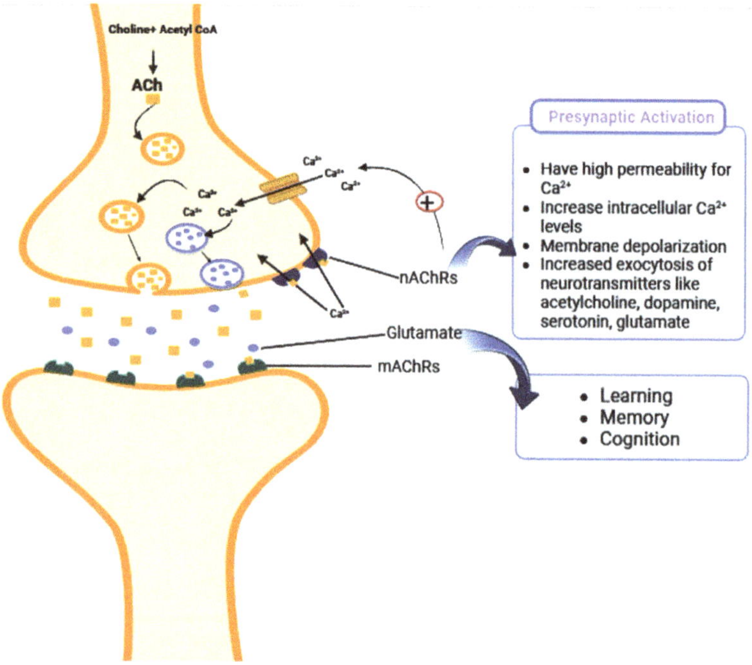

The latter refers to the brain's ability to alter the strength of neural connections, increasing or weakening them in response to activity. Due to its ability to adapt and efficiently store information, this dynamic process is essential for learning and memory. Glutamatergic transmission is associated with long-term potentiation (LTP). Long term potentiation is the long-lasting strengthening of neuronal connections that is essential for memory and learning. Strong synaptic activation that occurs repeatedly sets off processes like NMDA receptor activation. Synapses begin to change in strength and structure as a result, possibly involving increased receptor occupancy for better signaling. There are two different types of glutamate receptors: ionotropic and metabotropic receptors. The ionotropic receptors are ligand-gated cationic channels (LGICs) activated by specific agonists responsible for quick control of ion mobility (sodium, potassium, or sometimes calcium ions). Based on the agonist involved, these are sub-classified as AMPA, kainite and NMDA receptors. Cell electrical activity is impacted by this quick ion movement, which has an effect on processes including brain communication and sensory processing. Metabotropic receptors are G-protien coupled receptors (GPCRs) that are triggered by substances like neurotransmitters and open internal signaling channels. These have no direct impact on ion flow, in contrast to ionotropic receptors. Instead, these result in slower, longer-lasting biological reactions that affect activities including cell di-

Figure 4.

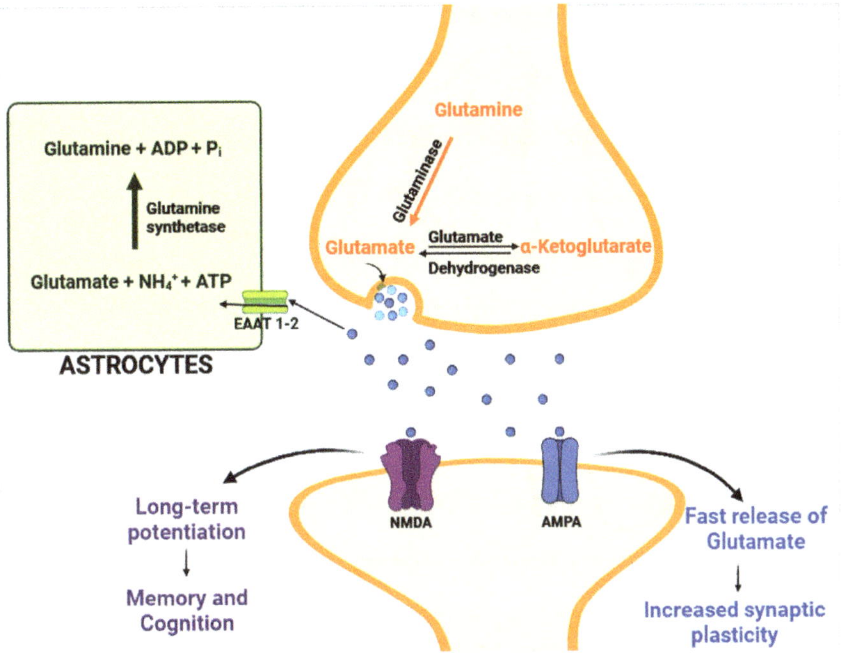

vision, gene expression, and neurotransmitter regulation). These are distributed on the presynaptic and postsynaptic sites of the neurons (Riedel, 2003). Amongst these, the ionotropic N-methyl-D-aspartate (NMDA) and the α-amino-3-hydroxy-5-methyl-4-isoxazolepropionic acid (AMPA) receptors have a significant role in enhancing cognitive performance. The release of glutamate by stimulation of neurons leads to the activation of the both NMDA and AMPA receptors. The AMPA receptor is responsible for synaptic transmission, whereas, the NMDA receptor is responsible for the memory and learning processes. Activation of the NMDA receptor initiates long-term potentiation which is responsible for enhancing cognition, whereas, the activation of AMPA receptors causes the fast release of glutamate, thus improving synaptic plasticity, and the resultant cognitive performance (Butterfield & Pocernich, 2003). The nootropic drug aniracetam, for instance, modifies glutamate receptors to enhance cognitive function. It is thought to boost receptor sensitivity, which improves memory and learning.

Serotonergic Pathway: Studies have also been directed towards the involvement of serotonin (5-hydroxytryptamine; 5-HT) as another plausible cognition-related biomarker (Figure. 5).

A part role of serotonin in the neurobiology of learning and memory has been linked with the complex interactions of the serotonergic system with other neurotransmitter systems such as the cholinergic, dopaminergic, GABAergic and glutamatergic receptor systems. The current evidence implies that reduced serotonergic neurotransmission negatively influences cognitive functions in certain conditions like depression and Alzheimer's disease and that normalization of serotonergic activity may have beneficial effects on cognition, however, this hypothesis needs to be confirmed through further clinical studies. The role of serotonin 5-HT_{2A} receptors in memory and cognition has been reviewed (Zhang & Stackman Jr, 2015). As an illustration, the serotonin precursor 5-HTP (5-Hydroxytryptophan) can be utilized to improve mood and possibly promote cognitive performance. 5-HTP may increase mental acuity and emotional well-being by raising serotonin levels.

Figure 5.

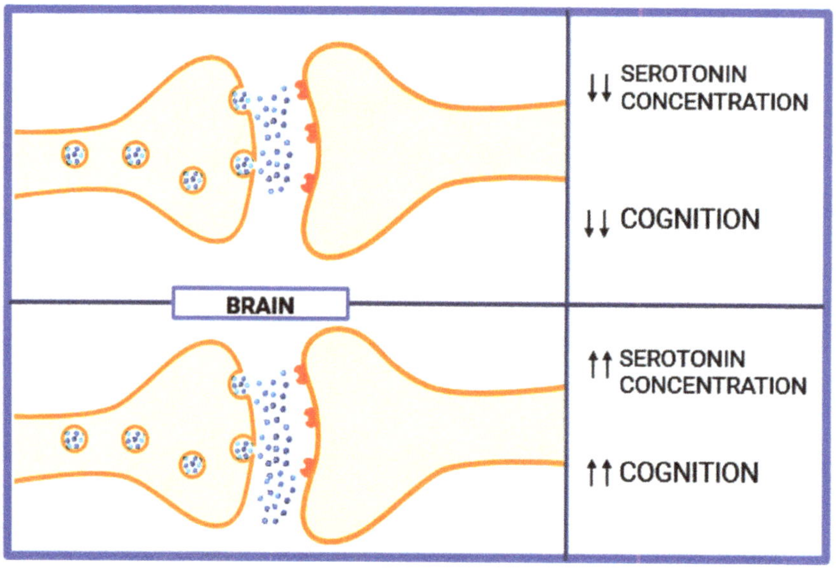

SMART DRUGS TARGETING THE CHOLINERGIC PATHWAY

Meclofenoxate

Meclofenoxate (Figure. 6), also known as centrophenoxine (Lucidril®, Amipolen®, Cellative®, Lucidryl®, Centrophenoxin, and several other brand names), is an extensively marketed smart drug that was developed at French National Scientific Research Center for the treatment of Alzheimer's disease in 1959. Chemically, this simple molecule is an ester of 4-chlorophenoxyacetic acid (a synthetic auxin) with another molecule having good cholinergic activity, i.e., dimethylaminoethanol or deanol (Durham, 1991). Meclofenoxate is well absorbed *via* the parenteral route and is known to increase CNS choline levels, which in turn increases the acetylcholine levels in the hippocampal areas of the brain. Its effects on choline and acetylcholine levels in the brain are similar to those of deanol, however, it is about twice as effective (Wood, 1982). It is a very useful therapeutic tool for potentiating depressed cholinergic neurons. Studies indicate that after crossing the blood-brain barrier (BBB), it either converts into an intermediary phospholipid that is needed to produce acetylcholine or breaks down into choline, which is subsequently turned to acetylcholine, increasing the amount of acetylcholine in synaptic vesicles. The drug exhibits a significant increase in memory, with concomitant reduction in neuronal damage, proinflammatory mediator levels, and oxidative stress. It has been found beneficial in cerebrovascular dementia (Hu et al., 2013) and also increases mental alertness and the combination of new information into long-term memory in elderly people (Marcer & Hopkins, 1977). Although the main mechanism for the nootropic action of meclofenoxate is by raising acetylcholine levels, it also improves cognition by other ways, such as, by making more oxygen and glucose available to the brain and by reducing cellular waste and toxins that have a negative impact on cognition. Centrophenoxine is also known to have anti-ageing effects.

Figure 6. Meclofenoxate

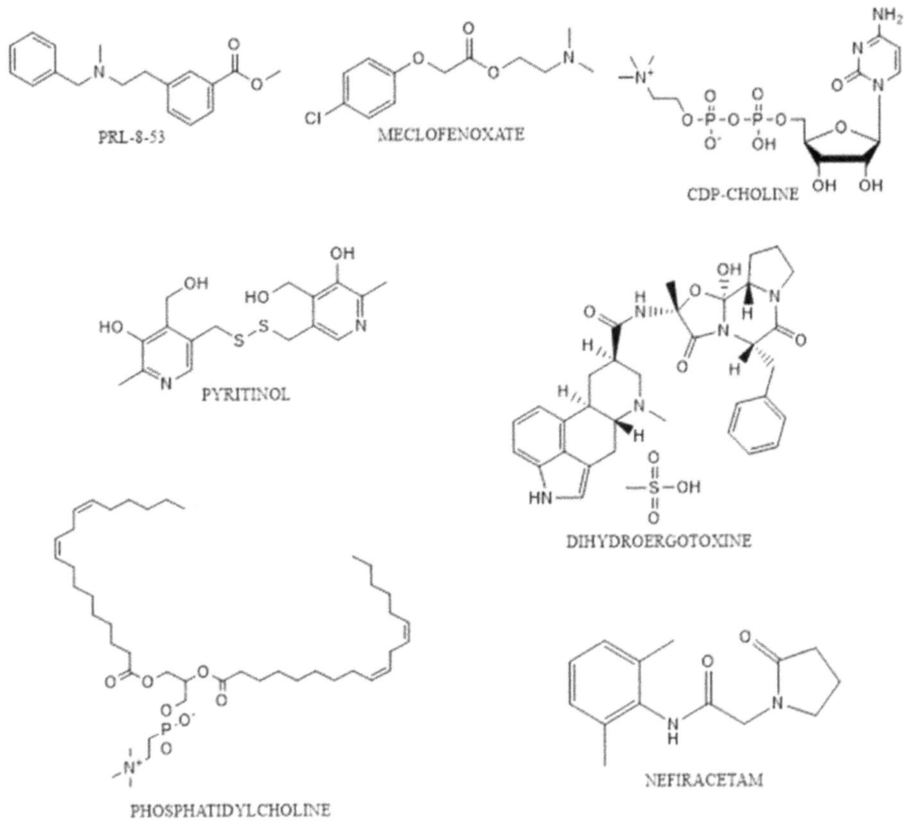

Citicoline (CDP Choline)

Citicoline (Figure. 6), or cytidine diphosphate-choline (CDP-Choline), is a potent neuroprotective substance that has been utilized for its capacity to improve memory and guard the brain against memory loss. The compound is an endogenous nucleoside, an intermediate in the generation of phosphatidyl choline (lecithin) from choline in cell membranes. Following absorption, CDP choline divides into choline and cytidine, crosses the blood-brain barrier and enters the neurological system (Weiss, 1995). Cytidine is subsequently changed into uridine, a nucleotide base essential for the formation of brain membranes and having a role in improving cognition. Raised choline levels in the brain enhances cholinergic function. The benefits of CDP choline go beyond just improved memory; they also include increased attention, mental endurance, faster reaction times, and improved thought clarity (Bruce et al., 2014). In addition to enhancing cognition, this drug also finds use in the treatment of glaucoma, age-related neurodegenerative illnesses such as Alzheimer's disease & Parkinson's disease, head traumas, stroke, cocaine addiction, and stroke. CDP choline functions in a number of ways, including restoring and fixing neuronal membranes increasing dopamine, adrenaline, and norepinephrine levels in the brain, and by increasing the synthesis of the "learning neurotransmitter," acetylcholine (Dixon et al., 1997). It synergistically enhances the effects of 'racetam' class of nootropics like piracetam and pramiracetam whilst reducing the incidence of headaches, a typical adverse effect of racetam intake.

Pyritinol

Pyritinol (or pyridoxine disulfide or pyrithioxin) (Figure. 6) (Cerbon 6®, Encefabol®), was first synthesized in 1961 by linking two molecules of vitamin B_6 (pyridoxine) *via* a disulfide bridge. Pyritinol passes through the blood-brain barrier and accumulates in major brain areas (Singh & Purohit, 2019) and demonstrated its effects on various neurotransmitters. It has been marketed as a nootropic dietary supplement for more than thirty years. It causes an increased choline acetyltransferase activity, leading to choline accumulation in the cholinergic neurons (Toledano & Bentura, 1994). In vivo studies have shown its supporting role in the recovery of age-related brain deficits. Pyritinol is able to scavenge oxygen free radicals, thus acting as an antioxidant and improving cerebral circulation, and also restores the levels of the age-related decrease in primary excitatory neurotransmitter N-methyl-D-aspartate (NMDA) (Hindmarch et al., 1990). Due to the status of pyritinol as a dietary supplement in some countries, it can be safely used as an adjunct to any standard treatment of the CNS diseases, such as developmental dysphasia and other cognitive disorders, for which current therapeutic options are limited. The minimum recommended daily dose is 300 mg, divided into three sub-doses. Pyritinol crosses the placental barrier, but animal studies have not shown any teratogenic or embryotoxic effects. Only a minimal amount of pyritinol is excreted in human milk but a careful evaluation should still be made before dosing during pregnancy and lactation (Nachbar et al., 1993).

Dihydroergotoxine Mesylate

Dihydroergotoxine mesylate (Figure. 6), also called ergoloid mesylate (Hydergine®), is a mixture of the methanesulfonate salts of dihydrogenated ergot alkaloids: dihydroergocornine (DHCO), dihydroergocristine (DHEC), α-dihydroergocryptine (α-DHC), and β-dihydroergocryptine (β-DHC). The drug, developed in the 1940s by Albert Hofmann (Stoll & Hofmann, 1943), it is one of the oldest nootropic drugs still in use. The drug was originally developed as an antihypertensive, but later, it was found to improve mental health when patients with Alzheimer's disease were treated for hypertension. Dihydroergotamine was found to enhance neuronal metabolism *via* stimulation of the local glucose utilization in the brain areas related to learning and memory. This effect has been utilized in geriatric patients with multi-infarct dementia (Nagasawa et al., 1990). It has been shown to reduce the levels of monoamine oxidase enzymes (responsible for depletion of catecholamine neurotransmitters like dopamine, norepinephrine, and adrenaline, thus, impairing mental function), which are generally elevated during the ageing process. It has been shown to regulate the release of the neurotransmitter acetylcholine from the hippocampus and increase the number of cholinergic receptors. Further, dihydroergotamine also protects the brain against hypoxia (Thompson et al., 1990). The combination of piracetam with xanthine derivatives, which have a bronchodilator and vasodilatory effect, increases the effect of the ergot alkaloids.

Phosphatidylcholine (Lecithin)

Phosphatidylcholine (Figure. 6) refers to a class of phospholipids containing choline, glycerophosphoric acid, with a variety of fatty acids, and this is the main liquid component of the cell membrane. A mixture of phospholipids in oil is called commercial lecithin, including phosphatidylcholine, followed by phosphatidylethanolamine, phosphatidylserine, phosphatidylinositol, and phosphatidylglycerols. The primary source of lecithin is soybean and sunflower oil. It is also found in some other nutrients such as egg yolk, fish liver oil, whole green products and nut kernels. The mechanism of action of lecithin is the release of choline from acetylcholine synthesis (Ladd et al., 1993). Animal studies have remission of dementia on administration of phosphatidylcholine, attrinuted to increasing acetylcholine concentrations in the brainresulting in improved memory. However, randomization of *in vivo* studies has also shown no benefit of lecithin in patients suffering from Alzheimer's and Parkinson's disease. Phosphatidylcholine supplements uniformly improve memory, suggesting that the dose and time parameters required to achieve a therapeutic effect may depend on variables intrinsic to individual subjects. The recommended normal dose of lecithin is 1200 mg, three times a day as a prophylactic, or 10-15 g/day or higher for therapeutic purpose (Wendel, 2000).

Nefiracetam

Nefiracetam (Figure. 6)), is another nootropic drug, pyrrolidinone derivative, belonging to the ácetam' family. Studies demonstrate that it reduces drug-induced, or age-related cognitive impairment (Woodruff-Pak, 1997). Nefiracetam use is hypothesised to involve the cholinergic system. By activating the protein kinase C, nefiracetam is thought to affect the phosphorylation of the nicotinic ACh receptor, facilitating the release of neurotransmitters from the presynaptic terminal (Nishizaki et al., 1999).

PRL-8-53

PRL-8-53 (Figure. 6), chemically, methyl 3-(2-(benzylmethylamino)ethyl)benzoate HCl, is a synthetic research chemical discovered and patented by the neuropharmacologist, Dr Nikolaus Hansl in 1970s. This simple molecule has drawn the attention of the nootropic community for many years. There has only been one small human trial on the substance that was funded by the patent holder, where the findings indicated that a single dose of PRL-8-53 could boost short-term memory by varying proportions in diverse types of subjects ranging from insignificant to as high as 150% (Hansl & Mead, 1978). PRL-8-53 has not been further developed since Hansl passed away in 2011, however, the substance is still sold in the market.

The drug was discovered in the 1970s, when preliminary animal experiments revealed the substance to be a safe nootropic, enhancing avoidance learning in rodents without any significant side effects. Later in 1978, the lone trial on human volunteers based on word retention scores showed that a single 5 mg dose of the compound could improve the memory nearly by two-fold. The increase in the rate of recall was found to be more significant (nearly double) in participants with poor initial word retention scores, compared to those with high initial scores (statistically insignificant). Although anecdotal data does seem to back up claims about its safety and nootropic efficacy, PRL-8-53 still remains an experimental substance on account of its insufficient testing. The precise mechanism of PRL-8-53 is not fully understood, however, it increases the response of the brain to acetylcholine, potentiates dopamine and somewhat recuces the biosynthesis of serotonin.

SMART DRUGS TARGETING DOPAMINERGIC PATHWAY

Mucuna

Mucuna puriens (Family Fabaceae), also known as Velvet bean, Bengal velvet bean or Monkey tamarind, is a nootropic of natural origin which contains 5% L-DOPA, a precursor of dopamine in the body. L-DOPA can quickly and easily penetrate the blood-brain barrier when one consumes mucuna, increasing the brain dopamine levels. For hundreds of years, Parkinson's disease and other ailments have been supported by the powdered form of mucuna in traditional Indian medicine. Mucuna can also boost the production of neurotransmitters like norepinephrine and serotonin, which are involved in mood regulation, in addition to dopamine levels. This plant extract has been investigated as a nootropic by virtue of its ability to raise dopamine levels. A modest improvement in cognition have still been shown by increasing L-DOPA even if dopamine levels are normal (Lampariello et al., 2012).

Figure 7. L-tyrosine

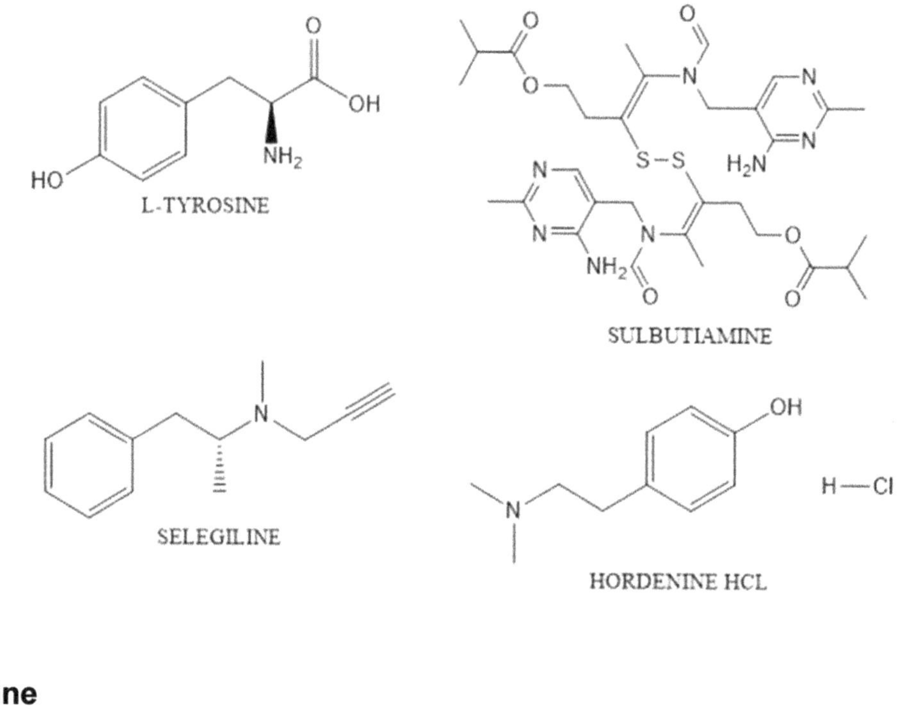

L-Tyrosine

L-tyrosine (Figure. 7) is a naturally occurring amino acid and acts as the precursor for three of the most significant neurotransmitters, dopamine, adrenaline, and norepinephrine having an important role in cognitive functions and memory. This is biosynthesized in the liver inside our body from another amino acid called L-phenylalanine, and can also be obtained through foods, dairy, meat, fish, and eggs (Hase et al., 2015). The L-tyrosine production by the body is adequate, so, supplementation is not required for good health. The body can use a buffer of L-tyrosine when it needs to make more catecholamines by having enough of the amino acid in the body. Under conditions of extreme physical or mental stress, the catecholamines like dopamine, norepinephrine, and adrenaline get swiftly depleted, and the body uses a buffer of L-tyrosine for producing more of these neurotransmitters by releasing an enzyme that transforms L-tyrosine first into the catecholamine precursor L-DOPA and then into dopamine, some of which is oxidized and transformed into norepinephrine and adrenaline (Hinz et al., 2011). This inbuilt regulatory system can be used to support the supplementing of L-tyrosine. By making sure there is enough tyrosine available during extended durations of stimulation, maintaining adequate amounts of L-tyrosine may boost catecholamine synthesis and release. L-tyrosine has been demonstrated to be safe and beneficial when taken in once daily dose of 500–2000 mg for improving the mental functions or to lessen the effects of stress. N-Acetyl-L-Tyrosine (NALT), is an acetylated derivative of L-tyrosine with increased bioavailability. NALT has a lower rate of conversion in the body than the normal form, making greater doses possible. The functions and advantages of the supplement are the same in both forms.

Sulbutiamine

Sulbutiamine (Enerion®, Arcalion®) (Figure. 7), a synthetic derivative of vitamin B_1 (thiamine), is formed by linking two thiamine molecules joined by a disulphide linkage. Thiamine deficiency has been associated with problems with memory and cognition, hence, thiamine and its analogs including sulbutiamine have been studied for age-relatted cognitive decline. Because of its structure, sulbutiamine is more accessible than normal thiamine and has improved blood-brain barrier penetration. In order to treat beriberi, a nervous illness brought on by thiamine deficiency, Japanese researchers initially developed the sulbutiamine molecule (Trovero et al., 2000). It effectively treats thiamine deficiency-related cognitive issues and may also improve memory and cognition in healthy individuals. It is also known to decrease both mental and physical exhaustion and might enhance long-term memory. Compared to racetams and ampakines, sulbutiamine operates by increasing the density of dopaminergic D_1 and D_2 receptors. Sulbutiamine has been shown to have some addictive potential despite being significantly safer than the majority of "smart drugs." Moreover, studies show that tolerance also develops on its use, which is common with most nootropics based on dopamine. Sulbutiamine also moderately affects the cholinergic pathway as it promotes the production of choline and inhibits acetylcholinesterase, an enzyme responsible for metabolism of the neurotransmitter acetylcholine. A daily dosage of about 600 mg was used in the clinical trials (Moerk, 2000) and a dose of 400–600 mg is recommended for adults, for a period of 4 weeks.

Selegiline (L-Deprenyl)

Selegiline or L-Deprenyl (Eldepryl®) (Figure. 7) is a prescription-only medication that has been employed as an antiparkinsonian and antidepressant drug. At lower doses, selegiline selectively and irreversibly inhibits the monoamine oxidase B (MAO-B) enzyme, thus, increasing dopamine brain levels. However, at higher doses, it also inhibits MAO-A enzyme, which results in an increase in brain serotonin and norepinephrine levels. Hence, the drug is regarded as a potent nootropic as it helps to stabilise memory and concentration levels by slowing down the breakdown of important neurotransmitters in the brain, including serotonin, dopamine, and norepinephrine. Though it is not FDA-approved for the purpose, it has also been used as an off-label preventive measure for dementia in Alzheimer's disease.

Hordenine HCl

Hordenine HCl (Figure. 7), or N,N-dimethyltyramine, is a naturally occurring substance present in a variety of plants, including germinated seeds of barley (*Hordeum vulgare*). It belongs to the category of phenethylamine type of alkaloids and thus, it shares its molecular structure with endogenous catecholamines like dopamine and norepinephrine. This dopaminergic nootropic, which is available over the counter in many nations, is supposed to moderately stimulate the mind. However, despite the fact that many nutritional supplements containing hordenine advertise as a fat burner, there is little to no proof that hordenine taken regularly *via* the oral route has the ability to speed up the body's metabolism. The drug is believed to act through modest inhibition of the MAO-B enzyme resulting in elevation of dopamine levels in the brain.

Figure 8. Aniracetam

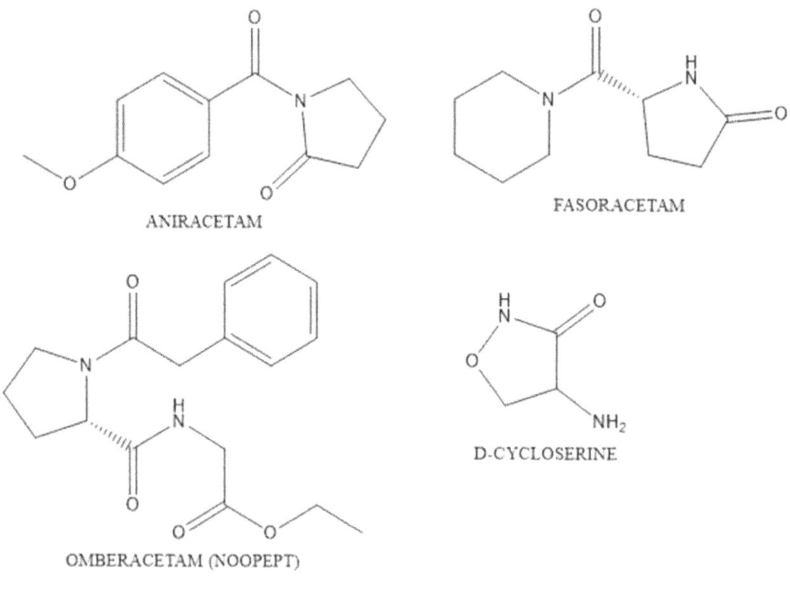

SMART DRUGS TARGETING GLUTAMATERGIC PATHWAY

Aniracetam

Aniracetam (Figure. 8) is a pyrrolidone derivative belonging to a general category of ácetam' drugs, typically associated with enhancement in cognitive function *via* tblockade of the AMPA receptors (Rao, 2001). It was first investigated in the 1980s as a more potent analog of piracetam for cognitive enhancement. It has been demonstrated that aniracetam, including other pyrrolidone derivatives, can lessen the cognitive impairment. The effects of aniracetam as a cognitive enhancer have been hypothesised to be a result of the slow rate of desensitisation and deactivation of the AMPA receptors. The participation of active hippocampus PKC and a sustained membrane/cytosolic PKC ratio has been hypothesised by other research. The phosphorylation of glutamate receptor subunits subsequently causes the increase of

PKC, which modifies the channel dynamics of the AMPA receptor (Gomes et al., 2003). As with other nootropics, antiracetam possesses neuroprotective actions and it also increases brain metabolism, especially under deficit conditions, including those resulting from hypoxia, chemical toxicity, or impaired cerebral circulation. Its use has been investigated for improving cerebral blood flow in Alzheimer's disease by Lee & Benfield (1994).

Fasoracetam

Fasoracetam (Figure. 8), the newest nootropic in the racetam family, is being researched as a possible non-stimulant treatment for attention deficit hyperactivity disorder (ADHD). It has been postulated to also increase memory, lessen anxiety and drug-withdrawal symptoms, according to animal studies. The drug is thought to control the synthesis and release of glutamate, GABA, and acetylcholine, having a predominant effect on glutamate (Harmsen et al., 2017). Despite its efficacy in animal studies and preliminary clinical studies, it had to be finally abandoned due to lack of efficacy in Phase 3 clinical trials. A critical excitatory neurotransmitter necessary for many aspects of brain function, glutamate, is modulated by fasoracetam in at least some of the glutamate receptors in the brain. Fasoracetam has been shown to target all the three types of metabotropic glutamate receptors (mGluRs), which have a range of functions in the body and brain but are majorly associated with learning, memory, and anxiety (Ogasawara et al., 1999).

Omberacetam (Noopept)

Omberacetam (GVS 111; Noopept®) (Figure. 8) is chemically, N-Phenylacetyl-l-prolylglycine ethyl ester. It is a powerful nootropic with mechanism of action similar to Piracetam but having a different chemical structure from racetam-drugs. It has a much higher absorption rate, higher potency (1000 times) along with an improved onset and duration of action compared to Piracetam (Vorobyov et al., 2011). As piracetam, noopept aids in memory development but it gives additional advantages like improved memory consolidation and retrieval. Noopept has been demonstrated to boost the expression of two crucial molecules involved in cognition, i.e., the Nerve Growth Factor (NGF) and Brain-Derived Neurotrophic Factor (BDNF), in animal tests. Diseases like dementia, schizophrenia, Alzheimer's, bipolar disorder, can precipitate on account of low levels of NGF. BDNF is one of the most significant molecules involved in memory, which plays a crucial role in the creation of both short-term and long-term memories. After entering the bloodstream, noopept works swiftly, increasing its concentration in about 15 to 20 minutes (Neznamov & Teleshova, 2009). Noopept also affects the activation of AMPA and NDMA receptors in the brain, thus, modulating the levels of glutamate. It reduces glutamate toxicity (responsible for neuronal damage), and also promotes neuroplasticity, which is essential for the development of long-term memories. The potent neuroprotective potential of nooopept is attributed to its antioxidant and anti-inflammatory properties.

D-Cycloserine

Cycloserine (Figure. 8), marketed as Seromycin®, is a nootropic medication with chemical name 4-amino-3-isoxazolidinone. It was earlier used as an antibiotic against the bacterium, *Mycobacterium tuberculosis*, the causative agent for tuberculosis. Preliminary studies suggest that D-cycloserine aids in learning by helping form stronger neural connections.NMDA stimulation is also touted as one of the mechanisms of D-cycloserine for produce its desired effects. D-cycloserine affects the glycine binding sites in these glutamatergic receptors which are crucial for activating NMDA channels. There is an increase in excitatory neurotransmission when the channels are open. Studies have shown that D-cycloserine is even more effective than the endogenous agonists like D-serine and glycine at the NR_1/NR_2 receptors (Baran & Kepplinger, 2014).

SMART DRUGS AFFECTING SEROTONERGIC PATHWAY

Tryptophan

Tryptophan (Figure. 9) is one of the precursors of serotonin, which is responsible for modulating mood, behaviour, and cognitive functions. Serotonin is often referred to as the "happy hormone". Tryptophan is a natural amino acid that our body needs in sufficient amounts in order to make serotonin. Tryptophan intake has been shown to increase brain serotonin metabolism which is considered to be responsible for enhancement in memory. Tryptophan administration results in an improvement in spatial learning and memory attributed to serotonin synthesis (Haider et al., 2007). Both short-term and long-term memory loss occurs in people with low tryptophan levels. Adults in good health have shown improved memory after taking tryptophan supplements. However, recently, a study has shown that memory improvement brought about by tryptophan is not dependent on its role as a precursor for serotonin and its effects are related to its interaction with microtubule proteins where it induces the rate of tubulin assembly with resultant increase in polymer mass (Yousefzadeh et al., 2020). Tryptophan is utilised as a nootropic supplement in patients with anxiety, depression, ADHD, insomnia, amnesia, and eating disorders.

Figure 9. Tryptophan

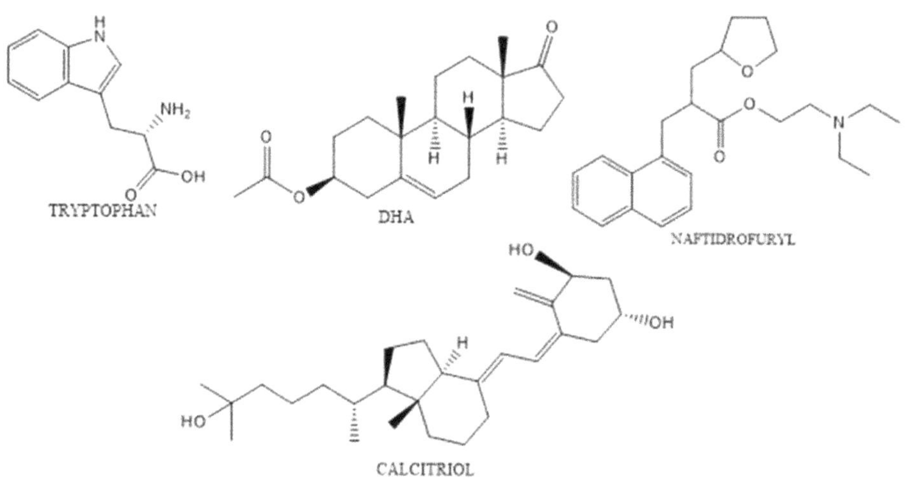

Omega-3 Fatty Acids

Omega-3 fatty acids are polyunsaturated fatty acids which cannot be synthesized by the body and are essential for a healthy brain and body. There are of three types, i.e., docosahexaenoic acid (DHA) (Figure. 9), eicosapentaenoic acid (EPA), and linolenic acid (ALA). The serotonin receptor in the brain is hypothesised to be activated by DHA, which makes it more susceptible to serotonin impulses and improves mood. EPA, on the other hand, is thought to encourage the release of serotonin in the brain. The importance of omega-3 fatty acids in the serotonin pathway is supported by this dual strategy. The reason for different responses in different people from supplementing is still a mystery given that a clear mechanism already exists. A positive association exists between the plasma and dietary intake levels of DHA and cognition. DHA, given alone or in combination with EPA, shows improvement in memory function in older patients (Yurko-Mauro et al., 2015).

Naftidrofuryl

Naftidrofuryl (Figure 9), a 2-(1-naphthalenylmethyl)-3-(2-oxolanyl) propanoic acid ester of 2-(diethyl-amino)ethanol, acts as a vasodilator showing a rheological effect on the blood. It is used for management of vascular disorders. It is a 5-HT_{2A} receptor antagonist and also inhibits serotonin-induced contractions in blood vessels. It has been shown t demonstrate a positive effect on neurons' energy metabolism. The drug has been used in senile dementia and for treatment of Alzheimer's disease (Goldsmith & Wellington, 2005).

Vitamin D

Vitamin D (Figure 9) is a fat-soluble vitamin which plays an important role in improving cognitive function. It is a neurosteroid hormone which is linked to regulation of neurotransmitters and neurotrophins. There are nootropic products available in the market, which contain vitamin B_{12} and vitamin D (OptiMind®). The brain's neurochemistry relies upon getting an adequate amount of daily vitamin D intake and this also aids in the body's synthesis of good neurochemicals. Reduced exposure to sunlight results in deficiency of Vitamin D, and numerous studies have linked this deficiency to low serotonin levels, sadness, and particularly seasonal affective disorder (SAD), and there is substantial evidence that supplementation may be helpful. Tryptophan, in combination with vitamin D, can help in the process of turning vitamin D into 5-hydroxytryptophan and subsequently, serotonin. Further, studies have shown that levels of vitamin D are significantly reduced in patients with Alzheimer's disease and cognitive impairment, when compared to healthy adults (Sultan et al., 2020).

SMART DRUGS TARGETING MULTIPLE PATHWAYS

Piracetam

Piracetam (Figure 10), chemically, 2-oxo-1-pyrrolidineacetamide, was the first synthetic smart drug to be developed. It is a cyclic derivative of the inhibitory neurotransmitter gamma-aminobutyric acid (GABA), and is a popular medication for treating Alzheimer's disease and senile dementia. Studies have shown that piracetam plays a role in improving memory and learning and that it works synergistically with choline to improve cognition. indicating that piracetam may have an effect on the NMDA receptors (Nickolson & Wolthuis, 1976). The mechanism of action of piracetam is not fully understood but it is believed to act through a combination of glutamatergic and cholinergic mechanisms. Treatment with piracetam results in AMPA receptor activation (it acts as a positive allosteric modulator), which stimulates the brain's Ca^{2+} influx and increases the density of AMPA receptors in the synaptic cleft. Piracetam enhances synaptic plasticity (the formation and maintenance of new connections between neurons), which is essential for learning, by enhancing blood flow inside the brain and by enhancing the production and activity of essential brain chemicals. It improves ACh function by acting on muscarinic cholinergic (ACh) receptors might increase acetylcholine production regulation. It is clinically useful in the treatment and prevention of mental dysfunction, and post-traumatic cognition and has been found to improve learning and memory in dyslexic children and Alzheimer's disease patients. Several structural analogues of piracetam are there such as oxiracetam, etiracetam, pramiracetam, nefiracetam, and aniracetam, which have similar actions but with varying efficacy. A single dose of 4800 mg of piracetam was shown to be the most beneficial for improving cognitive function in healthy persons. In another study, found that 1200 mg doses of piracetam given thrice a day, provided measurable benefits for cognition.

Figure 10. Piracetam

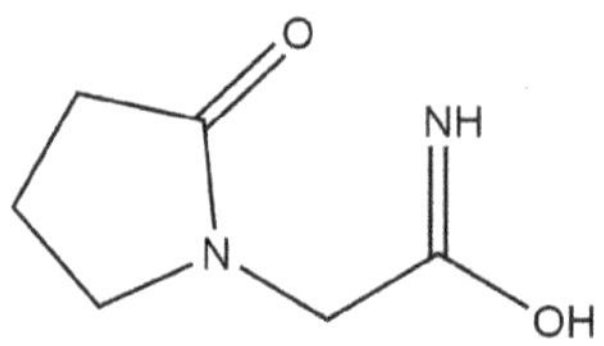

PIRACETAM

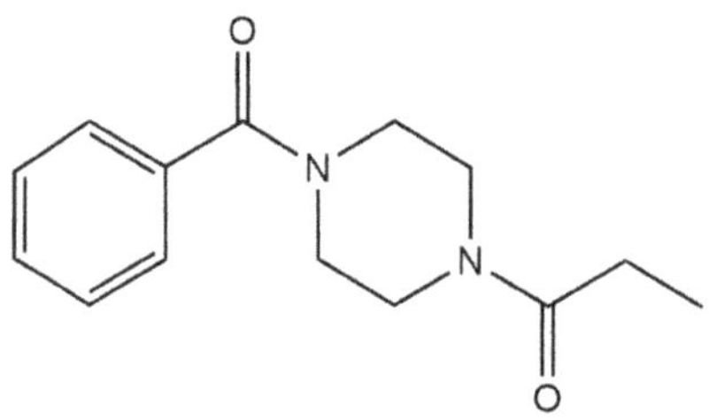

SUNIFIRAM

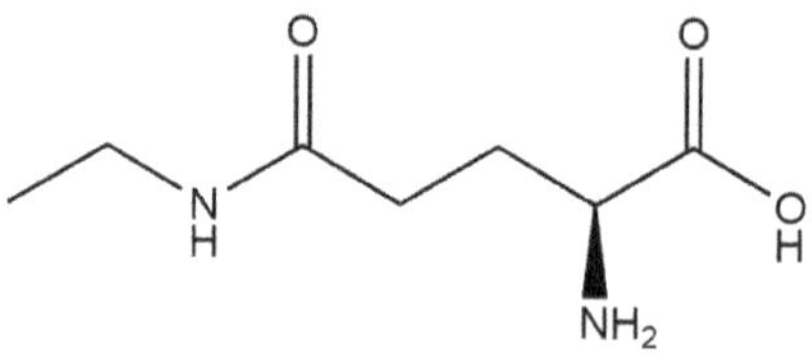

L-THEANINE

Sunifiram

Sunifiram (DM-235) (Figure 10), is an experimental drug developed as a derivative of piracetam which differs chemically from racetams. Although its exact mechanisms of action are not fully understood, it is believed to function by activating glutamate receptors and by enhancing the biosynthesis and release of the "learning neurotransmitter" acetylcholine. Sunifiram is thought to function primarily as an ampakine, which implies that after crossing the BBB, it binds to AMPA-type glutamate receptors in the brain (Romanelli et al., 2006). This promotes the synthesis of glutamate. Long-term potentiation, or the long-lasting improvement in signal strength between neurons, depends on adequate glutamate levels. Glutamate levels are particularly significant in brain hippocampus, which is primarily involved in spatial navigation as well as the development and storage of memory. It is thought that a large portion of sunifiram's ability to enhance cognition is ultimately achieved by strengthening neural signals in this area of the brain (Moriguchi et al., 2013). Ampakines generally operate as mild stimulants that boost alertness and energy without the restlessness and other negative side effects associated with coffee, methylphenidate (Ritalin), and amphetamines, in addition to improving memory formation and recall.

Sunifiram additionally increases acetylcholine release and synthesis. An Italian animal study found that experiments utilising piperazine compounds like sunifiram had a cholinergic impact similar to that of piracetam, despite the fact that the precise mechanism by which sunifiram modifies acetylcholine synthesis has not been determined.

L-Theanine

L-Theanine (Figure. 10), chemically γ-glutamylethylamide, is a non-protein amino acid which is found in tea plants. L-theanine has been shown to improve memory, concentration, and general cognitive function along with learning, focus, and attention. When combined with caffeine, the nootropic effect of L-theanine is especially potent. Studies show that combination of L-theanine with caffeine eliminates the vasoconstrictive and behavioural effects of caffeine, resulting in improved cerebral blood flow, enhanced cognition, and mood elevation (Giesbrecht et al., 2010). L-theanine primarily affects the activity of inhibitory neurotransmitters in the brain, raising their levels while actively suppressing the overproduction of excitatory neurotransmitters. L-theanine can bind to and inhibit glutamate receptors in the brain because its molecular structure is comparable to glutamate, the most significant excitatory neurotransmitter. This successfully reduces glutamate levels and stops the overactivity of glutamate receptors, which is often brought about by stress, leading to an improvement in calmness and relaxation. Theanine has been reported to raise levels of brain serotonin and dopamine, with possible improvement in specific memory and learning tasks (Nathan et al., 2006). A recent study has related green tea consumption to suppression of brain ageing (Keiko & Yoriyuki, 2021). The calming neurotransmitters dopamine, serotonin, and GABA, all of which are crucial for controlling mood, alertness, sleep, and energy, are reported to be increased by L-theanine.

SOME OTHER NATURAL SMART DRUGS

Ginkgo (*Ginkgo biloba* Fam. Ginkgoaceae)

Ginkgo biloba, also referred to as a living fossil, is one of the oldest plant species which has been used for treating various ailments ranging from infection to asthma. Numerous clinical studies indicate that *Ginkgo biloba* supplements significantly enhance memory, particularly in people with mild to severe age-related dementia. A six-week trial including middle-aged subjects with normal cognition found that daily *G. biloba* administration significantly improved information memory while leaving overall cognition unaltered. The mechanism of action of *ginkgo biloba* involves free radical scavenging antioxidant activity. There are also some other mechanisms of action such as antagonizing the effect of the platelet-activating factor which results in the reduction of blood viscosity and vasodilation (Diamond et al., 2000). *Ginkgo biloba* is also known to have anti-inflammatory properties and also acts *via* the cholinergic system (Kim et al., 2016). Triterpene Lactones (ginkgolide A, ginkgolide B, ginkgolide C and bilobalide) are present in the extract of *ginkgo biloba* which are responsible for the antioxidant, anti-inflammatory and neuroprotective effects. The extract also has an antagonizing effect on glycine and $GABA_A$ receptors (Nishimon et al., 2020). Clinical trials have shown the positive results of *G. biloba* extract in improving memory, speed of information processing, and cognitive functions, for treatment of cerebral dysfunction and neurological disorders. It is also known to cause blood thinning so it is contraindicated in patients taking anticoagulants (Ekor, 2014).

Ashwagandha (*Withania somnifera*, Fam. Solanaceae)

Ashwagandha, also known as Indian Winter Cherry or Indian Ginseng, is one of the most important herbs in the Indian traditional system of medicine. The root extract of *Withania somnifera,* has been shown to decrease lipid peroxidation, restore low glutathione levels, and reverse the drop in brain superoxide dismutase and catalase levels, brought about by injecting reserpine. *Withania somnifera* root extract has been shown to be an effective medication for treating tardive dyskinesia and cognitive dysfunction brought about by various drugs (Naidu et al., 2006). The nootropic effect of ashwagandha and its promise as a therapy for Alzheimer's disease has been validated by numerous studies in animal models (Schliebs et al., 1997). Ashwagandha is not advised in pregnant women and in women suffering from hyperthyroidism. Due to the herbal extract's potential to serve as a sedative in high amounts, ashwagandha is best consumed in the evening (Meher et al., 2016).

Water Hyssop (*Bacopa monnieri*, Fam. Plantaginaceae)

Water Hyssop, also known as Brahmi, has an important role in enhancing memory and cognition. Extract of *Bacopa monnieri* enhances protein kinase activity and shows an increase in protein in the hippocampus. Overall, the extract is proven to enhance learning ability by enhancing cognitive function and memory retention. The chemical compounds responsible for this facilitating effect have been identified as a mixture of two saponins, bacoside A and bacoside B (Dhawan & Singh, 1996). However, subsequent administration of an alcoholic extract of *Bacopa monnieri* reversed this effect and gradually. Experiments have shown that an alcoholic extract of *Bacopa monnieri* enhances the expression of choline acetyltransferase enzyme (responsible for biosynthesis of ACh) in the hippocampus, and also inhibits the degeneration of cholinergic neurons, thus reversing the induced cognitive decline due to cholinergic dysfunction. This suggests that the herb is a cognitive enhancer and a neuroprotectant, and it can be a potential adjunctive drug for the treatment of Alzheimer's disease (Le et al., 2013)

CONCLUSION

In an increasingly skill-driven and socially interactive world, memory is one of the keys to success. There are several disease states that are characterized by impaired cognition and memory, for example, Down's syndrome, stroke, senile dementia, Alzheimer's disease, etc. These are chronically debilitating conditions, therefore, there is a strong medical drive towards research for development of neuroprotection strategies, or at least to reduce the rate of decline in such disorders. Treatment of cognitive impairment associated with Alzheimer's disease has become an important target for research institutions and the pharmaceutical industry. As research on receptor pharmacology unravels new targets and sheds greater light on existing ones, better approaches can be directed towards the development of smart drugs with potential applications in various neurodegenerative disorders associated with memory loss. This chapter reviews the various types of smart drugs and their ability to enhance cognitive performance in various health disorders, their positive effects on brain health and synthetic biology approaches in their design with a focus on diverse pharmacological targets. This is one therapeutic class of agents which have got potential takers amongst normal individuals too, who wish to enhance or improve upon their working memory, performance and learning processes. In short, nootropics aim to improve IQ scores in children, professional performance in young adults, learning and memory in middle age subjects and senile dementia in geriatric patients suffering from chronic neurodegenerative disorders.

REFERENCES

Banga, S., Kumar, V., Suri, S., Kaushal, M., Prasad, R., & Kaur, S. (2019). Nutraceutical potential of Diet Drinks: A critical review on components, health effects, and Consumer Safety. *Journal of the American College of Nutrition*, *39*(3), 272–286. doi:10.1080/07315724.2019.1642811 PMID:31343956

Baran, H., & Kepplinger, B. (2014). D-cycloserine lowers kynurenic acid formation—New mechanism of action. *European Neuropsychopharmacology*, *24*(4), 639–644. doi:10.1016/j.euroneuro.2013.10.006 PMID:24189377

Benninghoff, J., & Perneczky, R. (2022). Anti-dementia medications and anti-alzheimer's disease drugs: Side effects, contraindications, and interactions. *NeuroPsychopharmacotherapy*, 1–10.

Bhattacharya, S. K., Kumar, A., & Ghosal, S. (1995). Effects of glycowithanolides from *withania somnifera* on an animal model of alzheimer's disease and perturbed central cholinergic markers of cognition in rats. *Phytotherapy Research*, *9*(2), 110–113. doi:10.1002/ptr.2650090206

Bleakney, T. L. (2008). Deconstructing an adaptogen. *Holistic Nursing Practice*, *22*(4), 220–224. doi:10.1097/01.HNP.0000326005.65310.7c PMID:18607235

Brinckmann, J., & Smith, E. (2004). Maca culture of the Junín Plateau. *Journal of Alternative and Complementary Medicine (New York, N.Y.)*, *10*(3), 426–430. doi:10.1089/1075553041323821 PMID:15253845

Brinkhaus, B., Lindner, M., Schuppan, D., & Hahn, E. G. (2000). Chemical, pharmacological and clinical profile of the East Asian Medical Plant *Centella Aslatica*. *Phytomedicine*, *7*(5), 427–448. doi:10.1016/S0944-7113(00)80065-3 PMID:11081995

Bruce, S. E., Werner, K. B., Preston, B. F., & Baker, L. M. (2014). Improvements in concentration, working memory and sustained attention following consumption of a natural citicoline–caffeine beverage. *International Journal of Food Sciences and Nutrition*, *65*(8), 1003–1007. doi:10.3109/09637486.2014.940286 PMID:25046515

Butterfield, D. A., & Pocernich, C. B. (2003). The glutamatergic system and alzheimer's disease. *CNS Drugs*, *17*(9), 641–652. doi:10.2165/00023210-200317090-00004 PMID:12828500

Chanana, P., & Kumar, A. (2016). Possible involvement of nitric oxide modulatory mechanisms in the neuroprotective effect of *centella asiatica* against sleep deprivation induced anxiety like behaviour, oxidative damage and neuroinflammation. *Phytotherapy Research*, *30*(4), 671–680. doi:10.1002/ptr.5582 PMID:26848139

Chaudhari, K. S., Tiwari, N. R., Tiwari, R. R., & Sharma, R. S. (2017). Neurocognitive effect of nootropic drug brahmi (*bacopa monnieri*) in alzheimer's disease. *Annals of Neurosciences*, *24*(2), 111–122. doi:10.1159/000475900 PMID:28588366

Chen, C.-L., Tsai, W.-H., Chen, C.-J., & Pan, T.-M. (2016). *Centella asiatica* extract protects against amyloid β1–40-induced neurotoxicity in neuronal cells by activating the Antioxidative Defence System. *Journal of Traditional and Complementary Medicine*, *6*(4), 362–369. doi:10.1016/j.jtcme.2015.07.002 PMID:27774420

Dhawan, B. N., & Singh, H. K. (1996). Pharmacological studies on *bacopa monniera*, an Ayurvedic nootropic agent. *European Neuropsychopharmacology*, *6*(1), 142–144. doi:10.1016/0924-977X(96)87969-7

Diamond, B. J., Shiflett, S. C., Feiwel, N., Matheis, R. J., Noskin, O., Richards, J. A., & Schoenberger, N. E. (2000). *Ginkgo biloba* extract: Mechanisms and clinical indications. *Archives of Physical Medicine and Rehabilitation*, *81*(5), 668–678. doi:10.1016/S0003-9993(00)90052-2 PMID:10807109

Dixon, C., Ma, X., & Marion, D. (1997). Effects of CDP-choline treatment on neurobehavioral deficits after TBI and on hippocampal and neocortical acetylcholine release. *Journal of Neurotrauma*, *14*(3), 161–169. doi:10.1089/neu.1997.14.161 PMID:9104933

Dormehi, I., & Jordaan, B. (1999). Spect monitoring of improved cerebral blood flow during long-term treatment of elderly patients with nootropic drugs. *Clinical Nuclear Medicine*, *24*(1), 29–34. doi:10.1097/00003072-199901000-00007 PMID:9890490

Durham, D. (1991). Dictionary of drugs—Chemical data, structures and bibliographies. Chapman & Hall.

Ekor, M. (2014). The growing use of herbal medicines: Issues relating to adverse reactions and challenges in monitoring safety. *Frontiers in Pharmacology*, *177*(4), 35–38. doi:10.3389/fphar.2013.00177 PMID:24454289

Espinola, E. B., Dias, R. F., Mattei, R., & Carlini, E. A. (1997). Pharmacological activity of Guarana (*Paullinia Cupana*) in laboratory animals. *Journal of Ethnopharmacology*, *55*(3), 223–229. doi:10.1016/S0378-8741(96)01506-1 PMID:9080343

Froestl, W., Muhs, A., & Pfeifer, A. (2012). Cognitive enhancers (nootropics). part 1: Drugs interacting with receptors. *Journal of Alzheimer's Disease*, *32*(4), 793–887. doi:10.3233/JAD-2012-121186 PMID:22886028

Giesbrecht, T., Rycroft, J. A., Rowson, M. J., & De Bruin, E. A. (2010). The combination of L-theanine and caffeine improves cognitive performance and increases subjective alertness. *Nutritional Neuroscience*, *13*(6), 283–290. doi:10.1179/147683010X12611460764840 PMID:21040626

Giurgea, C. (1973). The "nootropic" approach to the pharmacology of the integrative activity of the Brain 1, 2. *Conditional Reflex*, *8*(2), 108–115. doi:10.1007/BF03000311 PMID:4736348

Giurgea, C., & Salama, M. (1977). Nootropic drugs. *Progress in Neuro-Psychopharmacology*, *1*(34), 235–247. doi:10.1016/0364-7722(77)90046-7

Gohil, K. J., Patel, J. A., & Gajjar, A. K. (2010). Pharmacological review on Centella Asiatica: A potential herbal cure-all. *Indian Journal of Pharmaceutical Sciences*, *72*(5), 546. doi:10.4103/0250-474X.78519 PMID:21694984

Goldsmith, D. R., & Wellington, K. (2005). Naftidrofuryl. *Drugs & Aging*, *22*(11), 967–977. doi:10.2165/00002512-200522110-00006 PMID:16323973

Gomes, A. R., Correia, S. S., Carvalho, A. L., & Duarte, C. B. (2003)... *Neurochemical Research*, *28*(10), 1459–1473. doi:10.1023/A:1025610122776 PMID:14570391

Gonzales-Arimborgo, C., Yupanqui, I., Montero, E., Alarcón-Yaquetto, D., Zevallos-Concha, A., Caballero, L., Gasco, M., Zhao, J., Khan, I., & Gonzales, G. (2016). Acceptability, safety, and efficacy of oral administration of extracts of black or red maca (*lepidium meyenii*) in adult human subjects: A randomized, double-blind, placebo-controlled study. *Pharmaceuticals (Basel, Switzerland)*, *9*(3), 47–49. doi:10.3390/ph9030049 PMID:27548190

Haider, S., Khaliq, S., & Haleem, D. (2007). Enhanced serotonergic neurotransmission in the hippocampus following tryptophan administration improves learning acquisition and memory consolidation in rats. *Pharmacological Reports*, *59*, 53–57. PMID:17377206

Hansl, N. R., & Mead, B. T. (1978). PRL-8-53: Enhanced learning and subsequent retention in humans as a result of low oral doses of new psychotropic agent. *Psychopharmacology*, *56*(3), 249–253. doi:10.1007/BF00432846 PMID:418433

Harmsen, B., Robeyns, K., Wouters, J., & Leyssens, T. (2017). A study of Fasoracetam's Solid State forms: A potential Anti-Alzheimer Pharmaceutical. *Journal of Pharmaceutical Sciences*, *106*(5), 1317–1321. doi:10.1016/j.xphs.2017.01.016 PMID:28109793

Hase, A., Jung, S. E., & aan het Rot, M. (2015). Behavioral and cognitive effects of tyrosine intake in Healthy Human Adults. *Pharmacology, Biochemistry, and Behavior*, *133*, 1–6. doi:10.1016/j.pbb.2015.03.008 PMID:25797188

Hindmarch, I., Coleston, D. M., & Kerr, J. S. (1990). Psychopharmacological effects of pyritinol in normal volunteers. *Neuropsychobiology*, *24*(3), 159–164. doi:10.1159/000119478 PMID:2135070

Hinz, M., Stein, A., Neff, R., Weinberg, R., & Uncini, T. (2011). Treatment of attention deficit hyperactivity disorder with monoamine amino acid precursors and organic cation transporter assay interpretation. *Neuropsychiatric Disease and Treatment*, *32*(5), 29–31. doi:10.2147/NDT.S16270 PMID:21326653

Hu, X., Xie, C., He, S., Zhang, Y., Li, Y., & Jiang, L. (2013). Remifentanil post conditioning improves global cerebral ischemia-induced spatial learning and memory deficit in rats via inhibition of neuronal apoptosis through the PI3K signaling pathway. *Neurological Sciences*, *34*(11), 1955–1962. doi:10.100710072-013-1419-4 PMID:23609459

Huber, G., Bailly, Y., Martin, J. R., Mariani, J., & Brugg, B. (1997). Synaptic β-amyloid precursor proteins increase with learning capacity in rats. *Neuroscience*, *80*(2), 313–320. doi:10.1016/S0306-4522(97)00120-6 PMID:9284337

Ihl, R., & Kretschmar, C. (1997). Zur Nootropikabewertung für die praxis. *Der Nervenarzt*, *68*(11), 853–861. doi:10.1007001150050207 PMID:9732727

Kennedy, D. O., Haskell, C. F., Wesnes, K. A., & Scholey, A. B. (2004). Improved cognitive performance in human volunteers following administration of Guarana (*Paullinia Cupana*) extract: Comparison and interaction with panax ginseng. *Pharmacology, Biochemistry, and Behavior*, *79*(3), 401–411. doi:10.1016/j.pbb.2004.07.014 PMID:15582012

Kim, M. S., Bang, J. H., Lee, J., Han, J. S., Baik, T. G., & Jeon, W. K. (2016). *Ginkgo biloba L.* Extract protects against chronic cerebral hypo perfusion by modulating neuroinflammation and the cholinergic system. *Phytomedicine*, *23*(12), 1356–1364. doi:10.1016/j.phymed.2016.07.013 PMID:27765355

Kołodziej, B., Sugier, D., & Bielińska, E. (2013). The effect of leonardite application and various plantation modalities on yielding and quality of roseroot (*Rhodiola rosea L.*) and soil enzymatic activity. *Journal of Geochemical Exploration*, *129*(1), 64–69. doi:10.1016/j.gexplo.2012.10.014

Koncic, M., & Tomczyk, M. (2013). New insights into dietary supplements used in sport: Active substances, pharmacological and side effects. *Current Drug Targets*, *14*(9), 1079–1092. doi:10.2174/13894501113140900016 PMID:23574283

Ladd, S. L., Sommer, S. A., LaBerge, S., & Toscano, W. (1993). Brief report. *Clinical Neuropharmacology*, *16*(6), 540–549. doi:10.1097/00002826-199312000-00007 PMID:9377589

Lampariello, L. R., Cortelazzo, A., Guerranti, R., Sticozzi, C., & Valacchi, G. (2012). The magic velvet bean of *mucuna pruriens. Journal of Traditional and Complementary Medicine*, *2*(4), 331–339. doi:10.1016/S2225-4110(16)30119-5 PMID:24716148

Le, X. T., Pham, H. T., Do, P. T., Fujiwara, H., Tanaka, K., Li, F., Van Nguyen, T., Nguyen, K. M., & Matsumoto, K. (2013). *Bacopa monnieri* ameliorates memory deficits in olfactory bulbectomized mice: Possible involvement of glutamatergic and cholinergic systems. *Neurochemical Research*, *38*(10), 2201–2215. doi:10.100711064-013-1129-6 PMID:23949198

Lee, C. R., & Benfield, P. (1994). Aniracetam. *Drugs & Aging*, *4*(3), 257–273. doi:10.2165/00002512-199404030-00007 PMID:8199398

Magnusson. (2010). Selective vulnerabilities of N-methyl-D-aspartate (NMDA) receptors during brain aging. *Frontiers in Aging Neuroscience, 12*(3), 8-14.

Mahady, G. B., Gyllenhaal, C., Fong, H. H. S., & Farnsworth, N. R. (2000). Ginsengs: A Review of Safety and Efficacy. *Nutrition in Clinical Care*, *3*(2), 90–101. doi:10.1046/j.1523-5408.2000.00020.x

Malik, R., Sangwan, A., Saihgal, R., Paul Jindal, D., & Piplani, P. (2007). Towards better brain management: Nootropics. *Current Medicinal Chemistry*, *14*(2), 123–131. doi:10.2174/092986707779313408 PMID:17266573

Marcer, D., & Hopkins, S. A. N. D. R. A. M. (1977). The differential effects of Meclofenoxate on memory loss in the elderly. *Age and Ageing*, *6*(2), 123–131. doi:10.1093/ageing/6.2.123 PMID:329662

Marques, L. L., Klein, T., & de Mello, J. C. (2019). Guarana. *Nonvitamin and Nonmineral Nutritional*, (Supplements), 283–288.

Masteikova, R., Muselik, J., Bernatonienė, J., & Bernatonienė, R. (2007). Antioxidative activity of *ginkgo*, echinacea, and ginseng tinctures. *Medicina*, *43*(4), 306. doi:10.3390/medicina43040038 PMID:17485958

Meher, S. K., Das, B., Panda, P., Bhuyan, G. C., & Rao, M. M. (2016). Uses of *withania somnifera* (linn) Dunal (ashwagandha) in ayurveda and its pharmacological evidences. *Research Journal of Pharmacology and Pharmacodynamics*, *8*(1), 20–23. doi:10.5958/2321-5836.2016.00006.9

Moerk, K. (2000). The development of major depressive episodes during the course of dysthymic and Episodic Major Depressive Disorders: A retrospective examination of life events. *Journal of Affective Disorders*, *58*(2), 117–123. doi:10.1016/S0165-0327(99)00102-0 PMID:10781701

Moriguchi, S., Tanaka, T., Narahashi, T., & Fukunaga, K. (2013). Novel nootropic drug Sunifiram enhances hippocampal synaptic efficacy via glycine-binding site of *n*-methyl-d-aspartate receptor. *Hippocampus*, *23*(10), 942–951. doi:10.1002/hipo.22150 PMID:23733502

Müller, W. E., Koch, S., Scheuer, K., Rostock, A., & Bartsch, R. (1997). Effects of piracetam on membrane fluidity in the aged mouse, rat, and human brain. *Biochemical Pharmacology*, *53*(2), 135–140. doi:10.1016/S0006-2952(96)00463-7 PMID:9037245

Nachbar, F., Korting, H. C., & Vogl, T. (1993). Erythema multiforme-like eruption in association with severe headache following pyritinol. *Dermatology (Basel, Switzerland)*, *187*(1), 42–46. doi:10.1159/000247196 PMID:8324277

Nagasawa, H., Kogure, K., Kawashima, K., Ido, T., Itoh, M., & Hatazawa, J. (1990). Effects of co-dergocrine mesylate (Hydergine) in multi-infarct dementia as evaluated by positron emission tomography. *The Tohoku Journal of Experimental Medicine*, *162*(3), 225–233. doi:10.1620/tjem.162.225 PMID:2091324

Naidu, P. S., Singh, A., & Kulkarni, S. K. (2006). Effect of *withania somnifera* root extract on reserpine-induced orofacial dyskinesia and cognitive dysfunction. *Phytotherapy Research*, *20*(2), 140–146. doi:10.1002/ptr.1823 PMID:16444668

Napoletano, F., Schifano, F., Corkery, J. M., Guirguis, A., Arillotta, D., Zangani, C., & Vento, A. (2020). The psychonauts' world of cognitive enhancers. *Frontiers in Psychiatry*, ●●●, 11. PMID:33024436

Nathan, P., Lu, K., Gray, M., & Oliver, C. (2006). The neuropharmacology of *L*-theanine (N-ethyl-L-glutamine). *Journal of Herbal Pharmacotherapy*, *6*(2), 21–30. PMID:17182482

Chapter 11
Molecular Level Analogy and Approaches of Gibson Assembly in Genome-Wide Analysis

Venkata Satya Sai Nagasri Abhishek Chembuli
Northeastern University, USA

Ajaykumar Chittipolu
https://orcid.org/0000-0001-8243-3405
Osmania University, India

Haritha Katasani
Northeastern University, USA

ABSTRACT

DNA assembly has always been a vital foundational technology for synthetic biology and other bioengineering processes. The development of restriction digestion and ligation methods has played a significant role in developing better DNA assembly methods with higher fidelity, efficiency, and modularity. Daniel Gibson discovered an exonuclease-based robust cloning method to assemble the DNA in the correct sequence known as the Gibson assembly. This method has been widely adopted as a major workhouse for synthetic biology experiments that have provided a new platform for the rapid cloning of multiple DNA fragments into the vector with a minimal time frame under isothermal conditions utilizing an overhang generated by 5' exonuclease, polymerase, and a DNA ligase. The Gibson assembly technique is widely applied in assembling single fragments, multiple fragments, site-directed mutagenesis, shotgun cloning, and creating library construct. Several advancements were observed in terms of developing minimal cells and synthesizing the cells and combining them with CRISPR technology.

DOI: 10.4018/978-1-6684-6577-6.ch011

INTRODUCTION

Human life is greatly dependent on three major factors: environmental issues, deficiency of food, and health problems. Besides safety and cleanliness of the environment, food and health also play a vital role in the quality of human life. With the increasing world population's increasing demand for quality food and health is also increasing rapidly. It requires safe food and health at a reasonable price to reduce the enormous number of global deaths due to health and food-related problems. According to Whitman (2011), approximately 36 million deaths are observed each year from different non-communicable and communicable diseases like cancer, diabetes, AIDS/HIV, cardiovascular diseases, malaria, tuberculosis, etc. Despite extensive are made in technology and innovation, the current food production for the world population is much lower than its requirement. Health care facilities for the treatment are further below par in most third-world countries. With the rapid pace of industrialization, environmental pollution has soared with industrial wastes allowed to mix with the water which affects the water bodies and indirectly human beings resulting in the emergence of various new diseases. As a result, these issues need to be addressed through the latest innovations and modern technologies to improve the quality of life.

To overcome the issues related to health, environment, and agriculture by traditional medicines, pollutant degradants, and breeding, the technique of genetic engineering using modern tools and approaches has opened wide opportunities to combat such issues using molecular cloning, gene editing, recombinant plasmids and transformation which can be used the limited period and produce more reliable yield when compared to the conventional methods. Unlike conventional breeding methods where many specific and nonspecific genes are transferred to the host, genetic engineering focuses on transferring, only a small portion of desired genes to the target by various techniques like biolistic and agrobacterium-mediated transformation. These alterations are brought by either homologous recombinant dependent gene targeting or by nuclease mediated site-specific genome modifications. Besides, recombinase-mediated site-specific genome integration and oligonucleotide-directed mutagenesis are also used.

In 1971, Paul Berg from Stanford University showed a great interest in the genetics of microbes, leading him to understand the different techniques of animal cultures. His previous knowledge of Dulbecco has shown that some viruses can induce cancer in an infected cell by expressing their genetic material. The mechanism of the virus has made him curious about inserting foreign genes into a virus that can be carried into other cells as vectors. This curiosity has led him to succeed in the gene-splicing experiment, where a small part of bacterial virus DNA, known as lambda, spliced into the DNA of simian virus (SV40), a polyomavirus found in humans and monkeys.

The first step of Berg's experiment involved cutting the virus's loop using restriction enzymes such as *Eco RI,* which is called restriction digestion. The next step was to join the two types of DNA which are in linear form after restriction digestion. The final step is to join the ends of the linear DNA containing two different DNA again into a loop using enzymes like ligase which acts as a glue for sticking the ends of the DNA. This experiment of Paul Berg has finally resulted in the first-ever man-made recombinant DNA (rDNA). In 1980, Paul Berg, Walter Gilbert, and Frederick Sanger were awarded the Nobel Prize in the field of Chemistry for their work on recombinant DNA [12].

In 1973, Stanley Cohen, Herbert Boyer, and their colleagues from Stanford University and the University of California, San Francisco built on the work of Paul Berg and developed the basis of recombinant DNA technology which has led to the emergence of the biotechnology industry.

It was evident back in 1959 that bacteria contain extra DNA loops called plasmids along with their chromosome. These plasmids are beneficial for bacteria in transferring required genes that code for antibiotic resistance and other functions. In the early 1970s, scientists were successful in isolating the plasmids along with the special enzymes called restriction endonucleases which work as scissors that cut open the plasmid DNA loop.

At a conference in 1972, Stanley Cohen and Herbert Boyer whose expertise studies were in plasmids and restriction endonucleases respectively decided to combine and work together. In 1973, after initial experiments, the Boyer-Cohen team was successful in opening a plasmid loop from one species of bacteria, inserting a gene from different bacterial species, and closing the plasmid. This experiment resulted in the creation of the recombinant DNA molecule with a plasmid consisting of DNA from two different species of bacteria. Finally, they were able to demonstrate the expression of the recombinant plasmid by inserting it into the bacteria. This work by Boyer-Cohen and team have made a progress in creating the first genetically modified organism (GMO).

Figure 1. Bacteria with DNA and extra plasmid loops

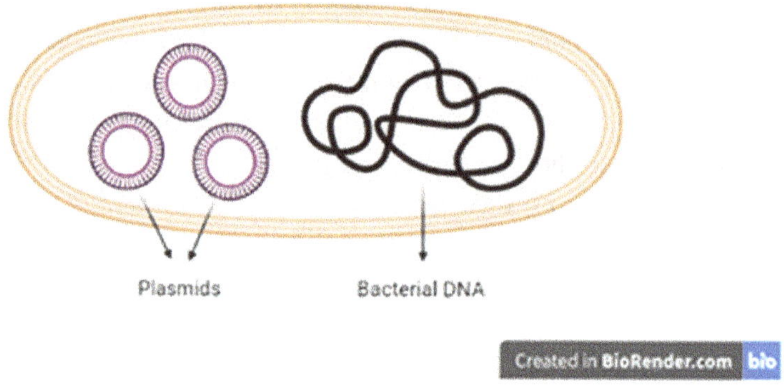

Figure 2. Insertion of rDNA into bacteria for replication

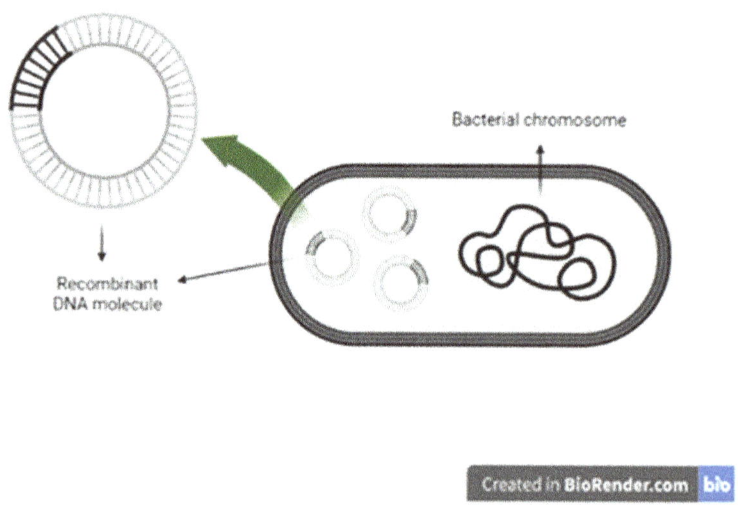

In 1972, Herbert Boyer and his graduate student, Bob Helling discovered a restriction enzyme called *Eco RI* that is produced in Escherichia coli. This enzyme can cut open the DNA and cut it into pieces with overlapping, single-stranded ends.

Figure 3. Event timeline

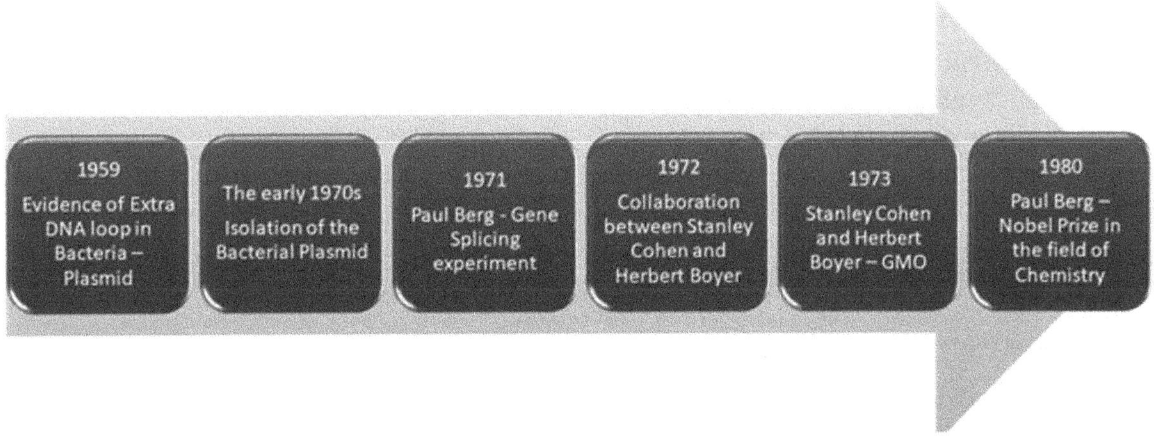

The development of novel medications and vaccines to treat disease and improve health conditions heavily relies on recombinant DNA technology. This technology also helps in improving various treatment strategies like monitoring devices, developing new diagnostic kits, and novel approaches to therapeutics. Leading examples of genetic engineering are the development of synthetic human insulin and erythropoietin using genetically modified bacteria and the production of mutant mouse types for experimental research [11]. Similarly, genetic engineering can play a key role in tackling environmental challenges by converting waste into biofuels and bioethanol. Besides, this technology can also be employed in the detection of arsenic and other contaminants in drinking water along with cleaning the oil spills and other toxic wastes from the water bodies. Genetically modified microorganisms are also used effectively in biomining and bioremediation.

METHODOLOGY

Isolation of DNA

The process of recombinant DNA technology is initiated by isolation of the desired DNA in its purest form, devoid of extraneous macromolecules such as RNA, carbohydrates, proteins, and triglycerides within the cell membrane, using enzymes like lysozymes, cellulase, chitinase, ribonuclease, and proteases.

In the presence of ethanol, DNA starts precipitating out as thin threads, and purified DNA can be spooled out.

Restriction Enzyme Digestion

- The restriction endonucleases identify and cleave the DNA strands at a sequence called the restriction site.
- The process of breaking DNA molecules into smaller units using restriction endonucleases is known as restriction digestion.
- The 'Agarose Gel Electrophoresis' method visualizes the process of restriction enzyme digestion.

When a current is supplied in this process, negatively charged DNA moves towards the positive electrode and is split into distinct sizes. This allows the digested DNA pieces to be sorted out, and the vector DNA is processed in the same way.

Type II restriction enzymes are extensively used in recombinant DNA technology.

Table 1. Types of Restriction endonucleases

Restriction Endonuclease	Function	Cofactor	Example
Type I	Cleaves DNA about 1000bp away from its restriction site and has both methylation and endonuclease activity.	ATP, Mg^{2+}, and S-adenosyl-L-methionine	*Eco*KI
Type II	It cleaves DNA at the restriction site.	Mg^{2+}	*Eco*RI, *Hind* III
Type III	It cuts DNA about 20–30 base pairs after the recognition site.	ATP and S-adenosyl-L-methionine	*EcoRI*
Type IV	It cuts the DNA close to or within the recognition sequence.	Mg^{2+}	McrBC, which cleaves 5mC-modified DNA
Type V	It can cut DNA if a suitable guide RNA is provided.	Mg^{2+}	cas9-gRNA complex

DNA Amplification Using Polymerase Chain Reaction (PCR)

Polymerase Chain Reaction is a technique that is used to replicate the segments of DNA exponentially. It is the safest known technique without any ethical concerns and controversies. The other technique involving embryos is much focused on ethical concerns, particularly those of human beings.

In the PCR process performed on 'thermal cyclers,' with several components like:

1. DNA template that must be amplified.
2. Oligonucleotide Primers that are complementary to a DNA region.
3. DNA polymerase enzyme synthesizing nucleotides complementary to the template strand.
4. Nucleotides to lengthen the primers.

The amplified DNA fragments can subsequently be ligated with the cut vector.

Figure 4. DNA fragment amplification

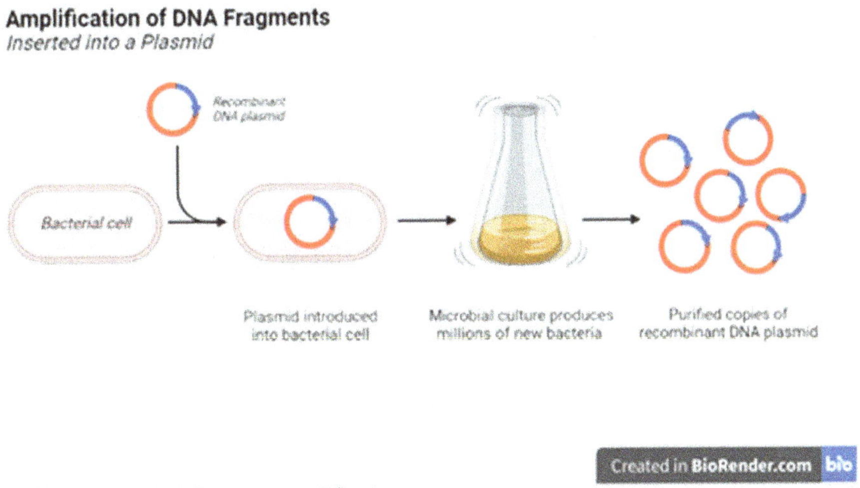

DNA Ligation

- The enzyme DNA ligase is used to link the vector with a segment of desired amplified DNA.
- DNA ligase is a type of enzyme that helps in the formation of the phosphodiester bonds and facilitates the joining of the DNA strands.
- This enzyme can repair and replicate single and double-stranded DNA. DNA ligase has an extensive application in molecular biology laboratories for recombinant DNA research experiments where it can be used for cloning a gene by joining DNA molecules.
- The resulting DNA molecule is a hybrid between the desired DNA and vector molecules. In genetics, recombination refers to the integration of distinct DNA strands.
- As a result, recombinant DNA molecules are produced, and this method is termed recombinant DNA technology.

Insertion of rDNA Into a Host

- In this step, recombinant DNA is added to the host cell, and the entire process is known as transformation.
- Microbial cells are resistant to foreign DNA. As a result, they are subjected to procedures such as thermal shock, Ca+2 ion therapy, and electroporation to enable them to receive new DNA.
- Under suitable conditions, the recombinant DNA replicates, and manifests as a manufactured protein after insertion.

Recombinant Cell Isolation

- Only the transformed host cells are screened during the selection phase from a heterogeneous population of cells.
- The plasmid vector's marker gene is used to distinguish recombinant cells from non-recombinant cells.

Since the invention of recombinant DNA technology, a series of changes have been seen in the field of biology. This technology has helped in innovating and producing a wide range of novel therapeutics with immediate effect in medical genetics and biomedicine by modifying the microorganisms, plants, and animals for the synthesis of medicinal value products. In the current market, most biotechnological pharmaceuticals are recombinant and act against lethal human diseases. The modified pharmaceutical products have shown such a significant impact on human life that the U.S. Food and Drug Administration (FDA) approved more recombinant DNA drugs in 1997 than in the previous years combined for the treatment of AIDS, anemia, genital warts, diabetic foot ulcers, multiple sclerosis, hepatitis B, hepatitis C, hereditary disorders (Turner's syndrome, SCID, Gaucher's disease, etc.) and different types of cancers like leukemia, kidney, colorectal and ovarian cancer. The development of multigene transfer, site-specific integration, and specifically regulated gene expression in the plants are crucial advancements in this technology along with some challenges like transcriptional regulation of endogenous genes, efficiency in the new location, and control over transgene expression that needs to be focused for further development.

Figure 5. Applications of recombinant DNA technology

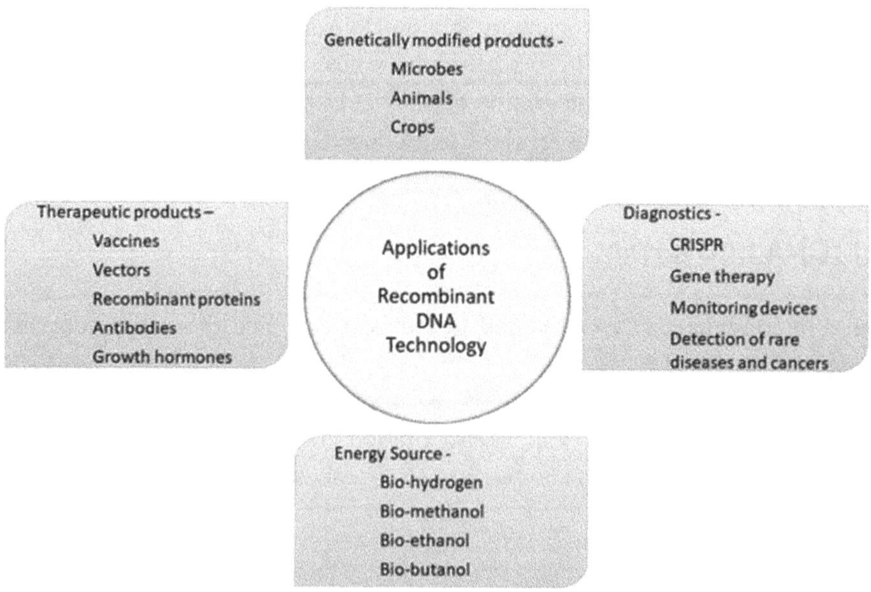

Table 2. Drugs based on recombinant DNA technology

Recombinant erythropoietin	i. Epoetin-α, β, ω ii. Darbepoetin-α
Recombinant blood coagulation modulators	• Coagulants – i. Factor-VII A ii. anti-hemophilic factor • Anticoagulants – i. Lepirudin ii. Desirudin iii. Drotecogin alfa
Recombinant enzymes	• Thrombolytic enzymes – i. Anistreplase ii. Duteplase iii. Reteplase • Therapeutic enzymes – i. Agalsidase β ii. Aeromonas aminopeptidase iii. Dornase α
Recombinant interferons	i. Interferon-gamma-1a ii. Interferon-gamma-1b iii. Interferons-ω iv. Consensus interferon
Recombinant interleukins	i. Interleukin-1 alpha and beta ii. Aldesleukin iii. Interleukin-3 iv. Oprelvekin
Recombinant anti-rheumatoid agents	i. Infliximab ii. Adalimumab iii. Basiliximab iv. Daclizumab
Recombinant hormones	• Insulin and its analogues – i. Human insulin ii. Insulin Aspart iii. Insulin glargine • Growth hormone and its analogues – i. Growth hormone ii. Growth hormone (ovine) iii. Growth hormone-binding protein • Gonadotropins – i. Recombinant follitropin-α and β ii. Luteinizing hormone

Several advancements were made for joining 2 or more DNA molecules using PCR and restriction enzymes since the invention of recombinant DNA technology. The development of a genetically identical copy or duplicate of any kind of biological material like cells or an organism for the purpose of scientific study can be achieved by a process called cloning. This process takes place in nature either naturally when a cell replicates asexually (prokaryotes) or by recombination technology or genetic alteration (eukaryotes). Bacteria are the most common microorganisms that can produce clones without any genetic alterations by binary fission or budding whereas in eukaryotes like human beings only certain cells like gastrointestinal tract cells can produce clones, but other cells need to be altered using recombination. There are several techniques of duplicating the cells or organisms that have altered DNA.

Figure 6. Steps in PCR

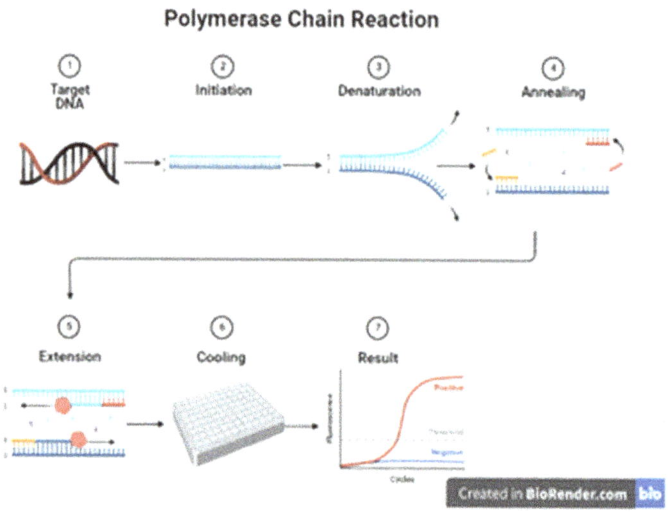

Gene cloning, reproductive cloning, and therapeutic cloning are the 3 different ways of cloning. Gene cloning consists of replicating segments of DNA using techniques like PCR; the process of developing a clone of an entire human or animal using an embryo is called reproductive cloning and therapeutic cloning is performed for creating embryonic stem cells which can be utilized for replacing damaged or diseased tissues in the body.

Human life is always threatened by a range of factors like limitations over food leading to malnutrition, various lethal diseases caused by different bacteria and viruses, environmental issues which are a result of extensive industrialization and urbanization, and many more. Genetic engineering was successful in replacing many conventional strategies and has shown a greater potential in overcoming such challenges. The current chapter emphasizes the recent technique called Gibson assembly which is an exonuclease-based cloning method for the assembly of the DNA in the correct sequence. A new platform for the rapid cloning of multiple DNA within a limited timeframe without restriction enzymes. In line with this, we will be discussing several types, and various applications of the Gibson assembly with their challenges and overcoming those for a better future for human beings.

Gibson Assembly

Gibson Assembly is an in vitro-based recombinant molecular cloning technique that Daniel G. Gibson, Chief Technology Officer, and Co-Founder of a synthetic biology business named Codex DNA, developed with his colleagues in 2009. This method makes it simple to link various DNA molecules into a circular plasmid in an isothermal setting. One tube may carry out the entire response and with the help of this technology, scarless DNA with a maximum size of 15 kb may be assembled.

This technology involves the process of designing and synthesizing the first few fragments with 15-80 bp that overlap with the adjacent DNA parts. After the synthesis of the DNA base pairs, the fragments are mixed with the reaction buffer consisting of 3 different types of enzymes: DNA polymerase, T5 exonuclease and DNA ligase.

Factors Involved in Gibson Assembly

- Phusion DNA Polymerase –

Phusion DNA polymerase is a unique Pyrococcus-like enzyme with a processivity-enhancing motif that creates correct PCR products at a faster pace than a single enzyme. This enzyme can proliferate long templates and its distinct structure and properties make it a preferable candidate for cloning, with excellent PCR performance criteria.

- T5 exonuclease –

T5 exonuclease is an enzyme that disrupts both single and double-stranded DNA in the 5' to 3' orientation. The enzyme transports the substrate to the 5' terminus of the DNA for disintegration and forms a gap or nick in the case of linear or circular ds DNA. T5 exonuclease cannot digest supercoiled ds DNA, making it suited for plasmid purification of linear or circular DNA from supercoiled plasmids. It can also be employed to remove the incompletely ligated ds DNA.

- Taq DNA Ligase –

Taq DNA Ligase is a thermostable NAD+-dependent enzyme from the thermophilic bacteria Thermus aquaticus. This enzyme facilitates the creation of phosphodiester linkages and enables the merging of DNA strands. This enzyme has the capacity to restore and reproduce single and double-stranded DNA. DNA ligase is frequently employed in molecular biology laboratories for recombinant DNA research activities, where it may be utilized to clone a gene by connecting DNA molecules.

How It Works

PCR amplification with primers containing the right equivalent sequences can be used to produce the affinity between two adjacent fragments. A suitable primer should have a melting temperature in the range of 48oC as well as an overlap of at least 15 - 40 base pairs. Consider a gene of interest that must be integrated into the vector with a large N-terminal tag. However, because the vector lacks a tag, a two-step approach is required, which will eventually create a gap between the two pieces when cleaved with restriction enzymes. Using Gibson assembly, the gene of interest and tag sequences may be efficiently inserted into the vector in one step without leaving any incisions. Primers are first intended to amplify the two segments, including sections of structural similarity to the vector or neighboring segment. The fragments and vectors are then amplified using PCR to ensure the purity and size of the DNA fragments.

Primer Designing

Designing acceptable similar common characteristics is crucial for Gibson Assembly cloning efficacy. Typically, 20-40 base pairs of the vector or insert homologous area must be added to the terminals of a standard-sized segment. In the case of several insertions, homologous ends are inserted into each adjacent fragment.

Figure 7. Gibson assembly protocol

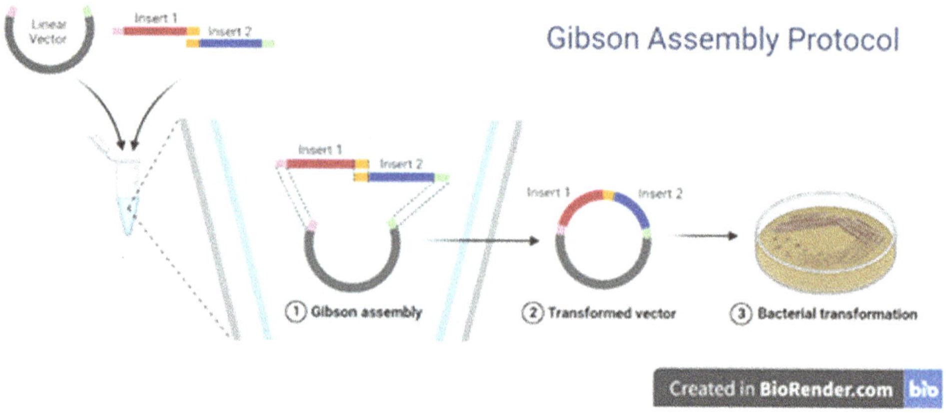

Table 3. Conditions of adding overlapping ends to the insert and vector

Addition of Overlapping Ends to the Insert	Addition of Overlapping Ends to the Vector
If insert PCR amplification is preferred, add the equivalent sequence to the vector to the ends of the insert	If vector PCR amplification is preferred, add the equivalent sequence to the insert to the ends of the linear vector
For small inserts < 2kb	For small vectors < 8kb
Utilizing one vector for cloning dissimilar inserts	Utilizing one insert to alternate between different vectors

Generating Homologous Overlaps

- The quantity and size of the segments have a significant correlation with the size of the adjacent region.
- Incorporating larger intersecting regions comes with advanced efficacy.
- Steer clear of tandem repetitions and guanine-cytosine concentrations less or above 40%.
- Insertion of a restriction enzyme site between the sequence-specific segment and the intersecting similar part.

Tool for Designing Gibson Assembly Primers: GAP Tool

- Kits that are used to facilitate the Gibson Assembly provide a free online tool called the Gibson Assembly Primer Design Tool that helps in generating primers for the synthesis of DNA.
- The vector and fragment sequence data are entered into the GAP tool along with the technique used to create DNA fragments.
- Using the GAP tool, the spherical structures are amplified by generating primers that add 30–40 base pairs of comparable sequences to the terminus of sequential loops.

Requirements of the Fragment for Using the GAP Tool

- The GAP Tool has been enhanced to work with DNA segments as small as 10 kb.
- For fragments less than 5.5 kb, the GAP tool offers primers for 1–15 fragments.
- To enable efficient PCR amplification for segments greater than 5.5 kb, the GAP tool constructs sub-assembly primers.

GAP Tool Protocol

Protocols

The identical overlaps can be introduced into the segments using PCR with the help of Gibson assembly tools. High-fidelity, Phusion DNA polymerase and fewer PCR cycles are performed to reduce the inefficiencies in the amplification. The fragments are examined on a gel for the confirmation of the insert size. Gel extraction is performed if several bands are identified on the gel.

PCR Preparation of DNA fragments for Gibson Assembly Cloning

Calculating the Amount of DNA to Use in a Gibson Assembly Reaction

pmol DNA = [ng DNA/ (660 x # of bases)] x 1000

ng of DNA = [pmol DNA x (660 x # of bases)]/1000

*pmol – picomolar (10^{-12})

*ng – nanogram (10^{-9})

Gibson Assembly SDM Reactions

Gibson Assembly HiFi 1-Step Kit

- Firstly, thaw the Gibson Assembly HiFi 1-Step master mix on the ice and combine the insert DNA with the vector DNA to a final volume of 5 µL for the 2X kit/7.5 µL for HC 4X kit. Mix the content by vertexing the reaction mixture.
- On the ice, 5 µL DNA and 5 µL master mix are added together for the 2X kit/ 7.5 µL of DNA and 2.5 µL master mix in the case of HC 4X kit.
- Mix the content and centrifuge for a short time.
- Incubate the reaction for 1 hour at 50°C before storing it at 20°C or proceed with the transformation.

Gibson Assembly Ultra Kit

In a total amount of 5 µL quantity add the insert and vector DNA after thawing Gibson Assembly ultra-master mix A on ice. Master Mix A (MM A) GA Ultra Vortex and quickly centrifuge. Then, mix 5 µL of DNA and 5 µL of MM A on ice. After a thorough mixing step, quickly centrifuge, then incubate.

Vortex GA Ultra Master Mix B (MM B) after it has been quickly centrifuged and frozen GA Ultra Master Mix B on ice. 10 µL of the reaction and 10 µL of MM B should be combined before being incubated at 45 °C for 15 minutes on ice. Reactions should be kept at 20°C or transformed.

Gibson Assembly Site-Directed Mutagenesis Kit

i. Thaw GA SDM PCR Mix on ice. Vortex and briefly centrifuge, SDM Forward Primer (2.5 µL), SDM Reverse Primer (2.5 µL), dsDNA template, GA SDM PCR Mix (12.5 µL), and PCR water are used to prepare the reaction mixture. Introduce mutations by heating them to 98 °C for one minute, then 72 °C for five minutes, then 4 °C. Purify after analyzing 5 L using gel electrophoresis.
ii. In a total volume of 5 µL, combine 1 to 5 SDM fragments and vector DNA after thawing GA SDM Assembly mix A on ice. Afterward, combine 5 l of DNA and 5 l of GA SDM Assembly mix A on ice. Thoroughly combine, quickly centrifuge, and anneal.
iii. On ice, thaw GA SDM Assembly combine B, then vortex and briefly centrifuge. 10 µL of reaction and 10 µL of GA SDM Assembly mix B should be combined on the ice. 15 minutes of incubation at 45 °C, followed by storage of by the reaction at -20 °C or transformation.

Transformation and Plating Guidelines

Gibson Assembly constructs may be effectively converted into a variety of competent cells with high efficacy. It is advised to prepare two plates for each transformation and plate two various quantities of the transformation mixture.

Calculating Cloning Efficiency

The presence of white colonies indicates successful assembly with the insert and blue colonies indicate the absence of insert and vector assembly.

Cloning efficiency is calculated using the following formula:

Cloning efficiency (%) = Number of white colonies/ Total colonies x 100

Applications for Gibson Assembly

A. Simple cloning method – One vector, One insert

An overlapping end of 20 to 40 bp double-strand DNA fragments are generated for the Gibson assembly cloning technique. For this process, the vector DNA along with the inserts are mixed and incubated with factors like exonuclease, polymerase, and ligase for the required amount of time. During this time, the factors initiate the generation of compatible nucleotides, anneal, extend, and ligate the plasmid for a fully assembled DNA construct.

B. Multiple Fragments cloning method – One vector, Many inserts

This technique enables a simultaneous assembly of the multiple fragments of DNA (inserts) which cuts down the time required to clone multiple fragments using traditional methods. The traditional method can be challenging in the identification of the recognition sites for multiple fragments, necessitating several subcloning steps. The generation of the blunt end fragments using PCR or restriction digestion can result in efficiency and require further screening and selection upon the orientation of the insert. Besides, in the process of multiple fragments cloning the possibility of ligating larger fragments is lesser when compared to the smaller fragments. Such shortcomings are forbidden by the Gibson Assembly by providing an efficient method of cloning multiple fragments of various sizes in a single round.

C. Site-directed Mutagenesis

Site-directed Mutagenesis (SDM) facilitates intentional, specific changes to the DNA sequence. These alterations often show an impact on the regulatory elements of the gene, RNA intermediates or on encoded proteins. SDM is often used in protein engineering and allows the examination of the biological activities of nucleic acids and proteins. A designed protein generated by SDM can be altered for improved or desired unique properties that make an appropriate protein for a specific function or application. SDM with the Gibson assembly method can initiate the point mutations, insertions, and deletions effectively and can induce multiple mutations to the targeted locations in both linear and circular templates of DNA.

D. Library Construction

The method of Gibson Assembly can result as an ideal option for creating a library of constructs. These libraries can facilitate the scanning of functional motifs or combinatorial site libraries for testing the genetic elements.

Preparation of the library vector for Gibson Assembly -

- The vector must be linearized, and 20 -40 bases must be identified from each end.
- These identified bases are added to the primers that are used for the amplification of the library and create homologous overlaps for assembly.
- The library is amplified to introduce the intended variations using PCR and homologous overlaps are added to the library inserts.
- The resulting library inserts are assembled using Gibson Assembly cloning with a vector.

E. Shotgun Cloning

A high throughput cloning application called shotgun cloning uses multiple DNA fragments having the same homologous ends and clones a pooled mixture of heterogenous DNA fragments with a single vector simultaneously. The combination of Gibson assembly and shotgun cloning along with the next-generation sequencing (NGS) can result in substantial yield and save a lot of resources.

Table 4. Comparison of cloning methods

Type	Gibson Assembly	Restriction Enzyme Digestion	PCR Fragment Cloning	Gateway Cloning	Golden Gate	Ligation Independent Cloning
Method	Depends on homologous overlap sequences	Restriction digestion	Single base overhang	Requires vectors and inserts along with recombination sequences (att sites)	Type IIS restriction enzymes cleave outside the recognition site	Creation of 3' to 5' exonuclease by T4 polymerase
Efficiency	>90%	Variable	Variable	Up to 95%	Variable	Variable
Insertion site after cloning	Seamless	May Leave a scar	Single base insertion	Att sites remain as a part of the construct	Seamless	Seamless
Multiple fragments cloned in a single round	Yes	No	No	Yes (limited)	Yes (limited)	Yes (limited)
Vector	Yes	Yes	No	Yes (att vectors only)	Yes	Yes
Direction oriented	Yes	Experiment dependent	No	Yes	Experiment dependent	Yes

Advantages of Gibson Assembly

I. Reduced time compared to the traditional method of cloning –
   - The assembly of the nucleotides is completed in 1 hour which cuts down time over the conventional methods. When performed using kits like the Gibson Assembly HiFi 1-Step Kit and the Gibson Assembly HiFi HC 1-Step Kit, the addition of nucleotides is completed in 1 hour under isothermal conditions in a single tube.
   - In robust conditions, a 2 – step process is completed in under 80 minutes when performed using Gibson Assembly Ultra Kit

II. No compatibility issues with restriction sites –
   - This method helps in joining any 2 fragments of DNA without any compatibility issues of restriction sites.
   - Eases the process of designing the construct by eliminating the compatible restriction site requirement.

III. Easy cloning of multiple fragments -
   - The ability to join multiple fragments with high efficiency reduces the need for multiple cloning rounds and screening.
     - Gibson Assembly HiFi 1-Step isothermal – 1 to 5 fragments can be cloned in a single round
     - Gibson Assembly Ultra two-step – 1 to 15 fragments of DNA can be cloned in a single round with robust conditions.

IV. This technique can be performed with minimal quantities like nanograms for the DNA to build the construct and can be cloned. This enhances the recovery of the construct with the limited starting material.

V. Gibson assembly enables the process of assembling a wide range of DNA fragments of various sizes from 100bp to 100kb to clone and utilized for various cloning experiments.
VI. This technique provides a new platform for cloning the simple to complex constructs of full lengths without any undesired sequences.
VII. Proofreading polymerase being included in the kits while performing the Gibson assembly significantly reduces the possibilities of errors in cloning and yields a high percentage of desired constructs.

Advancements

- Synthesis of cells (Synthetic cells)

The genomic DNA of several organisms has been sequenced for more than 2 decades and stored in the database [4]. Recent studies have shown the possibility to revert the process of sequencing and synthesizing the cells from digitized information. To prove the concept of synthetic cell generation, several tools with technologies were required to sequence, synthesize, and transplant the genome. The researchers of J. Craig Venter Institute (JCVI) along with Synthetic Genomics, Inc. (SGI), were successful in designing, synthesizing, and assembling the genome of 1.08 Mbp *Mycoplasma mycoides* JCVI-syn1.0 from the digitized genome sequence and transferring it into a *Mycoplasma capricolum* which is a recipient cell to control *Mycoplasma mycoides* with synthetic chromosomes. The organisms consist of designed DNA, along with watermark sequence and other mutational sequences like deletion, addition, or polymorphisms that are acquired during the process of synthesis. These cells were able self-replicate and exhibit desired phenotypic properties [4]. It is reported that it is one of the largest chemically synthesized genomic structures, it is almost double the size of *Mycoplasma genitalium* which is synthesized previously back in 2008. This proof of concept has helped in understanding and providing a new platform for creating synthetic cells with desired properties.

- Minimal Cells

Besides synthetic cells, researchers have also experimented on the specific functions of the gene in the organism. The objective of the study was to synthesize a minimal cell with limited genes that are useful for an independent life. Since we are capable of synthesizing synthetic DNA, the components of the gene can be altered and transplanted until only the essential genes are present in the organism. This helps in reducing the size of the synthetic genome and making it small as possible. This process can help in understanding the functions of every gene present in the organisms and modify them for the betterment of life.

- Gibson Assembly - CRISPR

Gibson Assembly can be combined with the more complicated cloning methods where the size of the vector is too large or has a high content of GC where it possesses several repeats which makes PCR difficult or no restriction site to linearize the vector. In such cases, Gibson Assembly can be combined with CRISPR technology as per Wang, et al. (2015). Despite using PCR or restriction enzymes to linearize the vector, the Cas9 enzyme along with specific genomic RNA can be used to cleave the vector. This technique can provide seamless and efficient results when no other possibility is viable. According to Jiang et al. (2015), the large fragments of the bacterial chromosome (up to 100kb) were specifically cut via the CRISPR technique and then assembled into a vector using Gibson assembly. In this process, the vector was amplified using PCR to contain homology regions with bacterial chromosome fragments. Repeated amplification is a challenging process and CRISPR cutting technique was used.

Variation of Gibson Assembly

- Gibson Assembly PBnJ Cloning

Gibson Assembly Primer-Bridge End Joining (PBnJ) cloning is a modified version of the Gibson Assembly cloning that attaches DNA fragments without any homologous overlaps. As this modification does not depend on overlapping ends to join the fragments, the assembly of the ends can be performed without generating overlapping regions using PCR. This technique uses a single or a pair of primers to assemble the non-overlapping ends of the DNA, typically with phosphorothioate-modified 3' ends. This modification eliminates the need for PCR amplification and increases flexibility.

Applications for Gibson Assembly PBnJ Cloning

- Gibson Assembly PBnJ™ Seamless Joining

This technique is applied to join multiple DNA fragments to construct a circular vector. A pair of primers are used to bridge non-homologous overlap regions of unrelated, non-homologous DNA fragments. The phosphorothioate-modified 3' ends in primers prevent the chew back from the 3' exonuclease. The primers anneal to the non-overlapping exposed template sequence on the template is chewed back. The primer sequence is extended during the incubation period leaving a non-homologous 3' single strands overhang. The HiFi 1-Step Master Mix mediates the chew back from the 5' end and allows annealing of the non-homologous bridge region of the primer to the template. Master Mix yields the extension and ligation by the polymerase for the final assembly product.

- Gibson Assembly PBnJ 3' Overhang Extension:

The 3' Overhang extension is accomplished with a single phosphorothioate-modified primer and the Gibson Assembly Ultra Kit. The process is initiated by combining a single phosphorothioate-modified primer with desired DNA fragment along with Master Mix A.

- Gibson Assembly PBnJ Sequence Insertion Cloning:

This technique is used to add any DNA sequence between two non-overlapping fragments during the assembly. This modified technique called Sequence Insertion Cloning is carried out using master mix A, phosphorothioate-modified primers and template DNA fragments. This technique can be applied to many types of downstream applications like in the studies of promoters, mutagenesis, and genome modifications.

Challenges

Gibson Assembly has very few drawbacks when compared to the traditional methods of cloning and assembling the DNA fragments.

- This technique can process best when there are more than 200 nucleotides in the fragment. The exonuclease used in the technique could chew the entire fragment which is shorter than 200 nucleotides much before annealing and polymerizing the DNA fragments.
- This technique does not work when the ends of the fragments have stable single-stranded DNA secondary structures like stem-loop or stem-loop loop which occur in a terminator sequence and compete directly with the required single-stranded annealing and priming of neighboring assembly fragments.

CONCLUSION

Genetic engineering has introduced several prospects that are less time-consuming and generate consistent yields than conventional approaches for resolving issues with health, the environment, and agriculture. To control the intended transformation in the host organism, these approaches typically only target a limited part of the targeted genes. The work of researchers like Paul Berg, Stanley Cohen, Herbert Boyer, and their colleagues laid the groundwork for recombinant DNA technology, which eventually gave rise to the biotechnology sector and the development of genetically altered species. Recombinant DNA technology involves the isolation of DNA from the desired host, which is subjected to restriction enzyme digestion and DNA amplification using polymerase chain reaction, a technique of replicating DNA fragments exponentially. PCR is followed by DNA ligation resulting in the production of a recombinant DNA molecule, which is subsequently inserted into the desired host cell using a vector. Recombinant DNA technology is vital for developing novel medications and vaccines that will enhance treatment prospects. Leading examples of genetic engineering include the use of genetically engineered microorganisms to create synthetic human erythropoietin and insulin. Daniel G. Gibson invented the Gibson Assembly technique, which enables rapid cloning of DNA fragments. This technology involves the process of synthesizing the first few fragments with 15-80 bp that overlap with the adjacent DNA parts. It is possible to create the affinity between two adjacent fragments using polymerase chain reaction using primers that contain comparable sequences. PCR is subsequently used to amplify the fragments and vectors, ensuring the DNA fragments' purity and size. Effective Gibson Assembly cloning depends on the design of acceptable comparable shared traits. A free online tool called the Gibson Assembly Primer Design Tool aids in creating primers for DNA synthesis. The same interaction pattern can be introduced into the sequence using PCR by using the Gibson assembly primer design tool. Gibson Assembly components may be effectively converted into a variety of functional cells. Gibson Assembly enables the precise assembly of many DNA adhesives, thus reducing the time required to clone various sections when using conventional methods. It is a straightforward cloning technique that offers an efficient way to clone several segments of different sizes in a single round, doing away with such inefficiencies. Site-Directed Mutagenesis (SDM) employing the Gibson Assembly approach may successfully introduce point mutations, insertions, and deletions as well as multiple mutations to the targeted locations in both linear and circular DNA templates. This method does not conflict with restriction sites and may end up being the best option for creating a library of constructs that make it simpler to filter combinatorial site libraries for biologic predisposition analysis. However, Gibson Assembly has very few drawbacks as compared to traditional cloning methods. When a consistent single-stranded DNA secondary structure is present at the extremities of the fragments, this approach fails but still works effectively for fragments with more than 200 nucleotides. Gibson Assembly involves developing synthetic cells, which help in deciphering the roles of each gene existing in organisms and in modifying them to create life. According to Wang et al., under these conditions, Gibson Assembly can be used with CRISPR technology. The Gibson Assembly procedure is good at assembling many DNA fragments in a single reaction since it does not rely on the existence of restriction sites inside a particular sequence that must be replicated.

ABBREVIATIONS

AIDS: Acquired immunodeficiency syndrome
CRISPR: Clustered Regularly Interspaced Short Palindromic Repeats
DNA: Deoxy ribonucleic acid
Eco RI: Escherichia coli RY13
FDA: Food and Drug Administration
GAP: Gibson Assembly Primers
GMO: Genetically modified organism
JCVI: J. Craig Venter Institute
NGS: Next-generation sequencing
PBnJ: Primer-Bridge End Joining
PCR: Polymerase Chain Reaction
rDNA: recombinant DNA
RNA: Ribo nucleic acid
SCID: Severe Combined Immunodeficiency
SDM: Site-Directed Mutagenesis
SGI: Synthetic Genomics, Inc.
SV40: simian virus 40

REFERENCES

Adivitiya, & Khasa, Y. P. (2016). The evolution of recombinant thrombolytics: Current status and future directions. *Bioengineered, 8*(4), 331–358. doi:10.1080/21655979.2016.1229718

Bloch, K. D., & Grossmann, B. (1995). Digestion of DNA with Restriction Endonucleases. *Current Protocols in Molecular Biology, 31*(1). Advance online publication. doi:10.1002/0471142727.mb0301s31 PMID:18265218

C, B. G. (2023). Mechanisms and types of restriction endonuclease enzymes. *The Science Notes*. https://thesciencenotes.com/restriction-endonuclease-enzymes-mechanisms-types/

Cardi, T., & Neal Stewart, C. Jr. (2016). Progress of targeted genome modification approaches in higher plants. *Plant Cell Reports*, *35*(7), 1401–1416. doi:10.100700299-016-1975-1 PMID:27025856

Cazzola, M., Mercuriali, F., & Brugnara, C. (1997). Use of recombinant human erythropoietin outside the setting of uremia. *Blood*, *89*(12), 4248–4267. doi:10.1182/blood.V89.12.4248 PMID:9192747

Chao, R., Yuan, Y., & Zhao, H. (2014). Recent advances in DNA assembly technologies. *FEMS Yeast Research*. doi:10.1111/1567-1364.12171 PMID:24903193

Eroshenko, N., & Church, G. M. (2013). Mutants of Cre recombinase with improved accuracy. *Nature Communications*, *4*(1), 2509. Advance online publication. doi:10.1038/ncomms3509 PMID:24056590

Festa, F., Steel, J. C., Bian, X., & LaBaer, J. (2013). High-throughput cloning and expression library creation for functional proteomics. *Proteomics*, *13*(9), 1381–1399. doi:10.1002/pmic.201200456 PMID:23457047

Galambos, L. (2011, December 13). Pharmaceutical Firms and the Transition to Biotechnology: A Study in Strategic Innovation. *Business History Review*. https://www.cambridge.org/core/journals/business-history-review/article/abs/pharmaceutical-firms-and-the-transition-to-biotechnology-a-study-in-strategic-innovation/18D8F37FD43C813B7984B06E58B733C0

Garcia, C. B., Chapman, I. F., Chen, S. H., Lazear, E., Lentz, T. B., Williams, C. L., Dums, J. T., Goller, C. C., & Robertson, S. D. (2020). Integrating research into a molecular cloning course to address the evolving biotechnology landscape. *Biochemistry and Molecular Biology Education*, *49*(1), 115–128. doi:10.1002/bmb.21402 PMID:33176069

Giansily-Blaizot, M., & Schved, J. (2017). Recombinant human factor VIIa (rFVIIa) in hemophilia: Mode of action and evidence to date. *Therapeutic Advances in Hematology*, *8*(12), 345–352. doi:10.1177/2040620717737701 PMID:29204261

Gibson, D. G. (2011). Enzymatic assembly of overlapping DNA fragments. In Methods in Enzymology (pp. 349–361). doi:10.1016/B978-0-12-385120-8.00015-2

Gibson, D. G., Benders, G. A., Andrews-Pfannkoch, C., Denisova, E. A., Baden-Tillson, H., Zaveri, J., Stockwell, T. B., Brownley, A., Thomas, D. W., Algire, M. A., Merryman, C., Young, L. S., Noskov, V. N., Glass, J. I., Venter, J. C., Hutchison, C. A. III, & Smith, H. O. (2008). Complete Chemical Synthesis, Assembly, and Cloning of a *Mycoplasma genitalium* Genome. *Science*, *319*(5867), 1215–1220. doi:10.1126cience.1151721 PMID:18218864

Gibson, D. G., Glass, J. I., Lartigue, C., Noskov, V. N., Chuang, R. Y., Algire, M. A., Benders, G. A., Montague, M. G., Ma, L., Moodie, M. M., Merryman, C., Vashee, S., Krishnakumar, R., Assad-Garcia, N., Andrews-Pfannkoch, C., Denisova, E. A., Young, L., Qi, Z. Q., Segall-Shapiro, T. H., ... Venter, J. C. (2010). Creation of a Bacterial Cell Controlled by a Chemically Synthesized Genome. *Science*, *329*(5987), 52–56. doi:10.1126cience.1190719 PMID:20488990

Gibson, D. G., Young, L. S., Chuang, R. Y., Venter, J. C., Hutchison, C. A. III, & Smith, H. O. (2009). Enzymatic assembly of DNA molecules up to several hundred kilobases. *Nature Methods*, *6*(5), 343–345. doi:10.1038/nmeth.1318 PMID:19363495

Gibson Assembly Cloning Guide (2nd ed.). (n.d.). BioCat. Retrieved August 2, 2022, from https://www.biocat.com/bc/files/Gibson_Guide_V2_101417_web_version_8.5_x_11_FINAL.pdf

Hutchison, C. A. III, Chuang, R. Y., Noskov, V. N., Assad-Garcia, N., Deerinck, T. J., Ellisman, M. H., Gill, J., Kannan, K., Karas, B. J., Ma, L., Pelletier, J. F., Qi, Z. Q., Richter, R. A., Strychalski, E. A., Sun, L., Suzuki, Y., Tsvetanova, B., Wise, K. S., Smith, H. O., ... Venter, J. C. (2016a). Design and synthesis of a minimal bacterial genome. *Science*, *351*(6280), aad6253. Advance online publication. doi:10.1126cience.aad6253 PMID:27013737

Hutchison, C. A. III, Chuang, R. Y., Noskov, V. N., Assad-Garcia, N., Deerinck, T. J., Ellisman, M. H., Gill, J., Kannan, K., Karas, B. J., Ma, L., Pelletier, J. F., Qi, Z. Q., Richter, R. A., Strychalski, E. A., Sun, L., Suzuki, Y., Tsvetanova, B., Wise, K. S., Smith, H. O., ... Venter, J. C. (2016b). Design and synthesis of a minimal bacterial genome. *Science*, *351*(6280), aad6253. Advance online publication. doi:10.1126cience.aad6253 PMID:27013737

Jiang, W., Bikard, D., Cox, D., Zhang, F., & Marraffini, L. A. (2013). RNA-guided editing of bacterial genomes using CRISPR-Cas systems. *Nature Biotechnology*, *31*(3), 233–239. doi:10.1038/nbt.2508 PMID:23360965

Jiang, W., Zhao, X., Gabrieli, T., Lou, C., Ebenstein, Y., & Zhu, T. (2015). Cas9-Assisted Targeting of CHromosome segments CATCH enables one-step targeted cloning of large gene clusters. *Nature Communications*, *6*(1), 8101. Advance online publication. doi:10.1038/ncomms9101 PMID:26323354

Jiang, W., Zhao, X., Gabrieli, T., Lou, C., Ebenstein, Y., & Zhu, T. F. (2015). Cas9-Assisted Targeting of CHromosome segments CATCH enables one-step targeted cloning of large gene clusters. *Nature Communications*, *6*(1), 8101. Advance online publication. doi:10.1038/ncomms9101 PMID:26323354

Juhas, M. (2015). On the road to synthetic life: The minimal cell and genome-scale engineering. *Critical Reviews in Biotechnology*, 1–8. doi:10.3109/07388551.2014.989423 PMID:25578717

Khan, S., Ullah, M. W., Siddique, R., Nabi, G., Manan, S., Yousaf, M., & Hou, H. (2016). Role of Recombinant DNA Technology to Improve Life. *International Journal of Genomics*, *2016*, 1–14. doi:10.1155/2016/2405954 PMID:28053975

Khattak, W. A., Ullah, M. W., Ul-Islam, M., Khan, S., Kim, M., Kim, Y., & Park, J. K. (2014). Developmental strategies and regulation of cell-free enzyme system for ethanol production: A molecular prospective. *Applied Microbiology and Biotechnology*, *98*(23), 9561–9578. doi:10.100700253-014-6154-0 PMID:25359472

Kishore, K., & Krishan, P. (2009). Pharmacology of recombinant or genetically engineered drugs. *Journal of Young Pharmacists : JYP*, *1*(2), 141. doi:10.4103/0975-1483.55747

Kumar, S., & Kumar, A. (2015). Role of genetic engineering in agriculture. *Plant Archives*, *15*(1), 1–6. http://plantarchives.org/PDF%2015%20-%201/1-6%20(2950).pdf

Lajoie, M. J., Kosuri, S., Mosberg, J. A., Gregg, C., Zhang, D., & Church, G. M. (2013). Probing the limits of genetic recoding in essential genes. *Science*, *342*(6156), 361–363. doi:10.1126cience.1241460 PMID:24136967

Lei, L., Jiang, W., & Lu, Y. (2018). A Modified Gibson Assembly Method for Cloning Large DNA Fragments with High GC Contents. In Methods in molecular biology (pp. 203–209). doi:10.1007/978-1-4939-7295-1_13

Liu, S., Sanchez, D. J., Aliyari, R., Sun, L., & Cheng, G. (2012). Systematic identification of type I and type II interferon-induced antiviral factors. *Proceedings of the National Academy of Sciences of the United States of America*, *109*(11), 4239–4244. doi:10.1073/pnas.1114981109 PMID:22371602

Liu, W., Yuan, J. S., & Stewart, C. N. Jr. (2013). Advanced genetic tools for plant biotechnology. *Nature Reviews. Genetics*, *14*(11), 781–793. doi:10.1038/nrg3583 PMID:24105275

Lomedico, P. T. (1982). Use of recombinant DNA technology to program eukaryotic cells to synthesize rat proinsulin: A rapid expression assay for cloned genes. *Proceedings of the National Academy of Sciences of the United States of America*, *79*(19), 5798–5802. doi:10.1073/pnas.79.19.5798 PMID:6764530

Mcpherson, J. D., Marra, M. A., Hillier, L. D. W., Waterston, R. H., Chinwalla, A. T., Wallis, J. W., Sekhon, M., Wylie, K. M., Mardis, E. R., Wilson, R. K., Fulton, R. S., Kucaba, T. A., Wagner-McPherson, C., Barbazuk, W. B., Gregory, S., Humphray, S., French, L., Evans, R. S., Bethel, G., ... Asakawa, S. (2001). A physical map of the human genome. *Nature*, *409*(6822), 934–941. doi:10.1038/35057157 PMID:11237014

Mrid, R. B., Bouchmaa, N., Ainani, H., Fatimy, R. E., Malka, G., & Mazini, L. (2022). Anti-rheumatoid drugs advancements: New insights into the molecular treatment of rheumatoid arthritis. *Biomedicine and Pharmacotherapy*, *151*, 113126. doi:10.1016/j.biopha.2022.113126 PMID:35643074

Paul Berg. (2022, March 29). *Science History Institute*. https://www.sciencehistory.org/historical-profile/paul-berg

Payal, S. (2016, October 20). *7 Main Stages of Recombinant DNA Technology*. Biology Discussion. https://www.biologydiscussion.com/dna/recombinant-dna-technology/7-main-stages-of-recombinant-dna-technology/56320

Pyhtila, B. (2016, March 1). *Plasmids 101: Gibson Assembly and Other Long-Homology Based Cloning Methods*. Addgene. https://blog.addgene.org/plasmids-101-gibson-assembly

Roberts, R. J., Belfort, M., Bestor, T. H., Bhagwat, A. S., Bickle, T. A., Bitinaite, J., Blumenthal, R., Degtyarev, S. K., Dryden, D. T. F., Dybvig, K., Firman, K., Gromova, E. S., Gumport, R. I., Halford, S. E., Hattman, S., Heitman, J., Hornby, D., Janulaitis, A., Jeltsch, A., ... Xu, S. (2003). A nomenclature for restriction enzymes, DNA methyltransferases, homing endonucleases and their genes. *Nucleic Acids Research*, *31*(7), 1805–1812. doi:10.1093/nar/gkg274 PMID:12654995

Rugnetta, M. (2020, April 15). Cloning. *Encyclopedia Britannica*. https://www.britannica.com/science/cloning

Sistla, S., & Rao, D. N. (2004). S-Adenosyl-L-methionine–Dependent Restriction Enzymes. *Critical Reviews in Biochemistry and Molecular Biology*, *39*(1), 1–19. doi:10.1080/10409230490440532 PMID:15121719

Steinberg, F. (1998, May 1). *Biotech pharmaceuticals and biotherapy: An overview*. Semantic Scholar. https://www.semanticscholar.org/paper/Biotech-pharmaceuticals-and-biotherapy%3A-an-Steinberg-Raso/93e6672461b240dc57fe4548d48c1e785d6d7df7

Stevenson, J., Krycer, J. R., Phan, L., & Brown, A. (2013). A practical comparison of Ligation-Independent cloning techniques. *PLoS One*, *8*(12), e83888. doi:10.1371/journal.pone.0083888 PMID:24376768

Thomas, S. R., Maynard, N. D., & Gill, J. (2015). DNA library construction using Gibson Assembly®. *Nature Methods*, *12*(11), i–ii. doi:10.1038/nmeth.f.384

Ulislam, M. (2014, January 1). *Yeast cell-free enzyme system for bio-ethanol production at elevated temperatures.* Academia. https://www.academia.edu/60123976/Yeast_cell_free_enzyme_system_for_bio_ethanol_production_at_elevated_temperatures?from_sitemaps=true&version=2

Ullah, M. W., Khattak, W. A., Ul-Islam, M., Khan, S., & Park, J. K. (2014). Bio-ethanol production through simultaneous saccharification and fermentation using an encapsulated reconstituted cell-free enzyme system. *Biochemical Engineering Journal*, *91*, 110–119. doi:10.1016/j.bej.2014.08.006

Ullah, M. W., Khattak, W. A., Ul-Islam, M., Khan, S., & Park, J. K. (2015). Encapsulated yeast cell-free system: A strategy for cost-effective and sustainable production of bio-ethanol in consecutive batches. *Biotechnology and Bioprocess Engineering; BBE*, *20*(3), 561–575. doi:10.100712257-014-0855-1

Venter, M. (2007). Synthetic promoters: Genetic control through cis engineering. *Trends in Plant Science*, *12*(3), 118–124. doi:10.1016/j.tplants.2007.01.002 PMID:17292658

Vidigal, J. A., & Ventura, A. (2015). Rapid and efficient one-step generation of paired gRNA CRISPR-Cas9 libraries. *Nature Communications*, *6*(1), 8083. Advance online publication. doi:10.1038/ncomms9083 PMID:26278926

Wang, J. W., Wang, A., Li, K., Wang, B., Jin, S., Reiser, M., & Lockey, R. F. (2015). CRISPR/Cas9 nuclease cleavage combined with Gibson assembly for seamless cloning. *BioTechniques*, *58*(4), 161–170. doi:10.2144/000114261 PMID:25861928

Weissmann, F., Petzold, G., VanderLinden, R. T., Veld, P. J. H. I., Brown, N. G., Lampert, F., Westermann, S., Stark, H., Schulman, B. A., & Peters, J. (2016). biGBac enables rapid gene assembly for the expression of large multisubunit protein complexes. *Proceedings of the National Academy of Sciences of the United States of America, 113*(19). 10.1073/pnas.1604935113

Whitman, E. (2011, September 21). *Corporate Profits Trumping Public Health*. Global Issues. https://www.globalissues.org/news/2011/09/21/11262

Chapter 12
Integrated Machine Learning Approach for Healthcare Disabilities Tracing and Detection:
Applied Intelligent Techniques

Raghuram Bhukya
https://orcid.org/0000-0003-4109-4974
Department of Computer Science and Engineering, Kakatiya Institute of Technologey and Science, India

Dler Salih Hasan
https://orcid.org/0009-0008-3212-5509
College of Science, University of Salahaddin, Erbil, Iraq

M. Sangeetha
Department of Data Science and Business Systems, SRM Institute of Science and Technology, India

L. Raja
Department of Electronics and Communication Engineering, Sri Eshwar College of Engineering, India

Abdul Hannan
Department of Computer Science and Engineering, Koneru Lakshmaiah Education Foundation, India

ABSTRACT

The convergence of internet of things (IoT) technology and machine learning has paved the way for innovative solutions in the healthcare domain. This chapter presents an integrated approach that leverages IoT devices and advanced machine learning techniques to trace and detect healthcare-related disabilities. By continuously monitoring physiological signals and activity patterns, the proposed system enables early intervention and personalized care. The authors outline the key components of this approach, including data collection from IoT devices, data preprocessing, feature extraction, and the application of machine learning algorithms. Furthermore, ethical and privacy considerations are discussed, along with the potential benefits and challenges of deploying such a system. This chapter sheds light on the potential of IoT and machine learning to revolutionize healthcare by providing proactive and data-driven solutions for disabilities tracing and detection.

DOI: 10.4018/978-1-6684-6577-6.ch012

INTRODUCTION

The development of the smart grid has resulted from rising worldwide electricity demand, as well as the need to address environmental issues and improve energy efficiency. The smart grid is a cutting-edge power infrastructure that includes cutting-edge technologies to improve energy monitoring and control capabilities. To provide real-time monitoring, efficient energy management, and effective control mechanisms, this revolutionary grid system incorporates numerous components such as advanced metering infrastructure (AMI), communication networks, intelligent sensors, and data analytics. Energy monitoring and control are crucial in the smart grid, enabling utilities, operators, and consumers to manage energy resources, optimize performance, and achieve energy efficiency goals(Avancini et al., 2021; Paul Divakar et al., 2020). By monitoring consumption patterns, identifying inefficiencies, and providing real-time information, these systems enable informed decision-making and proactive management strategies. Accurate data on energy usage, supply-demand dynamics, and grid conditions enables timely actions to mitigate peak loads, reduce wastage, and ensure grid stability.

Advancements in sensor technologies, communication protocols, and data analytics have revolutionized energy monitoring in smart grids. Intelligent meters with bidirectional communication enable real-time data collection and demand response programs. Smart sensors and IoT devices provide granular data on grid parameters, equipment performance, and environmental factors. Data analytics techniques like machine learning and artificial intelligence enable predictive maintenance, load forecasting, and energy resource optimization. Real-time monitoring and control systems are essential for efficient smart grid operations, collecting, processing, and analyzing data from various sources. SCADA systems monitor grid assets, ADMS optimize operations, and ADR systems enable dynamic load management based on grid conditions and consumer preferences. These systems enable stakeholders to monitor energy flows, detect anomalies, and respond quickly to grid events(Hartono et al., 2013).

Energy monitoring and control systems in the smart grid ecosystem offer numerous benefits, including optimizing energy generation, transmission, and distribution, reducing wastage, improving grid reliability, and providing real-time insights for consumers. These systems also enable informed decisions, reduced consumption, and lower energy bills. Additionally, they support renewable energy sources and facilitate efficient energy storage management, promoting a greener, more sustainable grid. Research on smart grid technology has gained attention due to the growing demand for efficient, sustainable energy systems. These systems integrate advanced technologies for intelligent monitoring, control, and management of energy resources. This section highlights key areas of focus and advancements in smart grid technology(Abo-Zahhad et al., 2016).

- Smart grid technology aims to improve power grid performance by upgrading aging infrastructure, enhancing resilience, and integrating renewable energy sources. Advanced sensors, communication networks, and control systems enable real-time monitoring, fault detection, localization, and grid stability.
- Advanced Metering Infrastructure (AMI) and smart meters are crucial areas of research in smart grid technology. Smart meters enable two-way communication between utilities and consumers, collecting real-time data on energy consumption, demand patterns, and power quality. Research focuses on efficient metering systems, data analytics, and consumer engagement strategies for energy conservation and load management.

- Smart grid technology research focuses on integrating distributed energy resources like solar panels, wind turbines, and energy storage systems. This involves developing algorithms and control strategies for optimal DER integration, managing bidirectional power flows, and enabling grid flexibility and resilience.
- Demand response programs are crucial in smart grid research, focusing on intelligent mechanisms and real-time data to optimize energy consumption, reduce peak demand, and promote grid stability through effective strategies.
- Smart grids require robust security protocols, intrusion detection systems, and data privacy frameworks to protect infrastructure from cyber threats and ensure consumer privacy. Research has been conducted to address these concerns in communication networks and data exchange.
- Data analytics and AI techniques are being utilized in smart grid technology to derive actionable insights for load forecasting, fault detection, asset management, and decision-making processes. Researchers explore machine learning algorithms, optimization techniques, and predictive analytics to improve efficiency and effectiveness.
- Research focuses on enhancing grid resilience and exploring microgrids, self-contained energy systems that can operate independently or in coordination with the main grid. Studies investigate their design, control, and operation, including integration with the main grid during normal and emergency conditions.

Research on smart grid technology covers modernization, advanced metering, distributed energy resource integration, demand response, cybersecurity, data analytics, and microgrids. It aims to overcome challenges, optimize performance, and create a sustainable, efficient energy future. Energy monitoring and control are crucial in smart grids, offering numerous benefits and advantages(Dalipi & Yayilgan, 2016; Olbrich, 2016).

- Energy monitoring helps utilities and consumers analyze real-time energy consumption patterns, enabling them to identify high usage areas, optimize consumption, and implement energy-saving measures. This information informs decisions on load management, peak shaving, and energy conservation, resulting in improved efficiency and cost savings.
- Energy monitoring systems offer valuable data on grid conditions, voltage levels, power quality, and demand-supply dynamics. These insights enable grid operators to monitor and manage performance, detect anomalies, and respond promptly to issues like voltage fluctuations, equipment failures, or congestion. This ensures uninterrupted power supply and minimizes blackout risks.
- Real-time energy monitoring enables demand response programs, allowing consumers to adjust energy consumption based on grid conditions and price signals. Grid operators can implement load management strategies, balance supply and demand, and reduce peak loads, optimizing grid operations, avoiding overloading, and avoiding infrastructure upgrades.
- Renewable energy sources like solar and wind increase variability and intermittency in energy generation. Energy monitoring enables grid operators to monitor generation levels, predict output fluctuations, and manage integration. This promotes sustainability and reduces reliance on fossil fuels, ensuring smooth integration and sustainable energy use(Boopathi, Kumar, et al., 2023; Boopathi & Davim, 2023; Dhanya et al., 2023; Hanumanthakari et al., 2023; Janardhana et al., 2023; Maguluri et al., 2023).

- Data-driven decision making in energy monitoring systems transforms vast amounts of data into valuable insights for utilities, grid operators, and consumers. This enables proactive maintenance, predictive analytics, accurate load forecasting, energy-saving opportunities identification, and demand-side management programs tailored to consumer needs.

IoT enhances smart grid capabilities and enabling remote monitoring and control(Hu et al., 2017; Saveen et al., 2018; Thakare et al., 2017).

- IoT devices and sensor networks enable real-time data collection from grid components like meters, sensors, transformers, and substations. This data captures energy consumption, grid parameters, equipment status, and environmental conditions, enabling comprehensive monitoring and situational awareness.
- IoT enables robust communication networks connecting smart devices, enabling seamless data exchange between grid assets, control systems, and stakeholders. This technology improves grid efficiency, remote monitoring, and decision-making, ensuring reliable and secure data transmission(Anitha et al., 2023; Vanitha, Radhika, et al., 2023; Venkateswaran, Vidhya, Naik, et al., 2023; Zekrifa et al., 2023).
- IoT integrates intelligent control systems in smart grids, optimizing operations, managing energy resources, and responding to events. These systems enable automated demand response, dynamic load balancing, fault detection, and self-healing, enhancing grid stability and efficiency.
- IoT-generated data and advanced analytics enable meaningful insights from grid data, identifying consumption patterns, anomalies, demand fluctuations, and optimizing energy management strategies. This data-driven intelligence enables proactive decision-making, efficient resource allocation, and predictive maintenance.
- IoT empowers consumers to participate in energy management through smart meters, devices, and user-friendly interfaces. This promotes energy conservation, load flexibility, and demand response, contributing to a more sustainable and efficient grid.

IoT technology significantly impacts smart grids by enabling real-time data collection, advanced communication infrastructure, intelligent control systems, data analytics, and consumer empowerment. This integration improves efficiency, reliability, and sustainability in energy systems.

MICROGRID IMPLEMENTATION PROCEDURE

The Microgrid Implementation Procedure is explained below (Ghasempour, 2019; Hartono et al., 2013; Katyara et al., 2019; Li et al., 2018; Olbrich, 2016).

- Conduct a feasibility study to evaluate microgrid implementation's viability and potential benefits, considering energy requirements, energy sources, grid connectivity, load profiles, and economic factors.
- Design a microgrid system based on project requirements, determining size, energy sources, and necessary control and monitoring systems.

- Select appropriate microgrid components based on efficiency, reliability, scalability, and compatibility with the architecture, considering generation sources, energy storage systems, and control systems.
- Integrate components for a cohesive microgrid system, ensuring proper electrical connections, protection mechanisms, and communication interfaces.
- Implement a microgrid control and management system to monitor, control, and coordinate energy generation, storage, and distribution, while ensuring load management, frequency and voltage control, and seamless transitions between grid-connected and islanded modes.
- Conduct comprehensive testing and commissioning of microgrid system to ensure proper functioning, adherence to design specifications, and verify component performance under various operating conditions.
- Ensure microgrid compliance with grid interconnection standards and regulations by obtaining necessary approvals, permits, and certifications from regulatory authorities.
- Establish a microgrid operation and maintenance plan, regularly monitor components, perform upgrades, and address issues promptly.

Microgrid components (Chandak & Rout, 2021; Ferahtia et al., 2022; Hartono et al., 2013; Saveen et al., 2018) work together to efficiently generate, store, and distribute energy, with primary components providing essential functions (Figure 1).

- Energy Generation Sources: Solar panels, wind turbines, and diesel generators generate electrical energy for microgrids, ranging from renewable to conventional, depending on project objectives and resources.
- Energy Storage Systems: Microgrid energy storage systems, like batteries or flywheels, store excess energy for future use, balancing supply and demand fluctuations and enhancing resilience.
- Power Converters/Inverters: Converters and inverters convert energy from DC to AC or vice versa for microgrids, ensuring compatibility with various energy sources and electrical infrastructure.
- Transformers: Transformers regulate voltage levels in microgrids, ensuring efficient transmission and distribution of electricity among components and loads.
- Protective Devices: Circuit breakers and relays are crucial for microgrid safety, reliability, fault detection, equipment protection, and system stability.
- Control and Monitoring Systems: Microgrid systems control and optimize energy flow, storage, load demand, and component operation by monitoring generation, storage levels, and load demand.

Microgrid focuses on efficient, sustainable energy solutions through localized energy generation, storage, and consumption using renewable or conventional sources like solar panels and wind turbines (Chandak & Rout, 2021; Hartono et al., 2013).

- Microgrid generates excess energy, stored in batteries, for high demand or insufficient primary energy sources during storage systems.
- Power converters and inverters ensure energy source compatibility and synchronization in microgrids, converting DC power from storage systems or renewables into AC for electrical appliances and equipment.

Figure 1. Micro-grid components

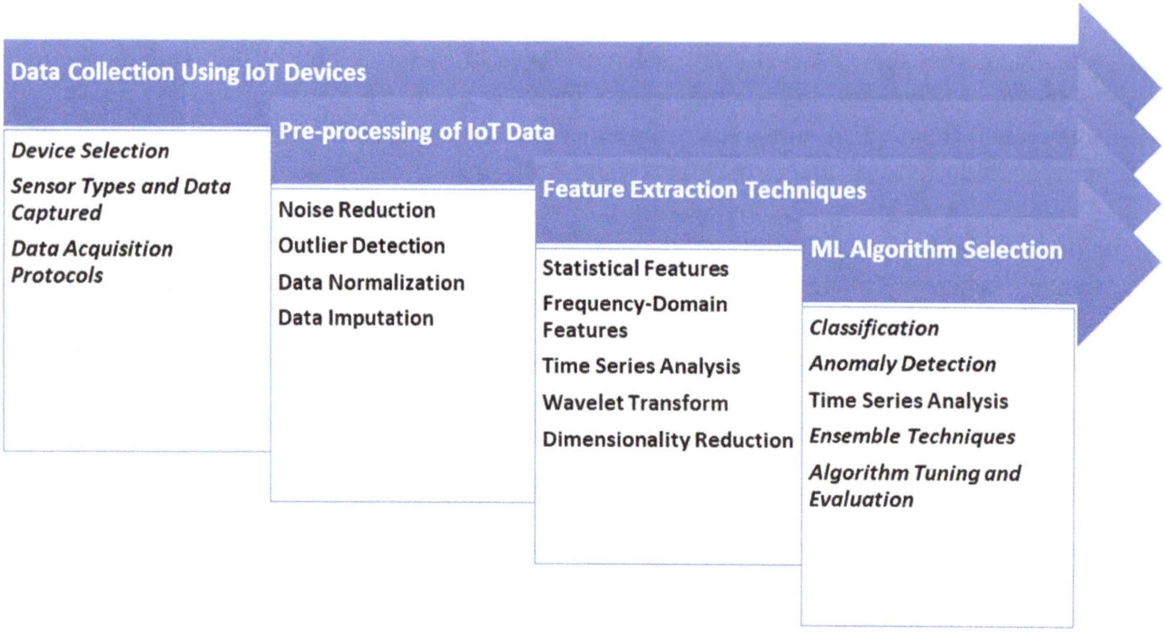

- Microgrid's load management system monitors energy demand, optimizes distribution, prioritizes critical loads, and balances energy supply, ensuring efficient and reliable power supply.
- Grid connectivity enables microgrids to exchange energy with the main electrical grid, enabling efficient resource utilization and grid stability by importing or exporting excess energy when demand exceeds generation capacity.
- Islanded operation allows microgrids to operate independently during grid outages or intentional islanding, using their own energy sources and storage systems. The control system balances generation and demand, ensuring a reliable, continuous power supply for critical loads.

IOT-INTEGRATED MICROGRID SYSTEM

Microgrids in Smart Grids

Microgrids are localized energy systems that operate independently or in coordination with the main grid, utilizing distributed energy resources like solar panels, wind turbines, energy storage, and controllable loads (Barman et al., 2019; Paul Divakar et al., 2020).

- Increased Resilience: Microgrids provide localized power generation and distribution, making them more resilient to grid disruptions and outages. In the event of a main grid failure, microgrids can continue to supply power to critical loads, such as hospitals, emergency services, or critical infrastructure, ensuring uninterrupted electricity supply.

- Integration of Renewable Energy: Microgrids facilitate the integration of renewable energy sources into the grid. They can efficiently harness solar or wind energy within a local area, reducing reliance on fossil fuels and decreasing greenhouse gas emissions. Microgrids enable the seamless integration of intermittent renewable resources by managing their variability and ensuring grid stability.
- Improved Energy Efficiency: Microgrids enable localized energy generation, which reduces transmission and distribution losses associated with long-distance power delivery. By generating power close to the point of consumption, microgrids minimize energy losses, improving overall energy efficiency and reducing energy costs.

Integration of IoT in Microgrids

IoT technologies improve microgrid functionality and performance by enabling real-time monitoring, control, and optimization of components (Avancini et al., 2021; Boopathi, Arigela, et al., 2023; Domakonda et al., 2023; Jeevanantham et al., 2023; Sathish et al., 2023; Subha et al., 2023; Vanitha, K., et al., 2023).

- Sensor Networks and Data Collection: IoT devices and sensors deployed in microgrids collect data on energy generation, consumption, grid parameters, and environmental conditions. These sensors provide real-time data, allowing operators to monitor and manage the microgrid's performance, detect anomalies, and optimize energy resources.
- Communication and Connectivity: IoT enables seamless communication and connectivity between microgrid components, such as DERs, energy storage systems, and control systems. This facilitates efficient coordination and control of distributed energy resources, load management, and grid operation in real-time (Hanumanthakari et al., 2023; Harikaran et al., 2023; Hema et al., 2023; Janardhana et al., 2023).
- Intelligent Control and Optimization: IoT integration enables intelligent control and optimization of microgrid operations. By leveraging data from IoT devices, machine learning algorithms, and advanced control strategies, microgrid operators can optimize energy generation, storage, and consumption, ensuring efficient utilization of resources and cost savings.

Benefits and Challenges of IoT-Integrated Microgrids

- Enhanced Efficiency and Reliability: IoT integration enables real-time monitoring, control, and optimization, leading to improved energy efficiency, reduced energy wastage, and enhanced grid reliability. Operators can identify and address issues promptly, optimize energy management strategies, and adapt to changing grid conditions in real-time.
- Increased Grid Flexibility: IoT integration enhances the flexibility of microgrids by enabling dynamic load management, demand response, and seamless integration of renewable resources. Microgrids can actively respond to grid conditions, adjust energy generation and consumption based on demand and supply fluctuations, and actively participate in grid-wide energy management programs.
- Cost Savings and Economic Benefits: IoT-integrated microgrids can optimize energy generation and consumption, reducing energy costs for consumers. By leveraging energy storage systems and intelligently managing energy resources, microgrids can potentially reduce peak demand charges,

optimize energy procurement, and participate in energy markets, leading to cost savings and economic benefits.

- IoT-integrated microgrids face challenges in data security, privacy, scalability, interoperability, and complexity. Data security and privacy are crucial, and robust cybersecurity measures are essential. Scaling up microgrids requires standardization and compatibility between different technologies. Complexity and management are also essential, as IoT-integrated microgrids involve multiple components, data streams, and control systems. Adequate expertise, monitoring systems, and management strategies are necessary for efficient operation and maintenance.

IoT integration in microgrids offers resilience, renewable energy integration, energy efficiency, and optimized grid operation. However, challenges like data security, scalability, and complexity must be addressed for successful implementation in smart grid systems (Dalipi & Yayilgan, 2016; Thakare et al., 2017).

SYSTEM IMPLEMENTATION FOR ENERGY MONITORING AND CONTROL

Hardware Components and Architecture

Implementing energy monitoring and control systems require hardware components and architecture, focusing on key aspects (Hang & Kim, 2018; Pawar et al., 2020; Saleem et al., 2021). The system implementation for energy monitoring and control is illustrated in Figure 2.

- Hardware Components: Smart meters, sensors, communication devices, data loggers, control devices, and computing infrastructure are essential hardware components for real-time energy measurement and data collection. Sensors monitor parameters like voltage, current, power quality, temperature, and environmental conditions, while communication devices facilitate data transmission. Data loggers store and record collected data, while control devices enable remote control and automation of energy systems.
- System Architecture: System architecture involves hardware components arrangement, interconnection, data flow, communication pathways, and control mechanisms for seamless energy monitoring and control. It may involve centralized or decentralized approaches depending on specific requirements.

Sensor Selection and Placement

Selecting and placing sensors is crucial for accurate energy monitoring and control, considering factors like size, shape, and placement (Padmanaban et al., 2019; Premkumar & Sigappi, 2021; Zhang et al., 2021).

- Monitor system parameters like voltage, current, power consumption, quality, temperature, humidity, and other relevant factors based on requirements.
- Select sensors with high accuracy and reliability, considering factors like measurement range, resolution, calibration, and longevity for data integrity.

Figure 2. System implementation for energy monitoring and control

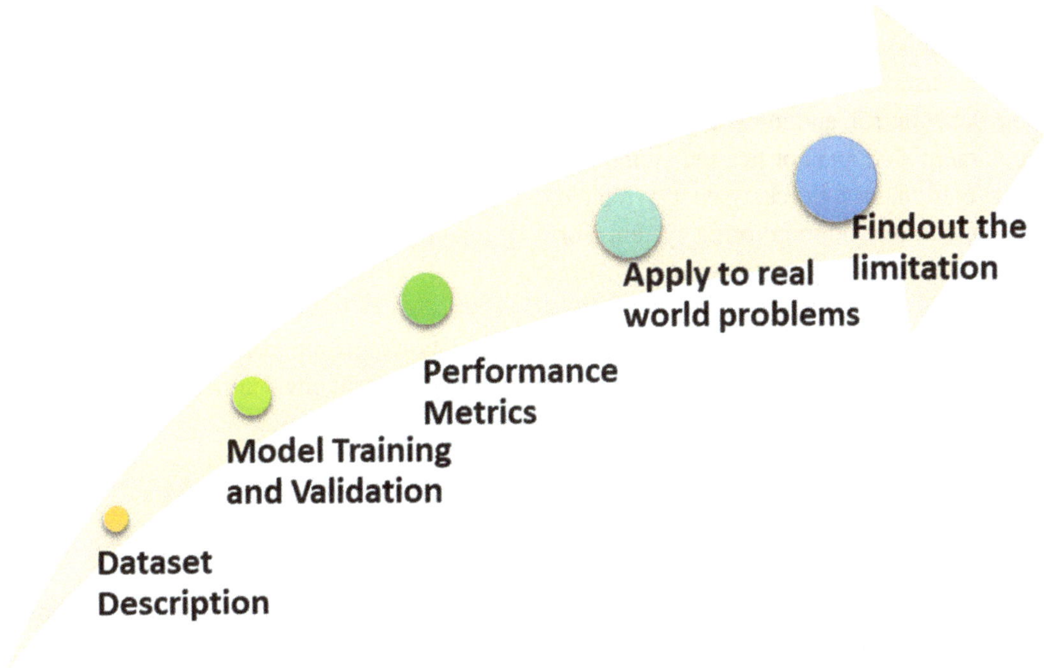

- Optimize sensor placement for representative data and comprehensive energy system view by considering accessibility, proximity to target parameters, and environmental factors affecting sensor performance.
- Incorporate redundancy in sensor placement for data reliability and fault tolerance, using duplicate sensors for backup measurements and identifying failures.

Data Acquisition and Communication Infrastructure

Efficient data acquisition and communication infrastructure are crucial for seamless data exchange.

- Implement a data acquisition system for collecting and processing data from sensors and smart meters using loggers, gateways, or edge computing devices.
- Develop a robust communication infrastructure, including wired or wireless technologies like Ethernet, Wi-Fi, Zigbee, or cellular networks, to ensure secure and reliable data transfer between system components.
- Adopt standardized protocols and communication standards for interoperability, compatibility, and seamless integration of devices and components in energy monitoring and control systems.

Control and Monitoring Systems

Control and monitoring systems manage and regulate energy resources using collected data.

- Develop control algorithms and strategies to optimize energy consumption, demand response, load balancing, and other control actions using advanced techniques like model predictive control, fuzzy logic, or machine learning.
- Create a user-friendly interface for system monitoring and control, enabling visualization, real-time data control, and informed decision-making.
- Implement alert and notification systems to promptly alert operators and users of abnormal conditions, equipment failures, or energy-related events.
- Design control and monitoring systems for easy integration with existing infrastructure and scalability for future expansions or modifications.

Implementing hardware components, sensor selection, data acquisition, communication infrastructure, and control systems for energy monitoring and control in smart grid environments effectively captures data and enables efficient resource management.

ANFIS FOR ENERGY MONITORING

Adaptive Neuro-Fuzzy Inference System (ANFIS) is a hybrid intelligent system that uses artificial neural networks (ANNs) and fuzzy logic to create a computational model. ANNs learn from data and capture complex patterns, while fuzzy logic handles linguistic variables and uncertainty reasoning. This system effectively models and analyzes complex relationships between input and output variables (Kampouropoulos et al., 2018; Suganthi et al., 2015; Suhail et al., 2021). In the field of energy monitoring, ANFIS has gained popularity due to its ability to handle the inherent complexity of energy systems. Energy consumption patterns often exhibit nonlinear behavior and are influenced by various factors such as weather conditions, occupancy, and time of day. Traditional modeling techniques may struggle to capture the intricate relationships between these variables. However, ANFIS excels in handling such complexities by providing a flexible and adaptive modeling framework (Kaur & Sood, 2019).

ANFIS consists of a layered structure that includes both fuzzy inference systems and neural networks. The fuzzy inference system is responsible for capturing linguistic rules and providing a fuzzy representation of the input-output relationships. Fuzzy sets and membership functions are used to define the linguistic variables and their relationships. The neural network component of ANFIS performs the task of parameter estimation and learning. It learns from the input-output data and adjusts the parameters of the fuzzy inference system to optimize the model's performance.

The training process of ANFIS involves two key steps: structure identification and parameter optimization. In the structure identification step, clustering algorithms are typically employed to determine the optimal number and configuration of fuzzy sets based on the input data. The clustering process identifies the appropriate number of fuzzy rules and determines their centers and spreads. Once the structure is identified, parameter optimization techniques such as least squares estimation or gradient descent are used to fine-tune the parameters of the ANFIS model (Khan & Rizwan, 2021a).

The resulting ANFIS model can accurately represent the complex relationships between input and output variables in energy monitoring applications. It can effectively capture the nonlinear behavior of energy consumption patterns and make accurate predictions based on the input variables. ANFIS models can be utilized for load forecasting, energy optimization, anomaly detection, and other energy management tasks.

ANFIS is a computational model that combines ANNs and fuzzy logic, making it ideal for energy monitoring applications. It handles complex relationships, enables accurate predictions, optimizes usage, and supports efficient decision-making in smart grid environments. ANFIS is a powerful tool for developing energy monitoring models that accurately predict energy consumption and parameters. It is used for load forecasting, optimization, and demand response in various energy management tasks (Khan & Rizwan, 2021b).

ANFIS models can accurately forecast future energy demand using historical energy consumption data and input variables like weather conditions and load patterns. This enables utilities and grid operators to efficiently plan, allocate resources, optimize energy generation and distribution, and ensure grid stability.

ANFIS models optimize energy usage in buildings, industrial processes, and smart grids by considering occupancy patterns, weather conditions, and equipment efficiency. This leads to energy savings, reduced peak loads, and improved efficiency, ultimately resulting in cost savings and environmental benefits.

ANFIS models are essential in demand response programs, managing energy consumption based on grid conditions and price signals. By analyzing historical data and real-time pricing information, utilities can predict consumer responses, incentivize peak usage reduction, balance supply and demand, and improve grid stability.

ANFIS models use historical data and input variables to predict energy consumption, load patterns, optimize usage, and support demand response initiatives. This insights enable efficient resource planning, improved grid operation, and enhanced energy efficiency, contributing to smarter, sustainable energy systems.

ANFIS Model Training and Optimization Techniques

ANFIS model training and optimization involve techniques for accurate predictions, using key steps and methods (Adedeji et al., 2022; Kampouropoulos et al., 2018; Khan & Rizwan, 2021a).

- Collect a representative dataset of historical energy consumption data and input variables like weather conditions, time, and occupancy patterns to capture system variability and cover various scenarios.
- ANFIS models utilize fuzzy logic to generate fuzzy rules, defining fuzzy sets and membership functions. Clustering techniques like subtractive, grid partitioning, or k-means determine the optimal number and shape of fuzzy sets based on the dataset.
- Parameter estimation is crucial for optimizing the ANFIS model after establishing fuzzy rules and membership functions. This involves determining centers, spreads, and weights to achieve the best fit with training data. Optimization techniques like least squares estimation, gradient descent, and genetic algorithms aim to minimize the difference between the model's predicted output and the actual output in the training dataset (Boopathi, 2022g, 2022a, 2022f, 2023b, 2023a, 2023c).
- Cross-validation is a technique used to evaluate the trained ANFIS model's generalization capability. It involves dividing the dataset into training and validation sets, training the model using the training set and testing against the validation set to detect overfitting and provide a more realistic estimate of its performance (Boopathi, 2022b, 2022d, 2022c, 2022e).
- Optimizing the ANFIS model involves fine-tuning parameters, adjusting fuzzy sets, and exploring different learning algorithms to improve accuracy and efficiency. The goal is to find the optimal configuration with the best performance metrics, such as MAE, RMSE, or R-squared.

- Validation and testing of the ANFIS model using independent datasets after training ensures its performance is not biased and provides a realistic assessment of predictive capabilities. Validation also assesses the model's robustness and generalization ability in handling new data.

Performance Evaluation and Results

Performance evaluation is crucial for assessing ANFIS-based energy monitoring models' accuracy and reliability using metrics like energy consumption prediction and target variables (Ghenai et al., 2022; Padmanaban et al., 2019).

- **Mean Absolute Error (MAE):** MAE measures the average absolute difference between predicted and actual values, indicating model accuracy; lower MAE indicates better performance.
- **Root Mean Square Error (RMSE):** RMSE measures the average squared difference between predicted and actual values, giving more weight to larger errors and indicating better accuracy.
- **Coefficient of Determination (R-squared):** R-squared measures the proportion of variance in energy consumption explained by independent variables, with a 1 value indicating higher correlation and predictive power.

ANFIS models' performance can be assessed through quantitative metrics and comparative analysis, comparing them with other prediction techniques or benchmarks. Performance evaluation should be conducted using independent datasets, ensuring fair assessment of generalization and handling unseen data. Graphical visualization techniques, such as line graphs or scatter plots, can help visualize predicted values and capture trends, patterns, and anomalies in energy consumption data (Boopathi, Pandey, et al., 2023; Boopathi & Kanike, 2023; Chandrika et al., 2023; Dhanya et al., 2023). Assessing ANFIS-based energy monitoring models using metrics and comparative analysis helps stakeholders evaluate effectiveness, identify improvement areas, and make informed decisions on smart grid energy management strategies (Mellit & Kalogirou, 2011; Mohammad Bilal, 2022).

- Energy Consumption Patterns: ANFIS models can accurately capture energy consumption patterns, allowing for a better understanding of energy usage trends over time. By analyzing the model's predictions, energy managers can identify peak demand periods, seasonal variations, and consumption patterns associated with specific activities or events.
- Peak Load Predictions: ANFIS models can forecast peak load periods, enabling utilities and grid operators to optimize energy generation and distribution resources accordingly. Accurate peak load predictions assist in reducing the strain on the grid during high-demand periods and implementing demand response strategies effectively.
- Energy Efficiency Improvements: ANFIS models help identify opportunities for energy efficiency improvements. By analyzing the relationships between input variables and energy consumption, the models can highlight factors that significantly impact energy usage. This information assists in identifying areas for energy-saving measures, equipment upgrades, and load management strategies to improve overall energy efficiency.
- Demand Response Strategies: ANFIS models play a crucial role in demand response programs by providing insights into consumer response patterns to pricing signals or incentive programs. By analyzing historical data and input variables, ANFIS can predict how consumers are likely to

respond to specific demand response initiatives. This information aids in designing effective demand response strategies and optimizing energy consumption during peak periods.

- Validation and Testing: Validation and testing of ANFIS models assess accuracy and reliability against independent datasets. These evaluations evaluate the model's generalization to unseen situations and ensure its performance is not limited to training data, providing accurate predictions for new scenarios.
- Sensitivity Analysis: Sensitivity analysis evaluates the impact of input variables on a model's predictions, helping energy managers identify significant factors affecting energy consumption. This helps prioritize resources and interventions for effective energy management.

ANFIS-based energy monitoring models offer valuable insights into energy consumption patterns, peak load predictions, efficiency improvements, and demand response strategies. Validation, testing, and sensitivity analysis improve models' accuracy and reliability, aiding energy management decision-making and fostering smarter, sustainable energy systems in smart grids.

Fuzzy Fruit Fly Optimization (FFO) Algorithm for IoT-Integrated Smart Grids

The Fuzzy Fruit Fly Optimization (FFO) algorithm is a nature-inspired optimization technique that combines fuzzy logic principles with fruit flies' behavior to solve optimization problems. It is ideal for IoT-integrated smart grids, where efficient control and optimization are crucial for managing energy resources. The FFO algorithm uses fuzzy logic to represent uncertainty and imprecision in smart grid control problems, offering a robust and adaptive approach for optimizing various aspects of smart grid operations (Ghenai et al., 2022).

The FFO algorithm is a population-based approach used in smart grid control to address challenges like load balancing, demand response, renewable energy integration, and power distribution optimization. It simulates the foraging behavior of fruit flies to find optimal control strategies that minimize energy consumption, maximize renewable energy utilization, and improve grid efficiency. The algorithm operates in iterations, maintaining a population of candidate solutions called fruit flies, which are evaluated using an objective function that captures the optimization goals of the smart grid control problem (Li et al., 2018).

The FFO algorithm uses fuzzy logic to manage uncertainty and imprecision in smart grid control variables. It uses fuzzy rules and membership functions to adapt to changing grid conditions. Through an iterative process, it converges towards optimal control solutions by dynamically adjusting population size, exploration parameters, and fuzzy rules (Hu et al., 2017). The Fuzzy Fruit Fly Optimization (FFO) algorithm offers several advantages in smart grid control. It provides a flexible, adaptive framework for handling complex systems, handles uncertainties and vagueness in control variables, and allows parallel processing and exploration of multiple control strategies. This population-based approach enhances efficiency and scalability in IoT-integrated systems, improving energy management, grid efficiency, and renewable energy resource integration.

FFO-Based Optimization of Energy Management Strategies

FFO-based optimization techniques improve energy management strategies in IoT-integrated smart grids by optimizing energy generation, distribution, and consumption, benefiting key areas like generation, distribution, and consumption (Hu et al., 2017; Li et al., 2018).

- Load Balancing: FFO optimizes load balancing in smart grids by dynamically allocating energy loads among sources and consumers, considering factors like demand, supply availability, and grid constraints, minimizing imbalances and ensuring efficient energy utilization.
- Demand Response: FFO-based optimization enhances demand response strategies by managing energy consumption based on grid conditions and pricing signals. It analyzes historical data, consumer behavior, and real-time pricing information to determine effective actions, such as shifting loads, adjusting consumption patterns, and incentivizing consumption reduction during peak periods.
- Renewable Energy Integration: FFO optimizes renewable energy integration in smart grid systems by considering factors like weather conditions, energy storage capacity, and demand patterns, maximizing clean and sustainable energy utilization while ensuring grid stability and reliability.
- Power Distribution Optimization: FFO-based optimization techniques improve power distribution in smart grids by considering line losses, voltage stability, and load demands, enhancing efficiency, minimizing losses, and reducing system failures.
- Energy Efficiency Improvements: FFO optimizes energy efficiency by analyzing consumption patterns, considering occupancy, weather, and equipment efficiency, recommending energy-saving measures, optimizing scheduling, and identifying load shedding or shifting opportunities.

FFO-based optimization in energy management strategies offers advantages such as adaptability, robustness, population-based approach, and integration of fuzzy logic techniques. This approach handles uncertainties and dynamic conditions in smart grids, allowing faster convergence and improved scalability. By utilizing FFO-based optimization techniques, smart grid systems can achieve improved energy efficiency, renewable energy utilization, grid stability, and reduced operational costs, contributing to the development of sustainable and intelligent energy systems in IoT-integrated smart grids.

Case Studies

Case studies demonstrate the effectiveness and benefits of FFO-based optimization techniques in energy management strategies, with examples of successful application in real-world scenarios (Kannan et al., 2020; Khalid & Javaid, 2020; Li et al., 2018; Premkumar & Sigappi, 2021).

- Load Balancing in a Microgrid: In a case study involving a microgrid with distributed energy resources, FFO was employed to optimize load balancing. The FFO algorithm considered factors such as energy demand, renewable energy generation, and battery storage capacity. The results showed significant improvements in load balancing, reducing grid imbalances and optimizing energy utilization across the microgrid.

- Demand Response Optimization: A case study focused on demand response optimization using FFO in a residential setting. Historical energy consumption data, weather conditions, and pricing signals were used as inputs to the FFO algorithm. The results demonstrated effective demand response strategies that incentivized consumers to reduce energy consumption during peak periods, resulting in reduced overall grid demand and improved grid stability.
- Renewable Energy Integration: FFO-based optimization was applied to optimize the integration of renewable energy sources into a smart grid system. The FFO algorithm considered factors such as solar and wind energy availability, demand patterns, and storage capacity. The results showcased improved utilization of renewable energy, reduced reliance on conventional energy sources, and enhanced grid resilience.
- Power Distribution Optimization: FFO was employed to optimize power distribution in a distribution network with multiple substations. The FFO algorithm considered parameters such as line losses, voltage stability, and load demands. The results demonstrated optimized power flow, reduced losses, and improved voltage stability, leading to increased grid efficiency and reduced operational costs.

Case studies demonstrate the effectiveness of FFO-based optimization techniques in improving energy management strategies. These techniques enhance load balancing, demand response, renewable energy integration, and power distribution optimization. However, specific results may vary depending on the system, data, and optimization objectives. Conducting case studies and performance evaluations is crucial for a comprehensive understanding of the benefits and effectiveness of FFO-based optimization techniques in energy management strategies.

INTEGRATION OF ANFIS AND FFO ALGORITHMS FOR ENERGY MONITORING AND CONTROL

The ANFIS-FFO hybrid system combines ANFIS's modeling capabilities with FFO's optimization capabilities for energy monitoring and control in smart grids. It leverages the strengths of both techniques, offering a powerful approach for smart grid energy management (Ghenai et al., 2022; Hu et al., 2017; Kannan et al., 2020; Mellit & Kalogirou, 2011). The ANFIS-FFO hybrid system combines ANFIS models for energy monitoring and control, while FFO optimizes ANFIS models for parameter fine-tuning and performance improvement. This comprehensive framework combines modeling and optimization capabilities, enhancing energy monitoring and control. ANFIS-FFO hybrid system architecture are depicted in Figure 3.

System Architecture and Implementation

ANFIS-FFO hybrid system architecture comprises components:

- Data Acquisition: The system collects energy-related data from various sources, such as smart meters, sensors, and weather stations. This data serves as input for the ANFIS models and optimization process.

Figure 3. ANFIS-FFO hybrid system architecture

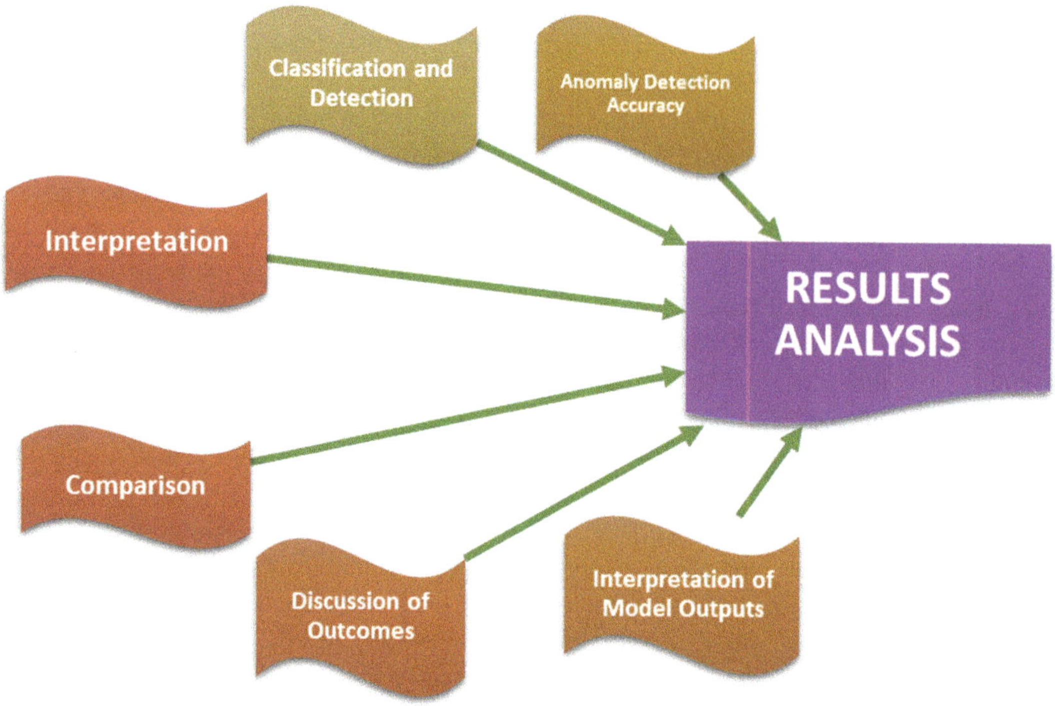

- ANFIS Modeling: ANFIS models are developed to capture the relationships between input variables (e.g., weather conditions, time of day, occupancy patterns) and output variables (e.g., energy consumption, load demand). The ANFIS models use fuzzy logic to handle uncertainty and linguistic variables.
- FFO Optimization: The FFO algorithm is applied to optimize the ANFIS models. FFO utilizes its population-based optimization approach to search for optimal parameter values for the ANFIS models. By fine-tuning the parameters, FFO aims to improve the accuracy and performance of the ANFIS models.
- Training and Optimization: The ANFIS models are trained and optimized using historical data. The FFO algorithm iteratively adjusts the parameters of the ANFIS models to minimize the objective function, such as mean absolute error (MAE) or root mean square error (RMSE), and enhance the accuracy of energy predictions.
- Control and Monitoring: The optimized ANFIS models are deployed in the control and monitoring system of the smart grid. They provide real-time predictions and insights into energy consumption, load demand, and other relevant parameters. This information is used to optimize energy management decisions, load balancing, demand response, and other control strategies.

The ANFIS-FFO hybrid system combines ANFIS modeling and FFO optimization, offering a comprehensive approach to energy monitoring and control in smart grids. It improves energy predictions accuracy, optimizes control strategies, and enhances overall energy management. Access to relevant data, modeling techniques, and computational resources is essential for training and optimization.

Experimental Setup and Evaluation

A comprehensive experimental setup is needed to evaluate the effectiveness of the ANFIS-FFO hybrid system for energy monitoring and control (Ghenai et al., 2022; Hu et al., 2017; Li et al., 2018; Mellit & Kalogirou, 2011).

a. Data Collection: Gather historical energy consumption data, weather data, occupancy patterns, and other relevant variables from smart meters, sensors, and other sources in the smart grid. This data will serve as the input for the ANFIS models and FFO optimization.
b. Data Preprocessing: Clean and preprocess the collected data to handle missing values, outliers, and normalization. Ensuring high-quality data is essential for accurate modeling and optimization.
c. ANFIS Model Development: Develop ANFIS models to capture the relationships between the input variables and energy consumption or other target variables. Train the ANFIS models using a portion of the collected data and validate them using another subset.
d. FFO Optimization: Apply the FFO algorithm to optimize the parameters of the ANFIS models. Utilize the validation data to fine-tune the ANFIS models and improve their performance.
    1. Control and Monitoring Integration: Integrate the optimized ANFIS models into the smart grid's control and monitoring system. This enables real-time predictions and monitoring of energy consumption, load demand, and other parameters.

Performance Analysis and Comparison

Conduct a performance analysis and comparison of the ANFIS-FFO hybrid system.

a) Metrics Selection: Choose appropriate performance metrics, such as mean absolute error (MAE), root mean square error (RMSE), coefficient of determination (R-squared), or others, to assess the accuracy and reliability of the ANFIS-FFO predictions.
b) Performance Evaluation: Evaluate the ANFIS-FFO hybrid system's performance using an independent test dataset that was not used during training and optimization. Calculate the selected metrics to quantify the accuracy of energy predictions and other control variables.
c) Comparative Analysis: Perform a comparative analysis to compare the performance of the ANFIS-FFO hybrid system with other energy monitoring and control methods. This may include comparing against traditional ANFIS models, standalone FFO optimization, or other prediction techniques.
d) Sensitivity Analysis: Conduct sensitivity analysis to evaluate the impact of different input variables on the ANFIS-FFO predictions. Identify the most significant factors affecting energy consumption and load demand.
e) Real-World Scenario Testing: If possible, validate the ANFIS-FFO hybrid system's performance in a real-world smart grid environment. Implement the system and observe its effectiveness in real-time energy monitoring and control.
f) Robustness and Scalability: Assess the robustness and scalability of the ANFIS-FFO hybrid system to handle varying data sizes, different smart grid configurations, and changing environmental conditions.

The ANFIS-FFO hybrid system's performance analysis and comparison showcase its advantages in energy monitoring and control, enhancing energy management and optimizing smart grid operations.

Case Study: ANFIS-FFO Hybrid System for Energy Monitoring and Control in a Smart Grid

This case study evaluates the effectiveness of the ANFIS-FFO hybrid system for energy monitoring and control in smart grid environments, evaluating energy predictions accuracy, control strategy optimization, and overall performance compared to traditional methods (Ghenai et al., 2022; Hu et al., 2017; Li et al., 2018; Mellit & Kalogirou, 2011).

Experimental Setup

- Data collection involves collecting historical energy consumption, weather conditions, occupancy patterns from smart meters and sensors in a real-world smart grid.
- Data preprocessing handles missing values, outliers, and normalization for consistent analysis.
- ANFIS models predict energy consumption using input variables like time, temperature, and occupancy, using fuzzy logic to handle uncertainties.
- FFO Optimization optimizes ANFIS models' parameters, enhancing accuracy and performance by fine-tuning parameters using Fuzzy Fruit Fly.
- Optimized ANFIS models integrated into smart grid control and monitoring system for real-time energy predictions.

Performance Evaluation

- Energy prediction accuracy evaluated using MAE, RMSE, and R-squared metrics for accuracy evaluation.
- ANFIS-FFO hybrid system's performance compared to traditional models, standalone optimization, and energy prediction techniques.
- Sensitivity analysis evaluates input variables' impact on ANFIS-FFO predictions, identifying influential factors affecting energy consumption.
- Real-world scenario testing evaluates ANFIS-FFO hybrid system's performance in real-time smart grid energy monitoring and control.

RESULTS AND FINDINGS

The ANFIS-FFO hybrid system offers more accurate energy predictions than traditional models, thanks to FFO optimization, reducing MAE and RMSE values. It outperforms standalone FFO optimization and other prediction techniques. The sensitivity analysis reveals the impact of weather conditions and occupancy patterns on energy consumption, aiding load forecasting and demand response strategies. The ANFIS-FFO hybrid system optimizes energy distribution, load balancing, and demand response, enhancing grid stability and efficiency.

The case study highlights the benefits of integrating ANFIS and FFO algorithms for energy monitoring and control in smart grids. This hybrid system offers accurate predictions and optimization, enhancing energy management and efficiency. It handles complex relationships between input and output variables, making it a valuable tool for sustainable and intelligent energy systems.

ENERGY MONITORING AND CONTROL IN RESIDENTIAL MICROGRIDS

Industrial IoT-enabled energy management systems optimize energy consumption and operational efficiency in industrial settings by collecting, analyzing, and acting on real-time data from devices and sensors. These systems offer numerous benefits (Barman et al., 2019; Ghenai et al., 2022; Li et al., 2018; Paul Divakar et al., 2020; Saveen et al., 2018).

- Data Acquisition and Monitoring: IoT devices and sensors are deployed throughout the industrial facility to capture energy-related data such as electricity usage, temperature, pressure, and other relevant parameters. These devices continuously monitor energy consumption and provide real-time data for analysis.
- Data Analytics and Insights: The collected data is analyzed using advanced analytics techniques, including machine learning and data mining algorithms. This enables the identification of energy consumption patterns, energy efficiency opportunities, and potential areas for improvement.
- Energy Optimization and Control: IoT-enabled energy management systems facilitate real-time energy optimization by intelligently adjusting energy-consuming devices, machinery, and processes. They can automate energy-saving measures, such as load shedding, demand response, and peak shaving, based on real-time data and predefined energy management strategies (Saravanan et al., 2022; Syamala et al., 2023; Venkateswaran, Vidhya, Ayyannan, et al., 2023; Vennila et al., 2022).
- Predictive Maintenance: Industrial IoT devices can monitor the health and performance of equipment and machinery, enabling predictive maintenance. By identifying potential equipment failures in advance, organizations can reduce downtime, improve energy efficiency, and extend the lifespan of critical assets.
- Integration with Enterprise Systems: Industrial IoT-enabled energy management systems can be integrated with enterprise resource planning (ERP) systems, asset management systems, and other business software to provide a holistic view of energy consumption and optimize resource allocation.
- Energy Reporting and Compliance: IoT-enabled energy management systems facilitate accurate and automated energy reporting, enabling organizations to meet regulatory compliance requirements. Real-time energy data can be used to generate reports, track key performance indicators (KPIs), and demonstrate adherence to sustainability goals.

Smart grids revolutionize energy management and distribution in urban environments by incorporating IoT technologies, intelligent sensors, and digital communication systems (Hema et al., 2023; Karthik et al., 2023; Koshariya et al., 2023; Kumara et al., 2023; Maguluri et al., 2023). Key applications include optimizing energy efficiency, ensuring grid stability, and integrating renewable energy sources (Barman et al., 2019; Paul Divakar et al., 2020; Saveen et al., 2018).

- Demand Response: Smart grids enable demand response programs that incentivize energy consumers to adjust their energy usage during peak demand periods. By participating in demand response initiatives, consumers can reduce their energy consumption and contribute to grid stability.
- Distributed Energy Resources (DERs) Integration: Smart grids facilitate the integration of distributed energy resources such as rooftop solar panels, wind turbines, and battery storage systems. These resources can feed surplus energy back into the grid or store excess energy for later use.
- Grid Monitoring and Control: Smart grid technologies enable real-time monitoring and control of the grid infrastructure. Intelligent sensors and advanced analytics provide insights into grid performance, voltage levels, power quality, and outage detection. This helps utilities proactively address issues and minimize downtime.
- Electric Vehicle (EV) Charging Infrastructure: Smart grids support the deployment of EV charging infrastructure in urban areas. They enable intelligent charging stations that optimize energy usage, manage load distribution, and support grid-friendly charging based on demand and available renewable energy.
- Microgrids: Microgrids are small-scale power systems that can operate independently or in coordination with the main grid. Smart grid technologies enable the efficient integration and management of microgrids, providing localized energy generation, storage, and consumption.
- Energy Management and Billing: Smart grids facilitate accurate energy metering, remote monitoring, and automated billing processes. Consumers have access to real-time energy usage data, enabling them to make informed decisions about their energy consumption and manage their energy costs effectively.

Industrial IoT-enabled energy management systems optimize energy in industrial settings, while smart grids revolutionize urban energy management by integrating advanced technologies for sustainable distribution and consumption. These technologies contribute to energy efficiency, sustainability, and grid resilience in both industrial and urban contexts.

CONCLUSION

Smart grids optimize energy efficiency, sustainability, and reliable supply by integrating IoT technologies like sensor networks, data analytics, and intelligent algorithms. This real-time monitoring, control, and optimization of energy generation, distribution, and consumption promotes energy-saving behaviors and optimizes energy management. IoT enhances interoperability, scalability, and resilience, supporting efficient energy management and grid optimization. Microgrids enable localized energy generation, storage, and distribution, ensuring reliable power supply, reducing transmission losses, and improving grid resilience.

Adaptive Neuro-Fuzzy Inference System (ANFIS) and Fuzzy Fruit Fly Optimization (FFO) algorithms are utilized in energy monitoring and control for modeling, prediction, and optimization. These algorithms combine artificial neural networks and fuzzy logic to handle complex relationships, while FFO offers optimization capabilities for energy management strategies. Implementing hardware components, sensor networks, data acquisition systems, and control systems is crucial for efficient deployment in smart grids. Integrating ANFIS and FFO algorithms enhances optimization and decision-making, resulting in more efficient energy management strategies.

Energy monitoring and control in smart grids face challenges such as data security, interoperability, scalability, grid integration issues, grid resilience, cybersecurity, cost management, stakeholder engagement, and consumer awareness. Future research will focus on advanced analytics, AI technologies, distributed energy resources management, demand-side management, decentralized energy markets, grid-edge intelligence, and self-healing capabilities. By leveraging IoT technologies, advanced algorithms, and intelligent systems, smart grids can optimize energy consumption, reduce environmental impact, and create a more resilient and reliable energy infrastructure.

LIST OF ACRONYMS

IoT - Internet of Things
ANFIS - Adaptive Neuro-Fuzzy Inference System
FFO - Fuzzy Fruit Fly Optimization
MAE - Mean Absolute Error
RMSE - Root Mean Square Error
R-squared - Coefficient of Determination
AI - Artificial Intelligence
ANN - Artificial Neural Network
DC - Direct Current
AC - Alternating Current
DE - Distributed Energy
DER - Distributed Energy Resources
EV - Electric Vehicle
GDP - Gross Domestic Product
ICT - Information and Communication Technology
ISO - International Organization for Standardization
PV - Photovoltaic
SCADA - Supervisory Control and Data Acquisition
SDN - Software-Defined Networking
V2G - Vehicle-to-Grid

REFERENCES

Abo-Zahhad, M., Ahmed, S. M., Farrag, M., Ahmed, M. F. A., & Ali, A. (2016). Design and implementation of building energy monitoring and management system based on wireless sensor networks. *Proceedings - 2015 10th International Conference on Computer Engineering and Systems, ICCES 2015*, 230–233. 10.1109/ICCES.2015.7393051

Adedeji, P. A., Akinlabi, S., Madushele, N., & Olatunji, O. O. (2022). Hybrid adaptive neuro-fuzzy inference system (ANFIS) for a multi-campus university energy consumption forecast. *International Journal of Ambient Energy*, *43*(1), 1685–1694. doi:10.1080/01430750.2020.1719885

Anitha, C., Komala, C., Vivekanand, C. V., Lalitha, S., & Boopathi, S. (2023). Artificial Intelligence driven security model for Internet of Medical Things (IoMT). *IEEE Explore*, 1–7.

Avancini, D. B., Rodrigues, J. J. P. C., Rabêlo, R. A. L., Das, A. K., Kozlov, S., & Solic, P. (2021). A new IoT-based smart energy meter for smart grids. *International Journal of Energy Research*, *45*(1), 189–202. doi:10.1002/er.5177

Barman, B. K., Yadav, S. N., Kumar, S., & Gope, S. (2019). IOT Based Smart Energy Meter for Efficient Energy Utilization in Smart Grid. *2nd International Conference on Energy, Power and Environment: Towards Smart Technology, ICEPE 2018*, 1–5. 10.1109/EPETSG.2018.8658501

Boopathi, S. (2022a). An experimental investigation of Quench Polish Quench (QPQ) coating on AISI 4150 steel. *Engineering Research Express*, *4*(4), 45009. doi:10.1088/2631-8695/ac9ddd

Boopathi, S. (2022b). An Extensive Review on Sustainable Developments of Dry and Near-Dry Electrical Discharge Machining Processes. *Journal of Manufacturing Science and Engineering*, *144*(5), 50801. doi:10.1115/1.4052527

Boopathi, S. (2022c). An investigation on gas emission concentration and relative emission rate of the near-dry wire-cut electrical discharge machining process. *Environmental Science and Pollution Research International*, *29*(57), 86237–86246. doi:10.100711356-021-17658-1 PMID:34837614

Boopathi, S. (2022d). Cryogenically treated and untreated stainless steel grade 317 in sustainable wire electrical discharge machining process: A comparative study. *Environmental Science and Pollution Research International*, 1–10. doi:10.100711356-022-22843-x PMID:36057706

Boopathi, S. (2022e). Effects of Cryogenically-treated Stainless Steel on Eco-friendly Wire Electrical Discharge Machining Process. *Preprint*, 1–18.

Boopathi, S. (2022f). Experimental investigation and multi-objective optimization of cryogenic Friction-stir-welding of AA2014 and AZ31B alloys using MOORA technique. *Materials Today. Communications*, *33*, 104937. doi:10.1016/j.mtcomm.2022.104937

Boopathi, S. (2022g). Performance Improvement of Eco-Friendly Near-Dry Wire-Cut Electrical Discharge Machining Process Using Coconut Oil-Mist Dielectric Fluid. *Journal of Advanced Manufacturing Systems*. Advance online publication. doi:10.1142/S0219686723500178

Boopathi, S. (2023a). An Investigation on Friction Stir Processing of Aluminum Alloy-Boron Carbide Surface Composite. In *Materials Horizons: From Nature to Nanomaterials* (pp. 249–257). Springer. doi:10.1007/978-981-19-7146-4_14

Boopathi, S. (2023b). Deep Learning Techniques Applied for Automatic Sentence Generation. In Promoting Diversity, Equity, and Inclusion in Language Learning Environments (pp. 255–273). IGI Global. doi:10.4018/978-1-6684-3632-5.ch016

Boopathi, S. (2023c). *Impact analysis of ceramic tile powder aggregates on self-compacting concrete.* Engineering Research Express.

Boopathi, S., Arigela, S. H., Raman, R., Indhumathi, C., Kavitha, V., & Bhatt, B. C. (2023). Prominent Rule Control-based Internet of Things: Poultry Farm Management System. *IEEE Explore*, 1–6. doi:10.1109/ICPECTS56089.2022.10047039

Boopathi, S., & Davim, J. P. (2023). Applications of Nanoparticles in Various Manufacturing Processes. In *Sustainable Utilization of Nanoparticles and Nanofluids in Engineering Applications* (pp. 1–31). IGI Global. doi:10.4018/978-1-6684-9135-5.ch001

Boopathi, S., & Kanike, U. K. (2023). Applications of Artificial Intelligent and Machine Learning Techniques in Image Processing. In *Handbook of Research on Thrust Technologies' Effect on Image Processing* (pp. 151–173). IGI Global. doi:10.4018/978-1-6684-8618-4.ch010

Boopathi, S., Kumar, P. K. S., Meena, R. S., Sudhakar, M., & Associates. (2023). Sustainable Developments of Modern Soil-Less Agro-Cultivation Systems: Aquaponic Culture. In Human Agro-Energy Optimization for Business and Industry (pp. 69–87). IGI Global.

Boopathi, S., Pandey, B. K., & Pandey, D. (2023). Advances in Artificial Intelligence for Image Processing: Techniques, Applications, and Optimization. In Handbook of Research on Thrust Technologies' Effect on Image Processing (pp. 73–95). IGI Global.

Chandak, S., & Rout, P. K. (2021). The implementation framework of a microgrid: A review. *International Journal of Energy Research*, *45*(3), 3523–3547. doi:10.1002/er.6064

Chandrika, V., Sivakumar, A., Krishnan, T. S., Pradeep, J., Manikandan, S., & Boopathi, S. (2023). Theoretical Study on Power Distribution Systems for Electric Vehicles. In *Intelligent Engineering Applications and Applied Sciences for Sustainability* (pp. 1–19). IGI Global. doi:10.4018/979-8-3693-0044-2.ch001

Dalipi, F., & Yayilgan, S. Y. (2016). Security and privacy considerations for IoT application on smart grids: Survey and research challenges. *Proceedings - 2016 4th International Conference on Future Internet of Things and Cloud Workshops, W-FiCloud 2016*, 63–68. 10.1109/W-FiCloud.2016.28

Dhanya, D., Kumar, S. S., Thilagavathy, A., Prasad, D., & Boopathi, S. (2023). Data Analytics and Artificial Intelligence in the Circular Economy: Case Studies. In Intelligent Engineering Applications and Applied Sciences for Sustainability (pp. 40–58). IGI Global.

Domakonda, V. K., Farooq, S., Chinthamreddy, S., Puviarasi, R., Sudhakar, M., & Boopathi, S. (2023). Sustainable Developments of Hybrid Floating Solar Power Plants. In *Human Agro-Energy Optimization for Business and Industry* (pp. 148–167). IGI Global. doi:10.4018/978-1-6684-4118-3.ch008

Ferahtia, S., Djeroui, A., Rezk, H., Houari, A., Zeghlache, S., & Machmoum, M. (2022). Optimal control and implementation of energy management strategy for a DC microgrid. *Energy*, *238*, 121777. doi:10.1016/j.energy.2021.121777

Ghasempour, A. (2019). Internet of things in smart grid: Architecture, applications, services, key technologies, and challenges. *Inventions (Basel, Switzerland)*, *4*(1), 22. doi:10.3390/inventions4010022

Ghenai, C., Al-Mufti, O. A. A., Al-Isawi, O. A. M., Amirah, L. H. L., & Merabet, A. (2022). Short-term building electrical load forecasting using adaptive neuro-fuzzy inference system (ANFIS). *Journal of Building Engineering*, *52*, 104323. doi:10.1016/j.jobe.2022.104323

Hang, L., & Kim, D. H. (2018). Enhanced model-based predictive control system based on fuzzy logic for maintaining thermal comfort in IoT smart space. *Applied Sciences (Basel, Switzerland)*, *8*(7), 1031. doi:10.3390/app8071031

Hanumanthakari, S., Gift, M. M., Kanimozhi, K., Bhavani, M. D., Bamane, K. D., & Boopathi, S. (2023). Biomining Method to Extract Metal Components Using Computer-Printed Circuit Board E-Waste. In *Handbook of Research on Safe Disposal Methods of Municipal Solid Wastes for a Sustainable Environment* (pp. 123–141). IGI Global. doi:10.4018/978-1-6684-8117-2.ch010

Harikaran, M., Boopathi, S., Gokulakannan, S., & Poonguzhali, M. (2023). Study on the Source of E-Waste Management and Disposal Methods. In *Sustainable Approaches and Strategies for E-Waste Management and Utilization* (pp. 39–60). IGI Global. doi:10.4018/978-1-6684-7573-7.ch003

Hartono, B. S., Budiyanto, & Setiabudy, R. (2013). Review of microgrid technology. *2013 International Conference on Quality in Research, QiR 2013 - In Conjunction with ICCS 2013: The 2nd International Conference on Civic Space*, 127–132. 10.1109/QiR.2013.6632550

Hema, N., Krishnamoorthy, N., Chavan, S. M., Kumar, N., Sabarimuthu, M., & Boopathi, S. (2023). A Study on an Internet of Things (IoT)-Enabled Smart Solar Grid System. In *Handbook of Research on Deep Learning Techniques for Cloud-Based Industrial IoT* (pp. 290–308). IGI Global. doi:10.4018/978-1-6684-8098-4.ch017

Hu, R., Wen, S., Zeng, Z., & Huang, T. (2017). A short-term power load forecasting model based on the generalized regression neural network with decreasing step fruit fly optimization algorithm. *Neurocomputing*, *221*, 24–31. doi:10.1016/j.neucom.2016.09.027

Janardhana, K., Singh, V., Singh, S. N., Babu, T. R., Bano, S., & Boopathi, S. (2023). Utilization Process for Electronic Waste in Eco-Friendly Concrete: Experimental Study. In Sustainable Approaches and Strategies for E-Waste Management and Utilization (pp. 204–223). IGI Global.

Jeevanantham, Y. A., A, S., V, V., J, S. Isaac., Boopathi, S., & Kumar, D. P. (2023). Implementation of Internet-of Things (IoT) in Soil Irrigation System. *IEEE Explore*, 1–5. doi:10.1109/ICPECTS56089.2022.10047185

Kampouropoulos, K., Andrade, F., Sala, E., Espinosa, A. G., & Romeral, L. (2018). Multiobjective optimization of multi-carrier energy system using a combination of ANFIS and genetic algorithms. *IEEE Transactions on Smart Grid*, *9*(3), 2276–2283. doi:10.1109/TSG.2016.2609740

Kannan, P., Venkataramanan, K., Sivaranjani, S., & Indhumathi, K. (2020)... *Analysis of ABC Algorithm Based ANFIS Controller for PV Integrated Low Voltage Weak Grid System.*, *3*(8), 196–201.

Karthik, S., Hemalatha, R., Aruna, R., Deivakani, M., Reddy, R. V. K., & Boopathi, S. (2023). Study on Healthcare Security System-Integrated Internet of Things (IoT). In Perspectives and Considerations on the Evolution of Smart Systems (pp. 342–362). IGI Global.

Katyara, S., Shah, M. A., Chowdhary, B. S., Akhtar, F., & Lashari, G. A. (2019). Monitoring, Control and Energy Management of Smart Grid System via WSN Technology Through SCADA Applications. *Wireless Personal Communications*, *106*(4), 1951–1968. doi:10.100711277-018-5738-x

Kaur, H., & Sood, S. K. (2019). Adaptive Neuro Fuzzy Inference System (ANFIS) based wildfire risk assessment. *Journal of Experimental & Theoretical Artificial Intelligence*, *31*(4), 599–619. doi:10.1080/0952813X.2019.1591523

Khalid, R., & Javaid, N. (2020). A survey on hyperparameters optimization algorithms of forecasting models in smart grid. *Sustainable Cities and Society*, *61*, 102275. doi:10.1016/j.scs.2020.102275

Khan, A., & Rizwan, M. (2021). ANN and ANFIS Based Approach for Very Short Term Load Forecasting: A Step Towards Smart Energy Management System. *Proceedings of the 8th International Conference on Signal Processing and Integrated Networks, SPIN 2021*, 464–468. 10.1109/SPIN52536.2021.9566146

Koshariya, A. K., Kalaiyarasi, D., Jovith, A. A., Sivakami, T., Hasan, D. S., & Boopathi, S. (2023). AI-Enabled IoT and WSN-Integrated Smart Agriculture System. In *Artificial Intelligence Tools and Technologies for Smart Farming and Agriculture Practices* (pp. 200–218). IGI Global. doi:10.4018/978-1-6684-8516-3.ch011

Kumara, V., Mohanaprakash, T., Fairooz, S., Jamal, K., Babu, T., & Sampath, B. (2023). Experimental Study on a Reliable Smart Hydroponics System. In *Human Agro-Energy Optimization for Business and Industry* (pp. 27–45). IGI Global. doi:10.4018/978-1-6684-4118-3.ch002

Li, D., Yin, H., & Zheng, B. (2018). An Annual Load Forecasting Model Based on Generalized Regression Neural Network With Multi-Swarm Fruit Fly Optimization Algorithm. *Dianwang Jishu/Power System Technology, 42*(2), 585–590. doi:10.13335/j.1000-3673.pst.2017.1403

Maguluri, L. P., Ananth, J., Hariram, S., Geetha, C., Bhaskar, A., & Boopathi, S. (2023). Smart Vehicle-Emissions Monitoring System Using Internet of Things (IoT). In Handbook of Research on Safe Disposal Methods of Municipal Solid Wastes for a Sustainable Environment (pp. 191–211). IGI Global.

Mellit, A., & Kalogirou, S. A. (2011). ANFIS-based modelling for photovoltaic power supply system: A case study. *Renewable Energy*, *36*(1), 250–258. doi:10.1016/j.renene.2010.06.028

Mohammad Bilal, M. Z. M. T. A. B. K. B. & O. K. (2022). Analysis and Validation of an Intelligent Energy Monitoring Framework for a Yacht Powered System Based on Adaptive Neuro-Fuzzy Inference System (Anfis). *International Journal of Robotics Research and Development, 12*(1), 1–10.

Olbrich, G. (2016). Energy management and smart cities. *Gis. Business (Atlanta, Ga.)*, *1*(4), 3.

Padmanaban, S., Priyadarshi, N., Bhaskar, M. S., Holm-Nielsen, J. B., Ramachandaramurthy, V. K., & Hossain, E. (2019). A Hybrid ANFIS-ABC Based MPPT Controller for PV System with Anti-Islanding Grid Protection: Experimental Realization. *IEEE Access : Practical Innovations, Open Solutions*, *7*, 103377–103389. doi:10.1109/ACCESS.2019.2931547

Paul Divakar, P., Vijaya Lakshmi, G., & Devi, L. (2020). Applications of Internet of things on Smart Grid. *IOP Conference Series. Materials Science and Engineering*, *906*(1), 13–17. doi:10.1088/1757-899X/906/1/012015

Pawar, P., TarunKumar, M., & Vittal K, P. (2020). An IoT based Intelligent Smart Energy Management System with accurate forecasting and load strategy for renewable generation. *Measurement*, *152*, 107187. doi:10.1016/j.measurement.2019.107187

Premkumar, S., & Sigappi, A. N. (2021). Functional framework for edge-based agricultural system. In AI, Edge and IoT-based Smart Agriculture (pp. 71–100). Elsevier. doi:10.1016/B978-0-12-823694-9.00029-3

Saleem, M. U., Usman, M. R., & Shakir, M. (2021). Design, Implementation, and Deployment of an IoT Based Smart Energy Management System. *IEEE Access : Practical Innovations, Open Solutions*, *9*, 59649–59664. doi:10.1109/ACCESS.2021.3070960

Saravanan, A., Venkatasubramanian, R., Khare, R., Surakasi, R., Boopathi, S., Ray, S., & Sudhakar, M. (2022). Policy trends of renewable energy and non. *Renewable Energy*.

Sathish, T., Sunagar, P., Singh, V., Boopathi, S., Al-Enizi, A. M., Pandit, B., Gupta, M., & Sehgal, S. S. (2023). Characteristics estimation of natural fibre reinforced plastic composites using deep multi-layer perceptron (MLP) technique. *Chemosphere*, *337*, 139346. doi:10.1016/j.chemosphere.2023.139346 PMID:37379988

Saveen, G., Prudhvi Raju, P., Manikanta, D. V., & Satya Praveen, M. (2018). Design and implementation of energy management system with fuzzy control for multiple microgrid. *Proceedings of the 2nd International Conference on Inventive Systems and Control, ICISC 2018, 28*(4), 1239–1244. 10.1109/ICISC.2018.8399003

Subha, S., Inbamalar, T. M., R, K. C., Suresh, L. R., Boopathi, S., & Alaskar, K. (2023, February). A Remote Health Care Monitoring system using internet of medical things (IoMT). *IEEE Explore*. doi:10.1109/ICIPTM57143.2023.10118103

Suganthi, L., Iniyan, S., & Samuel, A. A. (2015). Applications of fuzzy logic in renewable energy systems—A review. *Renewable & Sustainable Energy Reviews*, *48*, 585–607. doi:10.1016/j.rser.2015.04.037

Suhail, M., Akhtar, I., Kirmani, S., & Jameel, M. (2021). Development of Progressive Fuzzy Logic and ANFIS Control for Energy Management of Plug-In Hybrid Electric Vehicle. *IEEE Access : Practical Innovations, Open Solutions*, *9*, 62219–62231. doi:10.1109/ACCESS.2021.3073862

Syamala, M., Komala, C., Pramila, P., Dash, S., Meenakshi, S., & Boopathi, S. (2023). Machine Learning-Integrated IoT-Based Smart Home Energy Management System. In *Handbook of Research on Deep Learning Techniques for Cloud-Based Industrial IoT* (pp. 219–235). IGI Global. doi:10.4018/978-1-6684-8098-4.ch013

Thakare, S., Shriyan, A., Thale, V., Yasarp, P., & Unni, K. (2017). Implementation of an energy monitoring and control device based on IoT. *2016 IEEE Annual India Conference, INDICON 2016*, 1–6. 10.1109/INDICON.2016.7839066

Vanitha, S., Radhika, K., & Boopathi, S. (2023). Artificial Intelligence Techniques in Water Purification and Utilization. In *Human Agro-Energy Optimization for Business and Industry* (pp. 202–218). IGI Global. doi:10.4018/978-1-6684-4118-3.ch010

Vanitha, S. K. R., & Boopathi, S. (2023). Artificial Intelligence Techniques in Water Purification and Utilization. In *Human Agro-Energy Optimization for Business and Industry* (pp. 202–218). IGI Global. doi:10.4018/978-1-6684-4118-3.ch010

Venkateswaran, N., Vidhya, K., Ayyannan, M., Chavan, S. M., Sekar, K., & Boopathi, S. (2023). A Study on Smart Energy Management Framework Using Cloud Computing. In 5G, Artificial Intelligence, and Next Generation Internet of Things: Digital Innovation for Green and Sustainable Economies (pp. 189–212). IGI Global. doi:10.4018/978-1-6684-8634-4.ch009

Venkateswaran, N., Vidhya, R., Naik, D. A., Raj, T. M., Munjal, N., & Boopathi, S. (2023). Study on Sentence and Question Formation Using Deep Learning Techniques. In *Digital Natives as a Disruptive Force in Asian Businesses and Societies* (pp. 252–273). IGI Global. doi:10.4018/978-1-6684-6782-4.ch015

Vennila, T., Karuna, M., Srivastava, B. K., Venugopal, J., Surakasi, R., & Sampath, B. (2022). New Strategies in Treatment and Enzymatic Processes: Ethanol Production From Sugarcane Bagasse. In Human Agro-Energy Optimization for Business and Industry (pp. 219–240). IGI Global.

Zekrifa, D. M. S., Kulkarni, M., Bhagyalakshmi, A., Devireddy, N., Gupta, S., & Boopathi, S. (2023). Integrating Machine Learning and AI for Improved Hydrological Modeling and Water Resource Management. In *Artificial Intelligence Applications in Water Treatment and Water Resource Management* (pp. 46–70). IGI Global. doi:10.4018/978-1-6684-6791-6.ch003

Zhang, F., Sun, G., Zheng, B., & Dong, L. (2021). Design and implementation of energy management system based on spring boot framework. *Information (Basel)*, *12*(11), 457. doi:10.3390/info12110457

Chapter 13
Biofuel Production, Bioremediation, and the Frontier of Synthetic Cell-Free Processes:
IOT-AI Integration

Nalla Bhanu Teja
Department of Mechanical Engineering, Aditya College of Engineering, India

S. Bharathi
Department of Electrical and Electronics Engineering, S.A. Engineering College, India

Sivakumar Karthikeyan
https://orcid.org/0000-0001-8393-0083
Department of Mechanical Engineering, SRM Valliammai Engineering College, India

Rayavarapu Bhavani Sankar
https://orcid.org/0009-0005-5566-3703
Department of Computer Science and Engineering, Chirala Engineering College, India

Rohi Prasad
Department of Mechanical Engineering, Guru Gobind Singh Educational Society's Technical Campus, India

ABSTRACT

This chapter examines how synthetic cell-free processes, bioremediation, and biomass usage interact to promote sustainability and innovation. It emphasises biomass's potential as a source of renewable energy and for the generation of biofuels, concentrating on a variety of feedstocks and cutting-edge conversion methods. While synthetic cell-free methods offer customised biochemical reactions beyond biological cells, bioremediation employs microorganisms to reduce pollution. The chapter investigates how various disciplines might be combined to address environmental problems, emphasising moral questions, legal issues, and practical applications. The chapter seeks to shed light on a sustainable and peaceful future through case studies and technological breakthroughs. IoT-AI system streamlines biofuel production, bioremediation, and synthetic cell-free processes for efficient workflow.

DOI: 10.4018/978-1-6684-6577-6.ch013

INTRODUCTION

Biomass, as a renewable and abundant resource, has gained significant attention in recent years due to its potential to address pressing global challenges related to energy sustainability and environmental conservation. This chapter delves into the multifaceted domain of biomass utilization, exploring its pivotal roles in biofuel production, bioremediation, and the cutting-edge field of synthetic cell-free processes. By harnessing the inherent capabilities of biological systems, researchers and innovators are paving the way for sustainable energy solutions and advanced biotechnologies (Y. Li et al., 2020).

The utilization of biomass as a renewable and sustainable resource has garnered increasing attention in recent years due to its potential to mitigate the environmental impact of energy production and contribute to a more resilient energy landscape. This introduction provides a glimpse into the dynamic landscape of research and developments in biomass utilization, highlighting the strides made in harnessing biomass for energy generation, biofuel production, and innovative applications (Kwon et al., 2020).

As the global community grapples with the challenges posed by climate change and finite fossil fuel reserves, there is a growing imperative to transition towards cleaner and sustainable energy sources. Biomass, derived from organic materials such as plants, agricultural residues, and organic waste, offers a promising avenue to meet energy demands while reducing greenhouse gas emissions (Wang et al., 2020).

Research efforts have led to significant breakthroughs in converting biomass into various forms of bioenergy. Biomass can be utilized for electricity and heat generation through processes like combustion, gasification, and pyrolysis. Moreover, advancements in biofuel production have paved the way for converting biomass into liquid fuels, such as bioethanol and biodiesel, which can be seamlessly integrated into existing energy infrastructure (Huang et al., 2020).

The synergy between biomass utilization and cutting-edge technologies has accelerated progress in this field. Innovative techniques, such as biorefineries that extract multiple products from biomass, and advanced conversion processes that transform lignocellulosic biomass into valuable fuels and chemicals, are transforming the way we view biomass as a feedstock (P. Li et al., 2019). Beyond energy generation, biomass is finding novel applications in the creation of bio-based materials, contributing to the concept of a circular economy. From bioplastics and bio-composites to sustainable packaging solutions, researchers are exploring ways to substitute traditional petroleum-based materials with biomass-derived alternatives, reducing plastic waste and environmental harm (Yaashikaa et al., 2022). While the potential of biomass utilization is promising, challenges remain. Ensuring the sustainable sourcing of biomass, optimizing conversion technologies, and addressing environmental and social considerations are key areas of focus for researchers and policymakers. The future of biomass utilization lies in collaborative efforts that span scientific disciplines, industries, and international borders (Maguluri et al., 2023; Selvakumar, Shankar, et al., 2023; Sengeni et al., 2023).

The ever-increasing demand for energy, coupled with the urgency to mitigate climate change, has compelled a global shift towards renewable and sustainable energy sources. Among these, biofuels have emerged as promising alternatives to conventional fossil fuels due to their potential to reduce greenhouse gas emissions and dependence on finite resources. Biofuels, derived from biological materials such as biomass, represent a key component of the renewable energy landscape (Yaashikaa et al., 2022).

The modern world's insatiable energy appetite has driven the exploration of alternative energy sources. Fossil fuels, the primary energy providers for decades, are associated with environmental degradation, air pollution, and the accumulation of greenhouse gases in the atmosphere. As a result, the imperative to transition to cleaner and more sustainable energy options has become increasingly evident. Biofuels, originating from renewable biological feedstocks, offer an avenue to address both energy demands and environmental concerns. Biofuel production hinges on a wide variety of biomass sources (Saravanan, Deivayanai, et al., 2022). These include agricultural residues, forestry byproducts, algae, dedicated energy crops, and even organic waste materials. The versatility of biomass feedstocks provides flexibility in sourcing, reducing the competition between fuel and food crops. Harnessing these diverse feedstocks can not only lead to enhanced energy security but also contribute to rural development and job creation (Maguluri et al., 2023; Selvakumar, Shankar, et al., 2023; Sengeni et al., 2023).

The term "biofuel" encompasses several types, each with distinct production pathways and characteristics. First-generation biofuels, such as bioethanol and biodiesel, involve the conversion of food crops or vegetable oils into fuels. Second-generation biofuels, on the other hand, utilize lignocellulosic biomass—non-food plant materials like crop residues and wood—via advanced conversion techniques (Banerjee et al., 2020). Additionally, emerging third-generation biofuels focus on harnessing algae and other microorganisms to produce lipids and hydrocarbons. In the pursuit of efficient biofuel production, technological innovation plays a pivotal role. Advancements in biotechnology, catalysis, and chemical engineering have enabled the optimization of conversion processes. These include fermentation, enzymatic hydrolysis, pyrolysis, and gasification, among others. Each technology offers distinct advantages and challenges, influencing the choice of feedstock and final product (Chi et al., 2021).

While biofuels hold immense promise, they are not devoid of challenges. Concerns related to land use, food security, and biodiversity have prompted the need for sustainable and responsible biofuel production practices. Striking a balance between fuel production and environmental preservation requires careful consideration of feedstock selection, land management, and cultivation practices. The introduction of biofuels into the global energy mix represents a pivotal step towards a more sustainable and resilient future (Kelwick et al., 2020). By harnessing the potential of various biomass feedstocks and leveraging advancements in conversion technologies, biofuel production stands as a key driver in the pursuit of cleaner energy sources. As we navigate this evolving landscape, a holistic approach that encompasses economic viability, environmental responsibility, and social considerations will shape the trajectory of biofuel development and its integration into our energy systems (J. Li et al., 2021).

The interplay between environmental degradation and technological advancement has prompted a pressing need for innovative solutions to address pollution and restore ecological balance. Bioremediation, a natural or engineered process that employs microorganisms to remediate contaminants, and the emerging frontier of synthetic cell-free processes, which enable biochemical reactions outside living cells, exemplify two distinct yet interconnected pathways in the pursuit of sustainable environmental solutions (Zhu et al., 2020).

Human activities, industrial processes, and urbanization have led to the release of various pollutants into ecosystems, disrupting natural equilibrium and posing threats to human health and biodiversity. Bioremediation, leveraging the intrinsic capabilities of microorganisms to break down, transform, or sequester pollutants, presents an eco-friendly and efficient approach to tackling environmental contamination. From oil spills to heavy metal contamination, bioremediation showcases the potential of harnessing nature's remediation prowess for environmental restoration (Elani, 2021).

At the forefront of biotechnology, synthetic cell-free processes are driving transformative changes in our understanding of biochemical reactions and the manipulation of biological components. These processes involve isolating cellular machinery, such as enzymes and ribosomes, and orchestrating reactions in controlled environments, decoupled from living cells. This innovation offers unparalleled precision and control over biochemical pathways, enabling tailored synthesis, biomanufacturing, and even the creation of synthetic life (J. Li et al., 2023).

The convergence of bioremediation and synthetic cell-free processes heralds a new era in environmental remediation. Bioremediation's reliance on microorganisms aligns with synthetic cell-free processes' focus on molecular components, offering the potential to engineer or optimize microorganisms for enhanced pollutant degradation. Furthermore, the utilization of cell-free systems can enhance the production of enzymes and cofactors crucial for bioremediation, expediting and amplifying remediation efforts (Laohakunakorn et al., 2020).

The synergy between bioremediation and synthetic cell-free processes transcends mere technological innovation; it embodies a holistic approach to sustainability. By integrating nature's solutions with cutting-edge biotechnological tools, we are poised to address pollution challenges with unprecedented precision and efficiency. This interdisciplinary collaboration exemplifies the power of combining traditional ecological wisdom with the frontiers of synthetic biology to forge a more harmonious relationship between humanity and the environment (Sindhu et al., 2019).

As we embark on this exploration of bioremediation and the potential of synthetic cell-free processes, we delve into realms where scientific curiosity meets ecological urgency. The chapters that follow will unveil the intricacies of these fields, shedding light on their mechanisms, applications, challenges, and transformative potential. In an era where environmental stewardship and technological innovation are more crucial than ever, the fusion of bioremediation and synthetic cell-free processes stands as a beacon of hope, offering solutions that transcend disciplinary boundaries and cultivate a future defined by sustainable coexistence (Hoang et al., 2021).

Chapter explores biomass utilization, highlighting approaches, challenges, and achievements in this dynamic field. They highlight innovative conversion technologies, bioenergy solutions, and sustainable material production, transforming biomass into a cornerstone for a cleaner, more sustainable energy and materials landscape.

BACKGROUND AND SIGNIFICANCE

The world's escalating energy demand, coupled with concerns about depleting fossil fuel reserves and the environmental impacts of their combustion, has spurred intensive exploration of alternative energy sources. Biomass, derived from living or recently living organisms, holds immense promise as a renewable energy reservoir that can mitigate greenhouse gas emissions and promote energy security. Furthermore, as environmental pollution continues to threaten ecosystems and human health, the deployment of natural and engineered processes for bioremediation has gained prominence. Understanding the potential of microorganisms to degrade pollutants and restore ecological balance underscores the significance of bioremediation strategies (Kwon et al., 2020; Y. Li et al., 2020; Wang et al., 2020).

Emerging on the frontier of biotechnology, synthetic cell-free processes present a paradigm shift in manufacturing and biochemical production. By designing cellular processes that operate outside the confines of living cells, researchers are unlocking new avenues for tailoring biochemical pathways, optimizing production yields, and minimizing resource consumption. This novel approach has the potential to revolutionize industries ranging from pharmaceuticals to biofuels.

SCOPE AND OBJECTIVES

This chapter aims to provide a comprehensive overview of the intricate interplay between biomass utilization, biofuel production, bioremediation, and synthetic cell-free processes. Through a systematic exploration of these topics, the chapter intends to achieve the following objectives:

- **Contextualize Biomass**: Establish a foundational understanding of biomass as a versatile and renewable resource for sustainable energy and environmental applications.
- **Explore Biofuel Production**: Examine the various pathways and technologies involved in converting biomass into biofuels such as bioethanol, biodiesel, and biogas, highlighting advancements, challenges, and environmental considerations.
- **Unveil Bioremediation Strategies**: Investigate the principles and applications of bioremediation, showcasing how microorganisms can be harnessed to cleanse contaminated environments.
- **Probe Synthetic Cell-Free Processes**: Delve into the emerging realm of synthetic cell-free systems, elucidating their principles, applications in biochemical production, and implications for biomanufacturing.

By accomplishing these objectives, this chapter aims to contribute to the understanding of biomass's pivotal role in addressing energy and environmental challenges while shedding light on innovative approaches to harness its potential.

BIOMASS AS A RESOURCE

Biomass, in the context of energy and environmental applications, refers to organic matter derived from living or recently living organisms. It encompasses a wide array of materials, ranging from agricultural residues, forest residues, and dedicated energy crops to organic waste from industries and households. Biomass serves as a renewable and sustainable source of energy, carbon, and raw materials, offering diverse avenues for addressing energy demands and environmental concerns (Monlau et al., 2021; Ong et al., 2020; Sakarika & Kornaros, 2019).

Definition and Types of Biomasses

Biomass can be categorized into several types based on its origin and characteristics (Ong et al., 2020):

- **Woody Biomass**: This includes materials from trees and shrubs, such as wood chips, sawdust, and forest residues. Woody biomass is often utilized for heat and power generation.
- **Agricultural Biomass**: Crop residues, such as corn stover, wheat straw, and rice husks, constitute agricultural biomass. These materials can be converted into biofuels or used as feedstock for biorefineries.
- **Energy Crops**: Plants specifically grown for energy production, like switchgrass and miscanthus, fall under this category. They offer high biomass yields and can be used for bioenergy purposes.
- **Animal Waste**: Manure and other organic waste from livestock can be used to generate biogas through anaerobic digestion, providing both energy and waste management solutions.

- **Municipal Solid Waste (MSW)**: Organic components of household waste can be utilized for biogas production or composting, contributing to waste reduction and energy generation.
- **Algae**: Algae hold potential for biofuel production due to their rapid growth and high oil content. They can be cultivated in ponds or bioreactors.

ADVANTAGES AND CHALLENGES OF BIOMASS UTILIZATION

Advantages

- **Renewable Resource**: Biomass is replenishable over time, making it a sustainable alternative to finite fossil fuels.
- **Carbon Neutrality**: The carbon dioxide emitted during biomass combustion is roughly equal to the CO2 absorbed by the plants during their growth, resulting in a near-neutral carbon footprint.
- **Diverse Applications**: Biomass can be converted into various forms of energy, including heat, electricity, and biofuels, catering to different energy needs.
- **Waste Management**: Utilizing organic waste for energy production reduces landfill waste and associated environmental issues.
- **Rural Development**: Biomass utilization can stimulate rural economies by creating jobs related to cultivation, harvesting, and processing.

Challenges

Resource Competition: Biomass utilization can potentially compete with food production, leading to concerns about food security and land use conflicts.

Environmental Impact: Unsustainable harvesting practices or land-use changes for biomass cultivation can result in deforestation, loss of biodiversity, and soil degradation.

Energy Intensity: The energy required for cultivating, processing, and transporting biomass can offset the energy gained from its utilization.

Logistics and Storage: Biomass has lower energy density compared to fossil fuels, necessitating efficient storage and transportation solutions.

Feedstock Variability: Biomass feedstocks can vary in composition, affecting the efficiency and consistency of conversion processes.

In the pursuit of a more sustainable energy future, understanding the types, advantages, and challenges of biomass utilization is crucial for informed decision-making and the development of responsible and effective biomass-based solutions.

BIOFUEL PRODUCTION FROM BIOMASS

Bioethanol Production

Bioethanol, a renewable and widely used biofuel, is produced through the fermentation of sugars derived from biomass. This section explores the key aspects of bioethanol production, including fermentation processes, feedstock selection, and technological innovations (Saratale et al., 2022; Saravanan, Kumar, et al., 2022). The various factors considered for bio-ethanol production process are illustrated in Figure 1.

Fermentation Processes

The production of bioethanol involves microbial fermentation, where microorganisms convert sugars into ethanol and carbon dioxide. Two main fermentation processes are commonly employed (Hanumanthakari et al., 2023; Selvakumar, Adithe, et al., 2023):

- **Yeast Fermentation (Saccharification and Fermentation)**: This process involves breaking down complex sugars (polysaccharides) into simple sugars (monosaccharides) through enzymatic reactions, a step known as saccharification. The resulting sugars, such as glucose and fructose, are then fermented by yeast to produce ethanol and carbon dioxide. Saccharomyces cerevisiae is the most commonly used yeast due to its efficiency and ethanol tolerance.
- **Cellulolytic Microorganism Fermentation**: Some advanced processes utilize cellulolytic microorganisms or engineered microbes to directly convert complex plant cell wall components, like cellulose and hemicellulose, into ethanol. This approach holds promise for utilizing non-food biomass sources efficiently.

Figure 1. Bioethanol production factors

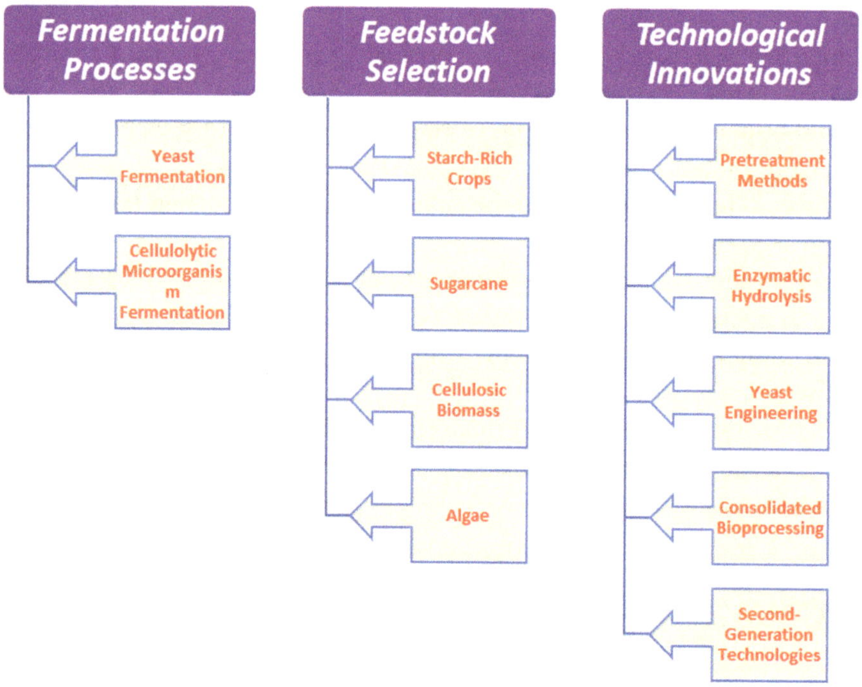

Feedstock Selection

The choice of feedstock significantly impacts bioethanol production's efficiency, sustainability, and economic viability. Feedstock options include (Sakarika & Kornaros, 2019):

Starch-Rich Crops: Crops such as corn, wheat, and barley contain high levels of starch, which can be easily converted into sugars for fermentation.

Sugarcane: Sugarcane is rich in sucrose, making it a favored feedstock in regions with suitable climates.

Cellulosic Biomass: Cellulosic feedstocks like agricultural residues, wood chips, and dedicated energy crops offer a more sustainable option, as they utilize non-edible parts of plants and reduce competition with food production.

Algae: Algal biomass, with its high sugar content and rapid growth, is being explored as a feedstock for bioethanol production (Babu et al., 2022; Gowri et al., 2023; Sampath, 2021).

Technological Innovations

Advancements in bioethanol production aim to enhance efficiency, reduce costs, and minimize environmental impacts (Hoang et al., 2021; Sindhu et al., 2019):

- **Pretreatment Methods**: Pretreatment processes break down plant cell walls, making cellulose more accessible for enzymes and microbes. Techniques include steam explosion, acid hydrolysis, and enzymatic pretreatment.
- **Enzymatic Hydrolysis**: Enzymes are used to convert complex sugars into fermentable sugars. Advances in enzyme engineering have led to more efficient and cost-effective enzymatic hydrolysis.
- **Yeast Engineering**: Genetic modification of yeast strains can enhance their ethanol production efficiency, stress tolerance, and substrate utilization.
- **Consolidated Bioprocessing**: This innovative approach combines multiple steps, such as pretreatment, hydrolysis, and fermentation, into a single process, reducing production steps and costs.
- **Second-Generation Technologies**: Processes that use lignocellulosic biomass as feedstock are considered second-generation bioethanol technologies and are seen as more sustainable alternatives.

Thus, bioethanol production involves intricate fermentation processes, careful feedstock selection, and continuous technological innovation. As researchers and engineers refine these aspects, bioethanol's potential as a cleaner and more sustainable alternative to fossil fuels continues to grow.

Biodiesel Production

Biodiesel, a renewable alternative to conventional diesel fuel, is derived from various biological sources through a process known as transesterification. This section explores the essential components of biodiesel production, including transesterification methods, feedstock diversity, and environmental implications (Monlau et al., 2021; Sakarika & Kornaros, 2019; Sharma et al., 2020).

Transesterification Methods

Transesterification is the chemical reaction in which triglycerides, found in oils and fats, are converted into biodiesel and glycerol. Two common methods of transesterification are widely employed in biodiesel production (Monlau et al., 2021; Sakarika & Kornaros, 2019):

Homogeneous Catalysis: In this method, an alkali catalyst (usually sodium or potassium hydroxide) is mixed with the feedstock oil or fat to initiate the transesterification reaction. This process is efficient but requires rigorous separation and neutralization steps to remove the catalyst and byproducts.

Heterogeneous Catalysis: In contrast, heterogeneous catalysis involves using solid acid or base catalysts, such as zeolites or metal oxides, to facilitate the reaction. This method offers the advantage of easier separation of the catalyst and reduced waste generation.

Feedstock Diversity

Biodiesel can be produced from a wide range of feedstock sources, each with its own advantages and challenges:

Vegetable Oils: Common vegetable oils like soybean, canola, and palm oil are frequently used feedstocks. They offer high oil yields and well-established processing techniques. However, concerns about deforestation and land use change associated with palm oil production have raised environmental and ethical considerations.

Animal Fats: Animal fats, such as tallow and lard, can also be used as feedstocks. Utilizing these fats can contribute to waste reduction in the meat processing industry.

Used Cooking Oil: Recycling used cooking oil from restaurants and households as a feedstock not only prevents disposal issues but also reduces the environmental impact associated with waste oil disposal.

Algae: Algal oil holds potential due to its rapid growth and high lipid content. Algae can be cultivated using non-arable land and wastewater, mitigating competition with food production.

Waste Oils: Various waste oils, such as trap grease and sewage sludge oil, can be repurposed for biodiesel production, promoting resource efficiency.

Environmental Implications

Biodiesel offers several environmental benefits, but its production and utilization also pose certain challenges (Banerjee et al., 2020):

Advantages

- **Reduced Greenhouse Gas Emissions**: Biodiesel combustion generates fewer greenhouse gases compared to conventional diesel, contributing to reduced climate impact.
- **Renewable Energy Source**: Biodiesel is derived from renewable feedstocks, which can help decrease reliance on finite fossil fuels.
- **Improved Air Quality**: Biodiesel emits fewer particulate matter and sulfur compounds, leading to improved local air quality and reduced health impacts.
- **Waste Utilization**: Utilizing waste oils and fats as feedstocks reduces waste disposal issues and potentially offsets the need for new crop cultivation.

Challenges

- **Land Use Competition**: The cultivation of biodiesel feedstock crops can compete with food crops, leading to concerns about food security and land degradation.
- **Deforestation and Habitat Loss**: The expansion of oil palm plantations, often used for biodiesel production, has been linked to deforestation and habitat destruction.
- **Biodiversity Impact**: Large-scale monoculture farming for biodiesel feedstocks can result in loss of biodiversity and disruption of ecosystems.
- **Indirect Land Use Change**: The shift of land use from food crops to biodiesel feedstocks can indirectly drive deforestation in other regions, leading to a complex and sometimes counterproductive environmental impact.

Hence, biodiesel production is a multifaceted process that involves transesterification, feedstock selection, and careful consideration of environmental implications. While biodiesel presents numerous environmental advantages, its sustainability depends on responsible feedstock sourcing, efficient production methods, and ongoing efforts to mitigate negative environmental impacts.

Biogas Generation

Biogas, a versatile renewable energy source, is produced through the anaerobic digestion of organic materials. This section delves into the key aspects of biogas generation, including anaerobic digestion processes, feedstock types, and energy recovery and applications (Monlau et al., 2021; Sharma et al., 2020).

Anaerobic Digestion

Anaerobic digestion is a biological process in which microorganisms break down organic matter in the absence of oxygen. This process occurs in several stages:

Hydrolysis: Complex organic compounds are broken down into simpler molecules, including sugars, amino acids, and fatty acids.

Acidogenesis: The breakdown products from hydrolysis are converted into volatile fatty acids and other organic acids by acidogenic bacteria.

Acetogenesis: Further conversion of these acids takes place, producing acetate, hydrogen, and carbon dioxide.

Methanogenesis: Methanogenic archaea utilize acetate, hydrogen, and carbon dioxide to produce methane gas (CH4) and carbon dioxide (CO2).

The resulting mixture of methane and carbon dioxide constitutes biogas.

Feedstock Types

A variety of organic materials can serve as feedstocks for biogas generation (Sakarika & Kornaros, 2019):

Agricultural Waste: Manure from livestock operations, crop residues, and other agricultural byproducts are commonly used feedstocks.

Food Waste: Organic waste from households, restaurants, and food processing facilities can be utilized.

Wastewater Treatment Sludge: Sewage treatment plants can employ anaerobic digestion to treat sludge and simultaneously generate biogas.

Industrial Residues: Organic waste from industries such as breweries, dairies, and pulp mills can be digested to produce biogas.

Energy Recovery and Applications

Biogas has several valuable applications for energy recovery (Mahla et al., 2021):

Electricity Generation: Biogas can be burned in gas turbines, internal combustion engines, or fuel cells to generate electricity.

Heat Production: The heat generated from biogas combustion can be used for space heating, water heating, and industrial processes.

Combined Heat and Power (CHP): CHP systems simultaneously produce electricity and heat from biogas, maximizing energy utilization and efficiency.

Vehicle Fuel: Biogas can be upgraded to biomethane by removing impurities and can be used as a clean alternative fuel for vehicles.

Injection into Natural Gas Grid: Biomethane can be injected into the natural gas grid, providing a renewable component to the gas supply.

Fertilizer Production: The nutrient-rich digestate left after anaerobic digestion can be used as organic fertilizer.

Benefits and Considerations

Advantages: Biogas is a sustainable energy source from organic waste, aiding in waste reduction and eco-friendly waste disposal. Anaerobic digestion captures and uses methane produced during digestion, reducing methane emissions from waste.

Challenges: Feedstock availability, digestate management, and process control are crucial for biogas production, ensuring consistent supply, preventing environmental issues, and maintaining optimal microbial activity conditions in anaerobic digestion.

Anaerobic digestion generates biogas, a sustainable energy solution for waste management, using diverse feedstocks and efficient conversion technologies, contributing to a more sustainable energy landscape.

BIOREMEDIATION: HARNESSING MICROBIAL POWER

Principles of Bioremediation

Bioremediation is a natural or engineered process that employs microorganisms to degrade, transform, or remove pollutants from contaminated environments. This section delves into the fundamental principles of bioremediation, the role of microorganisms, and its applications in soil remediation, water quality improvement, and air quality enhancement (Sarkar et al., 2022).

Microorganisms in Bioremediation

Microorganisms, including bacteria, fungi, and algae, play a crucial role in bioremediation due to their remarkable metabolic capabilities. These organisms have evolved to utilize various contaminants as energy sources, breaking down complex molecules into simpler forms. Bioremediation leverages these natural capabilities to facilitate the removal or transformation of pollutants (Verma et al., 2021).

Applications in Soil Remediation

Bioremediation holds significant promise for cleaning up contaminated soil. Some key applications include (Bhattacharya, Bose, Ganti, et al., 2023):

Hydrocarbon Degradation: Oil spills and hydrocarbon contamination from industrial activities can be mitigated by introducing hydrocarbon-degrading bacteria that break down oil compounds.

Heavy Metal Removal: Certain microorganisms can bind to heavy metals, immobilizing them and reducing their availability in the soil.

Volatile Organic Compounds (VOCs) Remediation: VOCs, commonly found in industrial waste, can be metabolized by specific microorganisms, reducing their concentration in the soil.

Nitrates and Phosphates Reduction: Microbes can facilitate the conversion of nitrates and phosphates, which contribute to water pollution, into harmless forms.

Water and Air Quality Improvement

Bioremediation also extends to improving water and air quality (Bharagava et al., 2020):

Water Remediation: In water bodies contaminated with organic pollutants, adding microbial cultures can enhance the breakdown of contaminants. Constructed wetlands utilize plants and microorganisms to filter and treat wastewater.

Wastewater Treatment: Microorganisms are used in wastewater treatment plants to break down organic matter and remove nutrients, producing cleaner effluent.

Air Quality Enhancement: Certain microbes can absorb or break down air pollutants like volatile organic compounds (VOCs), mitigating their impact on air quality.

Benefits and Considerations:

Advantages: Bioremediation utilizes natural microbial processes, reducing invasive techniques and minimizing environmental impact. It is often more cost-effective than traditional methods and produces fewer byproducts and disturbances.

Challenges: Bioremediation effectiveness depends on soil type, contaminant type, and microbial activity. It may be time-intensive, especially for complex contaminants, and requires close monitoring and regulation to ensure effective management of potential risks.

Bioremediation utilizes microorganisms' natural abilities to remediate contaminated environments, promoting sustainable solutions for environmental restoration and pollution control in soil, water, and air (Bhattacharya, Bose, Yadav, et al., 2023; Sahota & Sharma, 2020; Xu et al., 2023).

SYNTHETIC CELL-FREE PROCESSES: REVOLUTIONIZING BIOTECHNOLOGY

The field of synthetic cell-free processes is ushering in a new era in biotechnology by harnessing cellular components outside the confines of living cells (Figure 2). This section delves into the key aspects of synthetic cell-free systems, including understanding their principles and engineering cell-free reactions (Kelwick et al., 2020; Kruyer et al., 2021).

Figure 2. Synthetic cell-free processes: Revolutionizing biotechnology

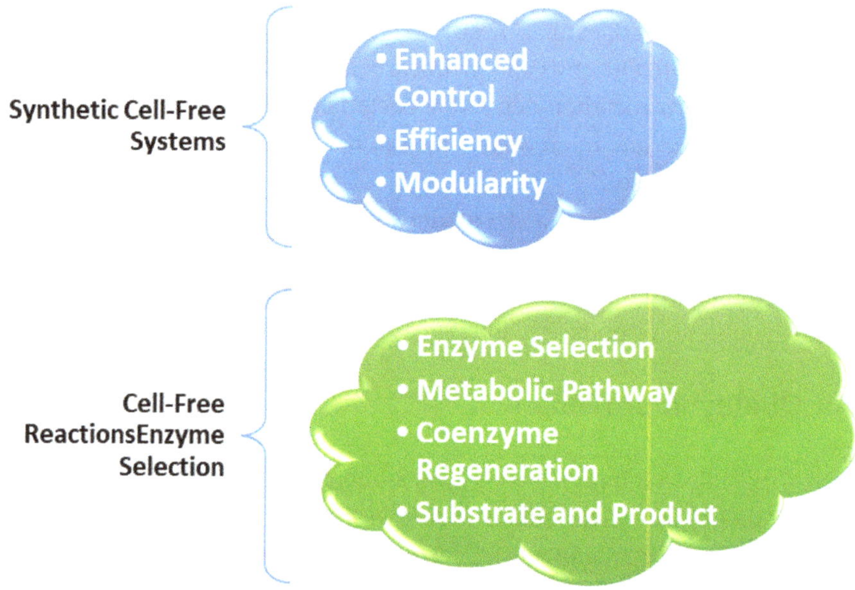

Understanding Synthetic Cell-Free Systems

Synthetic cell-free systems, often referred to as cell-free extracts or cell-free protein synthesis systems, involve extracting cellular components such as enzymes, ribosomes, and cofactors from living cells and utilizing them in vitro to perform biochemical reactions. This approach offers several advantages (Elani, 2021):

Enhanced Control: Researchers can precisely control reaction conditions, enabling tailored biochemical reactions and high-throughput experimentation.

Efficiency: Cell-free systems can bypass the energy requirements of cell growth and maintenance, resulting in higher yields of desired products.

Modularity: Components can be mixed and matched to create custom reaction pathways, facilitating the study of intricate biochemical processes.

Engineering Cell-Free Reactions

Engineering synthetic cell-free systems involves the manipulation of biological components and pathways to achieve desired outcomes (J. Li et al., 2021):

Enzyme Selection and Engineering: Choosing the right enzymes and engineering them for enhanced activity and stability is crucial for efficient cell-free reactions.

Metabolic Pathway Design: Synthetic cell-free systems can be programmed to perform multi-step reactions, enabling the synthesis of complex molecules.

Coenzyme Regeneration: The availability of cofactors can limit cell-free reactions. Strategies for regenerating cofactors are essential to sustain reactions.

Substrate and Product Engineering: Optimizing substrate concentration and product removal strategies can maximize reaction efficiency.

Benefits and Considerations:

Advantages

Customization: Synthetic cell-free systems enable the construction of tailor-made pathways for producing specific compounds.

Rapid Prototyping: Cell-free systems allow for rapid testing of novel pathways, reducing development time.

Minimal Resource Utilization: Since cell-free systems eliminate the need for cellular growth, resources are directed solely towards the desired product.

Challenges:

Complexity: Designing and engineering intricate cell-free reactions can be challenging due to the complexity of cellular processes.

Enzyme Compatibility: Not all cellular components are compatible with cell-free conditions, requiring careful selection and optimization.

Scalability: Transitioning from laboratory-scale reactions to industrial-scale production remains a challenge.

Future Directions

The field of synthetic cell-free processes is rapidly evolving and holds great promise for various applications (Laohakunakorn et al., 2020; J. Li et al., 2023; Yue et al., 2023):

Biomanufacturing: Cell-free systems can be used to produce valuable compounds, including pharmaceuticals, biofuels, and specialty chemicals.

Personalized Medicine: Customized protein synthesis using cell-free systems could enable personalized therapies and diagnostics.

Bioremediation: Cell-free systems might offer efficient ways to engineer enzymes for pollutant degradation.

Synthetic cell-free processes revolutionize biotechnology by providing control, efficiency, and customization. As research advances, these systems could revolutionize industries and drive innovation in bio-based product production.

Applications in Biochemical Production

Synthetic cell-free processes have opened up diverse applications in biochemical production, offering efficient and customizable platforms for synthesizing a wide range of compounds. This section explores key applications of synthetic cell-free systems in biochemical production (Hong & Fernandez-Baca, 2022; Lu, 2020; Yue et al., 2023).

Customized Compound Synthesis: Synthetic cell-free systems enable the rapid production of various biochemicals, including proteins, enzymes, metabolites, and even complex molecules like pharmaceuticals. These systems can be tailored to specific reaction pathways, allowing for the production of compounds that might be challenging to achieve through traditional methods.

Enzyme Production and Optimization: Cell-free systems are utilized to express and optimize enzymes for various applications. This includes enhancing enzyme activity, stability, and efficiency through directed evolution and rational design.

Biocatalysis and Chemoenzymatic Reactions: Synthetic cell-free platforms facilitate biocatalytic reactions, where enzymes drive chemical transformations. Chemoenzymatic synthesis can be achieved by coupling enzymatic and chemical steps in a controlled environment.

Pathway Screening and Design: Researchers can rapidly screen different metabolic pathways and optimize reaction conditions to identify the most efficient pathway for producing desired biochemicals.

Pharmaceutical and Drug Development: Synthetic cell-free systems enable rapid protein expression for drug discovery and testing, allowing researchers to produce therapeutic proteins for evaluation.

Biomanufacturing Advancements

The field of biomanufacturing is advancing with the integration of synthetic cell-free systems, offering novel solutions to traditional manufacturing challenges (Saravanan, Deivayanai, et al., 2022):

Faster Production: Cell-free systems eliminate the need for cell growth, enabling faster production of bio-based products.

Flexible Scale-Up: Synthetic cell-free processes can potentially be more scalable and adaptable to industrial-scale production compared to traditional cell-based approaches.

Resource Efficiency: As cell-free systems focus resources solely on the desired reactions, the process can be more resource-efficient.

Customized Products: Biomanufacturing using synthetic cell-free systems allows for on-demand and customized production of specific biochemicals.

Process Optimization: The controlled nature of synthetic cell-free systems enables precise optimization of reaction conditions and yields (Domakonda et al., 2022; Gunasekaran et al., 2022; Haribalaji et al., 2022; Jeevanantham et al., 2022).

Synthetic cell-free processes are reshaping biochemical production and biomanufacturing by offering efficient, customizable, and scalable platforms. As these technologies continue to advance and become integrated into industrial processes, they hold the potential to revolutionize the way we produce a wide range of biochemicals, contributing to a more sustainable and innovative future.

INTEGRATION OF BIOMASS TECHNOLOGIES FOR SUSTAINABILITY

Achieving a sustainable future requires the thoughtful integration of various biomass technologies (Boopathi & Davim, 2023b, 2023a; Sengeni et al., 2023). This section explores the synergies between biofuel production and bioremediation, showcasing how these two fields can work in harmony to address energy needs and environmental challenges (Ong et al., 2020; Rodionova et al., 2022; Sakarika & Kornaros, 2019).

Figure 3. Synergies between biofuel and bioremediation

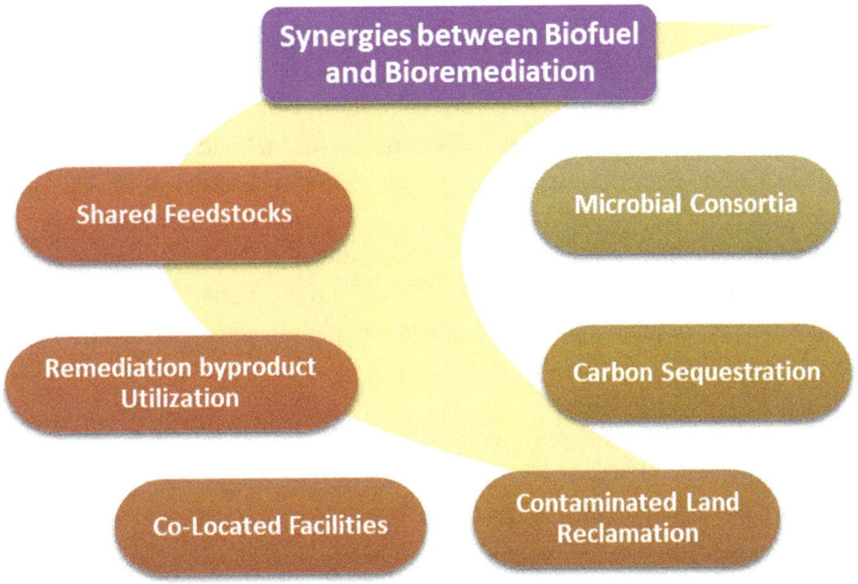

Synergies between Biofuel and Bioremediation

Synergies between biofuel and bioremediation is illustrated in Figure 3.

Shared Feedstocks: Many feedstocks used for biofuel production, such as agricultural residues and algae, can also be employed in bioremediation. This dual use maximizes resource utilization and reduces waste.

Remediation Byproduct Utilization: Effluents or byproducts from bioremediation processes, such as nutrient-rich sludge, can serve as valuable inputs for biofuel production, reducing waste and enhancing overall efficiency.

Co-Located Facilities: Integrated facilities that combine biofuel production and bioremediation processes can streamline resource sharing, reduce transportation costs, and enhance sustainability.

Contaminated Land Reclamation: Biofuel crops can be grown on contaminated lands, aiding in soil remediation while simultaneously yielding valuable biomass for biofuel production.

Carbon Sequestration: Certain biofuel crops can sequester carbon in the soil, contributing to carbon offsetting while producing feedstocks for biofuels.

Microbial Consortia: Microorganisms used in biofuel production and bioremediation can sometimes form synergistic relationships, enhancing the efficiency of both processes.

Benefits and Considerations

Advantages: Integrating biofuel and bioremediation processes improves resource efficiency, reduces waste, and enhances sustainability. Co-locating facilities and sharing resources lead to cost savings and improved economic viability. Simultaneous bioremediation and biofuel production contribute to pollution reduction, habitat restoration, and environmental health (Eskandar, 2023; Sánchez-Costa & López-Gallego, 2023; Yue et al., 2023).

Challenges: Complex logistics and technical compatibility are crucial for coordinating biofuel production and bioremediation processes, ensuring efficiency and avoiding interference.

Integrated systems combining biofuel and bioremediation can be pursued for future research and development. Governments can incentivize these technologies through policy measures and funding support. Raising awareness about the benefits of integrating biofuel and bioremediation can encourage adoption and collaboration (Hanumanthakari et al., 2023; Sathiyaseelan et al., 2023; Sengeni et al., 2023; Yatika et al., 2023). The integration of biofuel production and bioremediation holds immense potential for addressing both energy and environmental challenges. By designing systems that capitalize on the synergies between these two fields, we can achieve a more holistic and sustainable approach to utilizing biomass resources, reducing pollution, and advancing our journey towards a greener and more equitable future.

Synthetic Biology and Biomass Conversion

The marriage of synthetic biology and biomass conversion has opened up transformative possibilities in the realms of sustainable energy production and environmental management. This section explores the intersection of synthetic biology and biomass conversion, highlighting how synthetic biology's tools and principles are revolutionizing the way we harness and transform biomass resources (Laohakunakorn et al., 2020; J. Li et al., 2021; Yue et al., 2023). Synthetic biology and biomass conversion process is shown in Figure 4.

Genetic Engineering for Enhanced Biomass: Synthetic biology techniques enable the engineering of microorganisms to efficiently break down and utilize complex biomass materials. Genetic modifications can enhance the expression of enzymes involved in biomass degradation, thereby improving the efficiency of conversion processes.

Pathway Engineering for Biofuel Synthesis: Synthetic biology allows the design and assembly of metabolic pathways in microorganisms for the production of biofuels. This includes engineering microbes to produce bioethanol, biodiesel precursors, and other advanced biofuels from various biomass sources.

Designing Synthetic Microbes for Bioremediation: Synthetic biology can be used to create microorganisms capable of efficiently removing specific pollutants from the environment. Engineered bacteria can be designed to degrade contaminants in soil, water, and air, contributing to effective bioremediation.

Figure 4. Synthetic biology and biomass conversion process

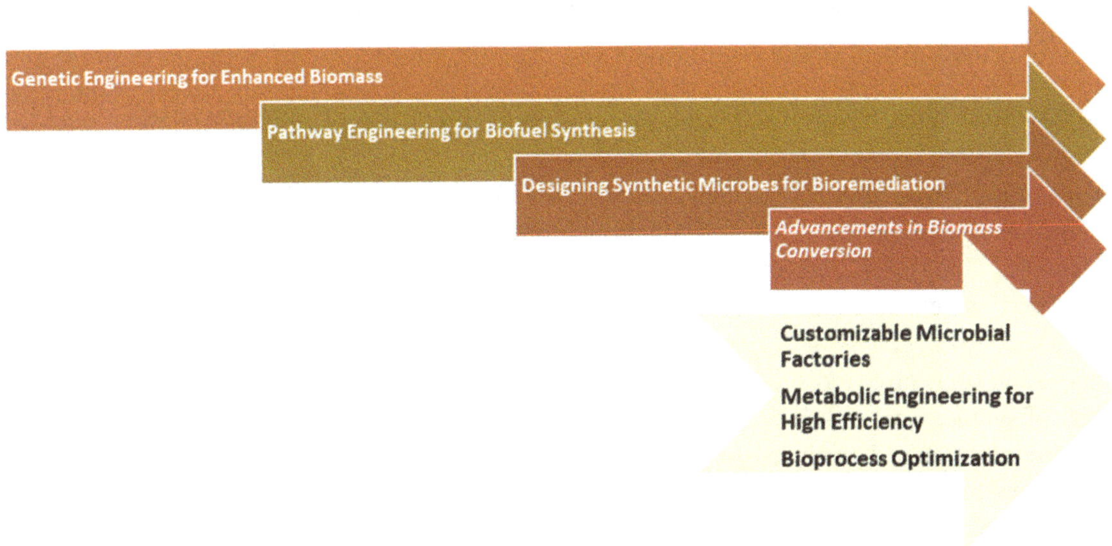

Advancements in Biomass Conversion

Customizable Microbial Factories: Synthetic biology enables the design of microbial cell factories tailored for specific biomass conversion tasks. These cell factories can be optimized to produce desired biofuels, chemicals, or enzymes with high yields and productivity (Bhattacharya, Bose, Ganti, et al., 2023; Saratale et al., 2022; Vandana et al., 2021; Xu et al., 2023).

Metabolic Engineering for High Efficiency: Through metabolic engineering, microorganisms can be modified to use a broader range of feedstocks, enhancing the flexibility and efficiency of biomass conversion processes.

Bioprocess Optimization: Synthetic biology tools aid in optimizing fermentation conditions, substrate utilization, and product formation, leading to more efficient and cost-effective biomass conversion.

Challenges and Considerations

Complexity of Biomass: Biomass is a heterogeneous and intricate mixture of materials, requiring sophisticated synthetic biology strategies to efficiently break it down and convert it into valuable products.

Scale-Up and Commercialization: Translating laboratory-scale synthetic biology approaches to industrial-scale biomass conversion processes requires addressing engineering, scalability, and economic challenges.

Ethical and Regulatory Aspects: The use of genetically modified microorganisms raises ethical considerations and requires adherence to regulatory frameworks to ensure safety and prevent unintended consequences.

The integration of synthetic biology and biomass conversion holds immense promise for sustainable energy production, waste reduction, and environmental restoration. As synthetic biology techniques become more advanced and accessible, they will play a pivotal role in developing innovative and eco-friendly solutions for biomass utilization, paving the way towards a greener and more sustainable future.

IOT- AI: BIOFUEL PRODUCTION, AND SYNTHETIC CELL PROCESS

In an era marked by escalating environmental concerns and a pressing need for sustainable solutions, the convergence of cutting-edge technologies holds the promise of transformative change. The integration of Internet of Things (IoT) and Artificial Intelligence (AI) has emerged as a powerful paradigm, offering unparalleled insights and control over intricate biological processes (Fowziya et al., 2023; Gnanaprakasam et al., 2023; Gowri et al., 2023; Hanumanthakari et al., 2023; Harikaran et al., 2023). This synthesis of innovation finds profound significance in the realms of biofuel production, bioremediation, and the frontier of synthetic cell-free processes – three pivotal domains that intersect at the crossroads of ecological preservation and renewable resource generation.

The quest for alternative and renewable energy sources has ignited interest in biofuel production, harnessing the inherent capabilities of microorganisms to convert organic matter into energy-rich compounds. Concurrently, bioremediation endeavors seek to remediate polluted environments by utilizing microorganisms' remarkable capacity to metabolize pollutants into benign byproducts. Meanwhile, the burgeoning field of synthetic cell-free processes ventures beyond traditional cell-based systems, exploring the orchestrated biochemical reactions that occur in the absence of intact cells.

The IoT-AI synergy enables real-time monitoring, data analysis, and dynamic control in diverse processes, enabling optimization of process parameters, resource wastage reduction, and precise biochemical reactions (Kumara et al., 2023; Maguluri et al., 2023; Maheswari et al., 2023; Mohanty, Jothi, et al., 2023; Mohanty, Venkateswaran, et al., 2023). This integration combines IoT-enabled sensors, data-driven AI algorithms, and smart actuators to enhance sustainability and efficiency in biofuel production, bioremediation, and synthetic cell-free processes. A comprehensive vision is proposed, combining biological intricacies with digital prowess. Sensors capture real-world data, while AI algorithms extract patterns and insights for decision-making. Actuators, connected to this cerebral AI, implement precise adjustments, bringing about tangible changes in process conditions (Palaniappan et al., 2023; Pramila et al., 2023; Rahamathunnisa et al., 2023; Ramudu et al., 2023; Reddy et al., 2023).

The IoT-AI connected ecosystem involves technology integration, data management, AI model development, and secure communication protocols. This endeavor aims to advance scientific frontiers and foster collaborative efforts across disciplines, bridging the gap between biology, engineering, and computer science. By embarking on a transformation, we navigate the uncharted territories of IoT-AI connected biofuel production, bioremediation, and synthetic cell-free processes, paving the way for a sustainable future combining human ingenuity and technological prowess (Sankar et al., 2023; Sathish et al., 2023; Sathiyaseelan et al., 2023; Selvakumar, Adithe, et al., 2023; Selvakumar, Shankar, et al., 2023).

IoT-AI integrated system for biofuel production, bioremediation, and synthetic cell-free processes combines technologies for efficient workflow, ensuring seamless and seamless processes (Veeranjaneyulu, Boopathi, Kumari, et al., 2023; Veeranjaneyulu, Boopathi, Narasimharao, et al., 2023; Venkateswaran, Vidhya, Ayyannan, et al., 2023; Venkateswaran, Vidhya, Naik, et al., 2023; Yatika et al., 2023; Yupapin et al., 2023).

- **Define Objectives:** Clearly define the goals of your project, such as optimizing biofuel production, enhancing bioremediation efficiency, and advancing synthetic cell-free processes. Understand the specific metrics you aim to improve, such as yield, speed, or resource efficiency.
- **System Architecture:** Design the overall architecture of your IoT-AI connected system. Consider the various components involved, including sensors, actuators, AI algorithms, databases, and communication protocols.
- **Sensor Integration:** Deploy sensors to monitor relevant parameters such as temperature, pH, nutrient levels, and gas emissions in the biofuel production and bioremediation processes. These sensors should be capable of real-time data collection and transmission.
- **Data Collection and Transmission:** Implement a data collection mechanism that gathers information from the deployed sensors. This data should be transmitted securely and efficiently to a central database or cloud platform. Use appropriate communication protocols such as MQTT or HTTPS.
- **Cloud Platform:** Choose a cloud platform to store, process, and analyze the collected data. Platforms like AWS, Azure, or Google Cloud provide tools for data storage, database management, and AI services.
- **Data Analytics and AI:** Develop AI algorithms to analyze the collected data. For example, AI can predict optimal biofuel production conditions based on historical data, recommend bioremediation strategies, and optimize synthetic cell-free processes. Machine learning techniques such as regression, clustering, and deep learning can be applied.
- **Decision System:** Integrate the AI analysis with a decision-making system. This system can automatically adjust process parameters based on AI recommendations, optimizing biofuel production and bioremediation in real time.
- **Actuator Control:** Implement actuators that can adjust process conditions based on decisions from the AI system. These could include adjusting nutrient levels, temperature, or flow rates in the bioreactors.
- **Feedback Loop:** Establish a feedback loop where data from actuators is collected and fed back into the system. This helps in verifying the effectiveness of the decisions made by the AI and improving the overall system over time.
- **Security and Privacy:** Ensure the security of data transmission and storage to protect sensitive information and prevent unauthorized access to the system. Implement encryption, authentication, and access controls.
- **User Interface:** Develop a user interface (UI) that provides a dashboard for monitoring the system's performance, visualizing data trends, and interacting with the AI decision-making process.
- **Testing and Optimization:** Thoroughly test the system in controlled environments to validate its performance. Fine-tune AI algorithms and system parameters to achieve the desired objectives efficiently.
- **Scalability:** Design the system with scalability in mind, allowing for the addition of more sensors, processes, and AI capabilities as needed.
- **Continuous Improvement:** Regularly update and improve the system based on new data insights, advancements in AI, and emerging technologies.
- **Documentation and Training:** Document the entire implementation process, including system architecture, algorithms used, and operational procedures. Provide training for personnel responsible for operating and maintaining the system.

CASE STUDIES AND REAL-WORLD IMPLEMENTATIONS

Successful Biomass-to-Biofuel Ventures: LanzaTech

LanzaTech, a pioneering bioenergy company, has developed a groundbreaking technology that converts industrial waste gases, such as carbon monoxide and carbon dioxide, into advanced biofuels and chemicals. By utilizing microorganisms that convert these gases into valuable products, LanzaTech demonstrates the feasibility of converting non-traditional biomass sources into biofuels. This innovative approach not only reduces greenhouse gas emissions but also produces valuable products from waste materials, showcasing the potential of biomass-to-biofuel technologies in a sustainable circular economy (Alishah Aratboni et al., 2019; Hoang et al., 2021; Saravanan, Kumar, et al., 2022; Sindhu et al., 2019).

CONCLUSION

The intersection of biomass utilization, bioremediation, synthetic biology, and innovative biotechnological processes is shaping a more sustainable and resilient future. Through the synergistic integration of these fields, we are unlocking novel pathways for addressing pressing environmental challenges and meeting growing energy demands. As we reflect on the multifaceted discussions presented in this book chapter, several key conclusions emerge:

Biomass, in its various forms, stands as a versatile resource that can be harnessed for energy production, biofuel synthesis, and environmental restoration. By exploring diverse biomass sources and refining conversion technologies, we can reduce our reliance on fossil fuels and minimize environmental impacts. The application of bioremediation technologies exemplifies the power of harnessing natural processes to restore polluted environments. Microorganisms, tailored or naturally occurring, possess the potential to alleviate environmental degradation, turning pollutants into harmless byproducts. Synthetic biology's tools and principles offer transformative opportunities for biotechnological advancement. From engineering microorganisms for targeted biomass conversion to constructing synthetic cell-free systems, this field is accelerating innovation across various sectors. The integration of biomass technologies, such as biofuel production and bioremediation, presents a holistic approach to sustainability. By leveraging shared resources, co-located facilities, and collaborative strategies, we can enhance resource efficiency and environmental benefits. As biotechnology progresses, ethical considerations and regulatory frameworks become essential in guiding responsible research, development, and deployment. Public engagement, transparency, and equitable access to benefits are crucial for navigating the complex ethical landscape. Case studies spotlight the tangible impact of these technologies. From LanzaTech's innovative waste gas-to-biofuel conversion to bioremediation success stories like the Exxon Valdez spill cleanup, real-world implementations underscore the transformative potential of these approaches.The advancement of biomass technologies is a collective endeavor that involves researchers, policymakers, industries, and the public. Collaborative efforts are necessary to address challenges, maximize benefits, and shape a sustainable future.

This book highlights the dynamic synergy between science, technology, and ethics, which can transform industries, redefine energy sources, and restore environments. The ethical and responsible application of these technologies is crucial for fostering a world thriving on innovation, environmental stewardship, and shared prosperity.

LIST OF ABBREVIATIONS

GMOs: Genetically Modified Organisms
VOCs: Volatile Organic Compounds
DNA: Deoxyribonucleic Acid
RNA: Ribonucleic Acid

REFERENCES

Alishah Aratboni, H., Rafiei, N., Garcia-Granados, R., Alemzadeh, A., & Morones-Ramírez, J. R. (2019). Biomass and lipid induction strategies in microalgae for biofuel production and other applications. *Microbial Cell Factories*, *18*(1), 1–17. doi:10.118612934-019-1228-4 PMID:31638987

Babu, B. S., Kamalakannan, J., Meenatchi, N., Karthik, S., & Boopathi, S. (2022). Economic impacts and reliability evaluation of battery by adopting Electric Vehicle. *IEEE Explore*, 1–6.

Banerjee, S., Banerjee, S., Ghosh, A. K., & Das, D. (2020). Maneuvering the genetic and metabolic pathway for improving biofuel production in algae: Present status and future prospective. *Renewable & Sustainable Energy Reviews*, *133*, 110155. doi:10.1016/j.rser.2020.110155

Bharagava, R. N., Saxena, G., & Mulla, S. I. (2020). *Introduction to industrial wastes containing organic and inorganic pollutants and bioremediation approaches for environmental management. In Bioremediation of Industrial Waste for Environmental Safety* (Vol. I). Industrial Waste and Its Management.

Bhattacharya, R., Bose, D., Ganti, P., Rizvi, A., Halder, G., & Sarkar, A. (2023). *Bioelectricity production and bioremediation potential of Withania somnifera in plant microbial fuel cells with food wastes as enrichment*. Academic Press.

Bhattacharya, R., Bose, D., Yadav, J., Sharma, B., Sangli, E., Patel, A., Mukherjee, A., & Singh, A. A. (2023). Bioremediation and bioelectricity from Himalayan rock soil in sediment-microbial fuel cell using carbon rich substrates. *Fuel*, *341*, 127019. doi:10.1016/j.fuel.2022.127019

Boopathi, S., & Davim, J. P. (2023a). Applications of Nanoparticles in Various Manufacturing Processes. In *Sustainable Utilization of Nanoparticles and Nanofluids in Engineering Applications* (pp. 1–31). IGI Global. doi:10.4018/978-1-6684-9135-5.ch001

Boopathi, S., & Davim, J. P. (2023b). *Sustainable Utilization of Nanoparticles and Nanofluids in Engineering Applications*. IGI Global. doi:10.4018/978-1-6684-9135-5

Chi, N. T. L., Anto, S., Ahamed, T. S., Kumar, S. S., Shanmugam, S., Samuel, M. S., Mathimani, T., Brindhadevi, K., & Pugazhendhi, A. (2021). A review on biochar production techniques and biochar based catalyst for biofuel production from algae. *Fuel*, *287*, 119411. doi:10.1016/j.fuel.2020.119411

Domakonda, V. K., Farooq, S., Chinthamreddy, S., Puviarasi, R., Sudhakar, M., & Boopathi, S. (2022). Sustainable Developments of Hybrid Floating Solar Power Plants: Photovoltaic System. In Human Agro-Energy Optimization for Business and Industry (pp. 148–167). IGI Global.

Elani, Y. (2021). Interfacing living and synthetic cells as an emerging frontier in synthetic biology. *Angewandte Chemie*, *133*(11), 5662–5671. doi:10.1002/ange.202006941 PMID:32909663

Eskandar, K. (2023). Revolutionizing biotechnology and bioengineering: Unleashing the power of innovation. *J Appl Biotechnol Bioeng*, *10*(3), 81–88. doi:10.15406/jabb.2023.10.00332

Fowziya, S., Sivaranjani, S., Devi, N. L., Boopathi, S., Thakur, S., & Sailaja, J. M. (2023). Influences of nano-green lubricants in the friction-stir process of TiAlN coated alloys. *Materials Today: Proceedings*. Advance online publication. doi:10.1016/j.matpr.2023.06.446

Gnanaprakasam, C., Vankara, J., Sastry, A. S., Prajval, V., Gireesh, N., & Boopathi, S. (2023). Long-Range and Low-Power Automated Soil Irrigation System Using Internet of Things: An Experimental Study. In Contemporary Developments in Agricultural Cyber-Physical Systems (pp. 87–104). IGI Global.

Gowri, N. V., Dwivedi, J. N., Krishnaveni, K., Boopathi, S., Palaniappan, M., & Medikondu, N. R. (2023). Experimental investigation and multi-objective optimization of eco-friendly near-dry electrical discharge machining of shape memory alloy using Cu/SiC/Gr composite electrode. *Environmental Science and Pollution Research International*, 1–19. doi:10.100711356-023-26983-6 PMID:37126160

Gunasekaran, K., Boopathi, S., & Sureshkumar, M. (2022). Analysis of a Cryogenically Cooled Near-Dry WEDM Process using Different Dielectrics. *Mater-Tehnol.Si. Materials Technology*, *56*(2), 179–186.

Hanumanthakari, S., Gift, M. M., Kanimozhi, K., Bhavani, M. D., Bamane, K. D., & Boopathi, S. (2023). Biomining Method to Extract Metal Components Using Computer-Printed Circuit Board E-Waste. In *Handbook of Research on Safe Disposal Methods of Municipal Solid Wastes for a Sustainable Environment* (pp. 123–141). IGI Global. doi:10.4018/978-1-6684-8117-2.ch010

Haribalaji, V., Venkatesan, G., Pandian, M., Subbiah, R., & Boopathi, S. (2022). Investigation on corrosion and tensile Characteristics: Friction stir welding of AA7075 and AA2014. *Materials Today: Proceedings*, *66*(3), 743–748. doi:10.1016/j.matpr.2022.04.037

Harikaran, M., Boopathi, S., Gokulakannan, S., & Poonguzhali, M. (2023). Study on the Source of E-Waste Management and Disposal Methods. In *Sustainable Approaches and Strategies for E-Waste Management and Utilization* (pp. 39–60). IGI Global. doi:10.4018/978-1-6684-7573-7.ch003

Hoang, A. T., Ong, H. C., Fattah, I. R., Chong, C. T., Cheng, C. K., Sakthivel, R., & Ok, Y. S. (2021). Progress on the lignocellulosic biomass pyrolysis for biofuel production toward environmental sustainability. *Fuel Processing Technology*, *223*, 106997. doi:10.1016/j.fuproc.2021.106997

Hong, S. H., & Fernandez-Baca, A. S. (2022). Cell-free synthetic biology as an emerging biotechnology. In *New Frontiers and applications of synthetic biology* (pp. 397–414). Elsevier. doi:10.1016/B978-0-12-824469-2.00016-6

Huang, Y., Li, B., Liu, D., Xie, X., Zhang, H., Sun, H., Hu, X., & Zhang, S. (2020). Fundamental advances in biomass autothermal/oxidative pyrolysis: A review. *ACS Sustainable Chemistry & Engineering*, *8*(32), 11888–11905. doi:10.1021/acssuschemeng.0c04196

Jeevanantham, Y. A., Saravanan, A., Vanitha, V., Boopathi, S., & Kumar, D. P. (2022). Implementation of Internet-of Things (IoT) in Soil Irrigation System. *IEEE Explore*, 1–5.

Kelwick, R. J., Webb, A. J., & Freemont, P. S. (2020). Biological materials: The next frontier for cell-free synthetic biology. *Frontiers in Bioengineering and Biotechnology*, *8*, 399. doi:10.3389/fbioe.2020.00399 PMID:32478045

Kruyer, N. S., Sugianto, W., Tickman, B. I., Alba Burbano, D., Noireaux, V., Carothers, J. M., & Peralta-Yahya, P. (2021). Membrane augmented cell-free systems: A new frontier in biotechnology. *ACS Synthetic Biology*, *10*(4), 670–681. doi:10.1021/acssynbio.0c00625 PMID:33749249

Kumara, V., Mohanaprakash, T., Fairooz, S., Jamal, K., Babu, T., & Sampath, B. (2023). Experimental Study on a Reliable Smart Hydroponics System. In *Human Agro-Energy Optimization for Business and Industry* (pp. 27–45). IGI Global. doi:10.4018/978-1-6684-4118-3.ch002

Kwon, G., Bhatnagar, A., Wang, H., Kwon, E. E., & Song, H. (2020). A review of recent advancements in utilization of biomass and industrial wastes into engineered biochar. *Journal of Hazardous Materials*, *400*, 123242. doi:10.1016/j.jhazmat.2020.123242 PMID:32585525

Laohakunakorn, N., Grasemann, L., Lavickova, B., Michielin, G., Shahein, A., Swank, Z., & Maerkl, S. J. (2020). Bottom-up construction of complex biomolecular systems with cell-free synthetic biology. *Frontiers in Bioengineering and Biotechnology*, *8*, 213. doi:10.3389/fbioe.2020.00213 PMID:32266240

Li, J., Castano, I., Liu, W.-Q., Oza, J. P., & Black, M. W., Jew, K., Smith, P. E., So, B., Kasman, J., & Oza, J. P. (2023). Cell-Free. *Synthetic Biology*, *II*, 87.

Li, J., Kwon, Y.-C., Lu, Y., & Moore, S. J. (2021). Cell-Free Synthetic Biology. *Frontiers in Bioengineering and Biotechnology*, *9*, 799122. doi:10.3389/fbioe.2021.799122 PMID:34912793

Li, P., Sakuragi, K., & Makino, H. (2019). Extraction techniques in sustainable biofuel production: A concise review. *Fuel Processing Technology*, *193*, 295–303. doi:10.1016/j.fuproc.2019.05.009

Li, Y., Xing, B., Ding, Y., Han, X., & Wang, S. (2020). A critical review of the production and advanced utilization of biochar via selective pyrolysis of lignocellulosic biomass. *Bioresource Technology*, *312*, 123614. doi:10.1016/j.biortech.2020.123614 PMID:32517889

Lu, Y. (2020). *Cell-Free Synthetic Biology*. Springer. doi:10.1007/978-981-13-1171-0

Maguluri, L. P., Ananth, J., Hariram, S., Geetha, C., Bhaskar, A., & Boopathi, S. (2023). Smart Vehicle-Emissions Monitoring System Using Internet of Things (IoT). In Handbook of Research on Safe Disposal Methods of Municipal Solid Wastes for a Sustainable Environment (pp. 191–211). IGI Global.

Maheswari, B. U., Imambi, S. S., Hasan, D., Meenakshi, S., Pratheep, V., & Boopathi, S. (2023). Internet of Things and Machine Learning-Integrated Smart Robotics. In Global Perspectives on Robotics and Autonomous Systems: Development and Applications (pp. 240–258). IGI Global. doi:10.4018/978-1-6684-7791-5.ch010

Mahla, S. K., Ardebili, S. M. S., Sharma, H., Dhir, A., Goga, G., & Solmaz, H. (2021). Determination and utilization of optimal diesel/n-butanol/biogas derivation for small utility dual fuel diesel engine. *Fuel*, *289*, 119913. doi:10.1016/j.fuel.2020.119913

Mohanty, A., Jothi, B., Jeyasudha, J., Ranjit, P., Isaac, J. S., & Boopathi, S. (2023). Additive Manufacturing Using Robotic Programming. In *AI-Enabled Social Robotics in Human Care Services* (pp. 259–282). IGI Global. doi:10.4018/978-1-6684-8171-4.ch010

Mohanty, A., Venkateswaran, N., Ranjit, P., Tripathi, M. A., & Boopathi, S. (2023). Innovative Strategy for Profitable Automobile Industries: Working Capital Management. In Handbook of Research on Designing Sustainable Supply Chains to Achieve a Circular Economy (pp. 412–428). IGI Global.

Monlau, F., Suarez-Alvarez, S., Lallement, A., Vaca-Medina, G., Giacinti, G., Munarriz, M., Urreta, I., Raynaud, C., Ferrer, C., & Castañón, S. (2021). A cascade biorefinery for the valorization of microalgal biomass: Biodiesel, biogas, fertilizers and high valuable compounds. *Algal Research, 59*, 102433. doi:10.1016/j.algal.2021.102433

Ong, H. C., Chen, W.-H., Singh, Y., Gan, Y. Y., Chen, C.-Y., & Show, P. L. (2020). A state-of-the-art review on thermochemical conversion of biomass for biofuel production: A TG-FTIR approach. *Energy Conversion and Management, 209*, 112634. doi:10.1016/j.enconman.2020.112634

Palaniappan, M., Tirlangi, S., Mohamed, M. J. S., Moorthy, R. S., Valeti, S. V., & Boopathi, S. (2023). Fused Deposition Modelling of Polylactic Acid (PLA)-Based Polymer Composites: A Case Study. In Development, Properties, and Industrial Applications of 3D Printed Polymer Composites (pp. 66–85). IGI Global.

Pramila, P., Amudha, S., Saravanan, T., Sankar, S. R., Poongothai, E., & Boopathi, S. (2023). Design and Development of Robots for Medical Assistance: An Architectural Approach. In Contemporary Applications of Data Fusion for Advanced Healthcare Informatics (pp. 260–282). IGI Global.

Rahamathunnisa, U., Sudhakar, K., Murugan, T. K., Thivaharan, S., Rajkumar, M., & Boopathi, S. (2023). Cloud Computing Principles for Optimizing Robot Task Offloading Processes. In *AI-Enabled Social Robotics in Human Care Services* (pp. 188–211). IGI Global. doi:10.4018/978-1-6684-8171-4.ch007

Ramudu, K., Mohan, V. M., Jyothirmai, D., Prasad, D., Agrawal, R., & Boopathi, S. (2023). Machine Learning and Artificial Intelligence in Disease Prediction: Applications, Challenges, Limitations, Case Studies, and Future Directions. In Contemporary Applications of Data Fusion for Advanced Healthcare Informatics (pp. 297–318). IGI Global.

Reddy, M. A., Reddy, B. M., Mukund, C., Venneti, K., Preethi, D., & Boopathi, S. (2023). Social Health Protection During the COVID-Pandemic Using IoT. In *The COVID-19 Pandemic and the Digitalization of Diplomacy* (pp. 204–235). IGI Global. doi:10.4018/978-1-7998-8394-4.ch009

Rodionova, M. V., Bozieva, A. M., Zharmukhamedov, S. K., Leong, Y. K., Lan, J. C.-W., Veziroglu, A., Veziroglu, T. N., Tomo, T., Chang, J.-S., & Allakhverdiev, S. I. (2022). A comprehensive review on lignocellulosic biomass biorefinery for sustainable biofuel production. *International Journal of Hydrogen Energy, 47*(3), 1481–1498. doi:10.1016/j.ijhydene.2021.10.122

Sahota, N. K., & Sharma, R. (2020). Bioremediation: Harnessing Natural Forces for Solid Waste Management. Handbook of Solid Waste Management: Sustainability through Circular Economy, 1–32.

Sakarika, M., & Kornaros, M. (2019). Chlorella vulgaris as a green biofuel factory: Comparison between biodiesel, biogas and combustible biomass production. *Bioresource Technology*, *273*, 237–243. doi:10.1016/j.biortech.2018.11.017 PMID:30447625

Sampath, B. (2021). *Sustainable Eco-Friendly Wire-Cut Electrical Discharge Machining: Gas Emission Analysis*. Academic Press.

Sánchez-Costa, M., & López-Gallego, F. (2023). *Solid-Phase Cell-Free Protein Synthesis and Its Applications in Biotechnology*. Springer. doi:10.1007/10_2023_226

Sankar, K. M., Booba, B., & Boopathi, S. (2023). Smart Agriculture Irrigation Monitoring System Using Internet of Things. In *Contemporary Developments in Agricultural Cyber-Physical Systems* (pp. 105–121). IGI Global. doi:10.4018/978-1-6684-7879-0.ch006

Saratale, R. G., Ponnusamy, V. K., Jeyakumar, R. B., Sirohi, R., Piechota, G., Shobana, S., Dharmaraja, J., Lay, C., Saratale, G. D., Shin, H. S., & ... (2022). Microalgae cultivation strategies using cost–effective nutrient sources: Recent updates and progress towards biofuel production. *Bioresource Technology*, *361*, 127691. doi:10.1016/j.biortech.2022.127691 PMID:35926554

Saravanan, A., Deivayanai, V., Kumar, P. S., Rangasamy, G., & Varjani, S. (2022). CO2 bio-mitigation using genetically modified algae and biofuel production towards a carbon net-zero society. *Bioresource Technology*, *363*, 127982. doi:10.1016/j.biortech.2022.127982 PMID:36126842

Saravanan, A., Kumar, P. S., Jeevanantham, S., Karishma, S., & Vo, D.-V. N. (2022). Recent advances and sustainable development of biofuels production from lignocellulosic biomass. *Bioresource Technology*, *344*, 126203. doi:10.1016/j.biortech.2021.126203 PMID:34710606

Sarkar, A., Umar, M. F., Fatima, U., Hossain, K., & Rafatullah, M. (2022). Modification of Bioanodes from Different Nanocomposite Materials for Wastewater Bioremediation through Microbial Fuel Cells. In Biorenewable Nanocomposite Materials, Vol. 2: Desalination and Wastewater Remediation (pp. 113–140). ACS Publications. doi:10.1021/bk-2022-1411.ch005

Sathish, T., Sunagar, P., Singh, V., Boopathi, S., Al-Enizi, A. M., Pandit, B., Gupta, M., & Sehgal, S. S. (2023). Characteristics estimation of natural fibre reinforced plastic composites using deep multi-layer perceptron (MLP) technique. *Chemosphere*, *337*, 139346. doi:10.1016/j.chemosphere.2023.139346 PMID:37379988

Sathiyaseelan, G., Srinath, P., Melwyn, J. G., Bhagyanathan, C., Socrates, S., & Boopathi, S. (2023). Tribology and hardness measurement of biosilica dispersed AA7075 metal matrix composites. *Materials Today: Proceedings*. Advance online publication. doi:10.1016/j.matpr.2023.02.440

Selvakumar, S., Adithe, S., Isaac, J. S., Pradhan, R., Venkatesh, V., & Sampath, B. (2023). A Study of the Printed Circuit Board (PCB) E-Waste Recycling Process. In Sustainable Approaches and Strategies for E-Waste Management and Utilization (pp. 159–184). IGI Global.

Selvakumar, S., Shankar, R., Ranjit, P., Bhattacharya, S., Gupta, A. S. G., & Boopathi, S. (2023). E-Waste Recovery and Utilization Processes for Mobile Phone Waste. In *Handbook of Research on Safe Disposal Methods of Municipal Solid Wastes for a Sustainable Environment* (pp. 222–240). IGI Global. doi:10.4018/978-1-6684-8117-2.ch016

Sengeni, D., Padmapriya, G., Imambi, S. S., Suganthi, D., Suri, A., & Boopathi, S. (2023). Biomedical Waste Handling Method Using Artificial Intelligence Techniques. In *Handbook of Research on Safe Disposal Methods of Municipal Solid Wastes for a Sustainable Environment* (pp. 306–323). IGI Global. doi:10.4018/978-1-6684-8117-2.ch022

Sharma, A., Singh, Y., Ansari, N. A., Pal, A., & Lalhriatpuia, S. (2020). Experimental investigation of the behaviour of a DI diesel engine fuelled with biodiesel/diesel blends having effect of raw biogas at different operating responses. *Fuel*, *279*, 118460. doi:10.1016/j.fuel.2020.118460

Sindhu, R., Binod, P., Pandey, A., Ankaram, S., Duan, Y., & Awasthi, M. K. (2019). Biofuel production from biomass: Toward sustainable development. In *Current developments in biotechnology and bioengineering* (pp. 79–92). Elsevier. doi:10.1016/B978-0-444-64083-3.00005-1

Vandana, U. K., Gulzar, A., Laskar, I. H., Meitei, L. R., & Mazumder, P. (2021). Role of microbes in bioremediation of radioactive waste. *Microbial Rejuvenation of Polluted Environment*, *1*, 329–352. doi:10.1007/978-981-15-7447-4_13

Veeranjaneyulu, R., Boopathi, S., Kumari, R. K., Vidyarthi, A., Isaac, J. S., & Jaiganesh, V. (2023). Air Quality Improvement and Optimisation Using Machine Learning Technique. *IEEE Explore*, 1–6.

Veeranjaneyulu, R., Boopathi, S., Narasimharao, J., Gupta, K. K., Reddy, R. V. K., & Ambika, R. (2023). Identification of Heart Diseases using Novel Machine Learning Method. *IEEE Explore*, 1–6.

Venkateswaran, N., Vidhya, K., Ayyannan, M., Chavan, S. M., Sekar, K., & Boopathi, S. (2023). A Study on Smart Energy Management Framework Using Cloud Computing. In 5G, Artificial Intelligence, and Next Generation Internet of Things: Digital Innovation for Green and Sustainable Economies (pp. 189–212). IGI Global. doi:10.4018/978-1-6684-8634-4.ch009

Venkateswaran, N., Vidhya, R., Naik, D. A., Raj, T. M., Munjal, N., & Boopathi, S. (2023). Study on Sentence and Question Formation Using Deep Learning Techniques. In *Digital Natives as a Disruptive Force in Asian Businesses and Societies* (pp. 252–273). IGI Global. doi:10.4018/978-1-6684-6782-4.ch015

Verma, J., Kumar, D., Singh, N., Katti, S. S., & Shah, Y. T. (2021). Electricigens and microbial fuel cells for bioremediation and bioenergy production: A review. *Environmental Chemistry Letters*, *19*(3), 2091–2126. doi:10.100710311-021-01199-7

Wang, G., Dai, Y., Yang, H., Xiong, Q., Wang, K., Zhou, J., Li, Y., & Wang, S. (2020). A review of recent advances in biomass pyrolysis. *Energy & Fuels*, *34*(12), 15557–15578. doi:10.1021/acs.energyfuels.0c03107

Xu, G., Zhao, S., Liu, J., & He, J. (2023). Bioremediation of organohalide pollutants: Progress, microbial ecology, and emerging computational tools. *Current Opinion in Environmental Science & Health*, *32*, 100452. doi:10.1016/j.coesh.2023.100452

Yaashikaa, P., Devi, M. K., & Kumar, P. S. (2022). Algal biofuels: Technological perspective on cultivation, fuel extraction and engineering genetic pathway for enhancing productivity. *Fuel*, *320*, 123814. doi:10.1016/j.fuel.2022.123814

Yatika, G., Kumar, S. R., Patel, P. B., Singh, D. P., Rajkamal, M., & Boopathi, S. (2023). Experimental investigation of RHA biochar and tamarind fibre epoxy composite. *Materials Today: Proceedings*. Advance online publication. doi:10.1016/j.matpr.2023.02.439

Yue, K., Chen, J., Li, Y., & Kai, L. (2023). Advancing synthetic biology through cell-free protein synthesis. *Computational and Structural Biotechnology Journal*, *21*, 2899–2908. doi:10.1016/j.csbj.2023.05.003 PMID:37216017

Yupapin, P., Trabelsi, Y., Nattappan, A., & Boopathi, S. (2023). Performance improvement of wire-cut electrical discharge machining process using cryogenically treated super-conductive state of Monel-K500 alloy. *Iranian Journal of Science and Technology. Transaction of Mechanical Engineering*, *47*(1), 267–283. doi:10.100740997-022-00513-0

Zhu, Y., Guo, X., Liu, J., Li, F., & Yang, D. (2020). Emerging advances of cell-free systems toward artificial cells. *Small Methods*, *4*(10), 2000406. doi:10.1002mtd.202000406

Chapter 14
Energy Monitoring and Control in the Smart Grid:
Integrated Intelligent IoT and ANFIS

C. R. Kavitha
https://orcid.org/0000-0003-1474-2243
Department of Computer Science Engineering, Amrita School of Computing, India

M. Varalatchoumy
Department of Computer Science and Engineering, Cambridge Institute of Technology, India

H. R. Mithuna
Department of Information Science and Engineering, Acharya Institute of Technology, India

K. Bharathi
Department of Thirumala Educational Trust, Sanford Public School, India

N. M. Geethalakshmi
Department of Information Science and Engineering, Acharya Institute of Technology, India

Sampath Boopathi
https://orcid.org/0000-0002-2065-6539
Muthayammal Engineering College, India

ABSTRACT

Monitoring and controlling energy use is critical for efficient power system management, particularly in smart grids. The internet of things (IoT) has compelled the development of intelligent systems such as the adaptive neuro-fuzzy inference system (ANFIS) and fuzzy fruit fly optimization to improve monitoring, optimise energy use, and enable demand response (FFO). This chapter examines the advantages and disadvantages of IoT-enabled microgrids, as well as system installation for energy monitoring and control. It focuses on the architecture, implementation, experimental setup, evaluation, and performance analysis of the ANFIS and FFO algorithms in smart grids. The chapter covers data security, privacy, interoperability, scalability, grid resilience, cost-effectiveness, stakeholder involvement, and AI breakthroughs in energy monitoring and control.

DOI: 10.4018/978-1-6684-6577-6.ch014

INTRODUCTION

The development of the smart grid has resulted from rising worldwide electricity demand, as well as the need to address environmental issues and improve energy efficiency. The smart grid is a cutting-edge power infrastructure that includes cutting-edge technologies to improve energy monitoring and control capabilities. To provide real-time monitoring, efficient energy management, and effective control mechanisms, this revolutionary grid system incorporates numerous components such as advanced metering infrastructure (AMI), communication networks, intelligent sensors, and data analytics. Energy monitoring and control are crucial in the smart grid, enabling utilities, operators, and consumers to manage energy resources, optimize performance, and achieve energy efficiency goals (Avancini et al., 2021; Paul Divakar et al., 2020). By monitoring consumption patterns, identifying inefficiencies, and providing real-time information, these systems enable informed decision-making and proactive management strategies. Accurate data on energy usage, supply-demand dynamics, and grid conditions enables timely actions to mitigate peak loads, reduce wastage, and ensure grid stability.

Advancements in sensor technologies, communication protocols, and data analytics have revolutionized energy monitoring in smart grids. Intelligent meters with bidirectional communication enable real-time data collection and demand response programs. Smart sensors and IoT devices provide granular data on grid parameters, equipment performance, and environmental factors. Data analytics techniques like machine learning and artificial intelligence enable predictive maintenance, load forecasting, and energy resource optimization. Real-time monitoring and control systems are essential for efficient smart grid operations, collecting, processing, and analyzing data from various sources. SCADA systems monitor grid assets, ADMS optimize operations, and ADR systems enable dynamic load management based on grid conditions and consumer preferences. These systems enable stakeholders to monitor energy flows, detect anomalies, and respond quickly to grid events (Hartono et al., 2013).

Energy monitoring and control systems in the smart grid ecosystem offer numerous benefits, including optimizing energy generation, transmission, and distribution, reducing wastage, improving grid reliability, and providing real-time insights for consumers. These systems also enable informed decisions, reduced consumption, and lower energy bills. Additionally, they support renewable energy sources and facilitate efficient energy storage management, promoting a greener, more sustainable grid. Research on smart grid technology has gained attention due to the growing demand for efficient, sustainable energy systems. These systems integrate advanced technologies for intelligent monitoring, control, and management of energy resources. This section highlights key areas of focus and advancements in smart grid technology (Abo-Zahhad et al., 2016).

- Smart grid technology aims to improve power grid performance by upgrading aging infrastructure, enhancing resilience, and integrating renewable energy sources. Advanced sensors, communication networks, and control systems enable real-time monitoring, fault detection, localization, and grid stability.
- Advanced Metering Infrastructure (AMI) and smart meters are crucial areas of research in smart grid technology. Smart meters enable two-way communication between utilities and consumers, collecting real-time data on energy consumption, demand patterns, and power quality. Research focuses on efficient metering systems, data analytics, and consumer engagement strategies for energy conservation and load management.

- Smart grid technology research focuses on integrating distributed energy resources like solar panels, wind turbines, and energy storage systems. This involves developing algorithms and control strategies for optimal DER integration, managing bidirectional power flows, and enabling grid flexibility and resilience.
- Demand response programs are crucial in smart grid research, focusing on intelligent mechanisms and real-time data to optimize energy consumption, reduce peak demand, and promote grid stability through effective strategies.
- Smart grids require robust security protocols, intrusion detection systems, and data privacy frameworks to protect infrastructure from cyber threats and ensure consumer privacy. Research has been conducted to address these concerns in communication networks and data exchange.
- Data analytics and AI techniques are being utilized in smart grid technology to derive actionable insights for load forecasting, fault detection, asset management, and decision-making processes. Researchers explore machine learning algorithms, optimization techniques, and predictive analytics to improve efficiency and effectiveness.
- Research focuses on enhancing grid resilience and exploring microgrids, self-contained energy systems that can operate independently or in coordination with the main grid. Studies investigate their design, control, and operation, including integration with the main grid during normal and emergency conditions.

Research on smart grid technology covers modernization, advanced metering, distributed energy resource integration, demand response, cybersecurity, data analytics, and microgrids. It aims to overcome challenges, optimize performance, and create a sustainable, efficient energy future. Energy monitoring and control are crucial in smart grids, offering numerous benefits and advantages (Dalipi & Yayilgan, 2016; Olbrich, 2016).

- Energy monitoring helps utilities and consumers analyze real-time energy consumption patterns, enabling them to identify high usage areas, optimize consumption, and implement energy-saving measures. This information informs decisions on load management, peak shaving, and energy conservation, resulting in improved efficiency and cost savings.
- Energy monitoring systems offer valuable data on grid conditions, voltage levels, power quality, and demand-supply dynamics. These insights enable grid operators to monitor and manage performance, detect anomalies, and respond promptly to issues like voltage fluctuations, equipment failures, or congestion. This ensures uninterrupted power supply and minimizes blackout risks.
- Real-time energy monitoring enables demand response programs, allowing consumers to adjust energy consumption based on grid conditions and price signals. Grid operators can implement load management strategies, balance supply and demand, and reduce peak loads, optimizing grid operations, avoiding overloading, and avoiding infrastructure upgrades.
- Renewable energy sources like solar and wind increase variability and intermittency in energy generation. Energy monitoring enables grid operators to monitor generation levels, predict output fluctuations, and manage integration. This promotes sustainability and reduces reliance on fossil fuels, ensuring smooth integration and sustainable energy use (Boopathi, Kumar, et al., 2023; Boopathi & Davim, 2023; Dhanya et al., 2023; Hanumanthakari et al., 2023; Janardhana et al., 2023; Maguluri et al., 2023).

- Data-driven decision making in energy monitoring systems transforms vast amounts of data into valuable insights for utilities, grid operators, and consumers. This enables proactive maintenance, predictive analytics, accurate load forecasting, energy-saving opportunities identification, and demand-side management programs tailored to consumer needs.

IoT enhances smart grid capabilities and enabling remote monitoring and control (Hu et al., 2017; Saveen et al., 2018; Thakare et al., 2017).

- IoT devices and sensor networks enable real-time data collection from grid components like meters, sensors, transformers, and substations. This data captures energy consumption, grid parameters, equipment status, and environmental conditions, enabling comprehensive monitoring and situational awareness.
- IoT enables robust communication networks connecting smart devices, enabling seamless data exchange between grid assets, control systems, and stakeholders. This technology improves grid efficiency, remote monitoring, and decision-making, ensuring reliable and secure data transmission (Anitha et al., 2023; Vanitha, Radhika, et al., 2023; Venkateswaran, Vidhya, Naik, et al., 2023; Zekrifa et al., 2023).
- IoT integrates intelligent control systems in smart grids, optimizing operations, managing energy resources, and responding to events. These systems enable automated demand response, dynamic load balancing, fault detection, and self-healing, enhancing grid stability and efficiency.
- IoT-generated data and advanced analytics enable meaningful insights from grid data, identifying consumption patterns, anomalies, demand fluctuations, and optimizing energy management strategies. This data-driven intelligence enables proactive decision-making, efficient resource allocation, and predictive maintenance.
- IoT empowers consumers to participate in energy management through smart meters, devices, and user-friendly interfaces. This promotes energy conservation, load flexibility, and demand response, contributing to a more sustainable and efficient grid.

IoT technology significantly impacts smart grids by enabling real-time data collection, advanced communication infrastructure, intelligent control systems, data analytics, and consumer empowerment. This integration improves efficiency, reliability, and sustainability in energy systems.

MICROGRID IMPLEMENTATION PROCEDURE

The Microgrid Implementation Procedure is explained below (Ghasempour, 2019; Hartono et al., 2013; Katyara et al., 2019; Li et al., 2018; Olbrich, 2016).

- Conduct a feasibility study to evaluate microgrid implementation's viability and potential benefits, considering energy requirements, energy sources, grid connectivity, load profiles, and economic factors.
- Design a microgrid system based on project requirements, determining size, energy sources, and necessary control and monitoring systems.

- Select appropriate microgrid components based on efficiency, reliability, scalability, and compatibility with the architecture, considering generation sources, energy storage systems, and control systems.
- Integrate components for a cohesive microgrid system, ensuring proper electrical connections, protection mechanisms, and communication interfaces.
- Implement a microgrid control and management system to monitor, control, and coordinate energy generation, storage, and distribution, while ensuring load management, frequency and voltage control, and seamless transitions between grid-connected and islanded modes.
- Conduct comprehensive testing and commissioning of microgrid system to ensure proper functioning, adherence to design specifications, and verify component performance under various operating conditions.
- Ensure microgrid compliance with grid interconnection standards and regulations by obtaining necessary approvals, permits, and certifications from regulatory authorities.
- Establish a microgrid operation and maintenance plan, regularly monitor components, perform upgrades, and address issues promptly.

Microgrid components (Chandak & Rout, 2021; Ferahtia et al., 2022; Hartono et al., 2013; Saveen et al., 2018) work together to efficiently generate, store, and distribute energy, with primary components providing essential functions (Figure 1).

- Energy Generation Sources: Solar panels, wind turbines, and diesel generators generate electrical energy for microgrids, ranging from renewable to conventional, depending on project objectives and resources.
- Energy Storage Systems: Microgrid energy storage systems, like batteries or flywheels, store excess energy for future use, balancing supply and demand fluctuations and enhancing resilience.
- Power Converters/Inverters: Converters and inverters convert energy from DC to AC or vice versa for microgrids, ensuring compatibility with various energy sources and electrical infrastructure.
- Transformers: Transformers regulate voltage levels in microgrids, ensuring efficient transmission and distribution of electricity among components and loads.
- Protective Devices: Circuit breakers and relays are crucial for microgrid safety, reliability, fault detection, equipment protection, and system stability.
- Control and Monitoring Systems: Microgrid systems control and optimize energy flow, storage, load demand, and component operation by monitoring generation, storage levels, and load demand.

Microgrid focuses on efficient, sustainable energy solutions through localized energy generation, storage, and consumption using renewable or conventional sources like solar panels and wind turbines (Chandak & Rout, 2021; Hartono et al., 2013).

- Microgrid generates excess energy, stored in batteries, for high demand or insufficient primary energy sources during storage systems.
- Power converters and inverters ensure energy source compatibility and synchronization in microgrids, converting DC power from storage systems or renewables into AC for electrical appliances and equipment.

Figure 1. Micro-grid components

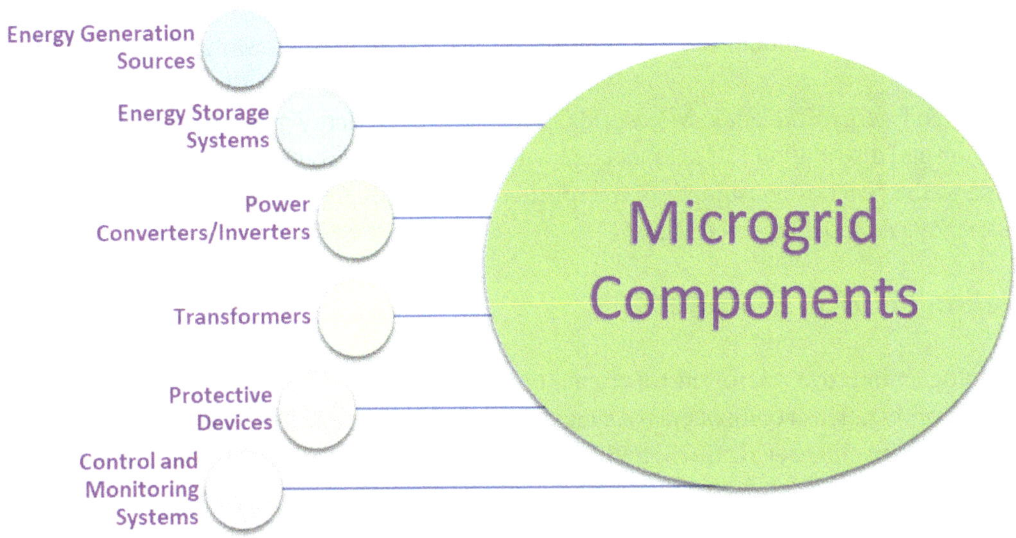

- Microgrid's load management system monitors energy demand, optimizes distribution, prioritizes critical loads, and balances energy supply, ensuring efficient and reliable power supply.
- Grid connectivity enables microgrids to exchange energy with the main electrical grid, enabling efficient resource utilization and grid stability by importing or exporting excess energy when demand exceeds generation capacity.
- Islanded operation allows microgrids to operate independently during grid outages or intentional islanding, using their own energy sources and storage systems. The control system balances generation and demand, ensuring a reliable, continuous power supply for critical loads.

IOT-INTEGRATED MICROGRID SYSTEM

Microgrids in Smart Grids

Microgrids are localized energy systems that operate independently or in coordination with the main grid, utilizing distributed energy resources like solar panels, wind turbines, energy storage, and controllable loads (Barman et al., 2019; Paul Divakar et al., 2020).

- Increased Resilience: Microgrids provide localized power generation and distribution, making them more resilient to grid disruptions and outages. In the event of a main grid failure, microgrids can continue to supply power to critical loads, such as hospitals, emergency services, or critical infrastructure, ensuring uninterrupted electricity supply.

- Integration of Renewable Energy: Microgrids facilitate the integration of renewable energy sources into the grid. They can efficiently harness solar or wind energy within a local area, reducing reliance on fossil fuels and decreasing greenhouse gas emissions. Microgrids enable the seamless integration of intermittent renewable resources by managing their variability and ensuring grid stability.
- Improved Energy Efficiency: Microgrids enable localized energy generation, which reduces transmission and distribution losses associated with long-distance power delivery. By generating power close to the point of consumption, microgrids minimize energy losses, improving overall energy efficiency and reducing energy costs.

Integration of IoT in Microgrids

IoT technologies improve microgrid functionality and performance by enabling real-time monitoring, control, and optimization of components (Avancini et al., 2021; Boopathi, Arigela, et al., 2023; Domakonda et al., 2023; Jeevanantham et al., 2023; Sathish et al., 2023; Subha et al., 2023; Vanitha, K., et al., 2023).

- Sensor Networks and Data Collection: IoT devices and sensors deployed in microgrids collect data on energy generation, consumption, grid parameters, and environmental conditions. These sensors provide real-time data, allowing operators to monitor and manage the microgrid's performance, detect anomalies, and optimize energy resources.
- Communication and Connectivity: IoT enables seamless communication and connectivity between microgrid components, such as DERs, energy storage systems, and control systems. This facilitates efficient coordination and control of distributed energy resources, load management, and grid operation in real-time (Hanumanthakari et al., 2023; Harikaran et al., 2023; Hema et al., 2023; Janardhana et al., 2023).
- Intelligent Control and Optimization: IoT integration enables intelligent control and optimization of microgrid operations. By leveraging data from IoT devices, machine learning algorithms, and advanced control strategies, microgrid operators can optimize energy generation, storage, and consumption, ensuring efficient utilization of resources and cost savings.

Benefits and Challenges of IoT-Integrated Microgrids

- Enhanced Efficiency and Reliability: IoT integration enables real-time monitoring, control, and optimization, leading to improved energy efficiency, reduced energy wastage, and enhanced grid reliability. Operators can identify and address issues promptly, optimize energy management strategies, and adapt to changing grid conditions in real-time.
- Increased Grid Flexibility: IoT integration enhances the flexibility of microgrids by enabling dynamic load management, demand response, and seamless integration of renewable resources. Microgrids can actively respond to grid conditions, adjust energy generation and consumption based on demand and supply fluctuations, and actively participate in grid-wide energy management programs.

- Cost Savings and Economic Benefits: IoT-integrated microgrids can optimize energy generation and consumption, reducing energy costs for consumers. By leveraging energy storage systems and intelligently managing energy resources, microgrids can potentially reduce peak demand charges, optimize energy procurement, and participate in energy markets, leading to cost savings and economic benefits.
- IoT-integrated microgrids face challenges in data security, privacy, scalability, interoperability, and complexity. Data security and privacy are crucial, and robust cybersecurity measures are essential. Scaling up microgrids requires standardization and compatibility between different technologies. Complexity and management are also essential, as IoT-integrated microgrids involve multiple components, data streams, and control systems. Adequate expertise, monitoring systems, and management strategies are necessary for efficient operation and maintenance.

IoT integration in microgrids offers resilience, renewable energy integration, energy efficiency, and optimized grid operation. However, challenges like data security, scalability, and complexity must be addressed for successful implementation in smart grid systems (Dalipi & Yayilgan, 2016; Thakare et al., 2017).

SYSTEM IMPLEMENTATION FOR ENERGY MONITORING AND CONTROL

Hardware Components and Architecture

Implementing energy monitoring and control systems require hardware components and architecture, focusing on key aspects (Hang & Kim, 2018; Pawar et al., 2020; Saleem et al., 2021). The system implementation for energy monitoring and control is illustrated in Figure 2.

- Hardware Components: Smart meters, sensors, communication devices, data loggers, control devices, and computing infrastructure are essential hardware components for real-time energy measurement and data collection. Sensors monitor parameters like voltage, current, power quality, temperature, and environmental conditions, while communication devices facilitate data transmission. Data loggers store and record collected data, while control devices enable remote control and automation of energy systems.
- System Architecture: System architecture involves hardware components arrangement, interconnection, data flow, communication pathways, and control mechanisms for seamless energy monitoring and control. It may involve centralized or decentralized approaches depending on specific requirements.

Sensor Selection and Placement

Selecting and placing sensors is crucial for accurate energy monitoring and control, considering factors like size, shape, and placement (Padmanaban et al., 2019; Premkumar & Sigappi, 2021; Zhang et al., 2021).

Figure 2. System implementation for energy monitoring and control

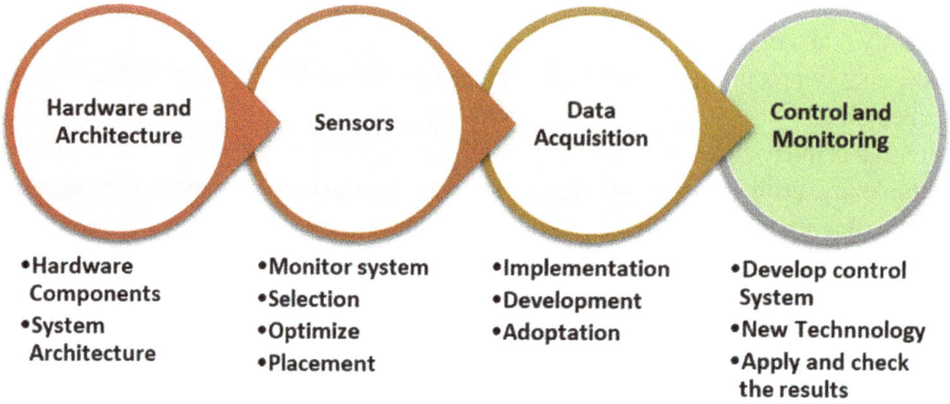

- Monitor system parameters like voltage, current, power consumption, quality, temperature, humidity, and other relevant factors based on requirements.
- Select sensors with high accuracy and reliability, considering factors like measurement range, resolution, calibration, and longevity for data integrity.
- Optimize sensor placement for representative data and comprehensive energy system view by considering accessibility, proximity to target parameters, and environmental factors affecting sensor performance.
- Incorporate redundancy in sensor placement for data reliability and fault tolerance, using duplicate sensors for backup measurements and identifying failures.

Data Acquisition and Communication Infrastructure

Efficient data acquisition and communication infrastructure are crucial for seamless data exchange.

- Implement a data acquisition system for collecting and processing data from sensors and smart meters using loggers, gateways, or edge computing devices.
- Develop a robust communication infrastructure, including wired or wireless technologies like Ethernet, Wi-Fi, Zigbee, or cellular networks, to ensure secure and reliable data transfer between system components.
- Adopt standardized protocols and communication standards for interoperability, compatibility, and seamless integration of devices and components in energy monitoring and control systems.

Control and Monitoring Systems

Control and monitoring systems manage and regulate energy resources using collected data.

- Develop control algorithms and strategies to optimize energy consumption, demand response, load balancing, and other control actions using advanced techniques like model predictive control, fuzzy logic, or machine learning.

- Create a user-friendly interface for system monitoring and control, enabling visualization, real-time data control, and informed decision-making.
- Implement alert and notification systems to promptly alert operators and users of abnormal conditions, equipment failures, or energy-related events.
- Design control and monitoring systems for easy integration with existing infrastructure and scalability for future expansions or modifications.

Implementing hardware components, sensor selection, data acquisition, communication infrastructure, and control systems for energy monitoring and control in smart grid environments effectively captures data and enables efficient resource management.

ANFIS FOR ENERGY MONITORING

Adaptive Neuro-Fuzzy Inference System (ANFIS) is a hybrid intelligent system that uses artificial neural networks (ANNs) and fuzzy logic to create a computational model. ANNs learn from data and capture complex patterns, while fuzzy logic handles linguistic variables and uncertainty reasoning. This system effectively models and analyzes complex relationships between input and output variables (Kampouropoulos et al., 2018; Suganthi et al., 2015; Suhail et al., 2021). In the field of energy monitoring, ANFIS has gained popularity due to its ability to handle the inherent complexity of energy systems. Energy consumption patterns often exhibit nonlinear behavior and are influenced by various factors such as weather conditions, occupancy, and time of day. Traditional modeling techniques may struggle to capture the intricate relationships between these variables. However, ANFIS excels in handling such complexities by providing a flexible and adaptive modeling framework (Kaur & Sood, 2019).

ANFIS consists of a layered structure that includes both fuzzy inference systems and neural networks. The fuzzy inference system is responsible for capturing linguistic rules and providing a fuzzy representation of the input-output relationships. Fuzzy sets and membership functions are used to define the linguistic variables and their relationships. The neural network component of ANFIS performs the task of parameter estimation and learning. It learns from the input-output data and adjusts the parameters of the fuzzy inference system to optimize the model's performance.

The training process of ANFIS involves two key steps: structure identification and parameter optimization. In the structure identification step, clustering algorithms are typically employed to determine the optimal number and configuration of fuzzy sets based on the input data. The clustering process identifies the appropriate number of fuzzy rules and determines their centers and spreads. Once the structure is identified, parameter optimization techniques such as least squares estimation or gradient descent are used to fine-tune the parameters of the ANFIS model (Khan & Rizwan, 2021a).

The resulting ANFIS model can accurately represent the complex relationships between input and output variables in energy monitoring applications. It can effectively capture the nonlinear behavior of energy consumption patterns and make accurate predictions based on the input variables. ANFIS models can be utilized for load forecasting, energy optimization, anomaly detection, and other energy management tasks.

ANFIS is a computational model that combines ANNs and fuzzy logic, making it ideal for energy monitoring applications. It handles complex relationships, enables accurate predictions, optimizes usage, and supports efficient decision-making in smart grid environments. ANFIS is a powerful tool for developing energy monitoring models that accurately predict energy consumption and parameters. It is used for load forecasting, optimization, and demand response in various energy management tasks (Khan & Rizwan, 2021b).

ANFIS models can accurately forecast future energy demand using historical energy consumption data and input variables like weather conditions and load patterns. This enables utilities and grid operators to efficiently plan, allocate resources, optimize energy generation and distribution, and ensure grid stability.

ANFIS models optimize energy usage in buildings, industrial processes, and smart grids by considering occupancy patterns, weather conditions, and equipment efficiency. This leads to energy savings, reduced peak loads, and improved efficiency, ultimately resulting in cost savings and environmental benefits.

ANFIS models are essential in demand response programs, managing energy consumption based on grid conditions and price signals. By analyzing historical data and real-time pricing information, utilities can predict consumer responses, incentivize peak usage reduction, balance supply and demand, and improve grid stability.

ANFIS models use historical data and input variables to predict energy consumption, load patterns, optimize usage, and support demand response initiatives. This insights enable efficient resource planning, improved grid operation, and enhanced energy efficiency, contributing to smarter, sustainable energy systems.

ANFIS Model Training and Optimization Techniques

ANFIS model training and optimization involve techniques for accurate predictions, using key steps and methods (Adedeji et al., 2022; Kampouropoulos et al., 2018; Khan & Rizwan, 2021a).

- Collect a representative dataset of historical energy consumption data and input variables like weather conditions, time, and occupancy patterns to capture system variability and cover various scenarios.
- ANFIS models utilize fuzzy logic to generate fuzzy rules, defining fuzzy sets and membership functions. Clustering techniques like subtractive, grid partitioning, or k-means determine the optimal number and shape of fuzzy sets based on the dataset.
- Parameter estimation is crucial for optimizing the ANFIS model after establishing fuzzy rules and membership functions. This involves determining centers, spreads, and weights to achieve the best fit with training data. Optimization techniques like least squares estimation, gradient descent, and genetic algorithms aim to minimize the difference between the model's predicted output and the actual output in the training dataset (Boopathi, 2022g, 2022a, 2022f, 2023b, 2023a, 2023c).
- Cross-validation is a technique used to evaluate the trained ANFIS model's generalization capability. It involves dividing the dataset into training and validation sets, training the model using the training set and testing against the validation set to detect overfitting and provide a more realistic estimate of its performance (Boopathi, 2022b, 2022d, 2022c, 2022e).
- Optimizing the ANFIS model involves fine-tuning parameters, adjusting fuzzy sets, and exploring different learning algorithms to improve accuracy and efficiency. The goal is to find the optimal configuration with the best performance metrics, such as MAE, RMSE, or R-squared.
- Validation and testing of the ANFIS model using independent datasets after training ensures its performance is not biased and provides a realistic assessment of predictive capabilities. Validation also assesses the model's robustness and generalization ability in handling new data.

Performance Evaluation and Results

Performance evaluation is crucial for assessing ANFIS-based energy monitoring models' accuracy and reliability using metrics like energy consumption prediction and target variables (Ghenai et al., 2022; Padmanaban et al., 2019).

- **Mean Absolute Error (MAE):** MAE measures the average absolute difference between predicted and actual values, indicating model accuracy; lower MAE indicates better performance.
- **Root Mean Square Error (RMSE):** RMSE measures the average squared difference between predicted and actual values, giving more weight to larger errors and indicating better accuracy.
- **Coefficient of Determination (R-squared):** R-squared measures the proportion of variance in energy consumption explained by independent variables, with a 1 value indicating higher correlation and predictive power.

ANFIS models' performance can be assessed through quantitative metrics and comparative analysis, comparing them with other prediction techniques or benchmarks. Performance evaluation should be conducted using independent datasets, ensuring fair assessment of generalization and handling unseen data. Graphical visualization techniques, such as line graphs or scatter plots, can help visualize predicted values and capture trends, patterns, and anomalies in energy consumption data (Boopathi, Pandey, et al., 2023; Boopathi & Kanike, 2023; Chandrika et al., 2023; Dhanya et al., 2023). Assessing ANFIS-based energy monitoring models using metrics and comparative analysis helps stakeholders evaluate effectiveness, identify improvement areas, and make informed decisions on smart grid energy management strategies (Mellit & Kalogirou, 2011; Mohammad Bilal, 2022).

- Energy Consumption Patterns: ANFIS models can accurately capture energy consumption patterns, allowing for a better understanding of energy usage trends over time. By analyzing the model's predictions, energy managers can identify peak demand periods, seasonal variations, and consumption patterns associated with specific activities or events.
- Peak Load Predictions: ANFIS models can forecast peak load periods, enabling utilities and grid operators to optimize energy generation and distribution resources accordingly. Accurate peak load predictions assist in reducing the strain on the grid during high-demand periods and implementing demand response strategies effectively.
- Energy Efficiency Improvements: ANFIS models help identify opportunities for energy efficiency improvements. By analyzing the relationships between input variables and energy consumption, the models can highlight factors that significantly impact energy usage. This information assists in identifying areas for energy-saving measures, equipment upgrades, and load management strategies to improve overall energy efficiency.
- Demand Response Strategies: ANFIS models play a crucial role in demand response programs by providing insights into consumer response patterns to pricing signals or incentive programs. By analyzing historical data and input variables, ANFIS can predict how consumers are likely to respond to specific demand response initiatives. This information aids in designing effective demand response strategies and optimizing energy consumption during peak periods.

- Validation and Testing: Validation and testing of ANFIS models assess accuracy and reliability against independent datasets. These evaluations evaluate the model's generalization to unseen situations and ensure its performance is not limited to training data, providing accurate predictions for new scenarios.
- Sensitivity Analysis: Sensitivity analysis evaluates the impact of input variables on a model's predictions, helping energy managers identify significant factors affecting energy consumption. This helps prioritize resources and interventions for effective energy management.

ANFIS-based energy monitoring models offer valuable insights into energy consumption patterns, peak load predictions, efficiency improvements, and demand response strategies. Validation, testing, and sensitivity analysis improve models' accuracy and reliability, aiding energy management decision-making and fostering smarter, sustainable energy systems in smart grids.

Fuzzy Fruit Fly Optimization (FFO) Algorithm for IoT-Integrated Smart Grids

The Fuzzy Fruit Fly Optimization (FFO) algorithm is a nature-inspired optimization technique that combines fuzzy logic principles with fruit flies' behavior to solve optimization problems. It is ideal for IoT-integrated smart grids, where efficient control and optimization are crucial for managing energy resources. The FFO algorithm uses fuzzy logic to represent uncertainty and imprecision in smart grid control problems, offering a robust and adaptive approach for optimizing various aspects of smart grid operations (Ghenai et al., 2022).

The FFO algorithm is a population-based approach used in smart grid control to address challenges like load balancing, demand response, renewable energy integration, and power distribution optimization. It simulates the foraging behavior of fruit flies to find optimal control strategies that minimize energy consumption, maximize renewable energy utilization, and improve grid efficiency. The algorithm operates in iterations, maintaining a population of candidate solutions called fruit flies, which are evaluated using an objective function that captures the optimization goals of the smart grid control problem (Li et al., 2018).

The FFO algorithm uses fuzzy logic to manage uncertainty and imprecision in smart grid control variables. It uses fuzzy rules and membership functions to adapt to changing grid conditions. Through an iterative process, it converges towards optimal control solutions by dynamically adjusting population size, exploration parameters, and fuzzy rules (Hu et al., 2017). The Fuzzy Fruit Fly Optimization (FFO) algorithm offers several advantages in smart grid control. It provides a flexible, adaptive framework for handling complex systems, handles uncertainties and vagueness in control variables, and allows parallel processing and exploration of multiple control strategies. This population-based approach enhances efficiency and scalability in IoT-integrated systems, improving energy management, grid efficiency, and renewable energy resource integration.

FFO-Based Optimization of Energy Management Strategies

FFO-based optimization techniques improve energy management strategies in IoT-integrated smart grids by optimizing energy generation, distribution, and consumption, benefiting key areas like generation, distribution, and consumption (Hu et al., 2017; Li et al., 2018).

- Load Balancing: FFO optimizes load balancing in smart grids by dynamically allocating energy loads among sources and consumers, considering factors like demand, supply availability, and grid constraints, minimizing imbalances and ensuring efficient energy utilization.
- Demand Response: FFO-based optimization enhances demand response strategies by managing energy consumption based on grid conditions and pricing signals. It analyzes historical data, consumer behavior, and real-time pricing information to determine effective actions, such as shifting loads, adjusting consumption patterns, and incentivizing consumption reduction during peak periods.
- Renewable Energy Integration: FFO optimizes renewable energy integration in smart grid systems by considering factors like weather conditions, energy storage capacity, and demand patterns, maximizing clean and sustainable energy utilization while ensuring grid stability and reliability.
- Power Distribution Optimization: FFO-based optimization techniques improve power distribution in smart grids by considering line losses, voltage stability, and load demands, enhancing efficiency, minimizing losses, and reducing system failures.
- Energy Efficiency Improvements: FFO optimizes energy efficiency by analyzing consumption patterns, considering occupancy, weather, and equipment efficiency, recommending energy-saving measures, optimizing scheduling, and identifying load shedding or shifting opportunities.

FFO-based optimization in energy management strategies offers advantages such as adaptability, robustness, population-based approach, and integration of fuzzy logic techniques. This approach handles uncertainties and dynamic conditions in smart grids, allowing faster convergence and improved scalability. By utilizing FFO-based optimization techniques, smart grid systems can achieve improved energy efficiency, renewable energy utilization, grid stability, and reduced operational costs, contributing to the development of sustainable and intelligent energy systems in IoT-integrated smart grids.

Case Studies

Case studies demonstrate the effectiveness and benefits of FFO-based optimization techniques in energy management strategies, with examples of successful application in real-world scenarios (Kannan et al., 2020; Khalid & Javaid, 2020; Li et al., 2018; Premkumar & Sigappi, 2021).

- Load Balancing in a Microgrid: In a case study involving a microgrid with distributed energy resources, FFO was employed to optimize load balancing. The FFO algorithm considered factors such as energy demand, renewable energy generation, and battery storage capacity. The results showed significant improvements in load balancing, reducing grid imbalances and optimizing energy utilization across the microgrid.
- Demand Response Optimization: A case study focused on demand response optimization using FFO in a residential setting. Historical energy consumption data, weather conditions, and pricing signals were used as inputs to the FFO algorithm. The results demonstrated effective demand response strategies that incentivized consumers to reduce energy consumption during peak periods, resulting in reduced overall grid demand and improved grid stability.
- Renewable Energy Integration: FFO-based optimization was applied to optimize the integration of renewable energy sources into a smart grid system. The FFO algorithm considered factors such as solar and wind energy availability, demand patterns, and storage capacity. The results show-

cased improved utilization of renewable energy, reduced reliance on conventional energy sources, and enhanced grid resilience.

- Power Distribution Optimization: FFO was employed to optimize power distribution in a distribution network with multiple substations. The FFO algorithm considered parameters such as line losses, voltage stability, and load demands. The results demonstrated optimized power flow, reduced losses, and improved voltage stability, leading to increased grid efficiency and reduced operational costs.

Case studies demonstrate the effectiveness of FFO-based optimization techniques in improving energy management strategies. These techniques enhance load balancing, demand response, renewable energy integration, and power distribution optimization. However, specific results may vary depending on the system, data, and optimization objectives. Conducting case studies and performance evaluations is crucial for a comprehensive understanding of the benefits and effectiveness of FFO-based optimization techniques in energy management strategies.

INTEGRATION OF ANFIS AND FFO ALGORITHMS FOR ENERGY MONITORING AND CONTROL

The ANFIS-FFO hybrid system combines ANFIS's modeling capabilities with FFO's optimization capabilities for energy monitoring and control in smart grids. It leverages the strengths of both techniques, offering a powerful approach for smart grid energy management (Ghenai et al., 2022; Hu et al., 2017; Kannan et al., 2020; Mellit & Kalogirou, 2011). The ANFIS-FFO hybrid system combines ANFIS models for energy monitoring and control, while FFO optimizes ANFIS models for parameter fine-tuning and performance improvement. This comprehensive framework combines modeling and optimization capabilities, enhancing energy monitoring and control. ANFIS-FFO hybrid system architecture are depicted in Figure 3.

System Architecture and Implementation

ANFIS-FFO hybrid system architecture comprises components:

- Data Acquisition: The system collects energy-related data from various sources, such as smart meters, sensors, and weather stations. This data serves as input for the ANFIS models and optimization process.
- ANFIS Modeling: ANFIS models are developed to capture the relationships between input variables (e.g., weather conditions, time of day, occupancy patterns) and output variables (e.g., energy consumption, load demand). The ANFIS models use fuzzy logic to handle uncertainty and linguistic variables.
- FFO Optimization: The FFO algorithm is applied to optimize the ANFIS models. FFO utilizes its population-based optimization approach to search for optimal parameter values for the ANFIS models. By fine-tuning the parameters, FFO aims to improve the accuracy and performance of the ANFIS models.

Figure 3. ANFIS-FFO hybrid system architecture

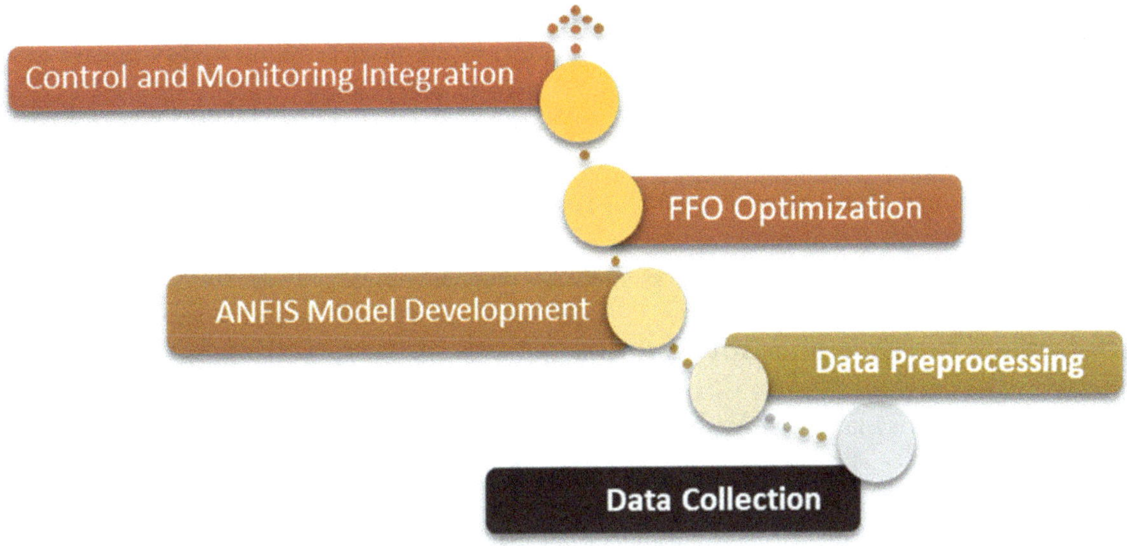

- Training and Optimization: The ANFIS models are trained and optimized using historical data. The FFO algorithm iteratively adjusts the parameters of the ANFIS models to minimize the objective function, such as mean absolute error (MAE) or root mean square error (RMSE), and enhance the accuracy of energy predictions.
- Control and Monitoring: The optimized ANFIS models are deployed in the control and monitoring system of the smart grid. They provide real-time predictions and insights into energy consumption, load demand, and other relevant parameters. This information is used to optimize energy management decisions, load balancing, demand response, and other control strategies.

The ANFIS-FFO hybrid system combines ANFIS modeling and FFO optimization, offering a comprehensive approach to energy monitoring and control in smart grids. It improves energy predictions accuracy, optimizes control strategies, and enhances overall energy management. Access to relevant data, modeling techniques, and computational resources is essential for training and optimization.

Experimental Setup and Evaluation

A comprehensive experimental setup is needed to evaluate the effectiveness of the ANFIS-FFO hybrid system for energy monitoring and control (Ghenai et al., 2022; Hu et al., 2017; Li et al., 2018; Mellit & Kalogirou, 2011).

a. Data Collection: Gather historical energy consumption data, weather data, occupancy patterns, and other relevant variables from smart meters, sensors, and other sources in the smart grid. This data will serve as the input for the ANFIS models and FFO optimization.

b. Data Preprocessing: Clean and preprocess the collected data to handle missing values, outliers, and normalization. Ensuring high-quality data is essential for accurate modeling and optimization.
c. ANFIS Model Development: Develop ANFIS models to capture the relationships between the input variables and energy consumption or other target variables. Train the ANFIS models using a portion of the collected data and validate them using another subset.
d. FFO Optimization: Apply the FFO algorithm to optimize the parameters of the ANFIS models. Utilize the validation data to fine-tune the ANFIS models and improve their performance.
    1. Control and Monitoring Integration: Integrate the optimized ANFIS models into the smart grid's control and monitoring system. This enables real-time predictions and monitoring of energy consumption, load demand, and other parameters.

Performance Analysis and Comparison

Conduct a performance analysis and comparison of the ANFIS-FFO hybrid system.

a) Metrics Selection: Choose appropriate performance metrics, such as mean absolute error (MAE), root mean square error (RMSE), coefficient of determination (R-squared), or others, to assess the accuracy and reliability of the ANFIS-FFO predictions.
b) Performance Evaluation: Evaluate the ANFIS-FFO hybrid system's performance using an independent test dataset that was not used during training and optimization. Calculate the selected metrics to quantify the accuracy of energy predictions and other control variables.
c) Comparative Analysis: Perform a comparative analysis to compare the performance of the ANFIS-FFO hybrid system with other energy monitoring and control methods. This may include comparing against traditional ANFIS models, standalone FFO optimization, or other prediction techniques.
d) Sensitivity Analysis: Conduct sensitivity analysis to evaluate the impact of different input variables on the ANFIS-FFO predictions. Identify the most significant factors affecting energy consumption and load demand.
e) Real-World Scenario Testing: If possible, validate the ANFIS-FFO hybrid system's performance in a real-world smart grid environment. Implement the system and observe its effectiveness in real-time energy monitoring and control.
f) Robustness and Scalability: Assess the robustness and scalability of the ANFIS-FFO hybrid system to handle varying data sizes, different smart grid configurations, and changing environmental conditions.

The ANFIS-FFO hybrid system's performance analysis and comparison showcase its advantages in energy monitoring and control, enhancing energy management and optimizing smart grid operations.

Case Study: ANFIS-FFO Hybrid System for Energy Monitoring and Control in a Smart Grid

This case study evaluates the effectiveness of the ANFIS-FFO hybrid system for energy monitoring and control in smart grid environments, evaluating energy predictions accuracy, control strategy optimization, and overall performance compared to traditional methods (Ghenai et al., 2022; Hu et al., 2017; Li et al., 2018; Mellit & Kalogirou, 2011).

Experimental Setup

- Data collection involves collecting historical energy consumption, weather conditions, occupancy patterns from smart meters and sensors in a real-world smart grid.
- Data preprocessing handles missing values, outliers, and normalization for consistent analysis.
- ANFIS models predict energy consumption using input variables like time, temperature, and occupancy, using fuzzy logic to handle uncertainties.
- FFO Optimization optimizes ANFIS models' parameters, enhancing accuracy and performance by fine-tuning parameters using Fuzzy Fruit Fly.
- Optimized ANFIS models integrated into smart grid control and monitoring system for real-time energy predictions.

Performance Evaluation

- Energy prediction accuracy evaluated using MAE, RMSE, and R-squared metrics for accuracy evaluation.
- ANFIS-FFO hybrid system's performance compared to traditional models, standalone optimization, and energy prediction techniques.
- Sensitivity analysis evaluates input variables' impact on ANFIS-FFO predictions, identifying influential factors affecting energy consumption.
- Real-world scenario testing evaluates ANFIS-FFO hybrid system's performance in real-time smart grid energy monitoring and control.

RESULTS AND FINDINGS

The ANFIS-FFO hybrid system offers more accurate energy predictions than traditional models, thanks to FFO optimization, reducing MAE and RMSE values. It outperforms standalone FFO optimization and other prediction techniques. The sensitivity analysis reveals the impact of weather conditions and occupancy patterns on energy consumption, aiding load forecasting and demand response strategies. The ANFIS-FFO hybrid system optimizes energy distribution, load balancing, and demand response, enhancing grid stability and efficiency.

The case study highlights the benefits of integrating ANFIS and FFO algorithms for energy monitoring and control in smart grids. This hybrid system offers accurate predictions and optimization, enhancing energy management and efficiency. It handles complex relationships between input and output variables, making it a valuable tool for sustainable and intelligent energy systems.

ENERGY MONITORING AND CONTROL IN RESIDENTIAL MICROGRIDS

Industrial IoT-enabled energy management systems optimize energy consumption and operational efficiency in industrial settings by collecting, analyzing, and acting on real-time data from devices and sensors. These systems offer numerous benefits (Barman et al., 2019; Ghenai et al., 2022; Li et al., 2018; Paul Divakar et al., 2020; Saveen et al., 2018).

- Data Acquisition and Monitoring: IoT devices and sensors are deployed throughout the industrial facility to capture energy-related data such as electricity usage, temperature, pressure, and other relevant parameters. These devices continuously monitor energy consumption and provide real-time data for analysis.
- Data Analytics and Insights: The collected data is analyzed using advanced analytics techniques, including machine learning and data mining algorithms. This enables the identification of energy consumption patterns, energy efficiency opportunities, and potential areas for improvement.
- Energy Optimization and Control: IoT-enabled energy management systems facilitate real-time energy optimization by intelligently adjusting energy-consuming devices, machinery, and processes. They can automate energy-saving measures, such as load shedding, demand response, and peak shaving, based on real-time data and predefined energy management strategies (Saravanan et al., 2022; Syamala et al., 2023; Venkateswaran, Vidhya, Ayyannan, et al., 2023; Vennila et al., 2022).
- Predictive Maintenance: Industrial IoT devices can monitor the health and performance of equipment and machinery, enabling predictive maintenance. By identifying potential equipment failures in advance, organizations can reduce downtime, improve energy efficiency, and extend the lifespan of critical assets.
- Integration with Enterprise Systems: Industrial IoT-enabled energy management systems can be integrated with enterprise resource planning (ERP) systems, asset management systems, and other business software to provide a holistic view of energy consumption and optimize resource allocation.
- Energy Reporting and Compliance: IoT-enabled energy management systems facilitate accurate and automated energy reporting, enabling organizations to meet regulatory compliance requirements. Real-time energy data can be used to generate reports, track key performance indicators (KPIs), and demonstrate adherence to sustainability goals.

Smart grids revolutionize energy management and distribution in urban environments by incorporating IoT technologies, intelligent sensors, and digital communication systems (Hema et al., 2023; Karthik et al., 2023; Koshariya et al., 2023; Kumara et al., 2023; Maguluri et al., 2023). Key applications include optimizing energy efficiency, ensuring grid stability, and integrating renewable energy sources (Barman et al., 2019; Paul Divakar et al., 2020; Saveen et al., 2018).

- Demand Response: Smart grids enable demand response programs that incentivize energy consumers to adjust their energy usage during peak demand periods. By participating in demand response initiatives, consumers can reduce their energy consumption and contribute to grid stability.
- Distributed Energy Resources (DERs) Integration: Smart grids facilitate the integration of distributed energy resources such as rooftop solar panels, wind turbines, and battery storage systems. These resources can feed surplus energy back into the grid or store excess energy for later use.
- Grid Monitoring and Control: Smart grid technologies enable real-time monitoring and control of the grid infrastructure. Intelligent sensors and advanced analytics provide insights into grid performance, voltage levels, power quality, and outage detection. This helps utilities proactively address issues and minimize downtime.

- Electric Vehicle (EV) Charging Infrastructure: Smart grids support the deployment of EV charging infrastructure in urban areas. They enable intelligent charging stations that optimize energy usage, manage load distribution, and support grid-friendly charging based on demand and available renewable energy.
- Microgrids: Microgrids are small-scale power systems that can operate independently or in coordination with the main grid. Smart grid technologies enable the efficient integration and management of microgrids, providing localized energy generation, storage, and consumption.
- Energy Management and Billing: Smart grids facilitate accurate energy metering, remote monitoring, and automated billing processes. Consumers have access to real-time energy usage data, enabling them to make informed decisions about their energy consumption and manage their energy costs effectively.

Industrial IoT-enabled energy management systems optimize energy in industrial settings, while smart grids revolutionize urban energy management by integrating advanced technologies for sustainable distribution and consumption. These technologies contribute to energy efficiency, sustainability, and grid resilience in both industrial and urban contexts.

CONCLUSION

Smart grids optimize energy efficiency, sustainability, and reliable supply by integrating IoT technologies like sensor networks, data analytics, and intelligent algorithms. This real-time monitoring, control, and optimization of energy generation, distribution, and consumption promotes energy-saving behaviors and optimizes energy management. IoT enhances interoperability, scalability, and resilience, supporting efficient energy management and grid optimization. Microgrids enable localized energy generation, storage, and distribution, ensuring reliable power supply, reducing transmission losses, and improving grid resilience.

Adaptive Neuro-Fuzzy Inference System (ANFIS) and Fuzzy Fruit Fly Optimization (FFO) algorithms are utilized in energy monitoring and control for modeling, prediction, and optimization. These algorithms combine artificial neural networks and fuzzy logic to handle complex relationships, while FFO offers optimization capabilities for energy management strategies. Implementing hardware components, sensor networks, data acquisition systems, and control systems is crucial for efficient deployment in smart grids. Integrating ANFIS and FFO algorithms enhances optimization and decision-making, resulting in more efficient energy management strategies.

Energy monitoring and control in smart grids face challenges such as data security, interoperability, scalability, grid integration issues, grid resilience, cybersecurity, cost management, stakeholder engagement, and consumer awareness. Future research will focus on advanced analytics, AI technologies, distributed energy resources management, demand-side management, decentralized energy markets, grid-edge intelligence, and self-healing capabilities. By leveraging IoT technologies, advanced algorithms, and intelligent systems, smart grids can optimize energy consumption, reduce environmental impact, and create a more resilient and reliable energy infrastructure.

LIST OF ACRONYMS

IoT - Internet of Things
ANFIS - Adaptive Neuro-Fuzzy Inference System
FFO - Fuzzy Fruit Fly Optimization
MAE - Mean Absolute Error
RMSE - Root Mean Square Error
R-squared - Coefficient of Determination
AI - Artificial Intelligence
ANN - Artificial Neural Network
DC - Direct Current
AC - Alternating Current
DE - Distributed Energy
DER - Distributed Energy Resources
EV - Electric Vehicle
GDP - Gross Domestic Product
ICT - Information and Communication Technology
ISO - International Organization for Standardization
PV - Photovoltaic
SCADA - Supervisory Control and Data Acquisition
SDN - Software-Defined Networking
V2G - Vehicle-to-Grid

REFERENCES

Abo-Zahhad, M., Ahmed, S. M., Farrag, M., Ahmed, M. F. A., & Ali, A. (2016). Design and implementation of building energy monitoring and management system based on wireless sensor networks. *Proceedings - 2015 10th International Conference on Computer Engineering and Systems, ICCES 2015*, 230–233. 10.1109/ICCES.2015.7393051

Adedeji, P. A., Akinlabi, S., Madushele, N., & Olatunji, O. O. (2022). Hybrid adaptive neuro-fuzzy inference system (ANFIS) for a multi-campus university energy consumption forecast. *International Journal of Ambient Energy*, *43*(1), 1685–1694. doi:10.1080/01430750.2020.1719885

Anitha, C., Komala, C., Vivekanand, C. V., Lalitha, S., & Boopathi, S. (2023). Artificial Intelligence driven security model for Internet of Medical Things (IoMT). *IEEE Explore*, 1–7.

Avancini, D. B., Rodrigues, J. J. P. C., Rabêlo, R. A. L., Das, A. K., Kozlov, S., & Solic, P. (2021). A new IoT-based smart energy meter for smart grids. *International Journal of Energy Research*, *45*(1), 189–202. doi:10.1002/er.5177

Barman, B. K., Yadav, S. N., Kumar, S., & Gope, S. (2019). IOT Based Smart Energy Meter for Efficient Energy Utilization in Smart Grid. *2nd International Conference on Energy, Power and Environment: Towards Smart Technology, ICEPE 2018*, 1–5. 10.1109/EPETSG.2018.8658501

Boopathi, S. (2022a). An experimental investigation of Quench Polish Quench (QPQ) coating on AISI 4150 steel. *Engineering Research Express*, *4*(4), 45009. doi:10.1088/2631-8695/ac9ddd

Boopathi, S. (2022b). An Extensive Review on Sustainable Developments of Dry and Near-Dry Electrical Discharge Machining Processes. *Journal of Manufacturing Science and Engineering*, *144*(5), 50801. doi:10.1115/1.4052527

Boopathi, S. (2022c). An investigation on gas emission concentration and relative emission rate of the near-dry wire-cut electrical discharge machining process. *Environmental Science and Pollution Research International*, *29*(57), 86237–86246. doi:10.100711356-021-17658-1 PMID:34837614

Boopathi, S. (2022d). Cryogenically treated and untreated stainless steel grade 317 in sustainable wire electrical discharge machining process: A comparative study. *Environmental Science and Pollution Research International*, 1–10. doi:10.100711356-022-22843-x PMID:36057706

Boopathi, S. (2022e). Effects of Cryogenically-treated Stainless Steel on Eco-friendly Wire Electrical Discharge Machining Process. *Preprint*, 1–18.

Boopathi, S. (2022f). Experimental investigation and multi-objective optimization of cryogenic Friction-stir-welding of AA2014 and AZ31B alloys using MOORA technique. *Materials Today. Communications*, *33*, 104937. doi:10.1016/j.mtcomm.2022.104937

Boopathi, S. (2022g). Performance Improvement of Eco-Friendly Near-Dry Wire-Cut Electrical Discharge Machining Process Using Coconut Oil-Mist Dielectric Fluid. *Journal of Advanced Manufacturing Systems*. Advance online publication. doi:10.1142/S0219686723500178

Boopathi, S. (2023a). An Investigation on Friction Stir Processing of Aluminum Alloy-Boron Carbide Surface Composite. In *Materials Horizons: From Nature to Nanomaterials* (pp. 249–257). Springer. doi:10.1007/978-981-19-7146-4_14

Boopathi, S. (2023b). Deep Learning Techniques Applied for Automatic Sentence Generation. In Promoting Diversity, Equity, and Inclusion in Language Learning Environments (pp. 255–273). IGI Global. doi:10.4018/978-1-6684-3632-5.ch016

Boopathi, S. (2023c). *Impact analysis of ceramic tile powder aggregates on self-compacting concrete.* Engineering Research Express.

Boopathi, S., Arigela, S. H., Raman, R., Indhumathi, C., Kavitha, V., & Bhatt, B. C. (2023). Prominent Rule Control-based Internet of Things: Poultry Farm Management System. *IEEE Explore*, 1–6. doi:10.1109/ICPECTS56089.2022.10047039

Boopathi, S., & Davim, J. P. (2023). Applications of Nanoparticles in Various Manufacturing Processes. In *Sustainable Utilization of Nanoparticles and Nanofluids in Engineering Applications* (pp. 1–31). IGI Global. doi:10.4018/978-1-6684-9135-5.ch001

Boopathi, S., & Kanike, U. K. (2023). Applications of Artificial Intelligent and Machine Learning Techniques in Image Processing. In *Handbook of Research on Thrust Technologies' Effect on Image Processing* (pp. 151–173). IGI Global. doi:10.4018/978-1-6684-8618-4.ch010

Boopathi, S., Kumar, P. K. S., Meena, R. S., Sudhakar, M., & Associates. (2023). Sustainable Developments of Modern Soil-Less Agro-Cultivation Systems: Aquaponic Culture. In Human Agro-Energy Optimization for Business and Industry (pp. 69–87). IGI Global.

Boopathi, S., Pandey, B. K., & Pandey, D. (2023). Advances in Artificial Intelligence for Image Processing: Techniques, Applications, and Optimization. In Handbook of Research on Thrust Technologies' Effect on Image Processing (pp. 73–95). IGI Global.

Chandak, S., & Rout, P. K. (2021). The implementation framework of a microgrid: A review. *International Journal of Energy Research*, *45*(3), 3523–3547. doi:10.1002/er.6064

Chandrika, V., Sivakumar, A., Krishnan, T. S., Pradeep, J., Manikandan, S., & Boopathi, S. (2023). Theoretical Study on Power Distribution Systems for Electric Vehicles. In *Intelligent Engineering Applications and Applied Sciences for Sustainability* (pp. 1–19). IGI Global. doi:10.4018/979-8-3693-0044-2.ch001

Dalipi, F., & Yayilgan, S. Y. (2016). Security and privacy considerations for IoT application on smart grids: Survey and research challenges. *Proceedings - 2016 4th International Conference on Future Internet of Things and Cloud Workshops, W-FiCloud 2016*, 63–68. 10.1109/W-FiCloud.2016.28

Dhanya, D., Kumar, S. S., Thilagavathy, A., Prasad, D., & Boopathi, S. (2023). Data Analytics and Artificial Intelligence in the Circular Economy: Case Studies. In Intelligent Engineering Applications and Applied Sciences for Sustainability (pp. 40–58). IGI Global.

Domakonda, V. K., Farooq, S., Chinthamreddy, S., Puviarasi, R., Sudhakar, M., & Boopathi, S. (2023). Sustainable Developments of Hybrid Floating Solar Power Plants. In *Human Agro-Energy Optimization for Business and Industry* (pp. 148–167). IGI Global. doi:10.4018/978-1-6684-4118-3.ch008

Ferahtia, S., Djeroui, A., Rezk, H., Houari, A., Zeghlache, S., & Machmoum, M. (2022). Optimal control and implementation of energy management strategy for a DC microgrid. *Energy*, *238*, 121777. doi:10.1016/j.energy.2021.121777

Ghasempour, A. (2019). Internet of things in smart grid: Architecture, applications, services, key technologies, and challenges. *Inventions (Basel, Switzerland)*, *4*(1), 22. doi:10.3390/inventions4010022

Ghenai, C., Al-Mufti, O. A. A., Al-Isawi, O. A. M., Amirah, L. H. L., & Merabet, A. (2022). Short-term building electrical load forecasting using adaptive neuro-fuzzy inference system (ANFIS). *Journal of Building Engineering*, *52*, 104323. doi:10.1016/j.jobe.2022.104323

Hang, L., & Kim, D. H. (2018). Enhanced model-based predictive control system based on fuzzy logic for maintaining thermal comfort in IoT smart space. *Applied Sciences (Basel, Switzerland)*, *8*(7), 1031. doi:10.3390/app8071031

Hanumanthakari, S., Gift, M. M., Kanimozhi, K., Bhavani, M. D., Bamane, K. D., & Boopathi, S. (2023). Biomining Method to Extract Metal Components Using Computer-Printed Circuit Board E-Waste. In *Handbook of Research on Safe Disposal Methods of Municipal Solid Wastes for a Sustainable Environment* (pp. 123–141). IGI Global. doi:10.4018/978-1-6684-8117-2.ch010

Harikaran, M., Boopathi, S., Gokulakannan, S., & Poonguzhali, M. (2023). Study on the Source of E-Waste Management and Disposal Methods. In *Sustainable Approaches and Strategies for E-Waste Management and Utilization* (pp. 39–60). IGI Global. doi:10.4018/978-1-6684-7573-7.ch003

Hartono, B. S., Budiyanto, & Setiabudy, R. (2013). Review of microgrid technology. *2013 International Conference on Quality in Research, QiR 2013 - In Conjunction with ICCS 2013: The 2nd International Conference on Civic Space*, 127–132. 10.1109/QiR.2013.6632550

Hema, N., Krishnamoorthy, N., Chavan, S. M., Kumar, N., Sabarimuthu, M., & Boopathi, S. (2023). A Study on an Internet of Things (IoT)-Enabled Smart Solar Grid System. In *Handbook of Research on Deep Learning Techniques for Cloud-Based Industrial IoT* (pp. 290–308). IGI Global. doi:10.4018/978-1-6684-8098-4.ch017

Hu, R., Wen, S., Zeng, Z., & Huang, T. (2017). A short-term power load forecasting model based on the generalized regression neural network with decreasing step fruit fly optimization algorithm. *Neurocomputing*, *221*, 24–31. doi:10.1016/j.neucom.2016.09.027

Janardhana, K., Singh, V., Singh, S. N., Babu, T. R., Bano, S., & Boopathi, S. (2023). Utilization Process for Electronic Waste in Eco-Friendly Concrete: Experimental Study. In Sustainable Approaches and Strategies for E-Waste Management and Utilization (pp. 204–223). IGI Global.

Jeevanantham, Y. A., A, S., V, V., J, S. Isaac., Boopathi, S., & Kumar, D. P. (2023). Implementation of Internet-of Things (IoT) in Soil Irrigation System. *IEEE Explore*, 1–5. doi:10.1109/ICPECTS56089.2022.10047185

Kampouropoulos, K., Andrade, F., Sala, E., Espinosa, A. G., & Romeral, L. (2018). Multiobjective optimization of multi-carrier energy system using a combination of ANFIS and genetic algorithms. *IEEE Transactions on Smart Grid*, *9*(3), 2276–2283. doi:10.1109/TSG.2016.2609740

Kannan, P., Venkataramanan, K., Sivaranjani, S., & Indhumathi, K. (2020).. . *Analysis of ABC Algorithm Based ANFIS Controller for PV Integrated Low Voltage Weak Grid System.*, *3*(8), 196–201.

Karthik, S., Hemalatha, R., Aruna, R., Deivakani, M., Reddy, R. V. K., & Boopathi, S. (2023). Study on Healthcare Security System-Integrated Internet of Things (IoT). In Perspectives and Considerations on the Evolution of Smart Systems (pp. 342–362). IGI Global.

Katyara, S., Shah, M. A., Chowdhary, B. S., Akhtar, F., & Lashari, G. A. (2019). Monitoring, Control and Energy Management of Smart Grid System via WSN Technology Through SCADA Applications. *Wireless Personal Communications*, *106*(4), 1951–1968. doi:10.100711277-018-5738-x

Kaur, H., & Sood, S. K. (2019). Adaptive Neuro Fuzzy Inference System (ANFIS) based wildfire risk assessment. *Journal of Experimental & Theoretical Artificial Intelligence*, *31*(4), 599–619. doi:10.1080/0952813X.2019.1591523

Khalid, R., & Javaid, N. (2020). A survey on hyperparameters optimization algorithms of forecasting models in smart grid. *Sustainable Cities and Society*, *61*, 102275. doi:10.1016/j.scs.2020.102275

Khan, A., & Rizwan, M. (2021). ANN and ANFIS Based Approach for Very Short Term Load Forecasting: A Step Towards Smart Energy Management System. *Proceedings of the 8th International Conference on Signal Processing and Integrated Networks, SPIN 2021*, 464–468. 10.1109/SPIN52536.2021.9566146

Koshariya, A. K., Kalaiyarasi, D., Jovith, A. A., Sivakami, T., Hasan, D. S., & Boopathi, S. (2023). AI-Enabled IoT and WSN-Integrated Smart Agriculture System. In *Artificial Intelligence Tools and Technologies for Smart Farming and Agriculture Practices* (pp. 200–218). IGI Global. doi:10.4018/978-1-6684-8516-3.ch011

Kumara, V., Mohanaprakash, T., Fairooz, S., Jamal, K., Babu, T., & Sampath, B. (2023). Experimental Study on a Reliable Smart Hydroponics System. In *Human Agro-Energy Optimization for Business and Industry* (pp. 27–45). IGI Global. doi:10.4018/978-1-6684-4118-3.ch002

Li, D., Yin, H., & Zheng, B. (2018). An Annual Load Forecasting Model Based on Generalized Regression Neural Network With Multi-Swarm Fruit Fly Optimization Algorithm. *Dianwang Jishu/Power System Technology, 42*(2), 585–590. doi:10.13335/j.1000-3673.pst.2017.1403

Maguluri, L. P., Ananth, J., Hariram, S., Geetha, C., Bhaskar, A., & Boopathi, S. (2023). Smart Vehicle-Emissions Monitoring System Using Internet of Things (IoT). In Handbook of Research on Safe Disposal Methods of Municipal Solid Wastes for a Sustainable Environment (pp. 191–211). IGI Global.

Mellit, A., & Kalogirou, S. A. (2011). ANFIS-based modelling for photovoltaic power supply system: A case study. *Renewable Energy, 36*(1), 250–258. doi:10.1016/j.renene.2010.06.028

Mohammad Bilal, M. Z. M. T. A. B. K. B. & O. K. (2022). Analysis and Validation of an Intelligent Energy Monitoring Framework for a Yacht Powered System Based on Adaptive Neuro-Fuzzy Inference System (Anfis). *International Journal of Robotics Research and Development, 12*(1), 1–10.

Olbrich, G. (2016). Energy management and smart cities. *Gis. Business (Atlanta, Ga.), 1*(4), 3.

Padmanaban, S., Priyadarshi, N., Bhaskar, M. S., Holm-Nielsen, J. B., Ramachandaramurthy, V. K., & Hossain, E. (2019). A Hybrid ANFIS-ABC Based MPPT Controller for PV System with Anti-Islanding Grid Protection: Experimental Realization. *IEEE Access : Practical Innovations, Open Solutions, 7*, 103377–103389. doi:10.1109/ACCESS.2019.2931547

Paul Divakar, P., Vijaya Lakshmi, G., & Devi, L. (2020). Applications of Internet of things on Smart Grid. *IOP Conference Series. Materials Science and Engineering, 906*(1), 13–17. doi:10.1088/1757-899X/906/1/012015

Pawar, P., TarunKumar, M., & Vittal K, P. (2020). An IoT based Intelligent Smart Energy Management System with accurate forecasting and load strategy for renewable generation. *Measurement, 152*, 107187. doi:10.1016/j.measurement.2019.107187

Premkumar, S., & Sigappi, A. N. (2021). Functional framework for edge-based agricultural system. In AI, Edge and IoT-based Smart Agriculture (pp. 71–100). Elsevier. doi:10.1016/B978-0-12-823694-9.00029-3

Saleem, M. U., Usman, M. R., & Shakir, M. (2021). Design, Implementation, and Deployment of an IoT Based Smart Energy Management System. *IEEE Access : Practical Innovations, Open Solutions, 9*, 59649–59664. doi:10.1109/ACCESS.2021.3070960

Saravanan, A., Venkatasubramanian, R., Khare, R., Surakasi, R., Boopathi, S., Ray, S., & Sudhakar, M. (2022). Policy trends of renewable energy and non. *Renewable Energy.*

Sathish, T., Sunagar, P., Singh, V., Boopathi, S., Al-Enizi, A. M., Pandit, B., Gupta, M., & Sehgal, S. S. (2023). Characteristics estimation of natural fibre reinforced plastic composites using deep multi-layer perceptron (MLP) technique. *Chemosphere*, *337*, 139346. doi:10.1016/j.chemosphere.2023.139346 PMID:37379988

Saveen, G., Prudhvi Raju, P., Manikanta, D. V., & Satya Praveen, M. (2018). Design and implementation of energy management system with fuzzy control for multiple microgrid. *Proceedings of the 2nd International Conference on Inventive Systems and Control, ICISC 2018, 28*(4), 1239–1244. 10.1109/ICISC.2018.8399003

Subha, S., Inbamalar, T. M., R, K. C., Suresh, L. R., Boopathi, S., & Alaskar, K. (2023, February). A Remote Health Care Monitoring system using internet of medical things (IoMT). *IEEE Explore*. doi:10.1109/ICIPTM57143.2023.10118103

Suganthi, L., Iniyan, S., & Samuel, A. A. (2015). Applications of fuzzy logic in renewable energy systems—A review. *Renewable & Sustainable Energy Reviews*, *48*, 585–607. doi:10.1016/j.rser.2015.04.037

Suhail, M., Akhtar, I., Kirmani, S., & Jameel, M. (2021). Development of Progressive Fuzzy Logic and ANFIS Control for Energy Management of Plug-In Hybrid Electric Vehicle. *IEEE Access : Practical Innovations, Open Solutions*, *9*, 62219–62231. doi:10.1109/ACCESS.2021.3073862

Syamala, M., Komala, C., Pramila, P., Dash, S., Meenakshi, S., & Boopathi, S. (2023). Machine Learning-Integrated IoT-Based Smart Home Energy Management System. In *Handbook of Research on Deep Learning Techniques for Cloud-Based Industrial IoT* (pp. 219–235). IGI Global. doi:10.4018/978-1-6684-8098-4.ch013

Thakare, S., Shriyan, A., Thale, V., Yasarp, P., & Unni, K. (2017). Implementation of an energy monitoring and control device based on IoT. *2016 IEEE Annual India Conference, INDICON 2016*, 1–6. 10.1109/INDICON.2016.7839066

Vanitha, S., Radhika, K., & Boopathi, S. (2023). Artificial Intelligence Techniques in Water Purification and Utilization. In *Human Agro-Energy Optimization for Business and Industry* (pp. 202–218). IGI Global. doi:10.4018/978-1-6684-4118-3.ch010

Vanitha, S. K. R., & Boopathi, S. (2023). Artificial Intelligence Techniques in Water Purification and Utilization. In *Human Agro-Energy Optimization for Business and Industry* (pp. 202–218). IGI Global. doi:10.4018/978-1-6684-4118-3.ch010

Venkateswaran, N., Vidhya, K., Ayyannan, M., Chavan, S. M., Sekar, K., & Boopathi, S. (2023). A Study on Smart Energy Management Framework Using Cloud Computing. In 5G, Artificial Intelligence, and Next Generation Internet of Things: Digital Innovation for Green and Sustainable Economies (pp. 189–212). IGI Global. doi:10.4018/978-1-6684-8634-4.ch009

Venkateswaran, N., Vidhya, R., Naik, D. A., Raj, T. M., Munjal, N., & Boopathi, S. (2023). Study on Sentence and Question Formation Using Deep Learning Techniques. In *Digital Natives as a Disruptive Force in Asian Businesses and Societies* (pp. 252–273). IGI Global. doi:10.4018/978-1-6684-6782-4.ch015

Vennila, T., Karuna, M., Srivastava, B. K., Venugopal, J., Surakasi, R., & Sampath, B. (2022). New Strategies in Treatment and Enzymatic Processes: Ethanol Production From Sugarcane Bagasse. In Human Agro-Energy Optimization for Business and Industry (pp. 219–240). IGI Global.

Zekrifa, D. M. S., Kulkarni, M., Bhagyalakshmi, A., Devireddy, N., Gupta, S., & Boopathi, S. (2023). Integrating Machine Learning and AI for Improved Hydrological Modeling and Water Resource Management. In *Artificial Intelligence Applications in Water Treatment and Water Resource Management* (pp. 46–70). IGI Global. doi:10.4018/978-1-6684-6791-6.ch003

Zhang, F., Sun, G., Zheng, B., & Dong, L. (2021). Design and implementation of energy management system based on spring boot framework. *Information (Basel)*, *12*(11), 457. doi:10.3390/info12110457

Chapter 15
Hospital Waste Management Using Internet of Things and Deep Learning:
Enhanced Efficiency and Sustainability

R. E. Ugandar
https://orcid.org/0000-0002-7014-0864
Department of Pharmacy Practice, Santhiram College of Pharmacy, India

U. Rahamathunnisa
School of Information Technology and Engineering, Vellore Institute of Technology, India

S. Sajithra
Department of Computer Science and Engineering, R.M.K. Engineering College, India

M. Beulah Viji Christiana
Department of Management Studies, Panimalar Engineering College, India

Basanta Kumar Palai
Department of Mechanical Engineering, GIET University, India

Sampath Boopathi
https://orcid.org/0000-0002-2065-6539
Muthayammal Engineering College, India

ABSTRACT

Hospital waste management is crucial for healthcare operations, ensuring safe disposal and handling of waste types. The integration of IoT and deep learning technologies offers a promising solution to address waste volumes and environmental concerns. This chapter presents an advanced research overview on hospital waste management using IoT and DL, exploring its potential benefits and applications. IoT-DL integration enables real-time monitoring of waste fill levels, temperature, and other parameters, while advanced DL techniques improve waste collection efficiency, segregation accuracy, and data-driven decision-making for optimized waste management practices. The chapter discusses challenges, benefits, challenges, and considerations in IoT implementation, waste sorting techniques, and ethical and environmental aspects, including sustainability and circular economy principles.

DOI: 10.4018/978-1-6684-6577-6.ch015

INTRODUCTION

Hospital waste management is a crucial aspect of modern healthcare operations that demands effective and sustainable solutions. The healthcare industry generates a significant amount of diverse waste, including hazardous materials, infectious substances, and pharmaceutical residues, posing environmental and public health risks if not managed appropriately. Conventional waste management practices often fall short in efficiently handling this complex waste stream, leading to potential pollution and health hazards (Harikaran, Boopathi, Gokulakannan, et al., 2023).

In recent years, there has been a growing interest in leveraging cutting-edge technologies to address the challenges of hospital waste management. Among these technologies, the integration of the Internet of Things (IoT) and Deep Learning (DL) has shown great promise in revolutionizing waste management practices in healthcare facilities. Conventional methods are often labour-intensive, error-prone, and lack real-time monitoring capabilities, leading to suboptimal waste disposal and increased risks of contamination. Additionally, inadequate waste segregation and classification can result in improper treatment, further exacerbating the environmental impact (Janardhana et al., 2023).

The advent of IoT has opened up new avenues for waste management by enabling continuous monitoring and tracking of waste generation, storage, and transportation processes. Simultaneously, DL techniques, particularly in image recognition and segmentation, offer powerful tools to automate waste sorting and classification, ensuring proper disposal and recycling of different waste categories (Boopathi, 2023a; Harikaran, Boopathi, Gokulakannan, et al., 2023).

Hospital waste management is a critical aspect of modern healthcare facilities that demands effective strategies to ensure the proper handling, disposal, and recycling of various types of waste generated within hospitals and other healthcare settings. Waste Segregation and Classification: One of the primary research aspects in hospital waste management is the development of efficient waste segregation and classification methods. This involves exploring technologies like deep learning and computer vision to automatically identify and sort different types of waste, facilitating proper disposal and recycling (Harikaran, Boopathi, Gokulakannan, et al., 2023; Koshariya, Khatoon, et al., 2023).

IoT-based Waste Monitoring and Tracking: Internet of Things (IoT) technologies offer opportunities for real-time waste monitoring and tracking throughout the waste management process. Research in this area focuses on the deployment of IoT sensors and devices to monitor waste generation, storage, transportation, and disposal, enabling data-driven decision-making and optimization (Janardhana et al., 2023; Selvakumar, Adithe, et al., 2023). Environmental and Health Impact Assessment: Understanding the environmental and health impacts of hospital waste is crucial for developing effective waste management strategies.

Investigating and promoting sustainable waste management practices in healthcare facilities is another key research aspect. This includes exploring the feasibility of waste-to-energy technologies, recycling methods, and waste reduction strategies to minimize the environmental footprint of hospital waste. Policy and Hospital waste management is subject to a range of local, national, and international regulations. Research in this aspect involves analyzing the existing policies and regulations related to hospital waste and proposing improvements or best practices to ensure compliance and better waste management outcomes (Boopathi, 2023f; Maguluri et al., 2023a; Subha et al., 2023).

Integrating advanced technologies like IoT and deep learning into hospital waste management systems presents various implementation challenges. Research in this aspect focuses on addressing issues related to data privacy, security, scalability, and cost-effectiveness to ensure successful technology adoption

(Reddy, Reddy, et al., 2023a; Syamala et al., 2023b). Public Awareness and Engagement: Public awareness and engagement play a crucial role in promoting responsible waste disposal practices and supporting hospital waste management initiatives.

Research in this aspect involves conducting cost-benefit analyses of implementing advanced waste management technologies, such as IoT and DL, in healthcare facilities. Evaluating the financial implications of adopting these technologies can help hospitals make informed decisions and prioritize their waste management investments (Hema et al., 2023a; Koshariya, Kalaiyarasi, et al., 2023a). Effective hospital waste management requires well-trained staff and personnel. Research in this area focuses on developing training programs and capacity-building initiatives to educate healthcare workers about waste segregation, handling, and disposal procedures, ensuring compliance with best practices (Boopathi, 2023b). Behavioral interventions can play a crucial role in improving waste management practices within hospitals. Research in this aspect involves understanding the behavioral factors influencing waste disposal behavior among healthcare workers and patients, and designing interventions to promote responsible waste management (Venkateswaran, Vidhya, Naik, et al., 2023a).

Investigating circular economy principles in hospital waste management is gaining attention. Research in this area explores ways to close the waste loop by converting waste into valuable resources, such as recycling medical equipment or using waste for energy generation (Sampath, Pandian, et al., 2022a). Conducting case studies on hospital waste management practices in different countries and healthcare systems offers valuable insights into regional challenges, best practices, and policy implications.

Life Cycle Assessment: Conducting life cycle assessments of various waste management techniques helps in understanding the environmental impacts of different disposal methods. Comparing the life cycle impacts of landfilling, incineration, recycling, and other approaches can inform more sustainable waste management decisions (Boopathi, 2023a; Sahal et al., 2022). Hospital waste management involves multiple stakeholders, including hospital administrators, waste management companies, regulators, and the public. Research in this aspect focuses on engaging these stakeholders in the decision-making process and fostering collaboration to improve waste management practices.

RESEARCH OBJECTIVES

The primary objectives of this advanced research on hospital waste management using IoT and DL are as follows:

- To explore the current challenges and limitations of hospital waste management practices, analyzing the environmental and health impacts of improper waste disposal.
- To investigate the fundamentals and applications of IoT and DL technologies in the healthcare sector, focusing on their potential contributions to waste management.
- To propose an integrated IoT-DL framework for hospital waste management, incorporating real-time waste monitoring, smart waste sorting systems, and data-driven optimization.
- To address data security, privacy, and ethical considerations associated with the implementation of IoT and DL in healthcare waste management.
- To present case studies and practical implementations of IoT-DL solutions in hospital waste management scenarios, assessing their effectiveness and benefits.

- To identify future trends and directions in this field, discussing advancements in IoT technologies, evolving DL techniques, and the potential integration of robotics and automation.

By accomplishing these research objectives, this study aims to provide valuable insights into the development and deployment of IoT and DL-based systems for advanced hospital waste management. Ultimately, the research endeavours to contribute to sustainable and responsible waste management practices in the healthcare industry, mitigating environmental impacts and safeguarding public health.

SCOPES

This book chapter explores IoT and Deep Learning technologies' potential in hospital waste management, focusing on their integration and potential transformation in healthcare facilities.

- Hospital Waste Types and Challenges: The chapter will provide an overview of the different types of hospital waste, including hazardous, infectious, pharmaceutical, and non-hazardous waste. It will also highlight the challenges and environmental/health implications associated with improper waste management.
- IoT Fundamentals and Applications in Healthcare: The scope encompasses the fundamental concepts of IoT, its relevance to the healthcare sector, and its potential applications in hospital waste management. Specific IoT-based solutions for waste monitoring, tracking, and collection will be explored.
- Deep Learning Techniques for Waste Sorting and Classification: This section will cover the basics of DL and its applications in waste sorting and classification. It will delve into various DL models and image recognition algorithms used to automate waste segregation.
- Integration of IoT and DL in Hospital Waste Management: The chapter will discuss the integration of IoT and DL technologies to develop a holistic waste management system. It will focus on real-time data processing, smart waste sorting systems, and the optimization of waste management processes.
- Data Security, Privacy, and Ethical Considerations: This part will address the importance of data security and privacy in healthcare IoT-DL systems. Ethical considerations related to the use of IoT and DL in managing hospital waste will also be explored.
- Case Studies and Practical Implementations: The scope includes presenting real-world case studies and practical implementations of IoT-DL solutions in hospital waste management scenarios. These case studies will demonstrate the effectiveness and benefits of such integrated systems.
- Future Trends and Directions: The chapter will discuss potential advancements in IoT technologies, emerging DL techniques, and their impact on the future of hospital waste management. It will explore opportunities for further research and development in this field.

HOSPITAL WASTE MANAGEMENT

Challenges and Current Practices

Hospital waste management faces several challenges, including the diverse nature of waste types, the potential for cross-contamination, and the need for specialized handling and disposal. Lack of awareness and training among healthcare personnel, insufficient resources, and regulatory compliance issues add to the complexity. Current practices often rely on manual sorting and segregation, leading to errors and suboptimal waste disposal (Harikaran, Boopathi, Gokulakannan, et al., 2023; Janardhana et al., 2023; Selvakumar, Adithe, et al., 2023).

Types of Hospital Waste

Hospital waste can be categorized into different types based on its nature and potential risks.

- General (Non-Hazardous) Waste: Non-infectious waste, such as paper, packaging, and food leftovers.
- Infectious Waste: Waste contaminated with infectious materials, such as discarded gloves, bandages, and laboratory cultures.
- Hazardous Waste: Chemicals, pharmaceuticals, and radioactive materials that pose a potential risk to human health and the environment.
- Sharps Waste: Medical sharps, like needles and syringes, which require safe disposal to prevent injuries and infections.

Environmental and Health Impacts of Improper Waste Management

Improper hospital waste management can lead to severe environmental and health impacts. Untreated or improperly disposed of waste may contaminate water bodies, soil, and air, releasing hazardous substances into the environment. This can result in pollution, the spread of infectious diseases, and harmful effects on flora and fauna. Additionally, healthcare workers and waste handlers are at risk of injury or infection if exposed to hazardous waste (Harikaran, Boopathi, Gokulakannan, et al., 2023; Harikaran, Boopathi, Rajkannan, et al., 2023; Janardhana et al., 2023; Selvakumar, Adithe, et al., 2023).

Existing Hospital Waste Management Practices

Common hospital waste management practices involve waste segregation at the point of generation, followed by storage and collection in color-coded bins. After collection, the waste undergoes treatment, such as incineration, autoclaving, or landfilling, based on the waste type. Some healthcare facilities have implemented waste-to-energy technologies to convert waste into usable energy.

Limitations of Conventional Approaches

Conventional hospital waste management practices have several limitations. Manual segregation can lead to errors, resulting in improper disposal. Furthermore, the high cost of advanced waste treatment technologies and the lack of resources in some healthcare settings may hinder the adoption of more sustainable methods. Additionally, conventional approaches may not address the need for real-time monitoring and data-driven optimization of waste management processes (Hanumanthakari et al., 2023; Harikaran, Boopathi, Gokulakannan, et al., 2023; Sengeni et al., 2023a). Addressing these challenges and limitations requires innovative solutions and the integration of advanced technologies like IoT and DL. Implementing more efficient waste segregation and classification methods, along with real-time monitoring and data analysis, can lead to improved hospital waste management practices, reducing environmental impacts and enhancing public health safety (Boopathi, 2022a; Koshariya, Khatoon, et al., 2023; Maguluri et al., 2023a; Selvakumar, Shankar, et al., 2023).

INTERNET OF THINGS (IOT) IN HOSPITAL WASTE MANAGEMENT

The Internet of Things (IoT) refers to a network of physical devices, vehicles, appliances, and other objects embedded with sensors, software, and connectivity, enabling them to collect and exchange data over the internet. IoT devices can interact with each other and with cloud-based platforms, allowing for seamless data transfer and remote monitoring.

- Connectivity: IoT devices communicate with each other and the internet through various communication protocols such as Wi-Fi, Bluetooth, Zigbee, or cellular networks.
- Sensors and Actuators: IoT devices are equipped with sensors to collect data from the environment, and actuators to perform specific actions based on received instructions.
- Data Analytics: IoT generates vast amounts of data, and data analytics play a crucial role in deriving insights, patterns, and trends from this data to make informed decisions.
- Cloud Computing: IoT devices often rely on cloud-based platforms for data storage, processing, and analysis, providing scalability and accessibility from anywhere.

IoT Applications in Healthcare

In the healthcare sector, IoT has a wide range of applications, including:

- Remote Patient Monitoring: IoT-enabled wearable devices and medical sensors allow continuous monitoring of patient vital signs, enabling remote healthcare monitoring and early detection of health issues.
- Telemedicine: IoT facilitates telemedicine services, connecting patients with healthcare providers remotely for virtual consultations and diagnoses.
- Asset Tracking: Hospitals use IoT to track medical equipment and assets, optimizing their utilization and ensuring timely maintenance.
- Medication Management: IoT-based medication adherence systems remind patients to take their medications on time, reducing medication errors and improving treatment outcomes.
- Smart Hospitals: IoT can create smart hospital environments, where systems are interconnected to improve patient care, energy efficiency, and overall operational management.

Iot Applications in Hospital Waste Management

In the context of hospital waste management, IoT applications include:

- Waste Monitoring: IoT sensors are deployed in waste bins to monitor waste levels and send real-time data to waste management systems. This enables timely waste collection and prevents overfilling.
- Smart Waste Collection: IoT-enabled waste collection vehicles can optimize routes based on real-time data, reducing fuel consumption and operational costs.
- Temperature Monitoring: IoT sensors can monitor the temperature of biomedical waste storage areas, ensuring compliance with regulations for waste storage and transportation.

Benefits of IoT in Waste Management

The integration of IoT in hospital waste management offers several benefits:

- Real-time Monitoring: IoT provides real-time data on waste generation, allowing for better planning and response to waste disposal needs.
- Efficient Resource Allocation: IoT data insights enable efficient allocation of waste management resources, optimizing waste collection schedules and routes.
- Improved Compliance: IoT ensures adherence to waste management regulations and standards, minimizing the risk of non-compliance and associated penalties.
- Sustainability: IoT-enabled waste management leads to more sustainable practices, reducing waste volumes, promoting recycling, and lowering environmental impact.
- Enhanced Safety: IoT helps reduce human intervention in waste handling, mitigating the risk of injuries and infections for waste management personnel.

Challenges and Considerations for IoT Implementation

Despite its benefits, IoT implementation in hospital waste management poses some challenges:

- Security and Privacy: IoT devices may be vulnerable to cyber threats, necessitating robust security measures to safeguard data and systems.
- Interoperability: Integrating different IoT devices and platforms may present interoperability challenges, requiring standardization and compatibility efforts.
- Data Management: Handling vast amounts of data from IoT devices demands efficient data storage, processing, and analytics capabilities.
- Cost: The initial investment and maintenance costs for IoT infrastructure can be substantial, especially for resource-constrained healthcare facilities.
- Data Ownership and Sharing: Issues related to data ownership and sharing need to be addressed to facilitate collaboration among stakeholders.
- Training and Adoption: Healthcare staff may require training and support to effectively use and embrace IoT technologies for waste management.

Healthcare facilities can optimize waste management using IoT, reducing waste and promoting a safer, sustainable environment by addressing challenges and considering IoT's potential.

DEEP LEARNING (DL) IN HOSPITAL WASTE SORTING AND CLASSIFICATION

Deep Learning (DL) has emerged as a powerful technique in computer vision and image recognition, making it an ideal approach for waste sorting and classification in hospital waste management. DL models, especially Convolutional Neural Networks (CNNs), have demonstrated remarkable performance in identifying and categorizing waste items based on their visual features (Anitha, R, et al., 2023).

Deep Learning Fundamentals

At the core of DL is the concept of artificial neural networks, inspired by the structure and function of the human brain. Neural networks consist of layers of interconnected nodes (neurons) that process and transmit information. Deep learning refers to the use of deep neural networks, which have multiple hidden layers, enabling them to learn complex patterns and hierarchies in data (Kenney et al., 2013; Nadif & Role, 2021).

DL Techniques for Waste Sorting and Classification

- Convolutional Neural Networks (CNNs): CNNs are the most widely used DL models for image-related tasks. They leverage convolutional layers to automatically extract hierarchical features from input images. In the context of waste sorting, CNNs can learn to distinguish different waste categories based on visual patterns, shapes, and textures (Bikash Chandra Saha, 2022; Sampath, Pandian, et al., 2022b; Sathish, Sunagar, Singh, Boopathi, Al-Enizi, et al., 2023; Syamala et al., 2023a; Venkateswaran, Vidhya, Naik, et al., 2023b).
- Transfer Learning: Transfer learning is a DL technique that involves utilizing pre-trained CNN models on large image datasets (e.g., ImageNet). The knowledge gained from these tasks can be transferred to waste sorting applications, enabling faster and more accurate model training even with limited waste-specific data (Boopathi, 2023c; Chandrika et al., 2023; Hema et al., 2023b; Syamala et al., 2023a; Venkateswaran, Vidhya, Naik, et al., 2023b).
- Image Recognition and Segmentation: DL models can perform both image recognition (classification) and image segmentation tasks. Image recognition involves assigning a specific label to the entire image, while image segmentation identifies and outlines individual objects or regions within an image. In waste sorting, image segmentation can be used to precisely identify and categorize waste items within complex scenes.
- Data Augmentation: To improve model generalization and reduce overfitting, data augmentation is employed. It involves generating variations of existing training images by applying transformations such as rotations, flips, and brightness adjustments. Data augmentation helps increase the diversity of the training data, leading to more robust waste sorting models.
- One-Shot Learning: One-shot learning aims to classify objects with only a single example of each category. This approach is useful in cases where waste items have limited examples or are difficult

to collect. One-shot learning techniques can be incorporated to improve the model's ability to recognize rare or infrequently occurring waste items.

- Ensemble Learning: Ensemble learning combines multiple DL models to make predictions. Each model may have its strengths and weaknesses, and ensemble learning leverages the diversity of these models to achieve more accurate and reliable waste sorting and classification results.

Utilizing DL techniques enhances waste management systems by ensuring precise sorting, reducing contamination risk, improving recycling rates, and enhancing hospital waste management processes.

Image Recognition and Segmentation in Waste Sorting

In waste sorting applications, image recognition and segmentation play crucial roles in accurately identifying and categorizing waste items from images. Image recognition involves assigning a specific label or class to the entire image, indicating the waste category it belongs to. On the other hand, image segmentation aims to precisely outline and identify individual waste items within the image, enabling more fine-grained classification (Abujar et al., 2019; Singh Grewal & Parkash Dhiman, 2019; Uçar et al., 2017).

DL models, particularly Convolutional Neural Networks (CNNs), have shown remarkable performance in both image recognition and segmentation tasks. For image recognition, CNNs analyze the entire image and predict the waste category, while for image segmentation, they produce pixel-level masks that highlight the boundaries of waste items.

DL Models for Hospital Waste Classification

Various DL models can be utilized for hospital waste classification. Some popular models include (Boopathi, Pandey, et al., 2023; Boopathi & Kanike, 2023; Chandrika et al., 2023; Dhanya et al., 2023; Gunasekaran & Boopathi, 2023):

- AlexNet: One of the early deep CNN models, AlexNet, is effective in image recognition tasks and can be adapted for waste classification.
- VGG (Visual Geometry Group)Net: VGGNet is known for its simplicity and effectiveness in image recognition, making it a reliable choice for waste classification.
- ResNet (Residual Network): ResNet introduces skip connections to tackle the vanishing gradient problem, enabling the training of much deeper networks. ResNet is suitable for more complex waste classification tasks.
- DenseNet: DenseNet emphasizes dense connections between layers, allowing for efficient information flow and parameter reduction. It is well-suited for tasks with limited data, making it useful in waste sorting scenarios with fewer samples.
- U-Net: U-Net is a popular model for image segmentation tasks. Its architecture consists of an encoder-decoder structure with skip connections, enabling accurate segmentation of waste items in images.

Performance Metrics and Evaluation

The evaluation of DL models for waste sorting and classification requires appropriate performance metrics (Boopathi, 2023g, 2023d; Boopathi, Pandey, et al., 2023; Boopathi & Kanike, 2023; Chandrika et al., 2023). Commonly used metrics include:

- Accuracy: Accuracy measures the proportion of correctly classified waste items in the dataset and is a general indicator of model performance.
- Precision and Recall: Precision measures the fraction of relevant instances among the items classified as a particular waste category. Recall, on the other hand, calculates the proportion of true positives correctly identified by the model.
- F1 Score: The F1 score is the harmonic mean of precision and recall, providing a balance between the two metrics. It is useful when there is an uneven class distribution in the dataset.
- Intersection over Union (IoU): IoU, also known as the Jaccard index, evaluates the overlap between the predicted segmentation masks and the ground truth masks. It is commonly used for image segmentation evaluation.
- Mean Intersection over Union (mIoU): The mIoU calculates the average IoU across all waste categories, providing an overall measure of segmentation performance.
- Receiver Operating Characteristic (ROC) Curve and Area Under the Curve (AUC): ROC curves illustrate the trade-off between true positive rate and false positive rate at different classification thresholds. AUC summarizes the overall performance of the classifier across all thresholds.

By assessing the performance of DL models using appropriate metrics, waste management systems can select the most effective models for waste sorting and classification, ensuring accurate and reliable waste categorization in hospital waste management processes.

INTEGRATION OF IOT AND DL FOR HOSPITAL WASTE MANAGEMENT

The integration of Internet of Things (IoT) and Deep Learning (DL) in waste management systems offers a powerful and transformative approach to optimize waste sorting and disposal processes (Anitha, Komala, et al., 2023; Nishanth et al., 2023; Ramudu et al., 2023; Sengeni et al., 2023b; Venkateswaran, Vidhya, Ayyannan, et al., 2023; Zekrifa et al., 2023). This integration leverages the real-time data collection capabilities of IoT sensors and combines them with the advanced image recognition and classification capabilities of DL models (Figure 1). By merging these technologies, waste management systems can achieve enhanced efficiency, accuracy, and sustainability in handling hospital waste (Boopathi, 2023b; Sathish, Sunagar, Singh, Boopathi, Sathyamurthy, et al., 2023).

IoT Sensors for Waste Monitoring and Collection

IoT sensors are deployed in waste bins, waste collection vehicles, and waste storage areas to monitor various parameters related to waste generation, storage, and collection (Anitha, R, et al., 2023; Jeevanantham et al., 2023; Subha et al., 2023).

Figure 1. Integration Of IOT and DL for hospital waste management

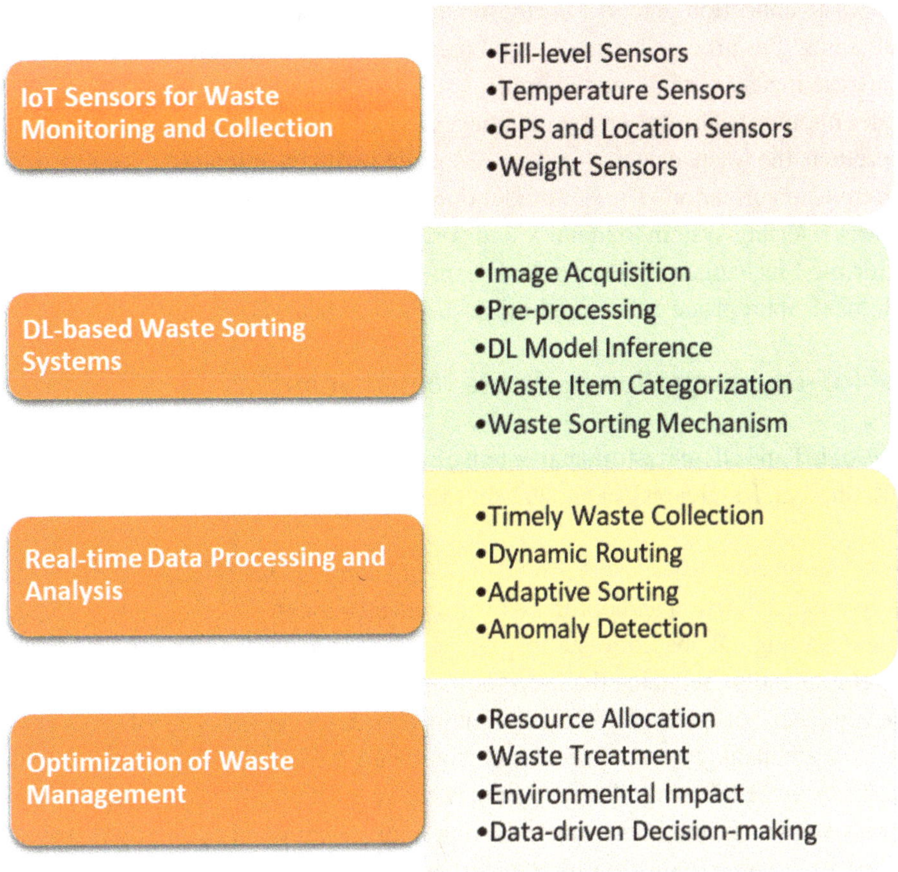

- Fill-level Sensors: Installed in waste bins, fill-level sensors measure the fill status of waste containers. They provide real-time data on waste levels, enabling optimized waste collection schedules and route planning.
- Temperature Sensors: Temperature sensors monitor waste storage areas to ensure compliance with regulations for temperature-sensitive biomedical waste.
- GPS and Location Sensors: Integrated into waste collection vehicles, GPS and location sensors track the location and movement of vehicles, enabling route optimization and real-time monitoring of waste collection activities.
- Weight Sensors: Weight sensors can be incorporated into waste bins or collection points to measure the weight of waste containers, providing additional data for waste management planning.

DL-Based Waste Sorting Systems

DL-based waste sorting systems use CNNs or other DL models to perform waste item recognition and classification based on visual characteristics (Boopathi, 2023a; Boopathi, Arigela, et al., 2023; Sahal et al., 2022). The workflow of a DL-based waste sorting system typically includes the following steps:

- Image Acquisition: Images of waste items are captured using cameras or other imaging devices, either at waste collection points or during waste sorting operations.
- Pre-processing: Pre-processing involves image resizing, normalization, and other transformations to prepare the images for DL model input.
- DL Model Inference: The pre-processed images are fed into the DL model for inference, where the model predicts the waste category or type associated with each image.
- Waste Item Categorization: The DL model outputs the waste category label for each item, allowing the waste sorting system to identify and sort waste items accordingly.
- Waste Sorting Mechanism: Based on the DL model's predictions, the waste sorting system directs waste items to appropriate collection bins or disposal processes.

Benefits of IoT-DL Integration in Waste Management

The integration of IoT and DL in waste management offers numerous benefits (Hema et al., 2023b; Karthik et al., 2023; Koshariya, Kalaiyarasi, et al., 2023b; Maguluri et al., 2023b; Reddy, Reddy, et al., 2023b):

- Real-time Monitoring: IoT sensors provide real-time data on waste levels and fill status, allowing for timely waste collection and optimized waste management.
- Automated Sorting: DL-based waste sorting systems automate the process of waste item recognition and classification, reducing the need for manual intervention and improving sorting accuracy.
- Data-driven Decision-making: IoT data combined with DL insights enable data-driven decision-making in waste management, optimizing resource allocation and waste disposal practices.
- Sustainability and Efficiency: The integration enhances waste management sustainability by reducing waste volumes and improving recycling rates, resulting in more efficient waste handling.
- Environmental Impact Reduction: Proper waste sorting and disposal, facilitated by IoT-DL integration, minimize environmental pollution and public health risks associated with improper waste management.

The combined capabilities of IoT and DL in waste management systems create a smarter and more effective approach to handling hospital waste, ensuring safer and more sustainable waste disposal practices.

Real-Time Data Processing and Analysis

IoT-enabled waste sorting systems can collect and process data in real-time using IoT sensors. The real-time data processing and analysis facilitate immediate decision-making and response to changing waste conditions (Bikash Chandra Saha, 2022; Jeevanantham et al., 2022; Reddy, Reddy, et al., 2023b; Samikannu et al., 2022; Syamala et al., 2023a).

- Timely Waste Collection: Real-time data on waste fill levels enable waste collection teams to prioritize and optimize collection schedules, avoiding unnecessary pickups or overflow situations.
- Dynamic Routing: Waste collection vehicles can be dynamically routed based on real-time data, reducing travel time and fuel consumption.

- Adaptive Sorting: Real-time data from sorting operations can be used to adjust the DL model's parameters or update classification rules on-the-fly, ensuring continuous improvement in sorting accuracy.
- Anomaly Detection: Real-time data analysis allows for the detection of anomalies in waste characteristics, such as unexpected hazardous waste items, which require immediate attention.

Optimization of Waste Management Processes

The integration of DL-based waste sorting systems with real-time data processing enables the optimization of various waste management processes (Boopathi, Kumar, et al., 2023; Boopathi, Pandey, et al., 2023; Boopathi & Sivakumar, 2013; Domakonda et al., 2022; Gowri et al., 2023; Kumara et al., 2023; Myilsamy & Sampath, 2017; Saravanan et al., 2022; Veeranjaneyulu, Boopathi, Kumari, et al., 2023):

- Resource Allocation: Real-time data on waste fill levels and collection needs help optimize the allocation of waste collection resources, minimizing unnecessary trips and costs.
- Waste Treatment: With accurate waste sorting, waste items can be directed to appropriate treatment facilities, such as recycling centers or hazardous waste disposal units, improving the overall efficiency of waste treatment.
- Environmental Impact: By reducing waste contamination and improving recycling rates, waste management processes can be optimized to reduce the environmental impact of hospital waste disposal.
- Data-driven Decision-making: Real-time data insights, combined with historical data, enable data-driven decision-making, leading to more informed and optimized waste management strategies.

DL-based waste sorting systems and real-time data processing enhance hospital waste management, efficiency, and sustainable practices, promoting efficient and eco-friendly waste disposal.

DATA SECURITY, PRIVACY, AND ETHICAL CONSIDERATIONS

Data Security in IoT-DL Systems

Data security is a critical concern in IoT-DL systems, especially in healthcare settings where sensitive and confidential information is involved (Hema et al., 2023b; Rahamathunnisa et al., 2023; Syamala et al., 2023a; Venkateswaran, Vidhya, Ayyannan, et al., 2023). Some key aspects of data security in these systems include:

- Encryption: Data transmitted between IoT devices and the central system should be encrypted to prevent unauthorized access or interception.
- Secure Authentication: Strong authentication mechanisms must be implemented to ensure that only authorized users can access the IoT-DL system and its data.
- Access Control: Role-based access control should be enforced to limit access to sensitive data and functionalities based on the user's role and privileges.

- Secure Communication: Secure communication protocols, such as HTTPS, should be used to protect data integrity during transmission.
- Regular Updates and Patches: IoT devices and software should be regularly updated with security patches to address known vulnerabilities.
- Data Storage and Retention: Sensitive data should be stored securely and only retained for as long as necessary, following data protection regulations.
- Secure Cloud Infrastructure: If the IoT-DL system relies on cloud-based services, ensuring the security of the cloud infrastructure and data storage is crucial.

Privacy Concerns in Healthcare IoT

Healthcare IoT systems often involve the collection and processing of personal health information, raising significant privacy concerns (Ramudu et al., 2023; Ravisankar et al., 2023; Reddy, Gaurav, et al., 2023; Reddy, Reddy, et al., 2023b). Key privacy considerations in healthcare IoT include:

- Informed Consent: Patients and users should be fully informed about the data collection and usage practices of the IoT system and give their informed consent.
- Data Anonymization: Whenever possible, personal health information should be anonymized or pseudonymized to protect the privacy of patients.
- Data Minimization: IoT systems should only collect and retain the minimum amount of personal data required for their intended purpose.
- Transparent Data Handling: Clear and transparent privacy policies should be in place, explaining how patient data is handled and who has access to it.
- Compliance with Regulations: Healthcare IoT systems must comply with data protection and privacy regulations, such as GDPR (General Data Protection Regulation) in the European Union and HIPAA (Health Insurance Portability and Accountability Act) in the United States.

Ethical Considerations in Hospital Waste Management

Ethical considerations in hospital waste management are essential to ensure responsible and sustainable waste handling (Boopathi, 2023g, 2023d; Durairaj et al., 2023; Karthik et al., 2023; Pramila et al., 2023). Some ethical considerations include:

- Environmental Impact: Hospital waste management should prioritize environmentally friendly practices, minimizing waste and promoting recycling.
- Health and Safety of Workers: Waste management personnel should be provided with proper training, protective equipment, and a safe working environment.
- Proper Waste Segregation: Ethical waste management involves accurate waste segregation to prevent cross-contamination and hazardous waste exposure.
- Responsible Waste Disposal: Hospital waste should be disposed of responsibly, complying with local regulations and minimizing potential harm to the environment and public health.

Figure 2. Case studies and practical implementations

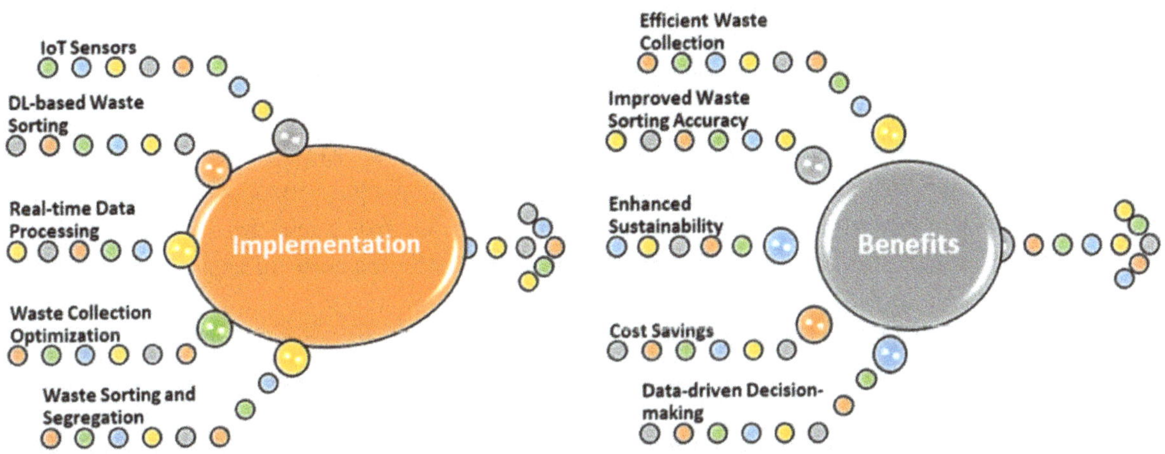

- Minimizing Waste Generation: Ethical waste management practices should focus on waste reduction strategies and promoting waste-to-energy or waste-to-resource approaches.
- Transparency and Accountability: Healthcare facilities should be transparent about their waste management practices and accountable for their environmental impact.

By addressing data security, privacy concerns, and ethical considerations, healthcare IoT-DL systems can uphold patient trust, ensure data protection, and promote responsible waste management practices in hospital settings.

CASE STUDIES AND PRACTICAL IMPLEMENTATIONS

Case Studies and Practical Implementations is illustrated in Figure 2.

Case Study 1: IoT-DL Integration in a Hospital Facility – Smart Waste Management System

Overview: A large hospital facility in a metropolitan area implemented an IoT-DL integrated smart waste management system to streamline waste sorting, improve waste collection efficiency, and enhance environmental sustainability. The system involved the deployment of IoT sensors and DL-based waste sorting algorithms (Boopathi, 2022b; Reddy, Reddy, et al., 2023a; Subha et al., 2023).

Implementation

- IoT Sensors: The hospital installed fill-level sensors in waste bins across various departments and waste collection areas. These sensors continuously monitored the fill levels of waste bins and transmitted real-time data to a centralized waste management platform.
- DL-based Waste Sorting: The hospital implemented a DL model, based on a pre-trained CNN architecture, for waste sorting and classification. The DL model was trained on a diverse dataset of waste images, including different waste categories commonly generated in healthcare settings.
- Real-time Data Processing: The IoT sensors' data was processed in real-time using cloud-based servers. The data included waste fill levels, timestamps, and location information of the waste bins.
- Waste Collection Optimization: The smart waste management system utilized the real-time data from the IoT sensors to optimize waste collection schedules. The system dynamically adjusted waste collection routes based on fill levels, ensuring timely and efficient waste pickup.
- Waste Sorting and Segregation: Waste items were collected, and images were captured using cameras at designated waste sorting stations. The images were then processed by the DL model for automatic waste sorting and segregation.
- Environmental Impact Analysis: The system provided data analytics on waste volumes, recycling rates, and environmental impact, enabling the hospital to assess its waste management practices' sustainability and make informed decisions.

Benefits: The implementation of the IoT-DL integrated smart waste management system resulted in several benefits for the hospital facility:

- Efficient Waste Collection: Real-time data on waste fill levels allowed waste collection teams to optimize routes, reducing fuel consumption and overall operational costs.
- Improved Waste Sorting Accuracy: The DL-based waste sorting system demonstrated high accuracy in classifying waste items, reducing the risk of improper waste disposal and cross-contamination.
- Enhanced Sustainability: The hospital observed a significant increase in recycling rates as waste sorting accuracy improved, promoting a more sustainable waste management approach.
- Cost Savings: The optimization of waste collection routes led to cost savings for the hospital, making waste management operations more cost-effective.
- Data-driven Decision-making: Data analytics provided valuable insights into waste generation patterns and trends, enabling data-driven decision-making in waste management strategies.

The IoT-DL integrated smart waste management system revolutionized waste management practices in the hospital facility. The combination of real-time data processing, IoT sensors, and DL-based waste sorting algorithms enhanced waste collection efficiency, improved recycling rates, and promoted environmentally responsible waste disposal. The success of this implementation serves as a model for other healthcare facilities seeking to implement advanced waste management solutions using IoT and DL technologies (Anitha, Komala, et al., 2023; Boopathi, Sureshkumar, Jeyakumar, et al., 2022; Boopathi, Sureshkumar, & Sathiskumar, 2022; Sampath, Pandian, et al., 2022b; Sampath, Yuvaraj, et al., 2022; Saravanan et al., 2022; Veeranjaneyulu, Boopathi, Kumari, et al., 2023; Veeranjaneyulu, Boopathi, Narasimharao, et al., 2023).

Case Study 2: IoT-DL Integration for Medical Waste Monitoring and Tracking

Overview: A mid-sized healthcare facility, comprising multiple clinics and medical centers, implemented an IoT-DL integrated system to monitor and track medical waste generated throughout its operations. The goal was to enhance waste segregation, optimize waste collection, and ensure compliance with regulatory requirements (Boopathi, 2023f, 2023e; Reddy, Reddy, et al., 2023a; Subha et al., 2023).

Implementation

- IoT Sensors: IoT sensors were installed in designated medical waste bins across various departments and medical centers. These sensors measured fill levels, temperature, and location information of the waste bins.
- DL-based Waste Segregation: The facility developed a DL model for waste segregation. The model was trained on a diverse dataset of medical waste images, including hazardous waste, sharps, pharmaceutical waste, and non-hazardous waste.
- Real-time Data Processing: The data from IoT sensors, including fill levels and temperature readings, were processed in real-time using edge computing devices located near the waste bins. Edge processing reduced latency and ensured swift response times.
- Waste Collection Optimization: The real-time data from IoT sensors and DL model predictions were fed into a centralized waste management platform. The platform dynamically optimized waste collection routes and schedules based on fill levels, minimizing waste handling and transportation costs.
- Temperature Monitoring: Temperature sensors in the medical waste bins monitored the storage conditions of hazardous waste, ensuring compliance with temperature-sensitive waste disposal regulations.
- Compliance Reporting: The integrated system provided automated compliance reporting, generating waste disposal manifests and other necessary documentation for regulatory authorities.

Benefits: The implementation of the IoT-DL integrated medical waste monitoring and tracking system yielded several advantages for the healthcare facility:

- Enhanced Waste Segregation: The DL-based waste segregation system accurately classified medical waste items, reducing the risk of improper disposal and cross-contamination.
- Regulatory Compliance: The automated compliance reporting ensured the facility met waste disposal regulations, minimizing the risk of fines and penalties.
- Optimal Waste Collection: The real-time data and optimization algorithms improved waste collection efficiency, reducing operational costs and carbon footprint associated with waste transportation.
- Timely Hazardous Waste Disposal: Temperature monitoring enabled timely disposal of temperature-sensitive hazardous waste, ensuring compliance with regulatory requirements.
- Improved Safety and Hygiene: With efficient waste segregation, the risk of exposure to hazardous waste for healthcare workers and waste handlers was significantly reduced.
- Data Analytics for Waste Management Planning: The system provided data analytics on waste generation patterns and trends, enabling better waste management planning and resource allocation.

The IoT-DL integrated medical waste monitoring and tracking system revolutionized waste management practices in the healthcare facility. The combination of IoT sensors, real-time data processing, and DL-based waste segregation ensured efficient waste collection, compliance with regulations, and improved safety standards. The successful implementation of this system serves as a model for other healthcare facilities seeking to optimize their medical waste management processes and enhance environmental sustainability.

Case Study 3: IoT-DL Integration for Smart Recycling in a Hospital Campus

Overview: A large hospital campus implemented an IoT-DL integrated smart recycling system to promote recycling practices among patients, visitors, and hospital staff. The system aimed to increase recycling rates, reduce landfill waste, and raise environmental awareness (Hanumanthakari et al., 2023; Sengeni et al., 2023a).

Implementation

- Smart Recycling Bins: IoT-enabled recycling bins were strategically placed throughout the hospital campus. These bins were equipped with fill-level sensors to monitor the levels of recyclable materials, such as paper, plastic, and glass.
- DL-based Material Recognition: The smart recycling system utilized a DL model to recognize and classify different recyclable materials. The DL model was trained on a diverse dataset of recyclable item images.
- Real-time Data Processing: The data from IoT sensors in the recycling bins were processed in real-time using edge computing devices. Real-time data analysis enabled immediate responses to changing waste conditions.
- Recycling Rewards System: To incentivize recycling, the hospital implemented a rewards system. Users who deposited recyclable materials in the smart recycling bins earned points that could be redeemed for discounts at the hospital cafeteria or gift shop.
- Waste Management Dashboard: The hospital established a waste management dashboard that provided real-time insights into recycling rates, waste diversion metrics, and environmental impact.
- Environmental Awareness Campaign: The smart recycling system was complemented by an environmental awareness campaign, including posters, educational materials, and digital screens displaying recycling statistics and environmental benefits.

Benefits: The implementation of the IoT-DL integrated smart recycling system resulted in several benefits for the hospital campus:

- Increased Recycling Rates: The rewards system and real-time data tracking encouraged more individuals to participate in recycling, leading to higher recycling rates.
- Waste Reduction: With improved recycling practices, the hospital achieved significant waste reduction and reduced the volume of waste sent to landfills.
- Sustainability and Environmental Impact: The smart recycling system positively impacted the hospital's environmental sustainability, promoting a culture of responsible waste management.

- Data-driven Decision-making: The waste management dashboard provided data-driven insights that guided waste management strategies, allowing the hospital to continuously optimize recycling efforts.
- Positive Public Perception: The implementation of smart recycling and the environmental awareness campaign enhanced the hospital's reputation as an environmentally responsible institution, garnering positive public perception.
- Employee Engagement: Hospital staff were actively involved in the recycling program, contributing to a sense of community and shared responsibility for environmental stewardship.

The IoT-DL integrated smart recycling system transformed hospital waste management practices by combining IoT sensors, DL-based material recognition, and environmental awareness initiatives. This led to increased recycling rates, reduced waste generation, and increased environmental consciousness among stakeholders. This case study demonstrates the potential of technology integration and environmental initiatives for sustainable waste management in healthcare settings.

FUTURE TRENDS AND DIRECTIONS IN HOSPITAL WASTE MANAGEMENT

Advancements in IoT Technologies

- Miniaturization and Energy Efficiency: IoT sensors and devices will become smaller, more energy-efficient, and cost-effective, enabling widespread deployment in waste management systems.
- Edge Computing: Edge computing will gain prominence in waste management, allowing real-time data processing and analysis at the source, reducing latency and dependence on cloud infrastructure.
- 5G Connectivity: The adoption of 5G networks will enable faster and more reliable data transmission, facilitating seamless communication between IoT devices in waste management systems.

Evolving DL Techniques for Waste Management

- Transfer Learning for Limited Data: DL models with transfer learning capabilities will become more prevalent, enabling accurate waste sorting even with limited waste-specific datasets.
- Few-Shot and Zero-Shot Learning: Few-shot and zero-shot learning techniques will be applied to waste sorting, allowing DL models to identify waste items with minimal training samples or even without direct examples.
- Active Learning: Active learning strategies will be employed to optimize data collection efforts, with DL models intelligently selecting the most informative samples for training.

Integration of Robotics and Automation

- Robotic Waste Sorting: Robotics will play a significant role in automating waste sorting processes, further reducing the need for manual intervention and improving sorting accuracy.
- Autonomous Waste Collection Vehicles: Self-driving waste collection vehicles will be developed, leveraging AI and robotics to optimize waste collection routes and reduce human labor.

Sustainability and Circular Economy Aspects

- Waste-to-Energy Conversion: More emphasis will be placed on waste-to-energy technologies, such as waste incineration and anaerobic digestion, to harness energy from organic waste.
- Closed-Loop Systems: Hospital waste management systems will evolve towards closed-loop systems, promoting recycling and reusing materials within the healthcare facility.
- Life Cycle Assessment: Hospitals will adopt life cycle assessment approaches to evaluate the environmental impact of waste management practices and make informed decisions to minimize their carbon footprint.

Smart Waste Management Integration

- Internet of Medical Things (IoMT): The integration of IoMT devices and sensors in medical equipment will provide valuable data on medical waste generation and improve waste management efficiency.
- Data Analytics and AI-driven Insights: Data analytics and AI-driven insights will be used to optimize waste management processes, predict waste generation patterns, and support evidence-based decision-making.

Public Awareness and Engagement

- Behavioral Nudges: Hospitals will employ behavioral nudge techniques to encourage waste segregation and recycling among patients, visitors, and staff.
- Gamification: Gamification elements, such as waste reduction challenges and rewards systems, will be integrated into waste management initiatives to promote active participation.

Hospital waste management will utilize IoT, advanced DL techniques, robotics, and automation for sustainability and circular economy principles, resulting in efficient, environmentally responsible, and technology-driven practices for a greener, healthier future.

CONCLUSION

The integration of IoT and Deep Learning (DL) technologies in hospital waste management has the potential to revolutionize waste handling practices, ensuring efficiency, accuracy, and sustainability. By combining real-time data processing from IoT sensors with advanced image recognition capabilities, hospitals can optimize waste collection schedules, improve waste sorting accuracy, and make data-driven decisions for better resource allocation. IoT-DL integration offers numerous benefits, including reduced operational costs, increased recycling rates, minimized environmental impact, and enhanced compliance with waste disposal regulations. It also promotes safer waste handling practices, mitigating the risk of contamination and infections for waste management personnel and healthcare workers. Future trends in IoT technologies, such as miniaturization, energy efficiency, and 5G connectivity, will make

IoT implementations more accessible and scalable. Evolving DL techniques, such as transfer learning, few-shot learning, and active learning, will enable more accurate waste sorting with limited data and promote continuous model improvement.

The integration of robotics and automation will enable autonomous waste collection vehicles and robotic waste sorting systems, reducing human labor and increasing efficiency. Hospitals will adopt sustainability principles, focusing on waste-to-energy conversion, closed-loop systems, and life cycle assessment to minimize environmental impact and promote a circular economy approach. Public awareness and engagement are crucial for driving positive behavioural change in waste segregation and recycling. Hospitals can foster responsible waste management through gamification and environmental awareness campaigns. The future of hospital waste management lies in combining IoT, DL, robotics, and sustainability principles, creating a greener, more efficient, and socially responsible waste management ecosystem. This contributes to a healthier environment and better healthcare outcomes for all.

LIST OF ACRONYMS

IoT - Internet of Things
DL - Deep Learning
CNN - Convolutional Neural Network
GPS - Global Positioning System
HIPAA - Health Insurance Portability and Accountability Act
GDPR - General Data Protection Regulation
mIoU - Mean Intersection over Union
ROC - Receiver Operating Characteristic
AUC - Area Under the Curve
IoMT - Internet of Medical Things
5G - Fifth Generation (of mobile communication)
AI - Artificial Intelligence
RFID - Radio Frequency Identification
U-Net - A type of deep learning architecture for image segmentation
VGGNet - Visual Geometry Group Network
ResNet - Residual Network
DLMS - Deep Learning Management System
SLAM - Simultaneous Localization and Mapping
GPU - Graphics Processing Unit
CPU - Central Processing Unit
RAM - Random Access Memory
AIoT - Artificial Intelligence of Things
NLP - Natural Language Processing
RNN - Recurrent Neural Network
LSTM - Long Short-Term Memory
GAN - Generative Adversarial Network

CAGR - Compound Annual Growth Rate
EMR - Electronic Medical Record
EHR - Electronic Health Record
PPE - Personal Protective Equipment

REFERENCES

Abujar, S., Hasan, M., & Hossain, S. A. (2019). Sentence similarity estimation for text summarization using deep learning. *Proceedings of the 2nd International Conference on Data Engineering and Communication Technology: ICDECT 2017*, 155–164. 10.1007/978-981-13-1610-4_16

Anitha, C., R, K. C., Vivekanand, C. V., Lalitha, S. D., Boopathi, S., & R, Revathi. (2023, February). Artificial Intelligence driven security model for Internet of Medical Things (IoMT). *IEEE Explore*. doi:10.1109/ICIPTM57143.2023.10117713

Anitha, C., Komala, C., Vivekanand, C. V., Lalitha, S., & Boopathi, S. (2023). Artificial Intelligence driven security model for Internet of Medical Things (IoMT). *IEEE Explore*, 1–7.

Boopathi, S. (2022a). An investigation on gas emission concentration and relative emission rate of the near-dry wire-cut electrical discharge machining process. *Environmental Science and Pollution Research International*, *29*(57), 86237–86246. doi:10.100711356-021-17658-1 PMID:34837614

Boopathi, S. (2022b). Cryogenically treated and untreated stainless steel grade 317 in sustainable wire electrical discharge machining process: A comparative study. *Environmental Science and Pollution Research International*, 1–10. doi:10.100711356-022-22843-x PMID:36057706

Boopathi, S. (2023a). Deep Learning Techniques Applied for Automatic Sentence Generation. In Promoting Diversity, Equity, and Inclusion in Language Learning Environments (pp. 255–273). IGI Global. doi:10.4018/978-1-6684-3632-5.ch016

Boopathi, S. (2023d). Internet of Things-Integrated Remote Patient Monitoring System: Healthcare Application. In *Dynamics of Swarm Intelligence Health Analysis for the Next Generation* (pp. 137–161). IGI Global. doi:10.4018/978-1-6684-6894-4.ch008

Boopathi, S. (2023f). Securing Healthcare Systems Integrated With IoT: Fundamentals, Applications, and Future Trends. In Dynamics of Swarm Intelligence Health Analysis for the Next Generation (pp. 186–209). IGI Global.

Boopathi, S., Arigela, S. H., Raman, R., Indhumathi, C., Kavitha, V., & Bhatt, B. C. (2023). Prominent Rule Control-based Internet of Things: Poultry Farm Management System. *IEEE Explore*, 1–6. doi:10.1109/ICPECTS56089.2022.10047039

Boopathi, S., & Kanike, U. K. (2023). Applications of Artificial Intelligent and Machine Learning Techniques in Image Processing. In *Handbook of Research on Thrust Technologies' Effect on Image Processing* (pp. 151–173). IGI Global. doi:10.4018/978-1-6684-8618-4.ch010

Boopathi, S., Kumar, P. K. S., Meena, R. S., Sudhakar, M., & Associates. (2023). Sustainable Developments of Modern Soil-Less Agro-Cultivation Systems: Aquaponic Culture. In Human Agro-Energy Optimization for Business and Industry (pp. 69–87). IGI Global.

Boopathi, S., Pandey, B. K., & Pandey, D. (2023). Advances in Artificial Intelligence for Image Processing: Techniques, Applications, and Optimization. In Handbook of Research on Thrust Technologies' Effect on Image Processing (pp. 73–95). IGI Global.

Boopathi, S., & Sivakumar, K. (2013). Experimental investigation and parameter optimization of near-dry wire-cut electrical discharge machining using multi-objective evolutionary algorithm. *International Journal of Advanced Manufacturing Technology*, *67*(9–12), 2639–2655. doi:10.100700170-012-4680-4

Boopathi, S., Sureshkumar, M., Jeyakumar, M., Kumar, R. S., & Subbiah, R. (2022). Influences of Fabrication Parameters on Natural Fiber Reinforced Polymer Composite (NFRPC) Material: A Review. *Materials Science Forum*, *1075*, 115–124. doi:10.4028/p-095f0t

Boopathi, S., Sureshkumar, M., & Sathiskumar, S. (2022). Parametric Optimization of LPG Refrigeration System Using Artificial Bee Colony Algorithm. *International Conference on Recent Advances in Mechanical Engineering Research and Development*, 97–105.

Chandrika, V., Sivakumar, A., Krishnan, T. S., Pradeep, J., Manikandan, S., & Boopathi, S. (2023). Theoretical Study on Power Distribution Systems for Electric Vehicles. In *Intelligent Engineering Applications and Applied Sciences for Sustainability* (pp. 1–19). IGI Global. doi:10.4018/979-8-3693-0044-2.ch001

Dhanya, D., Kumar, S. S., Thilagavathy, A., Prasad, D., & Boopathi, S. (2023). Data Analytics and Artificial Intelligence in the Circular Economy: Case Studies. In Intelligent Engineering Applications and Applied Sciences for Sustainability (pp. 40–58). IGI Global.

Domakonda, V. K., Farooq, S., Chinthamreddy, S., Puviarasi, R., Sudhakar, M., & Boopathi, S. (2022). Sustainable Developments of Hybrid Floating Solar Power Plants: Photovoltaic System. In Human Agro-Energy Optimization for Business and Industry (pp. 148–167). IGI Global.

Durairaj, M., Jayakumar, S., Karpagavalli, V., Maheswari, B. U., Boopathi, S., & ... (2023). Utilization of Digital Tools in the Indian Higher Education System During Health Crises. In *Multidisciplinary Approaches to Organizational Governance During Health Crises* (pp. 1–21). IGI Global. doi:10.4018/978-1-7998-9213-7.ch001

Gowri, N. V., Dwivedi, J. N., Krishnaveni, K., Boopathi, S., Palaniappan, M., & Medikondu, N. R. (2023). Experimental investigation and multi-objective optimization of eco-friendly near-dry electrical discharge machining of shape memory alloy using Cu/SiC/Gr composite electrode. *Environmental Science and Pollution Research International*, 1–19. doi:10.100711356-023-26983-6 PMID:37126160

Gunasekaran, K., & Boopathi, S. (2023). Artificial Intelligence in Water Treatments and Water Resource Assessments. In *Artificial Intelligence Applications in Water Treatment and Water Resource Management* (pp. 71–98). IGI Global. doi:10.4018/978-1-6684-6791-6.ch004

Hanumanthakari, S., Gift, M. D. M., Kanimozhi, K. V., Bhavani, M. D., Bamane, K. D., & Boopathi, S. (2023). Biomining Method to Extract Metal Components Using Computer-Printed Circuit Board E-Waste. In *Handbook of Research on Safe Disposal Methods of Municipal Solid Wastes for a Sustainable Environment* (pp. 123–141). IGI Global. doi:10.4018/978-1-6684-8117-2.ch010

Harikaran, M., Boopathi, S., Gokulakannan, S., & Poonguzhali, M. (2023). Study on the Source of E-Waste Management and Disposal Methods. In *Sustainable Approaches and Strategies for E-Waste Management and Utilization* (pp. 39–60). IGI Global. doi:10.4018/978-1-6684-7573-7.ch003

Harikaran, M., Boopathi, S., Rajkannan, M., & Gokulakannan, S. (2023). Impact analysis of ceramic tile powder aggregates on self-compacting concrete. *Engineering Research Express*, *5*(2), 025069. doi:10.1088/2631-8695/acde48

Hema, N., Krishnamoorthy, N., Chavan, S. M., Kumar, N., Sabarimuthu, M., & Boopathi, S. (2023). A Study on an Internet of Things (IoT)-Enabled Smart Solar Grid System. In *Handbook of Research on Deep Learning Techniques for Cloud-Based Industrial IoT* (pp. 290–308). IGI Global. doi:10.4018/978-1-6684-8098-4.ch017

Janardhana, K., Singh, V., Singh, S. N., Babu, T. S. R., Bano, S., & Boopathi, S. (2023). Utilization Process for Electronic Waste in Eco-Friendly Concrete: Experimental Study. In Sustainable Approaches and Strategies for E-Waste Management and Utilization (pp. 204–223). IGI Global.

Jeevanantham, Y. A., A, S., V, V., J, S. Isaac., Boopathi, S., & Kumar, D. P. (2023). Implementation of Internet-of Things (IoT) in Soil Irrigation System. *IEEE Explore*, 1–5. doi:10.1109/ICPECTS56089.2022.10047185

Jeevanantham, Y. A., Saravanan, A., Vanitha, V., Boopathi, S., & Kumar, D. P. (2022). Implementation of Internet-of Things (IoT) in Soil Irrigation System. *IEEE Explore*, 1–5.

Karthik, S., Hemalatha, R., Aruna, R., Deivakani, M., Reddy, R. V. K., & Boopathi, S. (2023). Study on Healthcare Security System-Integrated Internet of Things (IoT). In Perspectives and Considerations on the Evolution of Smart Systems (pp. 342–362). IGI Global.

Kenney, K. L., Smith, W. A., Gresham, G. L., & Westover, T. L. (2013). Understanding biomass feedstock variability. *Biofuels*, *4*(1), 111–127. doi:10.4155/bfs.12.83

Koshariya, A. K., Kalaiyarasi, D., Jovith, A. A., Sivakami, T., Hasan, D. S., & Boopathi, S. (2023a). AI-Enabled IoT and WSN-Integrated Smart Agriculture System. In *Artificial Intelligence Tools and Technologies for Smart Farming and Agriculture Practices* (pp. 200–218). IGI Global. doi:10.4018/978-1-6684-8516-3.ch011

Koshariya, A. K., Kalaiyarasi, D., Jovith, A. A., Sivakami, T., Hasan, D. S., & Boopathi, S. (2023b). AI-Enabled IoT and WSN-Integrated Smart Agriculture System. In *Artificial Intelligence Tools and Technologies for Smart Farming and Agriculture Practices* (pp. 200–218). IGI Global. doi:10.4018/978-1-6684-8516-3.ch011

Koshariya, A. K., Khatoon, S., Marathe, A. M., Suba, G. M., Baral, D., & Boopathi, S. (2023). Agricultural Waste Management Systems Using Artificial Intelligence Techniques. In *AI-Enabled Social Robotics in Human Care Services* (pp. 236–258). IGI Global. doi:10.4018/978-1-6684-8171-4.ch009

Kumara, V., Mohanaprakash, T., Fairooz, S., Jamal, K., Babu, T., & Sampath, B. (2023). Experimental Study on a Reliable Smart Hydroponics System. In *Human Agro-Energy Optimization for Business and Industry* (pp. 27–45). IGI Global. doi:10.4018/978-1-6684-4118-3.ch002

Maguluri, L. P., Ananth, J., Hariram, S., Geetha, C., Bhaskar, A., & Boopathi, S. (2023). Smart Vehicle-Emissions Monitoring System Using Internet of Things (IoT). In Handbook of Research on Safe Disposal Methods of Municipal Solid Wastes for a Sustainable Environment (pp. 191–211). IGI Global.

Myilsamy, S., & Sampath, B. (2017). Grey Relational Optimization of Powder Mixed Near-Dry Wire Cut Electrical Discharge Machining of Inconel 718 Alloy. *Asian Journal of Research in Social Sciences and Humanities*, *7*(3), 18–25. doi:10.5958/2249-7315.2017.00157.5

Nadif, M., & Role, F. (2021). Unsupervised and self-supervised deep learning approaches for biomedical text mining. *Briefings in Bioinformatics*, *22*(2), 1592–1603. doi:10.1093/bib/bbab016 PMID:33569575

Nishanth, J., Deshmukh, M. A., Kushwah, R., Kushwaha, K. K., Balaji, S., & Sampath, B. (2023). Particle Swarm Optimization of Hybrid Renewable Energy Systems. In *Intelligent Engineering Applications and Applied Sciences for Sustainability* (pp. 291–308). IGI Global. doi:10.4018/979-8-3693-0044-2.ch016

Pramila, P., Amudha, S., Saravanan, T., Sankar, S. R., Poongothai, E., & Boopathi, S. (2023). Design and Development of Robots for Medical Assistance: An Architectural Approach. In Contemporary Applications of Data Fusion for Advanced Healthcare Informatics (pp. 260–282). IGI Global.

Rahamathunnisa, U., Sudhakar, K., Murugan, T. K., Thivaharan, S., Rajkumar, M., & Boopathi, S. (2023). Cloud Computing Principles for Optimizing Robot Task Offloading Processes. In *AI-Enabled Social Robotics in Human Care Services* (pp. 188–211). IGI Global. doi:10.4018/978-1-6684-8171-4.ch007

Ramudu, K., Mohan, V. M., Jyothirmai, D., Prasad, D., Agrawal, R., & Boopathi, S. (2023). Machine Learning and Artificial Intelligence in Disease Prediction: Applications, Challenges, Limitations, Case Studies, and Future Directions. In Contemporary Applications of Data Fusion for Advanced Healthcare Informatics (pp. 297–318). IGI Global.

Ravisankar, A., Sampath, B., & Asif, M. M. (2023). Economic Studies on Automobile Management: Working Capital and Investment Analysis. In Multidisciplinary Approaches to Organizational Governance During Health Crises (pp. 169–198). IGI Global.

Reddy, M. A., Gaurav, A., Ushasukhanya, S., Rao, V. C. S., Bhattacharya, S., & Boopathi, S. (2023). Bio-Medical Wastes Handling Strategies During the COVID-19 Pandemic. In Multidisciplinary Approaches to Organizational Governance During Health Crises (pp. 90–111). IGI Global. doi:10.4018/978-1-7998-9213-7.ch006

Reddy, M. A., Reddy, B. M., Mukund, C., Venneti, K., Preethi, D., & Boopathi, S. (2023b). Social Health Protection During the COVID-Pandemic Using IoT. In *The COVID-19 Pandemic and the Digitalization of Diplomacy* (pp. 204–235). IGI Global. doi:10.4018/978-1-7998-8394-4.ch009

Reddy, M. A., Reddy, B. M., Mukund, C. S., Venneti, K., Preethi, D. M. D., & Boopathi, S. (2023a). Social Health Protection During the COVID-Pandemic Using IoT. In *The COVID-19 Pandemic and the Digitalization of Diplomacy* (pp. 204–235). IGI Global. doi:10.4018/978-1-7998-8394-4.ch009

Samikannu, R., Koshariya, A. K., Poornima, E., Ramesh, S., Kumar, A., & Boopathi, S. (2022). Sustainable Development in Modern Aquaponics Cultivation Systems Using IoT Technologies. In *Human Agro-Energy Optimization for Business and Industry* (pp. 105–127). IGI Global.

Sampath, B., Pandian, M., Deepa, D., & Subbiah, R. (2022). Operating parameters prediction of liquefied petroleum gas refrigerator using simulated annealing algorithm. *AIP Conference Proceedings*, *2460*(1), 70003. doi:10.1063/5.0095601

Sampath, B., Yuvaraj, T., & Velmurugan, D. (2022). Parametric analysis of mould sand properties for flange coupling casting. *AIP Conference Proceedings*, *2460*(1), 070002. doi:10.1063/5.0095599

Saravanan, M., Vasanth, M., Boopathi, S., Sureshkumar, M., & Haribalaji, V. (2022). Optimization of Quench Polish Quench (QPQ) Coating Process Using Taguchi Method. *Key Engineering Materials*, *935*, 83–91. doi:10.4028/p-z569vy

Sathish, T., Sunagar, P., Singh, V., Boopathi, S., Al-Enizi, A. M., Pandit, B., Gupta, M., & Sehgal, S. S. (2023). Characteristics estimation of natural fibre reinforced plastic composites using deep multi-layer perceptron (MLP) technique. *Chemosphere*, *337*, 139346. doi:10.1016/j.chemosphere.2023.139346 PMID:37379988

Sathish, T., Sunagar, P., Singh, V., Boopathi, S., Sathyamurthy, R., Al-Enizi, A. M., Pandit, B., Gupta, M., & Sehgal, S. S. (2023). Characteristics estimation of natural fibre reinforced plastic composites using deep multi-layer perceptron (MLP) technique. *Chemosphere*, *337*(June), 139346. doi:10.1016/j.chemosphere.2023.139346 PMID:37379988

Selvakumar, S., Adithe, S., Isaac, J. S., Pradhan, R., Venkatesh, V., & Sampath, B. (2023). A Study of the Printed Circuit Board (PCB) E-Waste Recycling Process. In Sustainable Approaches and Strategies for E-Waste Management and Utilization (pp. 159–184). IGI Global.

Selvakumar, S., Shankar, R., Ranjit, P. S., Bhattacharya, S., Gupta, A. S. G., & Boopathi, S. (2023). E-Waste Recovery and Utilization Processes for Mobile Phone Waste. In *Handbook of Research on Safe Disposal Methods of Municipal Solid Wastes for a Sustainable Environment* (pp. 222–240). IGI Global. doi:10.4018/978-1-6684-8117-2.ch016

Sengeni, D., Padmapriya, G., Imambi, S. S., Suganthi, D., Suri, A., & Boopathi, S. (2023). Biomedical Waste Handling Method Using Artificial Intelligence Techniques. In *Handbook of Research on Safe Disposal Methods of Municipal Solid Wastes for a Sustainable Environment* (pp. 306–323). IGI Global. doi:10.4018/978-1-6684-8117-2.ch022

Singh Grewal, G., & Parkash Dhiman, D. (2019). Effect of deep cryogenic treatment on copper electrode for non-traditional electric discharge machining (EDM). *Mechanical Sciences*, *10*(2), 413–427. doi:10.5194/ms-10-413-2019

Subha, S., Inbamalar, T. M., R, K. C., Suresh, L. R., Boopathi, S., & Alaskar, K. (2023, February). A Remote Health Care Monitoring system using internet of medical things (IoMT). *IEEE Explore*. doi:10.1109/ICIPTM57143.2023.10118103

Syamala, M., Komala, C., Pramila, P., Dash, S., Meenakshi, S., & Boopathi, S. (2023). Machine Learning-Integrated IoT-Based Smart Home Energy Management System. In *Handbook of Research on Deep Learning Techniques for Cloud-Based Industrial IoT* (pp. 219–235). IGI Global. doi:10.4018/978-1-6684-8098-4.ch013

Uçar, A., Demir, Y., & Güzeliş, C. (2017). Object recognition and detection with deep learning for autonomous driving applications. *Simulation*, *93*(9), 759–769. doi:10.1177/0037549717709932

Veeranjaneyulu, R., Boopathi, S., Kumari, R. K., Vidyarthi, A., Isaac, J. S., & Jaiganesh, V. (2023). Air Quality Improvement and Optimisation Using Machine Learning Technique. *IEEE Explore*, 1–6.

Veeranjaneyulu, R., Boopathi, S., Narasimharao, J., Gupta, K. K., Reddy, R. V. K., & Ambika, R. (2023). Identification of Heart Diseases using Novel Machine Learning Method. *IEEE Explore*, 1–6.

Venkateswaran, N., Vidhya, K., Ayyannan, M., Chavan, S. M., Sekar, K., & Boopathi, S. (2023). A Study on Smart Energy Management Framework Using Cloud Computing. In 5G, Artificial Intelligence, and Next Generation Internet of Things: Digital Innovation for Green and Sustainable Economies (pp. 189–212). IGI Global. doi:10.4018/978-1-6684-8634-4.ch009

Venkateswaran, N., Vidhya, R., Naik, D. A., Raj, T. F. M., Munjal, N., & Boopathi, S. (2023). Study on Sentence and Question Formation Using Deep Learning Techniques. In *Digital Natives as a Disruptive Force in Asian Businesses and Societies* (pp. 252–273). IGI Global. doi:10.4018/978-1-6684-6782-4.ch015

Zekrifa, D. M. S., Kulkarni, M., Bhagyalakshmi, A., Devireddy, N., Gupta, S., & Boopathi, S. (2023). Integrating Machine Learning and AI for Improved Hydrological Modeling and Water Resource Management. In *Artificial Intelligence Applications in Water Treatment and Water Resource Management* (pp. 46–70). IGI Global. doi:10.4018/978-1-6684-6791-6.ch003

Chapter 16
Bioremediation of Pharma Pollutants Using Synthetic Systems

Meghmala S. Waghmode
https://orcid.org/0000-0002-5635-134X
Department of Microbiology, PDEA's Annasaheb Magar Mahavidyalaya, Pune, India

Aparna Baban Gunjal
Department of Microbiology, Dr. D.Y. Patil, Arts, Commerce, and Science College, India

Manisha Gadekar
Department of Computer Science, PDEA's Annasaheb Magar Mahavidyalaya, Pune, India

Dipali Deepak Mali
https://orcid.org/0000-0002-6180-4155
Department of Computer Science, PDEA's Annasaheb Magar Mahavidyalaya, Pune, India

ABSTRACT

The pharmaceutical industry is responsible for the environmental pollution imposed by drugs, heavy metals, and surfactants. In-situ remediation of pharma pollutants with the cell-based system has some limitations like sensing of contaminants, efficacy, cost, concentration of pollutants, toxicity level, etc. Synthetic biology is important for gene mining, enzymes discovery, and biodegradation pathway of pollutants. The chapter is based on the synthetic systems for remediation of the pharma pollutants.

INTRODUCTION

Pharmaceutical wastes are one of the emerging pollutants to the ecosystem. Reports are available on their presence in water bodies, soil and WWTPs affecting the human and other life forms (Khetan and Collins, 2007; Kosjek and Krbavcic, 2005). Sources of pharmaceutical wastes are manufacturing industries, health care units, residential wastewater etc. (Gadepalli et al., 2014). Pharmaceutical wastes

DOI: 10.4018/978-1-6684-6577-6.ch016

involve dyes (azo, triphenylmethane dyes, xanthine dyes, and quinoline dyes), surfactants (sodium dodecyl sulphate (SDS) and cetrimonium salt (ATAC-C16)), antimicrobial compounds, synthetic musk's, hormones, preservatives, heavy metals (copper, cadmium, lead, nickel, chromium and zinc) (Chaturvedi et al., 2021; Chimezie et al., 2011; Gicevic et al., 2019). Azo dyes used in pharma industries are E133 Brilliant Blue, E102 Tartrazine, Ponceau 4R (Cochineal Red A), E110 Sunset Yellow FCF, Azorubine (Carmoisine), Amaranth, and E129 Allura Red, contains benzene rings which makes them harmful (Gicevic et al., 2019; Pawar and Kumar, 2002). Persistence of drugs in water is due to their high-water solubility, non-detectable effective concentration, and bioaccumulation. In waste water treatment plants, attention was given on the removal of physical objects and chemical residues. But with the recent research, concept of genomic surveillance of waste water treatment plants is emerging as a part of epidemiology. Brief outline of genomic surveillance of WWTP is given in Fig. 1. Genomic surveillance is important to find the chances of infection through antigens or mobile genetic elements as well as development of antimicrobial resistance in nosocomial pathogens (Rodriguez et al., 2021). During covid 19 pandemic situation, genomic surveillance was done via next generation sequencing to find out the antigenic variation in virus as well as effect of vaccine against this viral pathogen to control morbidity and mortality rates (Robishaw et al., 2021).

Pharma wastes removal from the waste water treatment plant has been carried out with the help of physico-chemical or biological methods. Biological methods are less costly and eco- friendly as compared to physico-chemical methods. In biological methods, aerobic and anaerobic microorganisms are used in free or immobilized form.

Bioremediation is the method which depends on the natural biological activity for the reduction in the

Figure 1. Genomic surveillance of waste water treatment plant

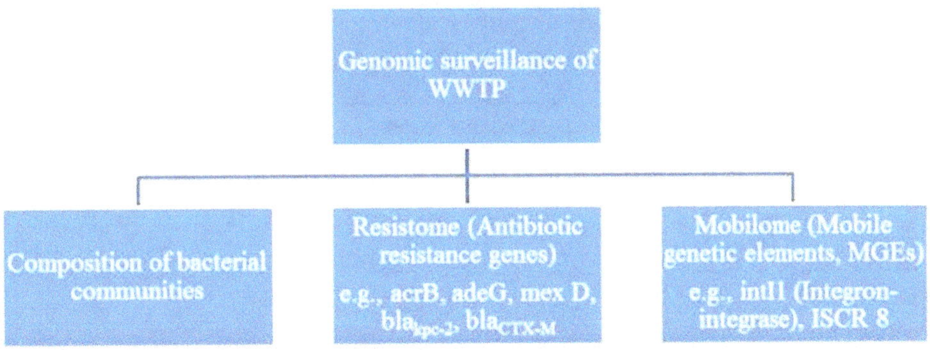

toxicity of harmful compounds (Vidali, 2001). Biological contribution for the remediation of toxicants is one of the extensively studied research topic due to the changes in the microbial behavior or alteration of the microbial community in presence of toxicant (Iwamoto and Nasu, 2001). With the progressive developments in the field of molecular biology, researchers are trying to explore the activity of culturable as well as unculturable bacteria. To find out the microbial load in the contaminated area, advanced molecular microbiological methods are required to optimize and validate the process of bioremediation (Iwamoto and Nasu, 2001). For the effective bioremediation process, some factors like viability of microorganisms, optimal concentration of pollutant, environmental conditions like pH, temperature, gases etc. are important. To circumvent the issues related with the environmental performance of bacteria, synthetic systems can be the promising technology.

POLLUTANTS RELEASED BY PHARMACEUTICAL INDUSTRIES

The pharma sectors release waste water during various process and manufacturing section. This waste water released from the pharmaceutical industries is polluted with heavy metals such as zinc (Zn), cadmium (Cd), lead (Pb), nickel (Ni), arsenic (As), various solids, etc. and hence needs treatment. The waste water released by the pharma sector hence needs to be treated to reduce the toxicity. Bioremediation of pharma pollutants using synthetic biology approach can be the best solution for this. Synthetic biology is contributing in the field of toxicology with the help of the subfield of this subject, "xenobiology" where synthetic systems are built as bio-orthogonal systems (Schmidt and Pei, 2011). The pollutants released by pharma sector and its waste water are represented in Fig. 2. The pH of pharmaceutical waste water is around 5.8-7.8, biological oxygen demand 20-620 mg/l, chemical oxygen demand (COD) 128-960 mg/l and total solids 880-2040 mg/l (Choudhary and Parmar, 2013). The methods for bioremediation are represented in Fig. 3. concept.

Figure 2. Pollutants from the pharma sector and its waste water

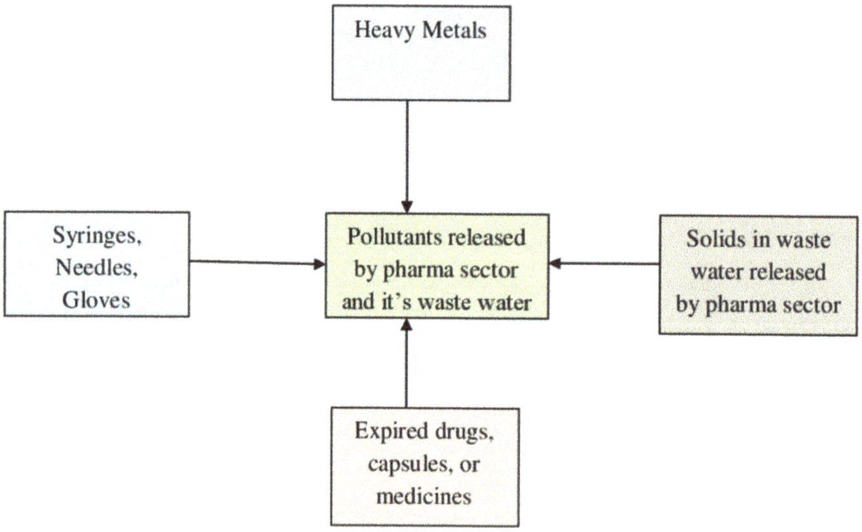

CONCEPT OF SYNTHETIC SYSTEMS

Development in the field of systems biology and synthetic biology, it's possible to explore the performance of cell factories by exploiting the potential of cell metabolism. Complete study of systems biology of plants, active pharmaceutical ingredients can be produced by microorganisms by using the synthetic biology approach (Xu et al., 2013). Metabolic network of the microorganisms can be influenced by collecting and assimilating the data of genomic, proteomic, metabolomic and transcriptomics. Synthetic biology is the engineering of complex biological units like protein, DNA, and gene networks, for finding the applications in the field of biotechnology (Heinemann and Panke, 2006). Application oriented synthetic systems are designed using the prior knowledge of synthetic biology (Purnick and Weiss, 2009). Synthetic biology and CRISPR has been important attributes for the survival of microorganisms living in hostile environmental conditions of space (Thombre et al., 2022). During space missions' nuclear energy is generally utilized which causes use of radioactive material. In BugNRG experiment, microbial fuel cell was developed with *Rhodoferax ferrireducens* to produce electricity (De Vet and Rutgers, 2007). Application based designed synthetic systems have gained importance in the field of remediation to sense and remediate the toxic pollutants from the environment.

Figure 3. Methods for bioremediation

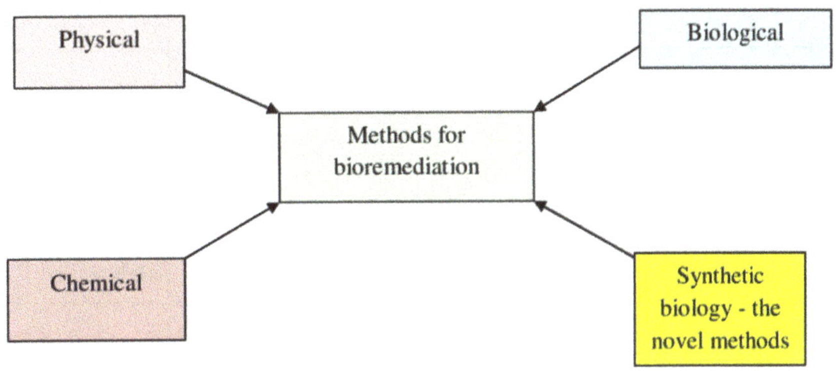

DATABASE AND TOOLS FOR REMEDIATION

For understanding and improvement in the process of remediation, the basic understanding of systems biology and synthetic biology is required. Systems biology deals with the cellular phenomena and integration of the network of enzymes, genes and metabolites using experimental and modeling tactics. For the bioremediation of recalcitrant or persistent compounds, studying as well as validating functional gene mining through the process of computational biology and multi omics approaches, gives new insights. Various databases are available to decode the biodegradation network as given in table stores manually curated information about proteins and genes directly implicated in the biodegradation metabolism. The databases for the elucidation of biodegradation network are given in Table 1.

BIOINFORMATIC TOOLS IN SYNTHETIC BIOLOGY

Bioinformatics tools are designed for DESIGN, BUILD as well as TEST stage. The DESIGN and BUILD stages includes bioinformatics tools for devices (pathways), selection, synthesis, systems (chassis), the optimization of parts (enzymes and regulatory elements) and assembly. The TEST stage includes bioinformatics tools for screening, identification, and metabolite's quantification for rapid prototyping.

Table 1. Databases for the elucidation of biodegradation network

Databases	Application	Reference
Bionemo (http://bionemo.bioinfo.cnio.es)	Gives information about the proteins, genes related data	Carbajosa et al., 2009
University of Minessota Biocatalysis / Biodegradation Database (UM-BBD) http://umbbd.ahc.umn.edu/	Contains the information about microbial catabolism of pollutants	Ellis et al., 2006
Metarouter http://pdg.cnb.uam.es/MetaRouter	Decoding pathway for biodegradation	Pazos et al., 2005
MOAtox	Database for predictive modelling about acute toxicity of chemicals based on QSAR	Barron et al., 2015
ECOTOX	Ecotoxicity of chemicals	EPA ECOTOX database, 2013
KEGG	Biodegradation pathway of pollutants	Kanehisa et al., 2004
Enzyme database ExPASy WWW server (http://www.expasy.ch/enzyme/).	Information about enzymes involved in catabolism.	Bairoch, 2000
BRENDA http://www.brenda.uni-koeln.de/	Information about enzymes	Schomburg et al., 2002
MetaCyc	Pathways, reaction, and enzymes	Karp, 2000
KAshinhou Tool for Ecotoxicity (KATE) system	Ecotoxicity based on QSAR	Furuhama et al., 2010
TOXNET	Ecotoxicity database	Hung and Nelson, 1997
VirtualTox lab www.biograf.ch	Prediction of pharmaceuticals in aquatic system.	Christen et al., 2010
Knowledge-based expert systems (KBES) [Assessment Tools for the Evaluation of Risk (ASTER), TOPKAT, Computer Assisted Evaluation of industrial chemical Substances According to Regulations (CAESAR), DEREK, OECD Tool box	Prediction about fate of chemicals using *in-silico* based methods.	Roy and Kar, 2016
Electronic Microbial Bioremediation (eMBR) web portal	Increases collaborative research activity in remediation by giving information about suitable microorganism, pathway etc.	Elekwachi, 2012

The biological databases consist of biological data in an organized manner such as DNA sequences, molecular structure, protein sequencing etc. Various computer tools are available for the manipulation the biological data. These databases are freely accessible to all. Depending upon the content stored in database, they are classified into primary, secondary, and composite databases (Kumari et al., 2016). As primary databases archive the experimental results submitted by the scientists so also known as an archival database. It is populated with experimentally derived data such as genome sequence, macro-molecular structure, etc. Examples of primary databases include Nucleic Acid Databases (GenBank and DDBJ) and protein databases (PDB, SwissProt, PIR, TrEMBL, Metacyc etc.). The primary database's analyzed result is the data stored in the secondary database. Inside the secondary database, informative and meaningful data is stored by applying computational algorithms on the primary database as well as the data are highly curated. As compare to primary database, a secondary database contains more valuable knowledge as well as is much better. Examples of secondary database include InterPro, Uniprot, Prosite (De Castro et al., 2006). In composite databases, data entered are first compared and then filtered depending on desired criteria. From primary database, initial data are taken and then based on certain condition it is merged. With the help of composite database, searching sequences becomes rapid. Data is non- redundant in Composite Databases. Examples of composite database include LIGAND (Sapundzhi and Dzimbova, 2022). The bioinformatics tools in synthetic biology are represented in Fig. 4.

Figure 4. Bioinformatics tools in synthetic biology

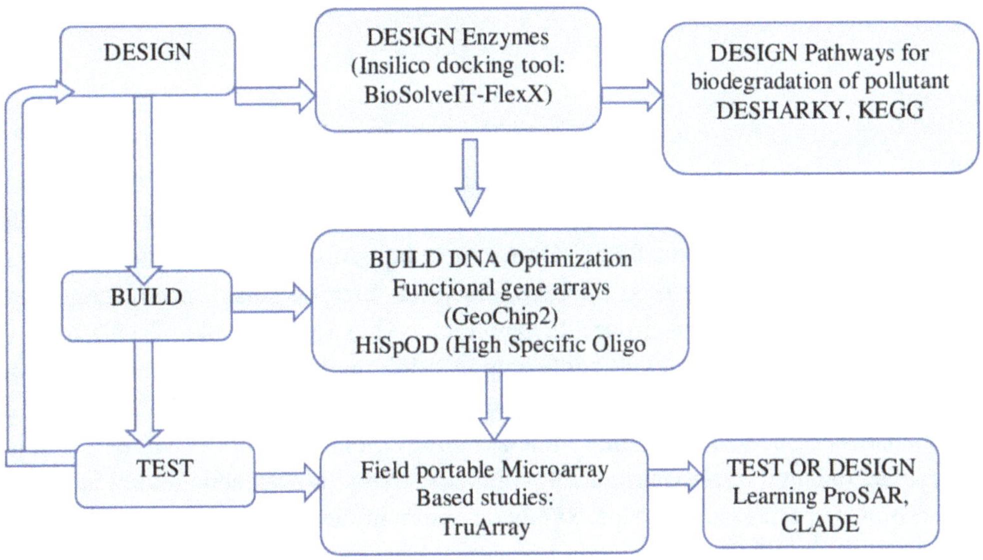

SYNTHETIC STRATEGIES FOR BIOREMEDIATION OF PHARMA POLLUTANTS

The synthetic strategies for bioremediation of pharm pollutants are represented in Fig. 5. Synthetic biology is the best solution for the bioremediation of pharma pollutants. The researchers are gaining immense interest in this area. The pharmaceutical contaminants accumulate on a large scale and cause pollution which affects the humans and environment. The synthetic biology methods are promising on a large-scale. The use of synthetic biology for the bioremediation of pharmaceutical pollutants will reduce the pollution which is very necessary. The synthetic biology methods for the bioremediation of pharmaceutical contaminants will be eco-friendly and effective. Omics technologies help researchers to construct the efficient microorganisms for the remediation application.

SYNTHETIC MICROBIAL CONSORTIA FOR BIOREMEDIATION OF PHARMA POLLUTANTS

A microbial consortium is combined effect of two or more microorganisms which work together (Khoo et al., 2020; Padmaperuma et al., 2020). The consortium of microorganisms is more effective when compared to single microorganism for the bioremediation (Bidja et al., 2019). The combination of al-

Figure 5. Synthetic strategies for bioremediation of pharma pollutants

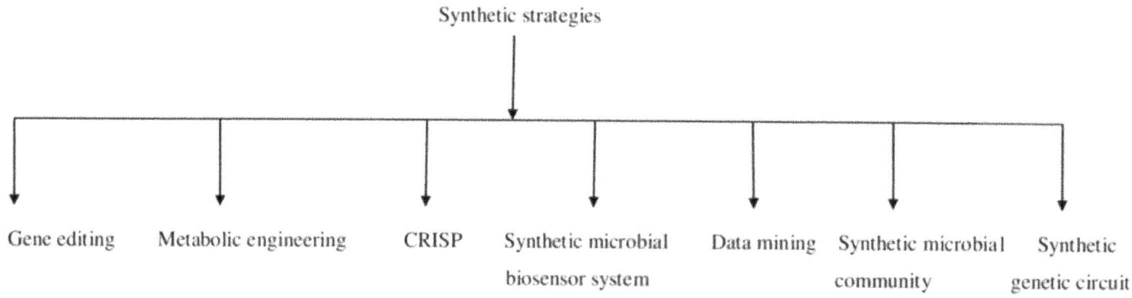

lochthonous microorganisms to prepare consortia must be used (Chen et al., 2020; Guo et al., 2017). Molecular tools along with systems biology will help to know the genome and various interactions among the cells (Jaiswal and Shukla, 2020). Omics tools help to know the metabolic pathways in the degradation process. With the help of synthetic biology, the microorganisms can be engineered which will be useful in bioremediation of pharmaceutical pollutants (Jaiswal and Shukla, 2020). In the consortia of microorganisms, the synthetic biology tools help to bring the interactions among microorganisms. The synthetic microbial consortia will be the best platform for bioremediation of pharma contaminants. The microorganisms engineered using the synthetic biology elements or tools will be a real promising solution for the degradation of pharmaceutical pollutants. Using the synthetic microbial consortia, the artificial microbiome species can be made. Synthetic communities can be formed by the use of two or more microorganisms under the same conditions which have proper interactions with each other and also function with each other (Sharma and Shukla, 2022). These interactions can be mutualism, competition and cooperation. These interactions are helpful for the bioremediation process. The community's dynamics and structure are dependent on this. The metabolism among these microorganisms facilitates the communication among them (Romero et al., 2019). Cooperation is the main factor in the microbial community structure. This cooperation is mainly influenced by synthetic communities (Che and Men, 2019). The deletion of genes can be useful for cooperation of microorganisms and can be used to engineer the microorganisms. These engineered microorganisms can also have various different interactions. The synthetic biology helps for long term existence of the microbes.

SYNTHETIC SYSTEMS FOR SENSING THE PHARMA POLLUTANTS

Quorum sensing operated cell coordination and building the microbial network with improved cell signaling pathway by gene editing tools is the first key step for the remediation of recalcitrant compounds (Jaiswal and Shukla, 2020).

SYNTHETIC GENETIC CIRCUIT AND MICROBIAL BIOSENSOR

The synthetic genetic circuit is important in approach of bioremediation of pharma pollutants. *Pseudomonas putida* is important model for synthetic biology chassis (Pabo and Nekludova, 2000). There are reports which show that this *P. putida* is the best model for use in bioremediation (Tanveer et al., 2018). The synthetic genetic circuit has been designed of P. putida for the bioremediation of pharma compounds (Adams, 2016). The genetic circuits can be helpful in the degradation of pharma products toxins (Checa et al., 2012; Tay et al., 2017). The two-component regulatory system (TCRS) based synthetic biology will be effective for bioremediation of pharmaceutical products. Synthetic biology application can be used for making whole cell biosensors. The whole cell biosensors can be beneficial to monitor the pharmaceutical pollutants. The biosensors will sense the pollutant, create signals, and detect the pharmaceutical pollutants. *P. fluorescens* was first used to make whole cell biosensor to be used in bioremediation. The synthetic systems and their applications are represented in Table 2.

Table 2. Synthetic systems and their applications

Synthetic Systems	Components	Application	Reference
Cas13-based SHERLOCK (Specific High Sensitivity Enzymatic Reporter Unlocking) technology	polymerase and caspase	Detection of RNA and DNA	Kellner et al., 2019
Organophosphates (OPs) hydrolyzing enzymatic systems	enzymes	For the remediation of OP compounds	Mali et al., 2022
Engineered enzymatic bacteriophage	enzymatically active DspB-expressing T7DspB phage	Dispersion of bacterial biofilm	Lu and Collins, 2007
Laccase mediated system	laccase and redox mediator	Degradation of synthetic fragrance	Vallecillos et al., 2017
Laccase system	laccase from *Trametes versicolor*	Removal of flumequine (FLU)	Ashrafi et al., 2016
Whole-cell oxido-reductase system	expression of rat NADPH-cytochrome P450 oxidoreductase (CPR, 77-kDa) on the surface of *Escherichia coli* with the help of ice-nucleation protein from *Pseudomonas syringae*	Bioremediation	Yim et al., 2006
HRP-immobilized cross-linked enzyme aggregates (HRP-CLEAs)	horse radish peroxidase, cross kinker	Degradation of dyes.	Bilal et al., 2017
DNA-editing All-in-one RNA-guided CRISPR-Cas Transposase (DART) systems	CRISPR	For targeted DNA insertion	Rubin et al., 2020
Cyt P450 based system	cytochrome p450	Dehalogenation and denitration of halogenated hydrocarbons and nitroamine explosives.	Behrendorff, 2021
Engineered P450 on Viral like particles (VLP)	-	*Insitu* detoxification of drugs in organ	Uchida et al., 2015
Display of Synthetic phytochelatins (ECs) on the surface of *Moraxella* sp.	-	Reduction of heavy metal	Bae et al., 2002

ENGINEERED MICROBES FOR BIOREMEDIATION

Microorganisms have the potential to degrade or completely mineralize the toxicants through the degradative pathway. Microorganisms are effective in the remediation due to their ecological adaptability, small size, speedy growth but some parameters like mixture of toxicants, environmental conditions, and incomplete biodegradation pathway. Such hurdles can be resolved with the genetically engineered microorganisms with functional biodegradative pathway (Yaashikaa et al., 2022). For bioremediation, developing genetically modified microorganisms with the help of systemic biology is the emerging trend in the field of biology. Synthetic biology adopts on the principle of "Design, Build, Check and Learn" where genetic materials is standardized to form functional unit (Capeness and Horsfall, 2020). With the help of synthetic biology, both culturable as well as unculturable bacteria can be effectively utilized by giving importance to the genes, proteins, or enzymes. Various extremophilic microorganisms have the potential to assist in the remediation process.

GENETIC AND METABOLIC ENGINEERING

Genome editing facilitates addition, deletion or replacement of fragments of DNA (Enriquez, 2016). Transcription-activators like effector nucleases (TALEN), clustered regularly interspaced short palindromic repeats (CRISPR-Cas) and zinc finger nucleases (ZFNs) are the recent gene editing tools. From these, CRISPR-Cas is most widely used gene editing tool (Kanchiswamy et al., 2016). These advanced gene editing tools are useful in bioremediation. TALEN has a DNA-binding modular which is sequence-specific for the host genome (Utturkar et al., 2013). These gene editing tools create the knock-in and knock-out and are under processing for implementation in bioremediation studies (Kumar et al., 2018). The model microorganisms used in CRISPR-Cas system are *Pseudomonas* sp. (Karimi et al., 2015; Nogales et al., 2020) or *E-coli* (Chen et al., 2018; Marshall et al., 2018; Pontrelli et al., 2018). Metabolic engineering leads to modification of the existing pathway for the betterment of the bioremediation process (Michel et al., 2007).

SOLUTIONS AND RECOMMENDATIONS

The bioremediation of pharma pollutants using synthetic approach will be very useful. The bioremediation of pharma pollutants will help in the clean-up of environment. The synthetic approach in bioremediation will involve various genetic aspects.

FUTURE RESEARCH DIRECTIONS

The bioremediation of various other wastes using synthetic approach needs to be studied in detail.

CONCLUSION

Bioremediation using synthetic approach for the degradation of pharma pollutants will be eco-friendly. This will be an advanced method for the clean-up of pollutants from pharmaceutical industries and can be used globally. This will reduce the pharma pollutants in the environment ensuring sustainability as well. Various other methods using synthetic biology must also be studied for the bioremediation of pollutants.

REFERENCES

Adams, B. L. (2016). The next generation of synthetic biology chassis: Moving synthetic biology from the laboratory to the field. *ACS Synthetic Biology*, *5*(12), 1328–1330. doi:10.1021/acssynbio.6b00256 PMID:27665861

Ashrafi, S. D., Nasseri, S., Alimohammadi, M., Mahvi, A. H., & Faramarzi, M. A. (2016). Optimization of the enzymatic elimination of flumequine by laccase-mediated system using response surface methodology. *Desalination and Water Treatment*, *57*(31), 14478–14487. doi:10.1080/19443994.2015.1063462

Bae, W., Mulchandani, A., & Chen, W. (2002). Cell surface display of synthetic phytochelatins using ice nucleation protein for enhanced heavy metal bioaccumulation. *Journal of Inorganic Biochemistry*, *88*(2), 223–227. doi:10.1016/S0162-0134(01)00392-0 PMID:11803043

Bairoch, A. (2000). The ENZYME database in 2000. *Nucleic Acids Research*, *28*(1), 304–305. doi:10.1093/nar/28.1.304 PMID:10592255

Barron, M. G., Lilavois, C. R., & Martin, T. M. (2015). MOAtox: A comprehensive mode of action and acute aquatic toxicity database for predictive model development. *Aquatic Toxicology (Amsterdam, Netherlands)*, *161*, 102–107. doi:10.1016/j.aquatox.2015.02.001 PMID:25700118

Behrendorff, J. B. (2021). Reductive cytochrome P450 reactions and their potential role in bioremediation. *Frontiers in Microbiology*, *12*, 649273. doi:10.3389/fmicb.2021.649273 PMID:33936006

Bidja Abena, M. T., Sodbaatar, N., Li, T., Damdinsuren, N., Choidash, B., & Zhong, W. (2019). Crude oil biodegradation by newly isolated bacterial strains and their consortium under soil microcosm experiment. *Applied Biochemistry and Biotechnology*, *189*(4), 1223–1244. doi:10.100712010-019-03058-2 PMID:31236896

Bilal, M., Iqbal, H. M., Hu, H., Wang, W., & Zhang, X. (2017). Development of horseradish peroxidase-based cross-linked enzyme aggregates and their environmental exploitation for bioremediation purposes. *Journal of Environmental Management*, *188*, 137–143. doi:10.1016/j.jenvman.2016.12.015 PMID:27978441

Capeness, M. J., & Horsfall, L. E. (2020). Synthetic biology approaches towards the recycling of metals from the environment. *Biochemical Society Transactions*, *48*(4), 1367–1378. doi:10.1042/BST20190837 PMID:32627824

Carbajosa, G., Trigo, A., Valencia, A., & Cases, I. (2009). Bionemo: Molecular information on biodegradation metabolism. *Nucleic Acids Research, 37*(Database, suppl_1), D598–D602. doi:10.1093/nar/gkn864 PMID:18986994

Chaturvedi, P., Shukla, P., Giri, B. S., Chowdhary, P., Chandra, R., Gupta, P., & Pandey, A. (2021). Prevalence and hazardous impact of pharmaceutical and personal care products and antibiotics in environment: A review on emerging contaminants. *Environmental Research, 194*, 110664. doi:10.1016/j.envres.2020.110664 PMID:33400949

Che, S., & Men, Y. (2019). Synthetic microbial consortia for biosynthesis and biodegradation: Promises and challenges. *Journal of Industrial Microbiology & Biotechnology, 46*(9-10), 1343–1358. doi:10.100710295-019-02211-4 PMID:31278525

Checa, S. K., Zurbriggen, M. D., & Soncini, F. C. (2012). Bacterial signaling systems as platforms for rational design of new generations of biosensors. *Current Opinion in Biotechnology, 23*(5), 766–772. doi:10.1016/j.copbio.2012.05.003 PMID:22658939

Chen, W., Kong, Y., Li, J., Sun, Y., Min, J., & Hu, X. (2020). Enhanced biodegradation of crude oil by constructed bacterial consortium comprising salt-tolerant petroleum degraders and biosurfactant producers. *International Biodeterioration & Biodegradation, 154*, 105047. doi:10.1016/j.ibiod.2020.105047

Chen, W., Zhang, Y., Zhang, Y., Pi, Y., Gu, T., Song, L., Wang, Y., & Ji, Q. (2018). CRISPR/Cas9-based genome editing in *Pseudomonas aeruginosa* and cytidine deaminase-mediated base editing in *Pseudomonas* species. *iScience, 6*, 222–231. doi:10.1016/j.isci.2018.07.024 PMID:30240613

Chimezie, A., Kenneth, N., Olufunsho, A., Chinwe, N., Mohsen, A., & Herbert, C. (2011). Concentrations of heavy metals in some pharmaceutical effluents in Lagos, Nigeria. *Journal of Environmental Chemistry and Ecotoxicology, 3*(2), 25–31.

Choudhary, S., & Parmar, N. (2013). Hazard assessment of liquid effluent treatment plant in pharmaceutical industry. *VSRD Int J Technol NonTechnol Res, 4*(9), 209–214.

Christen, V., Hickmann, S., Rechenberg, B., & Fent, K. (2010). Highly active human pharmaceuticals in aquatic systems: A concept for their identification based on their mode of action. *Aquatic Toxicology (Amsterdam, Netherlands), 96*(3), 167–181. doi:10.1016/j.aquatox.2009.11.021 PMID:20053463

De Castro, E., Sigrist, C. J., Gattiker, A., Bulliard, V., Langendijk-Genevaux, P. S., Gasteiger, E., & Hulo, N. (2006). Scan Prosite: Detection of PROSITE signature matches and ProRule-associated functional and structural residues in proteins. *Nucleic Acids Research, 34*(Web Server, suppl_2), W362–W365. doi:10.1093/nar/gkl124 PMID:16845026

De Vet, S. J., & Rutgers, R. (2007). From waste to energy: First experimental bacterial fuel cells on board the International Space Station. *Microgravity Science and Technology, 19*(5-6), 225–229. doi:10.1007/BF02919487

Elekwachi, C. O. (2012). *Bioinformatics resources to support bioremediation research* [Doctoral dissertation]. University of Nottingham.

Ellis, L. B., Roe, D., & Wackett, L. P. (2006). The University of Minnesota biocatalysis / biodegradation database: The first decade. *Nucleic Acids Research, 34*(suppl_1), D517-D521.

Enríquez, P. (2016). Genome editing and the jurisprudence of scientific empiricism. *Vand. J. Ent. & Tech. L., 19*, 603.

EPA ECOTOX database. (2013). https://www.epa.gov/ecotox/

Furuhama, A., Toida, T., Nishikawa, N., Aoki, Y., Yoshioka, Y., & Shiraishi, H. (2010). Development of an ecotoxicity QSAR model for the KAshinhou Tool for ecotoxicity (KATE) system, March 2009 version. *SAR and QSAR in Environmental Research, 21*(5-6), 403–413. doi:10.1080/1062936X.2010.501815 PMID:20818579

Gadipelly, C., Perez-Gonzalez, A., Yadav, G. D., Ortiz, I., Ibáñez, R., Rathod, V. K., & Marathe, K. V. (2014). Pharmaceutical industry waste water: Review of the technologies for water treatment and reuse. *Industrial & Engineering Chemistry Research, 53*(29), 11571–11592. doi:10.1021/ie501210j

Gicevic, A., Hindija, L., & Karacic, A. (2019, May). Toxicity of azo dyes in pharmaceutical industry. In *International Conference on Medical and Biological Engineering* (pp. 581-587). Springer.

Guo, G., Tian, F., Ding, K., Wang, L., Liu, T., & Yang, F. (2017). Effect of a bacterial consortium on the degradation of polycyclic aromatic hydrocarbons and bacterial community composition in Chinese soils. *International Biodeterioration & Biodegradation, 123*, 56–62. doi:10.1016/j.ibiod.2017.04.022

Heinemann, M., & Panke, S. (2006). Synthetic biology - Putting engineering into biology. *Bioinformatics (Oxford, England), 22*(22), 2790–2799. doi:10.1093/bioinformatics/btl469 PMID:16954140

Huber, S., Remberger, M., Kaj, L., Schlabach, M., Jörundsdóttir, H. O., Vester, J., Arnórsson, M., Mortensen, I., Schwartson, R., & Dam, M. (2016). A first screening and risk assessment of pharmaceuticals and additives in personal care products in waste water, sludge, recipient water and sediment from Faroe Islands, Iceland, and Greenland. *The Science of the Total Environment, 562*, 13–25. doi:10.1016/j.scitotenv.2016.03.063 PMID:27092417

Hung, O., & Nelson, L. S. (1997). Toxnet - A Guide to clinical toxicology resources. *Journal of Toxicology. Clinical Toxicology, 35*(6), 677–679. doi:10.3109/15563659709001254

Iwamoto, T., & Nasu, M. (2001). Current bioremediation practice and perspective. *Journal of Bioscience and Bioengineering, 92*(1), 1–8. doi:10.1016/S1389-1723(01)80190-0 PMID:16233049

Jaiswal, S., & Shukla, P. (2020). Alternative strategies for microbial remediation of pollutants via synthetic biology. *Frontiers in Microbiology, 11*, 808. doi:10.3389/fmicb.2020.00808 PMID:32508759

Kanchiswamy, C. N., Maffei, M., Malnoy, M., Velasco, R., & Kim, J. S. (2016). Fine-tuning next-generation genome editing tools. *Trends in Biotechnology, 34*(7), 562–574. doi:10.1016/j.tibtech.2016.03.007 PMID:27167723

Kanehisa, M., Goto, S., Kawashima, S., Okuno, Y., & Hattori, M. (2004). The KEGG resource for deciphering the genome. *Nucleic Acids Research, 32*(90001), D277–D280. doi:10.1093/nar/gkh063 PMID:14681412

Karimi, B., Habibi, M., & Esvand, M. (2015). Biodegradation of naphthalene using *Pseudomonas aeruginosa* by up flow anoxic–aerobic continuous flow combined bioreactor. *Journal of Environmental Health Science & Engineering*, *13*(1), 1–10. doi:10.118640201-015-0175-1 PMID:25859393

Karp, P. D., Riley, M., Saier, M., Paulsen, I. T., Paley, S. M., & Pellegrini-Toole, A. (2000). The ecocyc and metacyc databases. *Nucleic Acids Research*, *28*(1), 56–59. doi:10.1093/nar/28.1.56 PMID:10592180

Kellner, M. J., Koob, J. G., Gootenberg, J. S., Abudayyeh, O. O., & Zhang, F. (2019). SHERLOCK: Nucleic acid detection with CRISPR nucleases. *Nature Protocols*, *14*(10), 2986–3012. doi:10.103841596-019-0210-2 PMID:31548639

Khetan, S. K., & Collins, T. J. (2007). Human pharmaceuticals in the aquatic environment: A challenge to green chemistry. *Chemical Reviews*, *107*(6), 2319–2364. doi:10.1021/cr020441w PMID:17530905

Khoo, K. S., Chew, K. W., Yew, G. Y., Leong, W. H., Chai, Y. H., Show, P. L., & Chen, W. H. (2020). Recent advances in downstream processing of microalgae lipid recovery for biofuel production. *Bioresource Technology*, *304*, 122996. doi:10.1016/j.biortech.2020.122996 PMID:32115347

Kosjek, T., Heath, E., & Krbavcic, A. (2005). Determination of non-steroidal anti-inflammatory drug (NSAIDs) residues in water samples. *Environment International*, *31*(5), 679–685. doi:10.1016/j.envint.2004.12.001 PMID:15910964

Kumar, N. M., Muthukumaran, C., Sharmila, G., & Gurunathan, B. (2018). Genetically modified organisms and its impact on the enhancement of bioremediation. In S. J. Varjani, A. K. Agarwal, E. Gnansounou, & B. Gurunathan (Eds.), *Bioremediation: Applications for environmental protection and management* (pp. 53–76). Springer. doi:10.1007/978-981-10-7485-1_4

Kumari, A., Kanchan, S., Sinha, R. P., & Kesheri, M. (2016). Applications of biomolecular databases in bioinformatics. In *Medical Imaging in Clinical Applications* (pp. 329–351). Springer. doi:10.1007/978-3-319-33793-7_15

Lu, T. K., & Collins, J. J. (2007). Dispersing biofilms with engineered enzymatic bacteriophage. *Proceedings of the National Academy of Sciences of the United States of America*, *104*(27), 11197–11202. doi:10.1073/pnas.0704624104 PMID:17592147

Mali, H., Shah, C., Patel, D. H., Trivedi, U., & Subramanian, R. B. (2022). Biocatalytic system of metallohydrolases for remediation of neurotoxin organophosphates and applications with a future vision. *Journal of Inorganic Biochemistry*, *231*, 111771. doi:10.1016/j.jinorgbio.2022.111771 PMID:35278760

Marshall, R., Maxwell, C. S., Collins, S. P., Jacobsen, T., Luo, M. L., Begemann, M. B., Gray, B. N., January, E., Singer, A., He, Y., Beisel, C. L., & Noireaux, V. (2018). Rapid and scalable characterization of CRISPR technologies using an *E-coli* cell-free transcription-translation system. *Molecular Cell*, *69*(1), 146–157. doi:10.1016/j.molcel.2017.12.007 PMID:29304331

Michel, C., Jean, M., Coulon, S., Dictor, M. C., Delorme, F., Morin, D., & Garrido, F. (2007). Biofilms of As (III)-oxidizing bacteria: Formation and activity studies for bioremediation process development. *Applied Microbiology and Biotechnology*, *77*(2), 457–467. doi:10.100700253-007-1169-4 PMID:17846760

Nogales, J., Mueller, J., Gudmundsson, S., Canalejo, F. J., Duque, E., Monk, J., Feist, A. M., Ramos, J. L., Niu, W., & Palsson, B. O. (2020). High-quality genome-scale metabolic modeling of *Pseudomonas putida* highlights its broad metabolic capabilities. *Environmental Microbiology*, *22*(1), 255–269. doi:10.1111/1462-2920.14843 PMID:31657101

Pabo, C. O., & Nekludova, L. (2000). Geometric analysis and comparison of protein-DNA interfaces: Why is there no simple code for recognition? 1. *Journal of Molecular Biology*, *301*(3), 597–624. doi:10.1006/jmbi.2000.3918 PMID:10966773

Padmaperuma, G., Butler, T. O., Shuhaili, F. A. A., Almalki, W. J., & Vaidyanathan, S. (2020). Microbial consortia: Concept and application in fruit crop management. In *Fruit crops* (pp. 353–366). Elsevier. doi:10.1016/B978-0-12-818732-6.00025-3

Pawar, S., & Kumar, A. (2002). Issues in the formulation of drugs for oral use in children. *Paediatric Drugs*, *4*(6), 371–379. doi:10.2165/00128072-200204060-00004 PMID:12038873

Pazos, F., Guijas, D., Valencia, A., & De Lorenzo, V. (2005). MetaRouter: Bioinformatics for bioremediation. *Nucleic Acids Research*, *33*(Database issue, suppl_1), D588–D592. doi:10.1093/nar/gki068 PMID:15608267

Pontrelli, S., Chiu, T. Y., Lan, E. I., Chen, F. Y. H., Chang, P., & Liao, J. C. (2018). *Escherichia coli* as a host for metabolic engineering. *Metabolic Engineering*, *50*, 16–46. doi:10.1016/j.ymben.2018.04.008 PMID:29689382

Purnick, P. E., & Weiss, R. (2009). The second wave of synthetic biology: From modules to systems. *Nature Reviews. Molecular Cell Biology*, *10*(6), 410–422. doi:10.1038/nrm2698 PMID:19461664

Robishaw, J. D., Alter, S. M., Solano, J. J., Shih, R. D., DeMets, D. L., Maki, D. G., & Hennekens, C. H. (2021). Genomic surveillance to combat COVID-19: Challenges and opportunities. *The Lancet Microbe*, *2*(9), e481–e484. doi:10.1016/S2666-5247(21)00121-X PMID:34337584

Rodriguez, E. A., Ramirez, D., Balcazar, J. L., & Jimenez, J. N. (2021). Metagenomic analysis of urban waste water resistome and mobilome: A support for antimicrobial resistance surveillance in an endemic country. *Environmental Pollution*, *276*, 116736. doi:10.1016/j.envpol.2021.116736 PMID:33618114

Romero, M., Gallego, D., Blaz, J., Lechuga, A., Martínez, J. F., Barajas, H. R., & Alcaraz, L. D. (2019). Rhizosphere metagenomics of mine tailings colonizing plants: Assembling and selecting synthetic bacterial communities to enhance *in-situ* bioremediation. BioRxiv, 664805. doi:10.1101/664805

Roy, K., & Kar, S. (2016). In-silico models for ecotoxicity of pharmaceuticals. In *In-silico methods for predicting drug toxicity* (pp. 237–304). Humana Press. doi:10.1007/978-1-4939-3609-0_12

Rubin, B. E., Diamond, S., Cress, B. F., Crits-Christoph, A., He, C., Xu, M., & Doudna, J. A. (2020). Targeted genome editing of bacteria within microbial communities. bioRxiv. doi:10.1101/2020.07.17.209189

Sapundzhi, F. I., & Dzimbova, T. A. (2022). The importance of biological databases in modeling of structure-activity relationship. *Interpretation (Tulsa)*, *7*, 11.

Schmidt, M., & Pei, L. (2011). Synthetic toxicology: Where engineering meets biology and Toxicology? *Toxicological Sciences*, *120*(Supplement 1), S204–S224. doi:10.1093/toxsci/kfq339 PMID:21068213

Schomburg, I., Chang, A., Hofmann, O., Ebeling, C., Ehrentreich, F., & Schomburg, D. (2002). BRENDA: A resource for enzyme data and metabolic information. *Trends in Biochemical Sciences*, *27*(1), 54–56. doi:10.1016/S0968-0004(01)02027-8 PMID:11796225

Sharma, B., & Shukla, P. (2022). Futuristic avenues of metabolic engineering techniques in bioremediation. *Biotechnology and Applied Biochemistry*, *69*(1), 51–60. doi:10.1002/bab.2080 PMID:33242354

Tanveer, T., Shaheen, K., Parveen, S., Misbah, Z. T., Babar, M. M., & Gul, A. (2018). Omics-based bioengineering in environmental biotechnology. In D. Barh & V. Azevedo (Eds.), *Omics Technologies and Bio-Engineering* (pp. 353–364). Academic Press. doi:10.1016/B978-0-12-815870-8.00019-X

Tay, P. K. R., Nguyen, P. Q., & Joshi, N. S. (2017). A synthetic circuit for mercury bioremediation using self-assembling functional amyloids. *ACS Synthetic Biology*, *6*(10), 1841–1850. doi:10.1021/acssynbio.7b00137 PMID:28737385

Tecon, R., & Van der Meer, J. R. (2008). Bacterial biosensors for measuring availability of environmental pollutants. *Sensors (Basel)*, *8*(7), 4062–4080. doi:10.33908074062 PMID:27879922

Thombre, R. S., Kaur, K., Jagtap, S. S., Dixit, J., & Vaishampayan, P. V. (2022). *Microbial Life in Space*. New Frontiers in Astrobiology. doi:10.1016/B978-0-12-824162-2.00013-0

Uchida, M., LaFrance, B., Broomell, C. C., Prevelige, P. E. Jr, & Douglas, T. (2015). Higher order assembly of virus-like particles (VLPs) mediated by multi-valent protein linkers. *Small*, *11*(13), 1562–1570. doi:10.1002mll.201402067 PMID:25641768

Utturkar, S. M., Bollmann, A., Brzoska, R. M., Klingeman, D. M., Epstein, S. E., Palumbo, A. V., & Brown, S. D. (2013). Draft genome sequence for *Ralstonia* sp. *strain OR*214, a bacterium with potential for bioremediation. *Genome Announcements*, *1*(3), e00321–e13. doi:10.1128/genomeA.00321-13 PMID:23792748

Vallecillos, L., Sadef, Y., Borrull, F., Pocurull, E., & Bester, K. (2017). Degradation of synthetic fragrances by laccase-mediated system. *Journal of Hazardous Materials*, *334*, 233–243. doi:10.1016/j.jhazmat.2017.04.003 PMID:28415001

Vidali, M. (2001). Bioremediation: An overview. *Pure and Applied Chemistry*, *73*(7), 1163–1172. doi:10.1351/pac200173071163

Xu, P., Bhan, N., & Koffas, M. A. (2013). Engineering plant metabolism into microbes: From systems biology to synthetic biology. *Current Opinion in Biotechnology*, *24*(2), 291–299. doi:10.1016/j.copbio.2012.08.010 PMID:22985679

Yaashikaa, P. R., Devi, M. K., & Kumar, P. S. (2022). Engineering microbes for enhancing the degradation of environmental pollutants: A detailed review on synthetic biology. *Environmental Research*, *214*, 113868. doi:10.1016/j.envres.2022.113868 PMID:35835162

Yim, S. K., Jung, H. C., Pan, J. G., Kang, H. S., Ahn, T., & Yun, C. H. (2006). Functional expression of mammalian NADPH-cytochrome P450 oxido-reductase on the cell surface of *Escherichia coli*. *Protein Expression and Purification*, *49*(2), 292–298. doi:10.1016/j.pep.2006.05.013 PMID:16829136

KEY TERMS AND DEFINITIONS

Consortia: The consortia refer to the combination of more than two microorganisms.

Gene Editing: Gene editing is the technology that allows the genetic material of the microorganisms to be deleted, insert, or change at specific points in the genome.

Genomic Surveillance: It is monitoring of the pathogens and checking the similarities and differences between their genome.

Quorum Sensing: It is cell-cell communication among the microorganisms.

Chapter 17
Synthetic Biology for Waste Water to Energy Conversion:
IoT and AI Approaches

N. Venkateswaran
Department of Master of Business Administration, Panimalar Engineering College, India

S. Senthil Kumar
Department of Mechanical Engineering, R.M.K. College of Engineering and Technology, India

G. Diwakar
Department of Mechanical Engineering, Koneru Lakshmaiah Education Foundation, India

D. Gnanasangeetha
Department of Chemistry, PSNA College of Engineering and Technology (Autonomous), India

S. Boopathi
Muthayammal Engineering College, India

ABSTRACT

Waste-to-energy (WtE) is a process that generates electricity and heat from waste treatment or processing into fuel sources. WtE can reduce greenhouse gas emissions, landfill waste, and dependence on fossil fuels. However, it faces challenges such as high capital and operational costs, environmental and health impacts, and public acceptance. Synthetic biology (SynBio) can offer novel solutions for WtE by enhancing efficiency, sustainability, and diversity of biological processes. This chapter reviews the current state-of-the-art and future prospects of SynBio applications for WtE, focusing on the integration of IoT and AI technologies. IoT is a network of physical devices, sensors, and actuators that collect, transmit, and process data over the internet, while AI simulates human intelligence in machines. IoT and AI can enable smart monitoring, control, and optimization of SynBio systems for WtE, providing data-driven insights and decision support for stakeholders.

DOI: 10.4018/978-1-6684-6577-6.ch017

INTRODUCTION

The modern world faces a pressing need to address two intertwined challenges: effective waste management and the sustainable generation of energy. In response to these challenges, Waste-to-Energy (WtE) has emerged as a transformative process that not only disposes of waste responsibly but also harnesses its potential to generate electricity and heat. By converting waste into valuable fuel sources, WtE offers a two-fold solution, curbing greenhouse gas emissions, reducing landfill waste, and diminishing reliance on finite fossil fuels. However, the journey towards realizing the full potential of WtE is marked by intricate obstacles that range from economic feasibility to environmental concerns. In this pursuit of refining WtE processes, the realm of Synthetic Biology (SynBio) shines as a beacon of innovation (Afrane et al., 2021). SynBio presents an exciting frontier that holds the promise of enhancing the efficiency, sustainability, and adaptability of biological processes at the core of WtE. SynBio's ability to engineer organisms and pathways offers an avenue to reimagine waste transformation and energy generation, ushering in new possibilities that could reshape the energy landscape (Al-Ruzouq et al., 2022a).

A significant stride towards realizing these possibilities lies in the integration of two dynamic technological fields: the Internet of Things (IoT) and Artificial Intelligence (AI). The Internet of Things, with its interconnected network of physical devices, sensors, and data communication capabilities, empowers seamless data collection and transmission. On the other hand, Artificial Intelligence emulates human intelligence in machines, enabling predictive analytics, pattern recognition, and informed decision-making (Al-Ruzouq et al., 2022b). The convergence of SynBio with IoT and AI heralds a new era for Waste-to-Energy processes, marked by heightened precision, efficiency, and insights. Through smart monitoring, real-time control, and data-driven optimization, IoT and AI synergistically equip SynBio-based WtE systems to adapt, respond, and evolve. By offering stakeholders timely and informed insights, this integration redefines the paradigm of WtE, transforming it into a dynamic, intelligent, and sustainable endeavor (Beyene et al., 2018).

This chapter embarks on a journey through this transformative landscape, delving into the current state-of-the-art and future horizons of SynBio applications in Waste-to-Energy. With a keen focus on the integration of IoT and AI technologies, it will unravel the benefits, challenges, and opportunities presented by this convergence. Additionally, the chapter will delve into the ethical, social, and regulatory implications that underpin the responsible deployment of these technologies (Bhowmik et al., 2023). As we navigate through the complex interplay of innovation, ethics, and sustainability, a comprehensive understanding of these dynamics is vital for paving the path towards a cleaner, more resource-efficient, and ecologically balanced future powered by SynBio-based Waste-to-Energy.

BACKGROUND AND MOTIVATION OF WTE AND SYNBIO

Waste-to-energy (WtE) is the process of generating energy in the form of electricity and/or heat from the primary treatment of waste, or the processing of waste into a fuel source. WtE is a form of energy recovery that can reduce greenhouse gas emissions, landfill waste, and dependence on fossil fuels. According to the World Energy Council, WtE has the potential to provide up to 10% of global electricity demand by 2050. However, WtE faces several challenges, such as high capital and operational costs, environmental and health impacts, and public acceptance (Cernak et al., 2018).

Synthetic biology (SynBio) is the engineering of biological systems to perform new functions or improve existing ones. SynBio can offer novel solutions for WtE by enhancing the efficiency, sustainability, and diversity of biological processes for waste conversion. SynBio can also create new products and services from waste-derived biogas, biofuel, and bioelectricity. SynBio is expected to have a significant impact on the global bioeconomy, which is projected to reach $4.1 trillion by 2030. The integration of Internet of Things (IoT) and Artificial Intelligence (AI) technologies can further improve the performance and value of SynBio-based WtE systems. IoT is the network of physical devices, sensors, and actuators that can collect, transmit, and process data over the internet. AI is the simulation of human intelligence in machines that can learn, reason, and solve problems. IoT and AI can enable smart monitoring, control, and optimization of SynBio systems for WtE, as well as provide data-driven insights and decision support for stakeholders.

SCOPE AND OBJECTIVES OF THE CHAPTER

The scope of chapter is to review the current state-of-the-art and future prospects of synthetic biology applications for waste-to-energy conversion, focusing on the integration of Internet of Things and Artificial Intelligence technologies.

Objectives

- Provide an overview of the concepts and applications of synthetic biology for waste-to-energy conversion
- Discuss the technologies and solutions of Internet of Things and Artificial Intelligence for synthetic biology-based waste-to-energy systems
- Analyze the benefits, challenges, and opportunities of Internet of Things and Artificial Intelligence for synthetic biology-based waste-to-energy systems
- Examine the ethical, social, and regulatory implications of Internet of Things and Artificial Intelligence for synthetic biology-based waste-to-energy systems
- Summarize the main findings and recommendations for future research

SYNBIO FOR WTE: CONCEPTS AND APPLICATIONS

Definition and Principles of SynBio

Synthetic biology (SynBio) is a multidisciplinary field of science that focuses on living systems and organisms, and it applies engineering principles to develop new biological parts, devices, and systems or to redesign existing systems found in nature (Cernak et al., 2018).

Standardization: SynBio aims to create standardized biological parts and modules that can be easily assembled and reused in different contexts and applications. Standardization also facilitates the sharing and exchange of designs and data among synthetic biologists.

Modularity: SynBio uses modular design to break down complex biological systems into simpler components that can be independently characterized and optimized. Modularity also allows the integration of different parts and modules to create novel functions and behaviors.

Abstraction: SynBio employs abstraction to hide the details and complexity of lower-level biological parts and modules, and to focus on the higher-level functions and properties of the system. Abstraction also enables the use of mathematical models and computational tools to simulate and predict the behavior of synthetic biological systems.

Decoupling: SynBio separates the design and construction phases of synthetic biological systems, and uses well-defined interfaces and protocols to connect them. Decoupling also reduces the dependency and interference between different parts and modules of the system.

SynBio is an exciting and emerging field that has many potential applications in biotechnology, medicine, agriculture, energy, environment, and more. If you want to learn more about SynBio, you can check out these links:

SynBio Techniques and Tools for WtE

Synthetic Biology (SynBio) techniques and tools can be utilized in Waste-to-Energy (WtE) processes to enhance efficiency, reduce waste, and generate valuable products (Figure 1). WtE refers to technologies that convert various types of waste materials into energy, such as electricity or heat (Danchin, 2022).

- **Microbial Engineering**: Genetically modifying microorganisms like bacteria or yeast to enhance their ability to break down organic waste can improve the efficiency of waste digestion and biogas production. This involves introducing genes that encode for enzymes capable of breaking down complex organic molecules.
- **Metabolic Engineering**: Manipulating metabolic pathways within microorganisms to optimize their ability to produce desired products, such as biofuels or valuable chemicals, from waste materials. This could involve engineering pathways to maximize the production of specific metabolites or to increase the efficiency of energy extraction from waste.
- **Syngas Generation**: Using genetically engineered microbes to convert organic waste into synthesis gas (syngas), a mixture of carbon monoxide and hydrogen. Syngas can be used as a precursor for biofuels, chemicals, and other high-value products.
- **Algae and Cyanobacteria**: These photosynthetic microorganisms can be engineered to use waste nutrients (like nitrogen and phosphorus) to grow and produce biomass, which can then be converted into biofuels or other products. They also have the potential to capture carbon dioxide during their growth.
- **Enzyme Engineering**: Developing enzymes with enhanced activity or stability under harsh conditions (e.g., high temperatures) can improve the breakdown of waste materials during the WtE process. These enzymes could be produced by genetically modified microbes or used as additives in waste treatment systems.
- **Bioremediation**: Genetically engineering microbes to degrade specific contaminants or pollutants present in waste streams, thereby reducing environmental harm while generating useful byproducts.

Figure 1. SynBio techniques and tools for WtE

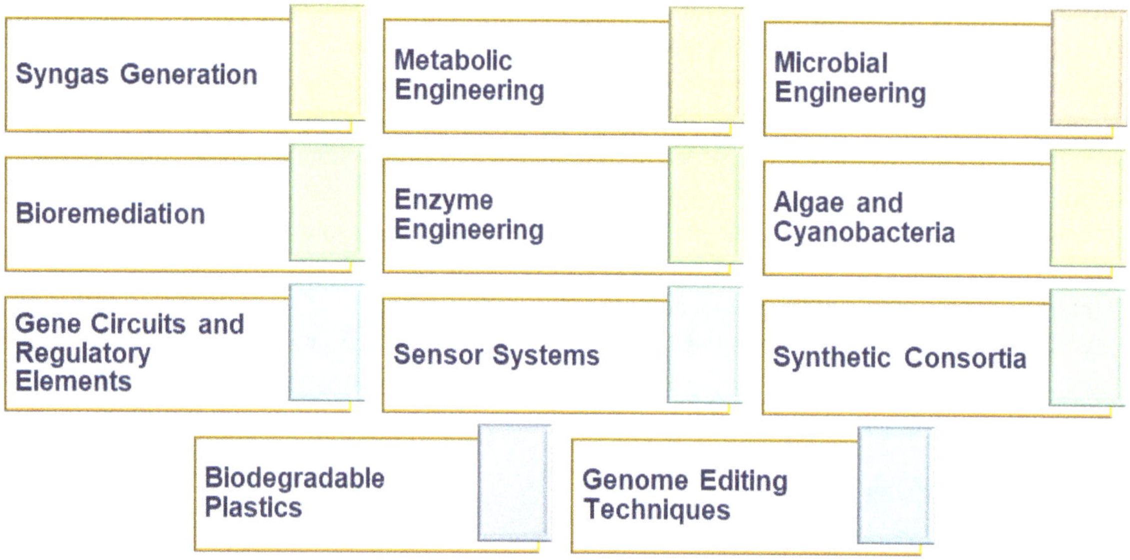

- **Synthetic Consortia**: Creating communities of engineered microorganisms that work together to efficiently break down waste materials and produce valuable products. This can mimic natural ecosystems and enhance overall efficiency.
- **Sensor Systems**: Developing synthetic biological sensors that can monitor the state of the waste conversion process, providing real-time data on parameters like pH, temperature, gas composition, and microbial activity. This data can help optimize the process conditions.
- **Gene Circuits and Regulatory Elements**: Designing genetic circuits that respond to specific environmental cues or waste composition, allowing for dynamic control of microbial processes to maximize efficiency.
- **Genome Editing Techniques**: Utilizing advanced genome editing tools like CRISPR-Cas9 to precisely modify the genomes of microorganisms, enhancing their waste-conversion capabilities or product yields.
- **Biodegradable Plastics**: Developing microbial strains that can digest and break down biodegradable plastics found in waste streams, helping to reduce plastic waste accumulation.

SynBio Platforms and Products for WtE

Synthetic Biology (SynBio) platforms and products can play a crucial role in Waste-to-Energy (WtE) processes by enabling the development of advanced technologies and solutions (de Lorenzo, 2022).

- **Bioconversion Platforms**: SynBio platforms can be used to engineer microorganisms for efficient bioconversion of waste materials into valuable products. This includes platforms like **Genomatica** that design microbes to produce chemicals from renewable feedstocks, potentially utilizing waste streams as inputs.

- **Biofuel Production**: Companies like **LanzaTech** employ SynBio to engineer microbes that can convert waste gases (such as carbon monoxide and carbon dioxide) into biofuels, such as ethanol and jet fuel.
- **Biogas Enhancement**: SynBio can be used to enhance the efficiency of anaerobic digestion processes in biogas production. Companies like **Biomethanol Chemie Nederland (BCN)** are working on engineering microbes to produce methane-rich biogas from various waste materials.
- **Waste-Derived Chemicals**: SynBio can enable the production of valuable chemicals from waste streams. Companies like **Renmatix** focus on converting waste biomass into useful intermediates for the chemical industry using engineered microbes.
- **Microbial Electrochemical Cells (MECs)**: These devices use microorganisms to break down organic matter and generate electricity. SynBio can enhance the performance of MECs by engineering microbes to efficiently metabolize waste materials.
- **Algae Cultivation**: SynBio can improve algae strains for wastewater treatment and biomass production. Companies like **Algenuity** work on developing genetically engineered algae for bioremediation and biofuel production.
- **Enzyme Products**: SynBio companies such as **Novozymes** develop engineered enzymes that can break down complex waste materials more efficiently, enhancing the overall effectiveness of waste digestion processes.
- **Bio-based Plastics**: SynBio can be used to engineer microbes for the production of biodegradable plastics from waste materials. Companies like **Bio-on** focus on producing bioplastics using engineered bacteria.
- **Carbon Capture and Utilization**: SynBio platforms can be applied to engineer microbes that capture carbon dioxide and convert it into useful products, such as biofuels or chemicals, using waste-derived resources.
- **Online Monitoring Tools**: SynBio can enable the development of biosensors and microbial indicators that monitor the progress of WtE processes in real-time, ensuring optimal conditions for waste conversion.
- **Metabolic Engineering Software**: Software platforms like **Synthetic Biology Studio (SBOL)** and **OptGene** aid in the design of genetic constructs and metabolic pathways, facilitating the engineering of microbes for waste conversion.
- **Custom Microbial Strains**: SynBio companies can develop custom microbial strains tailored for specific waste types and conditions, optimizing waste-to-energy conversion processes.

Examples and Case Studies of SynBio for WtE

Several examples and case studies illustrate how Synthetic Biology (SynBio) techniques have been applied to Waste-to-Energy (WtE) processes. These examples showcase the diverse ways in which SynBio can enhance waste conversion and energy generation (Grue et al., n.d.; Hoang et al., 2022):

- **LanzaTech's Gas Fermentation for Biofuels**: LanzaTech is a company that uses SynBio to engineer microbes capable of converting waste gases, such as carbon monoxide and carbon dioxide, into valuable biofuels and chemicals. Their technology has been applied in industrial settings, such as steel mills, where waste gases are generated during production. LanzaTech's engineered microbes ferment these waste gases to produce ethanol, which can be used as a biofuel.

- **Bio-on's Bioplastics from Waste Streams**: Bio-on is focused on producing bioplastics from waste materials using engineered bacteria. They have developed a SynBio platform to engineer microbes that can convert waste-derived sugars into biodegradable plastics. This approach not only addresses waste reduction but also provides an eco-friendly alternative to traditional plastics.
- **Synthetic Algae for Wastewater Treatment and Energy Generation**: Algenuity, a company specializing in microalgae-based solutions, utilizes SynBio to enhance the performance of algae strains for wastewater treatment and energy production. By engineering algae to efficiently remove nutrients and contaminants from wastewater, Algenuity aims to create a sustainable solution that simultaneously cleans water and generates bioenergy.
- **Genomatica's Bioconversion of Waste Streams**: Genomatica focuses on developing microbial strains for the production of chemicals from renewable feedstocks, including waste materials. They use SynBio techniques to engineer microorganisms capable of efficiently converting waste-derived feedstocks into valuable chemical intermediates, reducing reliance on fossil fuels.
- **Biomethanol Chemie Nederland's Biogas Production**: Biomethanol Chemie Nederland (BCN) employs SynBio to enhance the efficiency of biogas production from various waste materials. They aim to engineer microorganisms that can produce methane-rich biogas through anaerobic digestion processes. This technology has the potential to increase the energy yield from waste materials.
- **Microbial Electrochemical Cells (MECs) for Energy Generation**: Researchers are using SynBio to optimize the performance of MECs, which utilize microorganisms to convert organic waste into electricity. By engineering microbial communities to enhance waste degradation and electron transfer, MECs become more efficient energy generation systems.
- **Renmatix's Conversion of Waste Biomass**: Renmatix utilizes SynBio to convert waste biomass into valuable chemical intermediates. Their technology focuses on breaking down lignocellulosic materials from agricultural and forestry waste into sugars, which can be used as feedstock for biofuels and chemicals production.
- **Microbial Consortia for Waste Digestion**: Research studies have explored the use of engineered microbial consortia to efficiently break down complex waste materials. By designing microbial communities that work synergistically to degrade various waste components, researchers aim to enhance biogas production and waste conversion rates.

IOT AND AI FOR SYNBIO-BASED WTE: TECHNOLOGIES AND SOLUTIONS

Characteristics of IoT and AI

Internet of Things (IoT) and Artificial Intelligence (AI) are two transformative technologies that, when combined with Synthetic Biology (SynBio), can significantly enhance Waste-to-Energy (WtE) processes. Here's an overview of these technologies and their characteristics (Huang & Koroteev, 2021; Kasera et al., 2021; Khan et al., 2022):

Internet of Things (IoT): IoT refers to a network of interconnected devices and objects that can communicate, collect, and exchange data over the internet. These devices, often embedded with sensors and actuators, can interact with the physical world, gather real-time information, and transmit it to centralized systems for analysis and decision-making (Jeevanantham et al., 2022). Key characteristics of IoT include:

- **Connectivity**: IoT devices are connected to the internet, allowing seamless communication and data exchange between devices and with central servers.
- **Sensors and Actuators**: IoT devices are equipped with various sensors (temperature, humidity, pressure, etc.) to collect data from the environment. Actuators enable devices to perform actions based on received data (Boopathil et al., 2022; Jeevanantham et al., 2022).
- **Data Collection**: IoT devices continuously collect data from their surroundings, generating vast amounts of real-time information.
- **Data Transmission**: Collected data is transmitted to cloud-based platforms or local servers, where it can be stored, processed, and analyzed.
- **Remote Control and Monitoring**: IoT enables remote monitoring and control of devices and systems, providing real-time insights into operational status and performance.

Artificial Intelligence (AI): AI refers to the simulation of human intelligence in machines that are capable of learning from data, reasoning, problem-solving, and making informed decisions. AI technologies enable computers to perform tasks that typically require human intelligence (Kasera et al., 2021).

- **Machine Learning**: AI systems can learn from historical data and improve their performance over time without being explicitly programmed (Boopathi, 2022h, 2022e; Boopathi et al., 2021; Kannan et al., 2022).
- **Pattern Recognition**: AI can identify patterns, correlations, and trends in large datasets that may not be apparent to humans.
- **Data Analysis**: AI algorithms process and analyze data to extract valuable insights and make predictions or recommendations.
- **Decision-Making**: AI systems can make autonomous decisions based on the analysis of data and predefined rules.
- **Natural Language Processing**: AI can understand and interact with human language, enabling communication between humans and machines.

IoT and AI for SynBio-Based WtE: Technologies and Solutions: The integration of IoT and AI with SynBio in WtE processes (Figure 2) can yield several benefits (Kasera et al., 2021; Khan et al., 2022; Lee et al., 2017):

- **Real-Time Monitoring**: IoT sensors can provide real-time data on parameters such as waste composition, temperature, pH, and microbial activity. AI algorithms can analyze this data to optimize process conditions and predict potential issues.
- **Process Optimization**: AI can analyze complex biological and chemical interactions in SynBio-based WtE systems, suggesting optimal operating parameters for maximizing energy production and waste conversion efficiency (Boopathi, 2022g; Boopathi & Sivakumar, 2013).
- **Predictive Maintenance**: IoT-enabled devices can monitor equipment health, and AI can predict when maintenance is required, minimizing downtime and improving overall system reliability.
- **Data-Driven Insights**: AI can analyze historical data from WtE processes to identify trends, patterns, and correlations that guide decision-making and process improvement.
- **Energy Management**: IoT sensors can monitor energy consumption, and AI can suggest strategies for minimizing energy usage and optimizing resource allocation.

Figure 2. Benefits of IoT and AI for SynBio-based WtE

- **Smart Control Systems**: AI-driven control systems can adjust process parameters in real-time based on changing conditions, ensuring optimal waste conversion and energy generation.
- **Risk Assessment and Mitigation**: AI can assess potential risks associated with SynBio applications and provide recommendations for minimizing environmental, health, and safety risks.
- **Decision Support**: AI can assist in decision-making by providing data-driven insights and scenario analysis for choosing the best waste management strategies.

IoT and AI Techniques and Tools for SynBio-based WtE

Various IoT and AI techniques and tools for SynBio-based WtE are listed in Figure 3. These are explained as follows (Luo et al., 2015).

- parameters such as temperature, pH, gas composition, and microbial activity. These sensors provide real-time data that AI algorithms can analyze to optimize waste conversion processes (Saravanan et al., 2022; Subha et al., 2023; Trojovský et al., 2023; Yupapin et al., 2023).
- **Data Analytics Platforms**: AI-powered data analytics platforms process and analyze the vast amount of data collected from IoT sensors. These platforms can identify patterns, anomalies, and correlations, leading to improved process efficiency.
- **Predictive Analytics**: AI can use historical data and sensor inputs to predict potential issues, such as equipment failures or process deviations. This allows for proactive maintenance and process adjustments.

- **Machine Learning Models**: AI techniques like machine learning can develop models that correlate process parameters with energy generation, waste conversion rates, and product yields. These models can guide decision-making for optimal resource allocation.
- **Optimization Algorithms**: AI-driven optimization algorithms can determine the best operating conditions for maximizing energy production while minimizing waste generation. These algorithms consider various factors, including waste composition and microbial activity.

Figure 3. IoT and AI techniques and tools for SynBio-based WtE

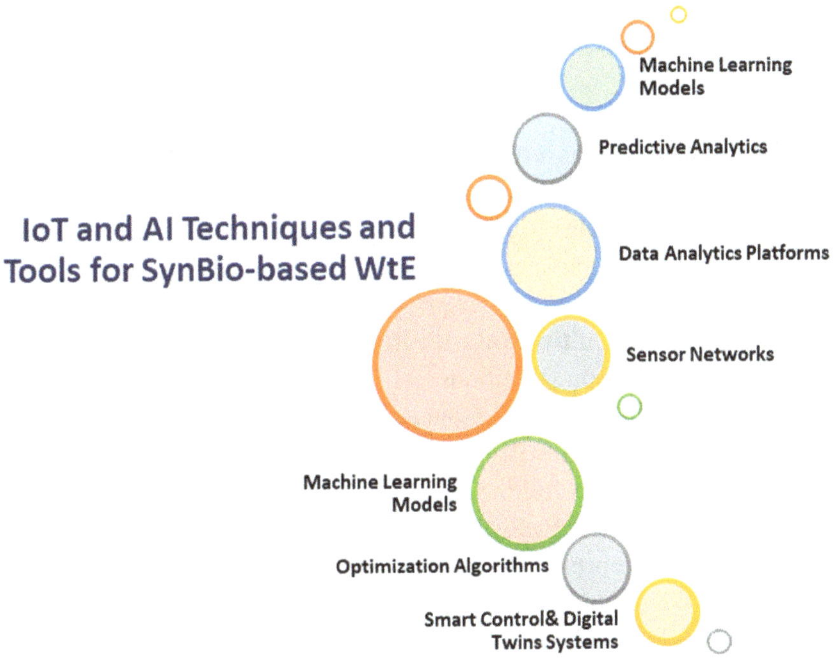

- **Sensor Networks**: IoT-enabled sensors can be placed throughout WtE facilities to monitor
- **Smart Control Systems**: IoT and AI can work together to create smart control systems that adjust process parameters in real-time based on sensor inputs. This ensures that the process operates within desired ranges for optimal performance.
- **Digital Twins**: IoT and AI can be used to create digital twins of WtE systems, which are virtual representations that mimic the behavior of the physical process. This enables testing and optimization of process changes without impacting the actual system.

IoT and AI Systems and Services for SynBio-Based WtE

- **Remote Monitoring and Control Systems**: IoT platforms offer remote monitoring and control of WtE processes, allowing operators to make real-time adjustments and receive alerts for anomalies (Jeevanantham et al., 2022; Jeyakymar et al., 2022; Kannan et al., 2022; Kavitha et al., 2023).

- **Predictive Maintenance Services**: IoT and AI-based predictive maintenance services can schedule maintenance tasks based on equipment health, reducing downtime and maintenance costs.
- **Data Analytics Services**: AI-driven data analytics services can process and interpret sensor data, providing insights and recommendations for process optimization.
- **Process Simulation Services**: AI-based simulation services can predict the outcome of process changes or the introduction of new microbial strains, helping to make informed decisions before implementation.
- **Energy Management Systems**: IoT and AI can optimize energy consumption by monitoring energy usage patterns and suggesting strategies for energy efficiency.

Examples and Case Studies of IoT and AI for SynBio-Based WtE

- **Echogen's Waste Heat Recovery with AI**: Echogen's heat recovery system captures waste heat from industrial processes and converts it into electricity. AI is used to optimize the system's efficiency by analyzing sensor data and adjusting parameters in real-time.
- **Bioneer's Smart Biogas System**: Bioneer developed an IoT-based biogas system that uses sensors to monitor key process parameters in biogas production. AI algorithms analyze the data to optimize gas production and ensure consistent performance.
- **Synthetic Genomics' Algae Cultivation with IoT**: Synthetic Genomics uses IoT sensors to monitor and control algae cultivation conditions. AI techniques analyze the sensor data to maintain optimal growth conditions for improved biomass yield.
- **Bioelektra Group's IoT and AI for Biogas Plants**: Bioelektra Group integrates IoT sensors with AI algorithms to monitor and control biogas production in their plants. The system adjusts input materials and process parameters to maximize gas output and energy efficiency.
- **Carbon Clean's Carbon Capture with AI**: Carbon Clean employs AI to optimize their carbon capture technology, which captures CO2 emissions from industrial processes. IoT sensors monitor system performance, and AI analyzes the data to enhance capture efficiency.

EXPERIMENTS AND CASE STUDIES OF IOT AND AI FOR SYNBIO-BASED WTE

Experimental Design and Methodology of IoT and AI for SynBio-Based WtE

Designing and implementing an experimental setup that integrates Internet of Things (IoT) and Artificial Intelligence (AI) for SynBio-based Waste-to-Energy (WtE) processes involves careful planning and coordination. Below is a general outline of the experimental design and methodology for such a project (Luo et al., 2015; Martin-Pascual et al., 2021; Naveenkumar et al., 2023):

- **Define Objectives and Scope**: Clearly outline the goals of the experiment. Determine what specific aspects of SynBio-based WtE you want to optimize using IoT and AI. Identify key process parameters, performance metrics, and targets for improvement.
- **Select Appropriate Sensors and Hardware**: Choose IoT sensors that can measure relevant parameters such as temperature, pH, gas composition, and microbial activity. Ensure compatibility

with the selected AI and data analysis platforms. Additionally, select appropriate hardware for data collection, transmission, and storage.

- **Data Collection Setup**: Install the selected sensors at strategic locations within the WtE system. Ensure sensors are properly calibrated and connected to IoT-enabled devices that can transmit data to a central data repository.
- **Data Transmission and Storage**: Set up a reliable data transmission system that securely sends sensor data to a cloud-based or on-site storage solution. Consider data security and privacy regulations in your setup.
- **AI Model Development**: Develop AI models using appropriate techniques such as machine learning or deep learning. Determine which AI algorithms will be suitable for analyzing the collected data and achieving the experiment's objectives.
- **Data Preprocessing**: Clean and preprocess the collected sensor data to remove noise, outliers, and irrelevant information. Ensure that the data is structured and ready for input into AI models.
- **AI Model Training and Validation**: Split the preprocessed data into training, validation, and testing sets. Train AI models using the training data and validate their performance using the validation set. Fine-tune the models as needed to achieve the desired accuracy.
- **Real-Time Analysis and Decision-Making**: Implement the trained AI models to analyze incoming sensor data in real-time. Based on the analysis, the AI system can make decisions or provide recommendations for adjusting process parameters to optimize WtE performance.
- **Feedback Loop and Adaptation**: Continuously monitor the performance of the AI system and the WtE process. Use feedback from the AI system's decisions to refine and adapt the models over time, improving their accuracy and effectiveness.
- **Performance Evaluation**: Evaluate the impact of the IoT and AI integration on the SynBio-based WtE process. Compare key performance indicators, such as energy generation, waste conversion rates, and product yields, with historical data or baseline performance.
- **Data Visualization and Reporting**: Develop dashboards or visualization tools to present real-time and historical data trends. Generate reports that communicate the experiment's outcomes and the benefits of using IoT and AI for SynBio-based WtE.
- **Ethical and Regulatory Considerations**: Ensure that the experimental setup adheres to ethical guidelines and regulatory requirements, especially regarding data privacy, environmental safety, and biosecurity.
- **Documentation and Knowledge Sharing**: Document the entire experimental process, including the setup, methodologies, AI models, and outcomes. Share this knowledge within your organization and the scientific community to contribute to the advancement of SynBio-based WtE technology.

This experimental design and methodology outline provides a framework for integrating IoT and AI into SynBio-based Waste-to-Energy processes. Adapt it according to the specific goals, resources, and technical requirements of your project.

Data Collection and Analysis of IoT and AI for SynBio-Based WtE

The data collection and analysis of IOT and AI (Vazana et al., 2013) for synbio-based WtE is illustrated in Figure 4 and explained below.

Figure 4. Data collection and analysis of IoT and AI for SynBio-based WtE

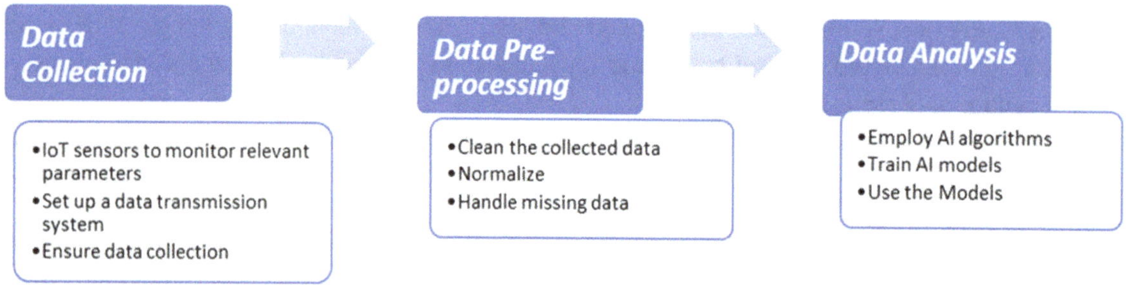

Data Collection

- Implement IoT sensors to monitor relevant parameters in the SynBio-based WtE process, such as waste composition, temperature, pH, gas production, and microbial activity.
- Set up a data transmission system to send sensor data to a central repository or cloud platform.
- Ensure data collection is continuous and synchronized with the process timeline.

Data Pre-Processing

- Clean the collected data by removing outliers, noise, and inconsistencies.
- Normalize or scale data to ensure uniformity across different sensor types.
- Handle missing data through imputation or interpolation techniques.

Data Analysis

- Employ AI algorithms (e.g., machine learning, neural networks) to analyze the collected sensor data.
- Train AI models to predict process outcomes, such as energy generation, waste conversion rates, or microbial growth.
- Use AI models to detect patterns, correlations, and anomalies in the data.

RESULTS AND DISCUSSION OF IOT AND AI FOR SYNBIO-BASED WTE

Performance Metrics: Present quantitative results related to process efficiency, energy production, waste reduction, or any other relevant metrics. Compare these results with baseline or historical data to showcase the improvements achieved through IoT and AI integration.

Real-time Monitoring: Discuss the benefits of real-time monitoring enabled by IoT sensors. Highlight how immediate data insights influenced operational decisions.

Predictive Abilities: Describe the accuracy of AI models in predicting process outcomes. Discuss instances where predictive capabilities aided in anticipating issues or optimizing process parameters.

Decision Support: Highlight how AI-driven recommendations influenced decisions related to process adjustments, resource allocation, or maintenance scheduling.

Comparison and Evaluation of IoT and AI for SynBio-Based WtE

Process Efficiency: Compare process efficiency metrics (e.g., energy output, waste conversion rates) before and after IoT and AI integration. Analyze how AI-driven process adjustments improved efficiency compared to traditional methods.

Resource Utilization: Evaluate resource utilization (e.g., raw materials, energy) and discuss how IoT and AI contributed to more efficient utilization.

Operational Flexibility: Compare the operational flexibility of the SynBio-based WtE process with and without IoT and AI integration. Discuss how real-time insights and AI recommendations allowed for agile process adjustments.

Economic Impact: Analyze the economic benefits of IoT and AI integration, including potential cost savings and increased revenue due to enhanced process performance.

Environmental Impact: Assess the environmental benefits of improved waste conversion and reduced emissions resulting from IoT and AI-enhanced SynBio-based WtE.

Future Potential: Discuss the scalability and potential for further improvements through continuous AI model refinement and increased IoT sensor deployment.

The study highlights the positive impact of IoT and AI on SynBio-based Waste-to-Energy processes. It discusses the implications for WtE, SynBio, and sustainable energy generation and waste management. Visual representations, technical details, and limitations of the experiment are essential for a comprehensive understanding of the results (Boopathi, 2019, 2022b, 2022a, 2022g, 2022e).

BENEFITS, CHALLENGES, AND OPPORTUNITIES OF IOT AND AI FOR SYNBIO-BASED WTE

Technical Benefits of IoT and AI for SynBio-based WtE (Boopathi et al., 2023; Boopathi & Davim, 2023b, 2023a; Fowziya et al., 2023)

- **Real-Time Monitoring**: IoT sensors provide continuous real-time data on process parameters, enabling immediate detection of deviations and anomalies.
- **Optimization and Efficiency**: AI algorithms analyze complex data patterns and optimize process parameters for improved energy generation and waste conversion efficiency.
- **Predictive Insights**: AI can predict process outcomes and potential issues, allowing operators to take proactive actions to prevent disruptions.
- **Adaptive Control**: IoT and AI enable dynamic adjustments to process conditions based on real-time data, leading to adaptive and responsive control systems.
- **Data-Driven Decision-Making**: Real time data analysis enhances decision-making by providing data-driven insights for optimizing resource allocation and process adjustments.
- **Resource Conservation**: IoT and AI integration can lead to reduced resource wastage by optimizing raw material usage and energy consumption.

- **Scalability**: IoT-enabled systems can be easily scaled up or replicated, and AI models can adapt to larger datasets without significantly increased computational resources.

Technical Challenges of IoT and AI for SynBio-based WtE (Ramudu et al., 2023)

- **Data Quality and Reliability**: Ensuring accurate and reliable data from IoT sensors is essential for accurate AI analysis. Sensor calibration and maintenance are critical.
- **Complex Data Integration**: SynBio-based WtE processes involve diverse data sources. Integrating and harmonizing data from different sensors and systems can be challenging.
- **Algorithm Selection and Training**: Selecting appropriate AI algorithms and training models require domain expertise. Overfitting and bias issues may arise during model development.
- **Data Security and Privacy**: Handling sensitive process data requires robust security measures to prevent unauthorized access and data breaches.
- **Energy Consumption**: IoT devices and AI algorithms can consume energy. Balancing energy efficiency with enhanced process performance is important.
- **Model Interpretability**: Complex AI models can be challenging to interpret, making it difficult to understand the reasoning behind their recommendations or decisions.
- **Integration Complexity**: Integrating IoT devices, data platforms, and AI models with existing WtE systems requires technical expertise and careful planning.
- **Maintenance and Upkeep**: IoT devices and AI models require continuous maintenance, updates, and troubleshooting to ensure reliable performance.
- **Regulatory Compliance**: Ensuring compliance with regulatory standards and biosecurity measures is crucial when dealing with SynBio-based processes.
- **Initial Investment**: Implementing IoT and AI technologies involves upfront costs for sensor deployment, hardware, software, and expertise.

Integrating IoT and AI in SynBio-based Waste-to-Energy processes requires a comprehensive approach addressing technical, operational, and ethical considerations. Thorough planning, implementation, and optimization are crucial for enhancing WtE performance (Hoang et al., 2022; Huang & Koroteev, 2021; Khan et al., 2022; Vazana et al., 2013; Wang et al., 2023).

Economic Benefits of IoT and AI for SynBio-based WtE (Babu et al., 2022; Dhanya et al., 2023; Mohanty et al., 2023; Ravisankar et al., 2023; Venkateswaran et al., 2023):

1. **Process Efficiency**: IoT and AI can optimize waste conversion, energy generation, and resource utilization, leading to increased overall process efficiency and higher yields.
2. **Reduced Operating Costs**: Improved process efficiency and predictive maintenance can lead to reduced operational costs by minimizing waste, energy consumption, and maintenance downtime.
3. **Increased Revenue**: Enhanced energy production and product yields can generate additional revenue streams, making the SynBio-based WtE process more economically viable.
4. **Data-Driven Decision-Making**: AI-driven insights enable informed decisions that maximize resource utilization, reducing waste and associated costs.
5. **Optimized Resource Allocation**: AI algorithms can suggest optimal resource allocation strategies, ensuring that inputs are used efficiently and effectively.

6. **Cost Savings in Waste Management**: Effective waste conversion through SynBio can lead to reduced waste disposal costs, as more waste is converted into valuable products.

Economic Challenges of IoT and AI for SynBio-based WtE (Boopathi, 2023; Hema et al., 2023; Karthik et al., 2023; Koshariya et al., 2023; Samikannu et al., 2022):

1. **Initial Investment**: Integrating IoT sensors and developing AI models involves upfront costs for hardware, software, expertise, and implementation.
2. **Maintenance Costs**: IoT devices and AI systems require ongoing maintenance, updates, and potential troubleshooting, which can incur additional expenses.
3. **Skill Gap**: Training personnel to manage IoT devices, AI models, and data analysis tools requires investment in education and training.
4. **Data Management Costs**: Storing and managing large amounts of data generated by IoT devices can lead to increased data storage costs.

Environmental Benefits of IoT and AI for SynBio-based WtE (Boopathi, 2022f, 2022c; Hanumanthakari et al., 2023; Maguluri et al., 2023; Selvakumar et al., 2023):

1. **Waste Reduction**: SynBio-based WtE processes, optimized by IoT and AI, can convert more waste into valuable products, reducing landfill and environmental pollution (Boopathi, 2022d, 2022e).
2. **Energy Generation**: Enhanced energy production from waste materials can contribute to a cleaner energy mix, reducing reliance on fossil fuels.
3. **Emissions Reduction**: Efficient waste conversion and energy generation can result in lower greenhouse gas emissions, contributing to environmental sustainability.
4. **Resource Conservation**: AI and IoT-driven optimization reduces resource wastage, conserving raw materials, energy, and water.

Environmental Challenges of IoT and AI for SynBio-based WtE

1. **Technological Footprint**: The production, deployment, and maintenance of IoT devices and AI systems may have environmental impacts, including resource consumption and electronic waste generation.
2. **Unintended Consequences**: AI and IoT interventions should be carefully designed to avoid unintended environmental consequences, such as changes in ecosystem dynamics.
3. **Biosecurity**: In SynBio-based processes, maintaining strict biosecurity measures is crucial to prevent accidental release of engineered organisms into the environment.
4. **Regulatory Compliance**: Compliance with environmental regulations and safety standards is essential to avoid negative impacts on ecosystems and public health.

Balancing Economic and Environmental Considerations

To maximize the economic and environmental benefits of integrating IoT and AI into SynBio-based Waste-to-Energy processes, it's important to adopt a holistic approach that considers both short-term gains and long-term sustainability. Proper planning, continuous monitoring, and adaptive strategies are key to navigating the challenges while leveraging the advantages of these technologies (Huang & Koroteev, 2021; Kasera et al., 2021; Khan et al., 2022; Lee et al., 2017).

Social Benefits of IoT and AI for SynBio-Based WtE

1. **Job Creation**: The integration of IoT and AI technologies into SynBio-based WtE processes can lead to the creation of new job roles related to data analysis, system maintenance, and technology management.
2. **Skill Development**: As industries adopt IoT and AI, there's an opportunity for skill development and training programs, empowering individuals to work with cutting-edge technologies.
3. **Community Engagement**: The use of advanced technologies can foster community engagement by showcasing innovative and sustainable approaches to waste management and energy generation.
4. **Education and Awareness**: IoT and AI applications can raise public awareness about waste-related challenges and solutions, promoting responsible waste management behaviors.
5. **Public Health**: Efficient waste conversion can reduce the environmental impact of waste disposal, leading to cleaner air and water and improved public health outcomes.

Social Challenges of IoT and AI for SynBio-Based WtE

1. **Technological Displacement**: The automation introduced by IoT and AI could lead to job displacement in certain sectors. Reskilling and retraining efforts are necessary to address this challenge.
2. **Digital Divide**: Unequal access to IoT and AI technologies could exacerbate existing societal inequalities, creating a digital divide.
3. **Privacy Concerns**: The collection and analysis of data using IoT and AI raise privacy concerns. Ensuring the responsible handling of sensitive information is crucial.
4. **Ethical Considerations**: The ethical implications of SynBio, IoT, and AI need careful consideration to ensure that engineered organisms and processes align with societal values.

Opportunities and Future Directions of IoT and AI for SynBio-Based WtE

1. **Advanced Monitoring and Control**: Further advancements in IoT can lead to more sophisticated monitoring systems that capture detailed data from multiple points within the WtE process.
2. **Hybrid Approaches**: Combining SynBio, IoT, and AI with other emerging technologies, such as nanotechnology and biotechnology, could lead to more efficient and sustainable WtE solutions.
3. **Decentralized Systems**: IoT and AI could enable the development of decentralized waste management systems that empower local communities to manage their waste and energy needs.
4. **Circular Economy Integration**: SynBio-based WtE processes can play a vital role in the circular economy by converting waste into valuable resources, and IoT and AI can facilitate tracking and optimization within this model.
5. **Data Sharing and Collaboration**: Collaborative platforms for sharing data and insights could lead to collective problem-solving and innovations in SynBio-based WtE.
6. **Education and Public Engagement**: IoT and AI-driven platforms can educate and engage the public about WtE, SynBio, and environmental sustainability.

7. **Regulatory Frameworks**: Developing regulatory frameworks that address the ethical, social, and environmental dimensions of IoT and AI in SynBio-based WtE is essential for responsible deployment.

The future of IoT and AI in SynBio-based Waste-to-Energy holds tremendous potential for revolutionizing waste management, energy generation, and sustainability (Chandrika et al., 2023; Dhanya et al., 2023; Nishanth et al., 2023). As these technologies continue to evolve, careful consideration of their societal impact, combined with responsible implementation, will be essential to harnessing their full potential for the benefit of society and the environment.

ETHICAL, SOCIAL, AND REGULATORY IMPLICATIONS OF IOT AND AI FOR SYNBIO-BASED WTE

Ethical Issues and Principles of IoT and AI for SynBio-Based WtE

Biosecurity: The release of engineered organisms into the environment through SynBio-based processes raises concerns about unintended ecological consequences and potential harm to ecosystems.

Informed Consent: Ethical considerations arise when using IoT and AI to monitor and manage processes involving biological materials. Ensuring informed consent from stakeholders is important.

Data Privacy: The collection, storage, and analysis of data generated by IoT sensors and AI systems require strict data privacy measures to protect sensitive information.

Equity and Access: Addressing inequalities in access to IoT and AI technologies is important to prevent creating a digital divide that excludes certain communities from the benefits.

Accountability: Ethical frameworks should establish clear lines of responsibility for managing IoT and AI technologies, including accountability for any adverse consequences.

Social Impacts and Responsibilities of IoT and AI for SynBio-Based WtE

Job Displacement: The introduction of IoT and AI may lead to job displacement. Society has a responsibility to provide retraining and reskilling opportunities for affected individuals.

Public Awareness and Engagement: Promoting public understanding and engagement about SynBio, IoT, and AI is essential for responsible technology deployment.

Community Consent: Local communities should have a say in the implementation of SynBio-based WtE processes that may affect their environment and well-being.

Transparency: Stakeholders should have access to transparent and understandable information about how IoT and AI are being used in WtE processes.

Social Inclusion: Ensuring that SynBio-based WtE processes benefiting from IoT and AI are inclusive and consider the needs of all segments of society.

Regulatory Frameworks and Policies of IoT and AI for SynBio-Based WtE

Biosecurity Regulations: Regulatory bodies should establish guidelines to prevent the unintended release of engineered organisms and ensure the containment of SynBio-based processes.

Data Privacy Laws: Stringent data privacy regulations must be in place to protect personal and sensitive data collected by IoT devices and AI systems.

Ethical Review Boards: Ethical review boards should assess the potential risks and benefits of SynBio-based WtE processes involving IoT and AI.

Standards and Compliance: Developing industry standards for the use of IoT and AI in SynBio-based WtE can ensure consistent and safe deployment.

Best Practices and Guidelines of IoT and AI for SynBio-Based WtE

Risk Assessment: Conduct thorough risk assessments to identify potential ethical, environmental, and safety concerns before implementing IoT and AI technologies.

Transparency: Communicate openly about the use of IoT and AI in SynBio-based WtE, ensuring stakeholders are informed about the technology's purpose and implications.

Interdisciplinary Collaboration: Engage experts from diverse fields, including ethics, law, biology, and technology, to ensure well-rounded decision-making.

Continuous Monitoring and Evaluation: Regularly monitor the impact of IoT and AI on SynBio-based WtE and adjust strategies based on feedback and emerging challenges.

Public Engagement: Involve local communities and the public in decision-making processes related to the implementation of SynBio-based WtE technologies.

Ethical, social, and regulatory considerations are critical to responsibly harnessing the benefits of IoT and AI in SynBio-based Waste-to-Energy processes. Balancing innovation with these considerations will be key to achieving sustainable and socially responsible outcomes.

CONCLUSION

In conclusion, the integration of Internet of Things (IoT) and Artificial Intelligence (AI) technologies with Synthetic Biology (SynBio) holds significant promise for advancing Waste-to-Energy (WtE) processes. This chapter has explored the multifaceted landscape of SynBio-based WtE enhanced by IoT and AI, highlighting its potential benefits, challenges, and ethical considerations.

The technical advantages of real-time monitoring, process optimization, predictive insights, and adaptive control underscore the transformative potential of IoT and AI in maximizing energy generation, waste conversion efficiency, and resource utilization. Despite the challenges posed by data quality, algorithm complexity, and privacy concerns, the economic benefits of improved process efficiency, reduced operating costs, and increased revenue demonstrate a compelling case for their integration.

Furthermore, the environmental benefits of waste reduction, energy generation, emissions reduction, and resource conservation emphasize the positive impact of IoT and AI on the sustainability of WtE processes. While acknowledging challenges related to the technological footprint, regulatory compliance, and potential unintended consequences, responsible deployment of IoT and AI can contribute to a cleaner, greener, and more sustainable environment.

On the social front, the potential for job creation, skill development, and community engagement underscores the positive role of IoT and AI in shaping the workforce of the future. However, considerations surrounding job displacement, digital divide, and ethical principles necessitate a holistic approach that prioritizes equitable access and responsible decision-making.

Regulatory frameworks and policies play a critical role in ensuring the safe and ethical deployment of IoT and AI technologies in SynBio-based WtE. Striking the right balance between innovation and adherence to biosecurity, data privacy, and environmental standards will pave the way for a secure and sustainable future.

As we venture into the future, opportunities abound to refine the application of IoT and AI in SynBio-based WtE. Advancements in monitoring and control systems, the integration of circular economy principles, and the convergence of various emerging technologies offer exciting prospects for enhancing efficiency and reducing environmental impact.

In embracing this technological evolution, it is imperative to maintain a steadfast commitment to ethical principles, community engagement, and regulatory compliance. By doing so, we can harness the transformative power of IoT and AI to drive sustainable progress in SynBio-based Waste-to-Energy processes, contributing to a cleaner environment, healthier communities, and a more prosperous future.

ABBREVIATIONS

IoT - Internet of Things
AI - Artificial Intelligence
SynBio - Synthetic Biology
WtE - Waste-to-Energy
AIoT - Artificial Intelligence of Things
ML - Machine Learning
DL - Deep Learning
pH - Potential of Hydrogen
CO2 - Carbon Dioxide
GHG - Greenhouse Gas
EaaS - Energy-as-a-Service
EROI - Energy Return on Investment
ESG - Environmental, Social, and Governance
LCA - Life Cycle Assessment
WtH - Waste-to-Heat
WtP - Waste-to-Power
HVAC - Heating, Ventilation, and Air Conditioning
BaaS - Biogas-as-a-Service
CCS - Carbon Capture and Storage
BMS - Building Management System
GDPR - General Data Protection Regulation
ROI - Return on Investment
KPI - Key Performance Indicator
O&M - Operations and Maintenance
R&D - Research and Development
API - Application Programming Interface
SEI - Sustainable Energy Initiative
SMART - Specific, Measurable, Achievable, Relevant, Time-bound

RIN - Renewable Identification Number
ISO - International Organization for Standardization
MSW - Municipal Solid Waste
RDF - Refuse-Derived Fuel
LFG - Landfill Gas
AD - Anaerobic Digestion
CHP - Combined Heat and Power

REFERENCES

Afrane, S., Ampah, J. D., Jin, C., Liu, H., & Aboagye, E. M. (2021). Techno-economic feasibility of waste-to-energy technologies for investment in Ghana: A multicriteria assessment based on fuzzy TOPSIS approach. *Journal of Cleaner Production, 318*, 128515. doi:10.1016/j.jclepro.2021.128515

Al-Ruzouq, R., Abdallah, M., Shanableh, A., Alani, S., Obaid, L., & Gibril, M. B. A. (2022). Waste to energy spatial suitability analysis using hybrid multi-criteria machine learning approach. *Environmental Science and Pollution Research International, 29*(2), 2613–2628. doi:10.100711356-021-15289-0 PMID:34374020

Babu, B. S., Kamalakannan, J., Meenatchi, N., Karthik, S., & Boopathi, S. (2022). Economic impacts and reliability evaluation of battery by adopting Electric Vehicle. *IEEE Explore*, 1–6.

Beyene, H. D., Werkneh, A. A., & Ambaye, T. G. (2018). Current updates on waste to energy (WtE) technologies: A review. *Renewable Energy Focus, 24*, 1–11. doi:10.1016/j.ref.2017.11.001

Bhowmik, D., Chetri, S., Enerijiofi, K. E., Naha, A., Kanungo, T. D., Shah, M. P., & Nath, S. (2023). Multitudinous approaches, challenges and opportunities of bioelectrochemical systems in conversion of waste to energy from wastewater treatment plants. *Cleaner and Circular Bioeconomy*, 100040.

Boopathi, S. (2019). Experimental investigation and parameter analysis of LPG refrigeration system using Taguchi method. *SN Applied Sciences, 1*(8), 892. Advance online publication. doi:10.100742452-019-0925-2

Boopathi, S. (2022a). An experimental investigation of Quench Polish Quench (QPQ) coating on AISI 4150 steel. *Engineering Research Express, 4*(4), 045009. Advance online publication. doi:10.1088/2631-8695/ac9ddd

Boopathi, S. (2022b). An Extensive Review on Sustainable Developments of Dry and Near-Dry Electrical Discharge Machining Processes. *Journal of Manufacturing Science and Engineering, 144*(5), 1–37. Advance online publication. doi:10.1115/1.4052527

Boopathi, S. (2022c). An investigation on gas emission concentration and relative emission rate of the near-dry wire-cut electrical discharge machining process. *Environmental Science and Pollution Research International, 29*(57), 86237–86246. doi:10.100711356-021-17658-1 PMID:34837614

Boopathi, S. (2022e). Cryogenically treated and untreated stainless steel grade 317 in sustainable wire electrical discharge machining process: A comparative study. *Environmental Science and Pollution Research International*. Advance online publication. doi:10.100711356-022-22843-x PMID:36057706

Boopathi, S. (2022g). Experimental investigation and multi-objective optimization of cryogenic Friction-stir-welding of AA2014 and AZ31B alloys using MOORA technique. *Materials Today. Communications*, *33*, 104937. Advance online publication. doi:10.1016/j.mtcomm.2022.104937

Boopathi, S. (2022h). Performance Improvement of Eco-Friendly Near-Dry Wire-Cut Electrical Discharge Machining Process Using Coconut Oil-Mist Dielectric Fluid. *Journal of Advanced Manufacturing Systems*. Advance online publication. doi:10.1142/S0219686723500178

Boopathi, S. (2023). Securing Healthcare Systems Integrated With IoT: Fundamentals, Applications, and Future Trends. In Dynamics of Swarm Intelligence Health Analysis for the Next Generation (pp. 186–209). IGI Global.

Boopathi, S., & Davim, J. P. (2023). Applications of Nanoparticles in Various Manufacturing Processes. In *Sustainable Utilization of Nanoparticles and Nanofluids in Engineering Applications* (pp. 1–31). IGI Global. doi:10.4018/978-1-6684-9135-5.ch001

Boopathi, S., Lewise, K. A. S., Subbiah, R., & Sivaraman, G. (2021). Near-dry wire-cut electrical discharge machining process using water-air-mist dielectric fluid: An experimental study. *Materials Today: Proceedings*, *49*, 1885–1890. doi:10.1016/j.matpr.2021.08.077

Boopathi, S., & Sivakumar, K. (2013). Experimental investigation and parameter optimization of near-dry wire-cut electrical discharge machining using multi-objective evolutionary algorithm. *International Journal of Advanced Manufacturing Technology*, *67*(9–12), 2639–2655. doi:10.100700170-012-4680-4

Boopathi, S., Umareddy, M., & Elangovan, M. (2023). Applications of Nano-Cutting Fluids in Advanced Machining Processes. In *Sustainable Utilization of Nanoparticles and Nanofluids in Engineering Applications* (pp. 211–234). IGI Global. doi:10.4018/978-1-6684-9135-5.ch009

Boopathil, S., Arigela, S. H., Raman, R., Indhumathi, C., Kavitha, V., & Bhatt, B. C. (2022). Prominent Rule Control-based Internet of Things: Poultry Farm Management System. *3rd International Conference on Power, Energy, Control and Transmission Systems, ICPECTS 2022 - Proceedings*. 10.1109/ICPECTS56089.2022.10047039

Cernak, P., Estrela, R., Poddar, S., Skerker, J. M., Cheng, Y.-F., Carlson, A. K., Chen, B., Glynn, V. M., Furlan, M., Ryan, O. W., Donnelly, M. K., Arkin, A. P., Taylor, J. W., & Cate, J. H. D. (2018). Engineering Kluyveromyces marxianus as a robust synthetic biology platform host. *mBio*, *9*(5), 10–1128. doi:10.1128/mBio.01410-18 PMID:30254120

Chandrika, V., Sivakumar, A., Krishnan, T. S., Pradeep, J., Manikandan, S., & Boopathi, S. (2023). Theoretical Study on Power Distribution Systems for Electric Vehicles. In *Intelligent Engineering Applications and Applied Sciences for Sustainability* (pp. 1–19). IGI Global. doi:10.4018/979-8-3693-0044-2.ch001

Danchin, A. (2022). In vivo, in vitro and in silico: An open space for the development of microbe-based applications of synthetic biology. *Microbial Biotechnology*, *15*(1), 42–64. doi:10.1111/1751-7915.13937 PMID:34570957 de Lorenzo, V. (2022). Environmental Galenics: Large-scale fortification of extant microbiomes with engineered bioremediation agents. *Philosophical Transactions of the Royal Society B*, *377*(1857), 20210395.

Dhanya, D., Kumar, S. S., Thilagavathy, A., Prasad, D., & Boopathi, S. (2023). Data Analytics and Artificial Intelligence in the Circular Economy: Case Studies. In Intelligent Engineering Applications and Applied Sciences for Sustainability (pp. 40–58). IGI Global.

Fowziya, S., Sivaranjani, S., Devi, N. L., Boopathi, S., Thakur, S., & Sailaja, J. M. (2023). Influences of nano-green lubricants in the friction-stir process of TiAlN coated alloys. *Materials Today: Proceedings*. Advance online publication. doi:10.1016/j.matpr.2023.06.446

Grue, B., Hamadache, S., Maddiboina, D., & Scott, B. (n.d.). *Invest in Canadian Synthetic Biology to Meet Commitments to Sustainable Development and Support Economic Recovery*. 10.38126/JSPG180106

Hanumanthakari, S., Gift, M. M., Kanimozhi, K., Bhavani, M. D., Bamane, K. D., & Boopathi, S. (2023). Biomining Method to Extract Metal Components Using Computer-Printed Circuit Board E-Waste. In *Handbook of Research on Safe Disposal Methods of Municipal Solid Wastes for a Sustainable Environment* (pp. 123–141). IGI Global. doi:10.4018/978-1-6684-8117-2.ch010

Hema, N., Krishnamoorthy, N., Chavan, S. M., Kumar, N., Sabarimuthu, M., & Boopathi, S. (2023). A Study on an Internet of Things (IoT)-Enabled Smart Solar Grid System. In *Handbook of Research on Deep Learning Techniques for Cloud-Based Industrial IoT* (pp. 290–308). IGI Global. doi:10.4018/978-1-6684-8098-4.ch017

Hoang, A. T., Varbanov, P. S., Nižetić, S., Sirohi, R., Pandey, A., Luque, R., Ng, K. H., & Pham, V. V. (2022). Perspective review on Municipal Solid Waste-to-energy route: Characteristics, management strategy, and role in circular economy. *Journal of Cleaner Production*, *359*, 131897. doi:10.1016/j.jclepro.2022.131897

Huang, J., & Koroteev, D. D. (2021). Artificial intelligence for planning of energy and waste management. *Sustainable Energy Technologies and Assessments*, *47*, 101426. doi:10.1016/j.seta.2021.101426

Jeevanantham, Y. A., Saravanan, A., Vanitha, V., Samson Isaac, J., Boopathi, S., & Kumar, D. P. (2022). Implementation of Internet-of Things (IoT) in Soil Irrigation System. *3rd International Conference on Power, Energy, Control and Transmission Systems, ICPECTS 2022 - Proceedings*. 10.1109/ICPECTS56089.2022.10047185

Jeyakymar, M., Munusami, V., Shanmugam, P., Boopalan, N., Boopathi, S., & Sureshkumar, M. (2022). An investigation on wear loss and hardness of nitro-carburizing coated stainless-steel grade-316. *Materials Today: Proceedings*, *66*, 1398–1404. doi:10.1016/j.matpr.2022.05.223

Kannan, E., Trabelsi, Y., Boopathi, S., & Alagesan, S. (2022). Influences of cryogenically treated work material on near-dry wire-cut electrical discharge machining process. *Surface Topography : Metrology and Properties*, *10*(1), 015027. Advance online publication. doi:10.1088/2051-672X/ac53e1

Karthik, S., Hemalatha, R., Aruna, R., Deivakani, M., Reddy, R. V. K., & Boopathi, S. (2023). Study on Healthcare Security System-Integrated Internet of Things (IoT). In Perspectives and Considerations on the Evolution of Smart Systems (pp. 342–362). IGI Global.

Kasera, O. A., Owiso, M. O., & Mburu, B. (2021). Governing Emerging Technologies: A Systematic Exploration of Kenya's Biotechnology and LMO-Specific Policy Documents for Adoption and Implementation of Synthetic Biology. *International Journal of Innovation Science, Research and Technology*, 946–974.

Kavitha, C., Geetha Malini, P. S., Charan Kantumuchu, V., Manoj Kumar, N., Verma, A., & Boopathi, S. (2023). An experimental study on the hardness and wear rate of carbonitride coated stainless steel. *Materials Today: Proceedings*, *74*, 595–601. doi:10.1016/j.matpr.2022.09.524

Khan, A. H., López-Maldonado, E. A., Alam, S. S., Khan, N. A., López, J. R. L., Herrera, P. F. M., Abutaleb, A., Ahmed, S., & Singh, L. (2022). Municipal solid waste generation and the current state of waste-to-energy potential: State of art review. *Energy Conversion and Management*, *267*, 115905. doi:10.1016/j.enconman.2022.115905

Koshariya, A. K., Kalaiyarasi, D., Jovith, A. A., Sivakami, T., Hasan, D. S., & Boopathi, S. (2023). AI-Enabled IoT and WSN-Integrated Smart Agriculture System. In *Artificial Intelligence Tools and Technologies for Smart Farming and Agriculture Practices* (pp. 200–218). IGI Global. doi:10.4018/978-1-6684-8516-3.ch011

Lee, U., Han, J., & Wang, M. (2017). Evaluation of landfill gas emissions from municipal solid waste landfills for the life-cycle analysis of waste-to-energy pathways. *Journal of Cleaner Production*, *166*, 335–342. doi:10.1016/j.jclepro.2017.08.016

Luo, Y., Zhang, L., Barton, K. W., & Zhao, H. (2015). Systematic identification of a panel of strong constitutive promoters from Streptomyces albus. *ACS Synthetic Biology*, *4*(9), 1001–1010. doi:10.1021/acssynbio.5b00016 PMID:25924180

Maguluri, L. P., Ananth, J., Hariram, S., Geetha, C., Bhaskar, A., & Boopathi, S. (2023). Smart Vehicle-Emissions Monitoring System Using Internet of Things (IoT). In Handbook of Research on Safe Disposal Methods of Municipal Solid Wastes for a Sustainable Environment (pp. 191–211). IGI Global.

Martin-Pascual, M., Batianis, C., Bruinsma, L., Asin-Garcia, E., Garcia-Morales, L., Weusthuis, R. A., van Kranenburg, R., & Dos Santos, V. A. M. (2021). A navigation guide of synthetic biology tools for Pseudomonas putida. *Biotechnology Advances*, *49*, 107732. doi:10.1016/j.biotechadv.2021.107732 PMID:33785373

Mohanty, A., Venkateswaran, N., Ranjit, P., Tripathi, M. A., & Boopathi, S. (2023). Innovative Strategy for Profitable Automobile Industries: Working Capital Management. In Handbook of Research on Designing Sustainable Supply Chains to Achieve a Circular Economy (pp. 412–428). IGI Global.

Naveenkumar, R., Iyyappan, J., Pravin, R., Kadry, S., Han, J., Sindhu, R., Awasthi, M. K., Rokhum, S. L., & Baskar, G. (2023). A strategic review on sustainable approaches in municipal solid waste management and energy recovery: Role of artificial intelligence, economic stability and life cycle assessment. *Bioresource Technology*, *379*, 129044. doi:10.1016/j.biortech.2023.129044 PMID:37044151

Nishanth, J., Deshmukh, M. A., Kushwah, R., Kushwaha, K. K., Balaji, S., & Sampath, B. (2023). Particle Swarm Optimization of Hybrid Renewable Energy Systems. In *Intelligent Engineering Applications and Applied Sciences for Sustainability* (pp. 291–308). IGI Global. doi:10.4018/979-8-3693-0044-2.ch016

Ramudu, K., Mohan, V. M., Jyothirmai, D., Prasad, D., Agrawal, R., & Boopathi, S. (2023). Machine Learning and Artificial Intelligence in Disease Prediction: Applications, Challenges, Limitations, Case Studies, and Future Directions. In Contemporary Applications of Data Fusion for Advanced Healthcare Informatics (pp. 297–318). IGI Global.

Ravisankar, A., Sampath, B., & Asif, M. M. (2023). Economic Studies on Automobile Management: Working Capital and Investment Analysis. In Multidisciplinary Approaches to Organizational Governance During Health Crises (pp. 169–198). IGI Global.

Samikannu, R., Koshariya, A. K., Poornima, E., Ramesh, S., Kumar, A., & Boopathi, S. (2022). Sustainable Development in Modern Aquaponics Cultivation Systems Using IoT Technologies. In *Human Agro-Energy Optimization for Business and Industry* (pp. 105–127). IGI Global.

Saravanan, M., Vasanth, M., Boopathi, S., Sureshkumar, M., & Haribalaji, V. (2022). Optimization of Quench Polish Quench (QPQ) Coating Process Using Taguchi Method. *Key Engineering Materials*, *935*, 83–91. doi:10.4028/p-z569vy

Selvakumar, S., Shankar, R., Ranjit, P., Bhattacharya, S., Gupta, A. S. G., & Boopathi, S. (2023). E-Waste Recovery and Utilization Processes for Mobile Phone Waste. In *Handbook of Research on Safe Disposal Methods of Municipal Solid Wastes for a Sustainable Environment* (pp. 222–240). IGI Global. doi:10.4018/978-1-6684-8117-2.ch016

Subha, S., Inbamalar, T. M., R, K. C., Suresh, L. R., Boopathi, S., & Alaskar, K. (2023). A Remote Health Care Monitoring system using internet of medical things (IoMT). *IEEE Explore*. doi:10.1109/ICIPTM57143.2023.10118103

Trojovský, P., Dhasarathan, V., & Boopathi, S. (2023). Experimental investigations on cryogenic friction-stir welding of similar ZE42 magnesium alloys. *Alexandria Engineering Journal*, *66*(1), 1–14. doi:10.1016/j.aej.2022.12.007

Vazana, Y., Barak, Y., Unger, T., Peleg, Y., Shamshoum, M., Ben-Yehezkel, T., Mazor, Y., Shapiro, E., Lamed, R., & Bayer, E. A. (2013). A synthetic biology approach for evaluating the functional contribution of designer cellulosome components to deconstruction of cellulosic substrates. *Biotechnology for Biofuels*, *6*(1), 1–18. doi:10.1186/1754-6834-6-182 PMID:24341331

Venkateswaran, N., Vidhya, K., Ayyannan, M., Chavan, S. M., Sekar, K., & Boopathi, S. (2023). A Study on Smart Energy Management Framework Using Cloud Computing. In 5G, Artificial Intelligence, and Next Generation Internet of Things: Digital Innovation for Green and Sustainable Economies (pp. 189–212). IGI Global. doi:10.4018/978-1-6684-8634-4.ch009

Wang, G., Kong, Q., Wang, D., & Asmi, F. (2023). Ethical and social insights into synthetic biology: Predicting research fronts in the post-COVID-19 era. *Frontiers in Bioengineering and Biotechnology*, *11*, 1085797. doi:10.3389/fbioe.2023.1085797 PMID:37274167

Yupapin, P., Trabelsi, Y., Nattappan, A., & Boopathi, S. (2023). Performance Improvement of Wire-Cut Electrical Discharge Machining Process Using Cryogenically Treated Super-Conductive State of Monel-K500 Alloy. *Iranian Journal of Science and Technology. Transaction of Mechanical Engineering*, *47*(1), 267–283. doi:10.100740997-022-00513-0

Compilation of References

Abaandou, L., Quan, D., & Shiloach, J. (2021). Affecting HEK293 Cell Growth and Production Performance by Modifying the Expression of Specific Genes. *Cells*, *10*(7), 1667. Advance online publication. doi:10.3390/cells10071667 PMID:34359846

Abo-Zahhad, M., Ahmed, S. M., Farrag, M., Ahmed, M. F. A., & Ali, A. (2016). Design and implementation of building energy monitoring and management system based on wireless sensor networks. *Proceedings - 2015 10th International Conference on Computer Engineering and Systems, ICCES 2015*, 230–233. 10.1109/ICCES.2015.7393051

Abujar, S., Hasan, M., & Hossain, S. A. (2019). Sentence similarity estimation for text summarization using deep learning. *Proceedings of the 2nd International Conference on Data Engineering and Communication Technology: ICDECT 2017*, 155–164. 10.1007/978-981-13-1610-4_16

Acharya, U. R., Fujita, H., Lih, O. S., Hagiwara, Y., Tan, J. H., & Adam, M. (2017). Automated detection of arrhythmias using different intervals of tachycardia ECG segments with convolutional neural network. *Information Sciences*, *405*, 81–90. doi:10.1016/j.ins.2017.04.012

Acharya, U. R., Fujita, H., Oh, S. L., Hagiwara, Y., Tan, J. H., Adam, M., & San Tan, R. (2019). Deep convolutional neural network for the automated diagnosis of congestive heart failure using ecg signals. *Applied Intelligence*, *49*(1), 16–27. doi:10.100710489-018-1179-1

Acharya, U. R., Fujita, H., Oh, S. L., Raghavendra, U., Tan, J. H., Adam, M., Gertych, A., & Hagiwara, Y. (2018). Automated identification of shockable and non-shockable life-threatening ventricular arrhythmias using convolutional neural network. *Future Generation Computer Systems*, *79*, 952–959. doi:10.1016/j.future.2017.08.039

Acharya, U. R., Oh, S. L., Hagiwara, Y., Tan, J. H., Adam, M., Gertych, A., & San Tan, R. (2017c). A deep convolutional neural network model to classify heart- beats. *Computers in Biology and Medicine*, *89*, 389–396. doi:10.1016/j.compbiomed.2017.08.022 PMID:28869899

Adams, B. L. (2016). The next generation of synthetic biology chassis: Moving synthetic biology from the laboratory to the field. *ACS Synthetic Biology*, *5*(12), 1328–1330. doi:10.1021/acssynbio.6b00256 PMID:27665861

Adams, R. L., Knowler, J. T., & Leader, D. P. (1992). Degradation and modification of nucleic acids. In *The biochemistry of the nucleic acids* (pp. 97–133). Springer. doi:10.1007/978-94-011-2290-0_4

Adedeji, P. A., Akinlabi, S., Madushele, N., & Olatunji, O. O. (2022). Hybrid adaptive neuro-fuzzy inference system (ANFIS) for a multi-campus university energy consumption forecast. *International Journal of Ambient Energy*, *43*(1), 1685–1694. doi:10.1080/01430750.2020.1719885

Adivitiya, & Khasa, Y. P. (2016). The evolution of recombinant thrombolytics: Current status and future directions. *Bioengineered, 8*(4), 331–358. doi:10.1080/21655979.2016.1229718

Adlakha, N., Pfau, T., Ebenhöh, O., & Yazdani, S. S. (2015). Insight into metabolic pathways of the potential biofuel producer, *Paenibacillus polymyxa* ICGEB2008. *Biotechnology for Biofuels*, *8*(1), 1–10. doi:10.118613068-015-0338-4 PMID:26413158

Adrio, J. L. (2017). Oleaginous yeasts: Promising platforms for the production of oleochemicals and biofuels. *Biotechnology and Bioengineering*, *114*(9), 1915–1920. doi:10.1002/bit.26337 PMID:28498495

Afrane, S., Ampah, J. D., Jin, C., Liu, H., & Aboagye, E. M. (2021). Techno-economic feasibility of waste-to-energy technologies for investment in Ghana: A multicriteria assessment based on fuzzy TOPSIS approach. *Journal of Cleaner Production*, *318*, 128515. doi:10.1016/j.jclepro.2021.128515

Agrawal, A. V., Pitchai, R., Senthamaraikannan, C., Balaji, N. A., Sajithra, S., & Boopathi, S. (2023). Digital Education System During the COVID-19 Pandemic. In Using Assistive Technology for Inclusive Learning in K-12 Classrooms (pp. 104–126). IGI Global. doi:10.4018/978-1-6684-6424-3.ch005

Agren, R., Liu, L., Shoaie, S., Vongsangnak, W., Nookaew, I., & Nielsen, J. (2013). The RAVEN Toolbox and Its Use for Generating a Genome-scale Metabolic Model for *Penicillium chrysogenum*. *PLoS Computational Biology*, *9*(3), e1002980. Advance online publication. doi:10.1371/journal.pcbi.1002980 PMID:23555215

Ajjolli Nagaraja, A., Charton, P., Cadet, X., Fontaine, N., Delsaut, M., Wiltschi, B., Voit, A., Offmann, B., Damour, C., Grondin-Perez, B., & Cadet, F. (2020). A Machine Learning Approach for Efficient Selection of Enzyme Concentrations and Its Application for Flux Optimization. *Catalysts*, *10*(3), 291. doi:10.3390/catal10030291

Ajmal, N., Khan, S., & Shaikh, R. (2019). Polycystic ovary syndrome (PCOS) and genetic predisposition: A review article. *European Journal of Obstetrics, Gynecology, and Reproductive Biology*, *3*, 1–6. PMID:31403134

Akash, M. S., Rehman, K., & Chen, S. (2014, July 1). Effects of coffee on type 2 diabetes mellitus. *Nutrition (Burbank, Los Angeles County, Calif.)*, *30*(7-8), 755–763. doi:10.1016/j.nut.2013.11.020 PMID:24984989

Akash, M. S., Rehman, K., Sun, H., & Chen, S. (2014, May 1). Assessment of release kinetics, stability and polymer interaction of poloxamer 407-based thermosensitive gel of interleukin-1 receptor antagonist. *Pharmaceutical Development and Technology*, *19*(3), 278–284. doi:10.3109/10837450.2013.775158 PMID:23506246

Akash, M. S., Rehman, K., Tariq, M., & Chen, S. (2015). Development of therapeutic proteins: Advances and challenges. *Turkish Journal of Biology*, *39*(3), 343–358. doi:10.3906/biy-1411-8

Al Rahhal, M. M., Bazi, Y., Al Zuair, M., Othman, E., & BenJdira, B. (2018). Convo- lutional neural networks for electrocardiogram classification. *Journal of Medical and Biological Engineering*, *38*(6), 1014–1025. doi:10.100740846-018-0389-7

Alam, I., Alam, G., Ayub, S., & Siddiqui, A. A. (2019). Assessment of bio-medical waste management in different hospitals in Aligarh city. *Advances in Waste Management: Select Proceedings of Recycle*, *2016*, 501–510. doi:10.1007/978-981-13-0215-2_36

Ali, A., Zafar, H., Zia, M., Ul Haq, I., Phull, A. R., Ali, J. S., & Hussain, A. (2016). Synthesis, characterization, applications, and challenges of iron oxide nanoparticles. *Nanotechnology, Science and Applications*, *9*, 49–67. doi:10.2147/NSA.S99986 PMID:27578966

Alishah Aratboni, H., Rafiei, N., Garcia-Granados, R., Alemzadeh, A., & Morones-Ramírez, J. R. (2019). Biomass and lipid induction strategies in microalgae for biofuel production and other applications. *Microbial Cell Factories*, *18*(1), 1–17. doi:10.118612934-019-1228-4 PMID:31638987

Alper, H., Jin, Y. S., Moxley, J. F., & Stephanopoulos, G. (2005). Identifying gene targets for the metabolic engineering of lycopene biosynthesis in *Escherichia coli*. *Metabolic Engineering*, *7*(3), 155–164. doi:10.1016/j.ymben.2004.12.003 PMID:15885614

Al-Ruzouq, R., Abdallah, M., Shanableh, A., Alani, S., Obaid, L., & Gibril, M. B. A. (2022). Waste to energy spatial suitability analysis using hybrid multi-criteria machine learning approach. *Environmental Science and Pollution Research International*, *29*(2), 2613–2628. doi:10.100711356-021-15289-0 PMID:34374020

Alvarez Blasco, F., Martinez Garcia, M. A., Luque Ramirez, M., Parraza, N., & Millan, J. (2009). Role of Haptoglobin in Polycystic ovary syndrome (PCOS), Obesity and Disorders of glucose tolerance in premenopausal women. *PLoS One*, *4*(5), 1–7. doi:10.1371/journal.pone.0005606 PMID:19440331

Ambekar, A. S., Kelkar, D. S., Pinto, S. M., Sharma, R., Hinduja, I., Zaveri, K., Pandey, A., Prasad, K., Gowda, H., & Mukherjee, S. (2015). Protcomics of follicular fluid from women with polycystic ovary syndrome suggests molecular defects in follicular development. *The Journal of Clinical Endocrinology and Metabolism*, *100*(2), 744–753. doi:10.1210/jc.2014-2086 PMID:25393639

Anastasina, M., Terenin, I., Butcher, S. J., & Kainov, D. E. (2014). A technique to increase protein yield in a rabbit reticulocyte lysate translation system. *BioTechniques*, *56*(1), 36–39. doi:10.2144/000114125 PMID:24447137

Andersen, R. S., Peimankar, A., & Puthusserypady, S. (2019). A deep learning ap- proach for real-time detection of atrial fibrillation. *Expert Systems with Applications*, *115*, 465–473. doi:10.1016/j.eswa.2018.08.011

Andrea, B., Karin, S., Diana, T., Klaus, H., Marjori, A. M., & Antionius, J. M. M. (1986). The expression of a nopaline synthase - human growth hormone chimeric gene in transformed tobacco and sunflower callus tissue. *Plant Molecular Biology*.

Andreotti, F., Carr, O., Pimentel, M. A., Mahdi, A., & De Vos, M. (2017). Comparing feature-based classifiers and convolutional neural networks to detect ar- rhythmia from short segments of ecg. In 2017 computing in cardiology (CINC) (pp. 1–4). IEEE.

Anitha, C., Komala, C., Vivekanand, C. V., Lalitha, S., & Boopathi, S. (2023). Artificial Intelligence driven security model for Internet of Medical Things (IoMT). *IEEE Explore*, 1–7.

Anitha, C., R, K. C., Vivekanand, C. V., Lalitha, S. D., Boopathi, S., & R, Revathi. (2023, February). Artificial Intelligence driven security model for Internet of Medical Things (IoMT). *IEEE Explore*. doi:10.1109/ICIPTM57143.2023.10117713

Ari, Z., Michael, W., Betina, H., & Gregory, M. P. (2018). Taligiucerase alfa: Safety and efficacy across clinical studies in adults and children with Gaucher disease. *Orphanet Journal of Rare Diseases*, *13*(1), 36. Advance online publication. doi:10.118613023-018-0776-8

Arunprasad, R., & Boopathi, S. (2019). Chapter-4 Alternate Refrigerants for Minimization Environmental Impacts: A Review. In Advances in Engineering Technology (p. 75). AkiNik Publications.

Arya, R., Bhattacharya, A., & Saini, K. S. (2008). Dictyostelium discoideum—A promising expression system for the production of eukaryotic proteins. *The FASEB Journal*, *22*(12), 4055–4066. doi:10.1096/fj.08-110544 PMID:18714070

Asadollahi, M. A., Maury, J., Patil, K. R., Schalk, M., Clark, A., & Nielsen, J. (2009). Enhancing sesquiterpene production in *Saccharomyces cerevisiae* through in silico driven metabolic engineering. *Metabolic Engineering*, *11*(6), 328–334. doi:10.1016/j.ymben.2009.07.001 PMID:19619667

Ashford, M. B., England, R. M., & Akhtar, N. (2021, May). Highway to success—Developing advanced polymer therapeutics. *Advanced Therapeutics*, *4*(5), 2000285. doi:10.1002/adtp.202000285

Ashrafi, S. D., Nasseri, S., Alimohammadi, M., Mahvi, A. H., & Faramarzi, M. A. (2016). Optimization of the enzymatic elimination of flumequine by laccase-mediated system using response surface methodology. *Desalination and Water Treatment*, *57*(31), 14478–14487. doi:10.1080/19443994.2015.1063462

Astuti, Y., Ahda, Y., & Putri, D. H. (2021). Development of Bioremediation Learning Materials based on Contextual Teaching and Learning (CTL). *Journal of Physics: Conference Series*, *1940*(1), 12126. doi:10.1088/1742-6596/1940/1/012126

Atanasov, A. G., Zotchev, S. B., Dirsch, V. M., Orhan, I. E., Banach, M., Rollinger, J. M., & … . (2021). Natural products in drug discovery: Advances and opportunities. *Nature Reviews. Drug Discovery*, *20*(3), 200–216. doi:10.103841573-020-00114-z PMID:33510482

Atiomo, W., Khalid, S., Parameshweran,S., Houda,M., Layfield,R. (2009). Proteomic Biomarkers for the diagnosis and risk stratification of polycystic ovary syndrome: a systematic review. *BJOG: An International Journal of Obstetrics and Gynecology, 116*(2), 137-143.

Au, H. H., & Jan, E. (2014). Novel viral translation strategies. *Wiley Interdisciplinary Reviews. RNA*, *5*(6), 779–801. doi:10.1002/wrna.1246 PMID:25045163

Avancini, D. B., Rodrigues, J. J. P. C., Rabêlo, R. A. L., Das, A. K., Kozlov, S., & Solic, P. (2021). A new IoT-based smart energy meter for smart grids. *International Journal of Energy Research*, *45*(1), 189–202. doi:10.1002/er.5177

Axup, J. Y., Bajjuri, K. M., Ritland, M., Hutchins, B. M., Kim, C. H., Kazane, S. A., ... Stafin, K. (2012). Synthesis of site-specific antibody-drug conjugates using unnatural amino acids. *Proceedings of the National Academy of Sciences of the United States of America*, *109*(40), 16101–16106. doi:10.1073/pnas.1211023109 PMID:22988081

Aziz, R. K., Bartels, D., Best, A. A., Dejongh, M., Disz, T., Edwards, R. A., Formsma, K., Gerdes, S., Glass, E. M., Kubal, M., Meyer, F., Olsen, G. J., Olson, R., Osterman, A. L., Overbeek, R. A., Mcneil, L. K., Paarmann, D., Paczian, T., Parrello, B., ... Zagnitko, O. (2008). The RAST Server: Rapid Annotations using Subsystems Technology. *BMC Genomics*, *9*, 75.

Azziz, R., Bradley, E. L., Potter, H. D., & Boots, L. R. (1995). Adrenal androgen excess in women: Lack of a role for 17-hydroxylase and 17,20-lyase dysregulation. *The Journal of Clinical Endocrinology and Metabolism*, *80*(2), 400–405. PMID:7852496

Babic, A., & Sepehri, A. A. (2019a). Extraction of diagnostic infor- mation from phonocardiographic signal using time-growing neural network. In L. Lhotska, L. Sukupova, I. Lackovi ́c, & G. S. Ibbott (Eds.), *World congress on med- ical physics and biomedical engineering 2018* (pp. 849–853). Singapore: Springer Singapore.

Babu, B. S., Kamalakannan, J., Meenatchi, N., Karthik, S., & Boopathi, S. (2022). Economic impacts and reliability evaluation of battery by adopting Electric Vehicle. *IEEE Explore*, 1–6.

Babu, B. S., Kamalakannan, J., Meenatchi, N., M, S. K. S., S, K., & Boopathi, S. (2023). Economic impacts and reliability evaluation of battery by adopting Electric Vehicle. *IEEE Explore*, 1–6. doi:10.1109/ICPECTS56089.2022.10046786

Bae, W., Mulchandani, A., & Chen, W. (2002). Cell surface display of synthetic phytochelatins using ice nucleation protein for enhanced heavy metal bioaccumulation. *Journal of Inorganic Biochemistry*, *88*(2), 223–227. doi:10.1016/S0162-0134(01)00392-0 PMID:11803043

Bahera, B. K., Prasad, R., & Behera, S. (2021). *Therapeutic Proteins: Production and Delivery. InLife Sciences Industry*. Springer.

Bairoch, A. (2000). The ENZYME database in 2000. *Nucleic Acids Research*, *28*(1), 304–305. doi:10.1093/nar/28.1.304 PMID:10592255

Balamurugan, S., Kaewta, R., Suwimon, M., Arunee, T., & Waranyoo, P. (2020, November). Monoclonal Antibodies B38 and H4 Produced in Nicotiana benthamiana Neutralize SARS-CoV-2 in vitro. *Frontiers in Plant Science*. PMID:33329653

Baldwin, G., Bayer, T., Dickinson, R., Ellis, T., Freemont, P. S., Kitney, R. I., Polizzi, K., & Stan, G.-B. (2015). Modeling Synthetic Biology Systems. In Synthetic Biology — A Primer (pp. 89–108). Imperial College Press. doi:10.1142/9781783268801_0006

Banerjee, S., Banerjee, S., Ghosh, A. K., & Das, D. (2020). Maneuvering the genetic and metabolic pathway for improving biofuel production in algae: Present status and future prospective. *Renewable & Sustainable Energy Reviews*, *133*, 110155. doi:10.1016/j.rser.2020.110155

Banga, S., Kumar, V., Suri, S., Kaushal, M., Prasad, R., & Kaur, S. (2019). Nutraceutical potential of Diet Drinks: A critical review on components, health effects, and Consumer Safety. *Journal of the American College of Nutrition*, *39*(3), 272–286. doi:10.1080/07315724.2019.1642811 PMID:31343956

Baran, H., & Kepplinger, B. (2014). D-cycloserine lowers kynurenic acid formation—New mechanism of action. *European Neuropsychopharmacology*, *24*(4), 639–644. doi:10.1016/j.euroneuro.2013.10.006 PMID:24189377

Barman, B. K., Yadav, S. N., Kumar, S., & Gope, S. (2019). IOT Based Smart Energy Meter for Efficient Energy Utilization in Smart Grid. *2nd International Conference on Energy, Power and Environment: Towards Smart Technology, ICEPE 2018*, 1–5. 10.1109/EPETSG.2018.8658501

Barron, M. G., Lilavois, C. R., & Martin, T. M. (2015). MOAtox: A comprehensive mode of action and acute aquatic toxicity database for predictive model development. *Aquatic Toxicology (Amsterdam, Netherlands)*, *161*, 102–107. doi:10.1016/j.aquatox.2015.02.001 PMID:25700118

Bateman, A., Martin, M. J., Orchard, S., Magrane, M., Agivetova, R., Ahmad, S., Alpi, E., Bowler-Barnett, E. H., Britto, R., Bursteinas, B., Bye-A-Jee, H., Coetzee, R., Cukura, A., da Silva, A., Denny, P., Dogan, T., Ebenezer, T. G., Fan, J., Castro, L. G., ... Teodoro, D. (2021). UniProt: The universal protein knowledgebase in 2021. *Nucleic Acids Research*, *49*(D1), D480–D489. doi:10.1093/nar/gkaa1100 PMID:33237286

Baumschlager, A., & Khammash, M. (2021). Synthetic Biological Approaches for Optogenetics and Tools for Transcriptional Light-Control in Bacteria. *Advanced Biology*, *5*(5), 2000256. doi:10.1002/adbi.202000256 PMID:34028214

Becker, J., Zelder, O., Häfner, S., Schröder, H., & Wittmann, C. (2011). From zero to hero—Design-based systems metabolic engineering of *Corynebacterium glutamicum* for l-lysine production. *Metabolic Engineering*, *13*(2), 159–168. doi:10.1016/j.ymben.2011.01.003 PMID:21241816

Behrendorff, J. B. (2021). Reductive cytochrome P450 reactions and their potential role in bioremediation. *Frontiers in Microbiology*, *12*, 649273. doi:10.3389/fmicb.2021.649273 PMID:33936006

Bekri, S. (2006). Importance of glycosylation in enzyme replacement therapy. *Fabry disease: Perspectives from 5 years of FOS*.

Benjamin, E. J., Virani, S. S., Callaway, C. W., Chamberlain, A. M., Chang, A. R., Cheng, S., & Deo, R. (2018). Heart disease and stroke statistics-2018 up- date: A report from the american heart association. *Circulation*, *137*(12), e67. doi:10.1161/CIR.0000000000000558 PMID:29386200

Benninghoff, J., & Perneczky, R. (2022). Anti-dementia medications and anti-alzheimer's disease drugs: Side effects, contraindications, and interactions. *NeuroPsychopharmacotherapy*, 1–10.

Beyene, H. D., Werkneh, A. A., & Ambaye, T. G. (2018). Current updates on waste to energy (WtE) technologies: A review. *Renewable Energy Focus*, *24*, 1–11. doi:10.1016/j.ref.2017.11.001

Bharagava, R. N., Saxena, G., & Mulla, S. I. (2020). *Introduction to industrial wastes containing organic and inorganic pollutants and bioremediation approaches for environmental management. In Bioremediation of Industrial Waste for Environmental Safety* (Vol. I). Industrial Waste and Its Management.

Bhattacharya, R., Bose, D., Ganti, P., Rizvi, A., Halder, G., & Sarkar, A. (2023). *Bioelectricity production and bioremediation potential of Withania somnifera in plant microbial fuel cells with food wastes as enrichment.* Academic Press.

Bhattacharya, R., Bose, D., Yadav, J., Sharma, B., Sangli, E., Patel, A., Mukherjee, A., & Singh, A. A. (2023). Bioremediation and bioelectricity from Himalayan rock soil in sediment-microbial fuel cell using carbon rich substrates. *Fuel, 341*, 127019. doi:10.1016/j.fuel.2022.127019

Bhattacharya, S. K., Kumar, A., & Ghosal, S. (1995). Effects of glycowithanolides from *withania somnifera* on an animal model of alzheimer's disease and perturbed central cholinergic markers of cognition in rats. *Phytotherapy Research, 9*(2), 110–113. doi:10.1002/ptr.2650090206

Bhowmik, D., Chetri, S., Enerijiofi, K. E., Naha, A., Kanungo, T. D., Shah, M. P., & Nath, S. (2023). Multitudinous approaches, challenges and opportunities of bioelectrochemical systems in conversion of waste to energy from wastewater treatment plants. *Cleaner and Circular Bioeconomy*, 100040.

Bidja Abena, M. T., Sodbaatar, N., Li, T., Damdinsuren, N., Choidash, B., & Zhong, W. (2019). Crude oil biodegradation by newly isolated bacterial strains and their consortium under soil microcosm experiment. *Applied Biochemistry and Biotechnology, 189*(4), 1223–1244. doi:10.100712010-019-03058-2 PMID:31236896

Bilal, M., Iqbal, H. M., Hu, H., Wang, W., & Zhang, X. (2017). Development of horseradish peroxidase-based cross-linked enzyme aggregates and their environmental exploitation for bioremediation purposes. *Journal of Environmental Management, 188*, 137–143. doi:10.1016/j.jenvman.2016.12.015 PMID:27978441

Bizopoulos, P., & Koutsouris, D. (2018). Deep learning in cardiology. *IEEE Reviews in Biomedical Engineering, 12*, 168–193.

Blanken, D., Foschepoth, D., Serrão, A. C., & Danelon, C. (2020). Genetically controlled membrane synthesis in liposomes. *Nature Communications, 11*(1), 1–13. doi:10.103841467-020-17863-5 PMID:32859896

Bleakney, T. L. (2008). Deconstructing an adaptogen. *Holistic Nursing Practice, 22*(4), 220–224. doi:10.1097/01.HNP.0000326005.65310.7c PMID:18607235

Bloch, K. D., & Grossmann, B. (1995). Digestion of DNA with Restriction Endonucleases. *Current Protocols in Molecular Biology, 31*(1). Advance online publication. doi:10.1002/0471142727.mb0301s31 PMID:18265218

Blow, J. J., & Laskey, R. A. (1986). Initiation of DNA replication in nuclei and purified DNA by a cell-free extract of Xenopus eggs. *Cell, 47*(4), 577–587. doi:10.1016/0092-8674(86)90622-7 PMID:3779837

Boopathi, S. (2022). Performance Improvement of Eco-Friendly Near-Dry wire-Cut Electrical Discharge Machining Process Using Coconut Oil-Mist Dielectric Fluid. *World Scientific: Journal of Advanced Manufacturing Systems.*

Boopathi, S. (2022e). Effects of Cryogenically-treated Stainless Steel on Eco-friendly Wire Electrical Discharge Machining Process. *Preprint*, 1–18.

Boopathi, S. (2023). Securing Healthcare Systems Integrated With IoT: Fundamentals, Applications, and Future Trends. In Dynamics of Swarm Intelligence Health Analysis for the Next Generation (pp. 186–209). IGI Global.

Boopathi, S. (2023c). Securing Healthcare Systems Integrated With IoT: Fundamentals, Applications, and Future Trends. In Dynamics of Swarm Intelligence Health Analysis for the Next Generation (pp. 186–209). IGI Global.

Boopathi, S. (2023f). Securing Healthcare Systems Integrated With IoT: Fundamentals, Applications, and Future Trends. In Dynamics of Swarm Intelligence Health Analysis for the Next Generation (pp. 186–209). IGI Global.

Boopathi, S., Arigela, S. H., Raman, R., Indhumathi, C., Kavitha, V., & Bhatt, B. C. (2023). Prominent Rule Control-based Internet of Things: Poultry Farm Management System. *IEEE Explore*, 1–6. doi:10.1109/ICPECTS56089.2022.10047039

Boopathi, S., Gavaskar, T., Dogga, A. D., Mahendran, R. K., Kumar, A., Kathiresan, G., N., V., Ganesan, M., Ishwarya, K. R., & Ramana, G. V. (2021). *Emergency medicine delivery transportation using unmanned aerial vehicle (Patent Grant)*. Academic Press.

Boopathi, S., Khare, R., Jaya Christiyan, K. G., Muni, T. V., & Khare, S. (2023). Additive manufacturing developments in the medical engineering field. In Development, Properties, and Industrial Applications of 3D Printed Polymer Composites (pp. 86–106). IGI Global. doi:10.4018/978-1-6684-6009-2.ch006

Boopathi, S., Khare, R., KG, J. C., Muni, T. V., & Khare, S. (2023). Additive Manufacturing Developments in the Medical Engineering Field. In Development, Properties, and Industrial Applications of 3D Printed Polymer Composites (pp. 86–106). IGI Global.

Boopathi, S., Kumar, P. K. S., Meena, R. S., Sudhakar, M., & Associates. (2023). Sustainable Developments of Modern Soil-Less Agro-Cultivation Systems: Aquaponic Culture. In Human Agro-Energy Optimization for Business and Industry (pp. 69–87). IGI Global.

Boopathi, S., Pandey, B. K., & Pandey, D. (2023). Advances in Artificial Intelligence for Image Processing: Techniques, Applications, and Optimization. In Handbook of Research on Thrust Technologies' Effect on Image Processing (pp. 73–95). IGI Global.

Boopathi, S., Thillaivanan, A., Mohammed, A. A., Shanmugam, P., & VR, P. (2022). Experimental investigation on Abrasive Water Jet Machining of Neem Wood Plastic Composite. *IOP: Functional Composites and Structures, 4*, 025001.

Boopathi, S. (2019). Experimental investigation and parameter analysis of LPG refrigeration system using Taguchi method. *SN Applied Sciences*, *1*(8), 892. Advance online publication. doi:10.100742452-019-0925-2

Boopathi, S. (2022a). An experimental investigation of Quench Polish Quench (QPQ) coating on AISI 4150 steel. *Engineering Research Express*, *4*(4), 45009. doi:10.1088/2631-8695/ac9ddd

Boopathi, S. (2022b). An Extensive Review on Sustainable Developments of Dry and Near-Dry Electrical Discharge Machining Processes. *Journal of Manufacturing Science and Engineering*, *144*(5), 50801. doi:10.1115/1.4052527

Boopathi, S. (2022c). An investigation on gas emission concentration and relative emission rate of the near-dry wire-cut electrical discharge machining process. *Environmental Science and Pollution Research International*, *29*(57), 86237–86246. doi:10.100711356-021-17658-1 PMID:34837614

Boopathi, S. (2022d). Cryogenically treated and untreated stainless steel grade 317 in sustainable wire electrical discharge machining process: A comparative study. *Environmental Science and Pollution Research International*, 1–10. doi:10.100711356-022-22843-x PMID:36057706

Boopathi, S. (2022f). Experimental investigation and multi-objective optimization of cryogenic Friction-stir-welding of AA2014 and AZ31B alloys using MOORA technique. *Materials Today. Communications*, *33*, 104937. doi:10.1016/j.mtcomm.2022.104937

Boopathi, S. (2022g). Performance Improvement of Eco-Friendly Near-Dry Wire-Cut Electrical Discharge Machining Process Using Coconut Oil-Mist Dielectric Fluid. *Journal of Advanced Manufacturing Systems*. Advance online publication. doi:10.1142/S0219686723500178

Boopathi, S. (2023a). An Investigation on Friction Stir Processing of Aluminum Alloy-Boron Carbide Surface Composite. In *Advances in Processing of Lightweight Metal Alloys and Composites* (pp. 249–257). Springer. doi:10.1007/978-981-19-7146-4_14

Boopathi, S. (2023a). *Deep Learning Techniques Applied for Automatic Sentence Generation*. doi:10.4018/978-1-6684-3632-5.ch016

Boopathi, S. (2023b). Internet of Things-Integrated Remote Patient Monitoring System: Healthcare Application. In *Dynamics of Swarm Intelligence Health Analysis for the Next Generation* (pp. 137–161). IGI Global. doi:10.4018/978-1-6684-6894-4.ch008

Boopathi, S. (2023c). *Impact analysis of ceramic tile powder aggregates on self-compacting concrete*. Engineering Research Express.

Boopathi, S., Alqahtani, A. S., Mubarakali, A., & Panchatcharam, P. (2023). Sustainable developments in near-dry electrical discharge machining process using sunflower oil-mist dielectric fluid. *Environmental Science and Pollution Research International*, 1–20. doi:10.100711356-023-27494-0 PMID:37199846

Boopathi, S., & Davim, J. P. (2023a). Applications of Nanoparticles in Various Manufacturing Processes. In *Sustainable Utilization of Nanoparticles and Nanofluids in Engineering Applications* (pp. 1–31). IGI Global. doi:10.4018/978-1-6684-9135-5.ch001

Boopathi, S., & Davim, J. P. (2023b). *Sustainable Utilization of Nanoparticles and Nanofluids in Engineering Applications*. IGI Global. doi:10.4018/978-1-6684-9135-5

Boopathi, S., & Kanike, U. K. (2023). Applications of Artificial Intelligent and Machine Learning Techniques in Image Processing. In *Handbook of Research on Thrust Technologies' Effect on Image Processing* (pp. 151–173). IGI Global. doi:10.4018/978-1-6684-8618-4.ch010

Boopathi, S., Lewise, K. A. S., Subbiah, R., & Sivaraman, G. (2021). Near-dry wire-cut electrical discharge machining process using water-air-mist dielectric fluid: An experimental study. *Materials Today: Proceedings*, *49*, 1885–1890. doi:10.1016/j.matpr.2021.08.077

Boopathi, S., & Sivakumar, K. (2013). Experimental investigation and parameter optimization of near-dry wire-cut electrical discharge machining using multi-objective evolutionary algorithm. *International Journal of Advanced Manufacturing Technology*, *67*(9–12), 2639–2655. doi:10.100700170-012-4680-4

Boopathi, S., & Sivakumar, K. (2016). Optimal parameter prediction of oxygen-mist near-dry wire-cut EDM. *Inderscience: International Journal of Manufacturing Technology and Management*, *30*(3–4), 164–178. doi:10.1504/IJMTM.2016.077812

Boopathi, S., Sureshkumar, M., Jeyakumar, M., Kumar, R. S., & Subbiah, R. (2022). Influences of Fabrication Parameters on Natural Fiber Reinforced Polymer Composite (NFRPC) Material: A Review. *Materials Science Forum*, *1075*, 115–124. doi:10.4028/p-095f0t

Boopathi, S., Sureshkumar, M., & Sathiskumar, S. (2022). Parametric Optimization of LPG Refrigeration System Using Artificial Bee Colony Algorithm. *International Conference on Recent Advances in Mechanical Engineering Research and Development*, 97–105.

Boopathi, S., Umareddy, M., & Elangovan, M. (2023). Applications of Nano-Cutting Fluids in Advanced Machining Processes. In *Sustainable Utilization of Nanoparticles and Nanofluids in Engineering Applications* (pp. 211–234). IGI Global. doi:10.4018/978-1-6684-9135-5.ch009

Borkowski, O., Koch, M., Zettor, A., Pandi, A., Batista, A. C., Soudier, P., & Faulon, J. L. (2020). Large scale active-learning-guided exploration for in vitro protein production optimization. *Nature Communications*, *11*(1), 1872. doi:10.103841467-020-15798-5 PMID:32312991

BoydM. A.ThavarajahW.LucksJ. B.KamatN. P. (2022). Robust and tunable performance of a cell-free biosensor encapsulated in lipid vesicles. bioRxiv, 2022.2003.2002.482665. doi:10.1101/2022.03.02.482665

Brinckmann, J., & Smith, E. (2004). Maca culture of the Junín Plateau. *Journal of Alternative and Complementary Medicine (New York, N.Y.)*, *10*(3), 426–430. doi:10.1089/1075553041323821 PMID:15253845

Brinkhaus, B., Lindner, M., Schuppan, D., & Hahn, E. G. (2000). Chemical, pharmacological and clinical profile of the East Asian Medical Plant *Centella Aslatica. Phytomedicine*, *7*(5), 427–448. doi:10.1016/S0944-7113(00)80065-3 PMID:11081995

Brooks, S. M., & Alper, H. S. (2021, March 2). Applications, challenges, and needs for employing synthetic biology beyond the lab. *Nature Communications*, *12*(1), 1–6. doi:10.103841467-021-21740-0 PMID:33654085

Brookwell, A., Oza, J. P., & Caschera, F. (2021). Biotechnology Applications of Cell-Free Expression Systems. *Life (Basel, Switzerland)*, *11*(12), 1367. Advance online publication. doi:10.3390/life11121367 PMID:34947898

Bruce, S. E., Werner, K. B., Preston, B. F., & Baker, L. M. (2014). Improvements in concentration, working memory and sustained attention following consumption of a natural citicoline–caffeine beverage. *International Journal of Food Sciences and Nutrition*, *65*(8), 1003–1007. doi:10.3109/09637486.2014.940286 PMID:25046515

Buchan, B. W., & Ledeboer, N. A. (2014). Emerging technologies for the clinical microbiology laboratory. *Clinical Microbiology Reviews*, *27*(4), 783–822. doi:10.1128/CMR.00003-14 PMID:25278575

Buchner, E. (1897). Alcoholic fermentation without yeast cells. *Cornish-Bowden*, 25-31.

Buntru, M., Vogel, S., Spiegel, H., & Schillberg, S. (2014). Tobacco BY-2 cell-free lysate: An alternative and highly-productive plant-based in vitro translation system. *BMC Biotechnology*, *14*(1), 37. doi:10.1186/1472-6750-14-37 PMID:24886601

Burgard, A. P., Pharkya, P., & Maranas, C. D. (2003). Optknock: A bilevel programming framework for identifying gene knockout strategies for microbial strain optimization. *Biotechnology and Bioengineering*, *84*(6), 647–657. doi:10.1002/bit.10803 PMID:14595777

Burnett, M. J., & Burnett, A. C. (2020, March). Therapeutic recombinant protein production in plants: Challenges and opportunities. *Plants, People, Planet*, *2*(2), 121–132. doi:10.1002/ppp3.10073

Butreddy, A., Gaddam, R. P., Kommineni, N., Dudhipala, N., & Voshavar, C. (2021, August 18). PLGA/PLA-Based long-acting injectable depot microspheres in clinical use: Production and characterization overview for protein/peptide delivery. *International Journal of Molecular Sciences*, *22*(16), 8884. doi:10.3390/ijms22168884 PMID:34445587

Butterfield, D. A., & Pocernich, C. B. (2003). The glutamatergic system and alzheimer's disease. *CNS Drugs*, *17*(9), 641–652. doi:10.2165/00023210-200317090-00004 PMID:12828500

C, B. G. (2023). Mechanisms and types of restriction endonuclease enzymes. *The Science Notes*. https://thesciencenotes.com/restriction-endonuclease-enzymes-mechanisms-types/

Calhoun, K. A., & Swartz, J. R. (2005). An Economical Method for Cell-Free Protein Synthesis using Glucose and Nucleoside Monophosphates. *Biotechnology Progress*, *21*(4), 1146–1153. doi:10.1021/bp050052y PMID:16080695

Camacho-Zaragoza, J. M., Hernández-Chávez, G., Moreno-Avitia, F., Ramírez-Iñiguez, R., Martínez, A., Bolívar, F., & Gosset, G. (2016). Engineering of a microbial coculture of *Escherichia coli* strains for the biosynthesis of resveratrol. *Microbial Cell Factories*, *15*(1), 163. doi:10.118612934-016-0562-z PMID:27680538

Cameron, D. E., Bashor, C. J., & Collins, J. J. (2014). A brief history of synthetic biology. *Nature Reviews. Microbiology*, *12*(5), 381–390. doi:10.1038/nrmicro3239

Campbell, K., Xia, J., & Nielsen, J. (2017). The Impact of Systems Biology on Bioprocessing. *Trends in Biotechnology*, *35*(12), 1156–1168. doi:10.1016/j.tibtech.2017.08.011 PMID:28987922

Campo, A. M. (n.d.). *NMR Metabolomics for Optimizing Cell-Free Protein Synthesis* [Doctoral dissertation]. Wright State University.

Can, T. (2020). Gating Creates Slow Modes And Controls Phase-space Complexity In GRUs And LSTMs. *PMLR Proceedings*. http://proceedings.mlr.press/v107/can20a.

Capeness, M. J., & Horsfall, L. E. (2020). Synthetic biology approaches towards the recycling of metals from the environment. *Biochemical Society Transactions*, *48*(4), 1367–1378. doi:10.1042/BST20190837 PMID:32627824

Carbajosa, G., Trigo, A., Valencia, A., & Cases, I. (2009). Bionemo: Molecular information on biodegradation metabolism. *Nucleic Acids Research*, *37*(Database, suppl_1), D598–D602. doi:10.1093/nar/gkn864 PMID:18986994

Carbonell, P., Radivojevic, T., & García Martín, H. (2019). Opportunities at the Intersection of Synthetic Biology, Machine Learning, and Automation. In ACS Synthetic Biology (Vol. 8, Issue 7, pp. 1474–1477). ACS Publications. doi:10.1021/acssynbio.8b00540

Carbonell, P., Jervis, A. J., Robinson, C. J., Yan, C., Dunstan, M., Swainston, N., Vinaixa, M., Hollywood, K. A., Currin, A., Rattray, N. J. W., Taylor, S., Spiess, R., Sung, R., Williams, A. R., Fellows, D., Stanford, N. J., Mulherin, P., Le Feuvre, R., Barran, P., ... Scrutton, N. S. (2018). An automated Design-Build-Test-Learn pipeline for enhanced microbial production of fine chemicals. *Communications Biology*, *1*(1), 66. doi:10.103842003-018-0076-9 PMID:30271948

Cardi, T., & Neal Stewart, C. Jr. (2016). Progress of targeted genome modification approaches in higher plants. *Plant Cell Reports*, *35*(7), 1401–1416. doi:10.100700299-016-1975-1 PMID:27025856

Carlson, E. D., Gan, R., Hodgman, C. E., & Jewett, M. C. (2012). Cell-free protein synthesis: Applications come of age. *Biotechnology Advances*, *30*(5), 1185–1194. doi:10.1016/j.biotechadv.2011.09.016 PMID:22008973

Caschera, F., Bedau, M. A., Buchanan, A., Cawse, J., de Lucrezia, D., Gazzola, G., Hanczyc, M. M., & Packard, N. H. (2011). Coping with complexity: Machine learning optimization of cell-free protein synthesis. *Biotechnology and Bioengineering*, *108*(9), 2218–2228. doi:10.1002/bit.23178 PMID:21520017

Caschera, F., Karim, A. S., Gazzola, G., d'Aquino, A. E., Packard, N. H., & Jewett, M. C. (2018). High-Throughput Optimization Cycle of a Cell-Free Ribosome Assembly and Protein Synthesis System. *ACS Synthetic Biology*, *7*(12), 2841–2853. doi:10.1021/acssynbio.8b00276 PMID:30354075

Caschera, F., & Noireaux, V. (2014). Synthesis of 2.3 mg/ml of protein with an all Escherichia coli cell-free transcription-translation system. *Biochimie*, *99*, 162–168. doi:10.1016/j.biochi.2013.11.025 PMID:24326247

Caspi, R., Altman, T., Billington, R., Dreher, K., Foerster, H., Fulcher, C. A., Holland, T. A., Keseler, I. M., Kothari, A., Kubo, A., Krummenacker, M., Latendresse, M., Mueller, L. A., Ong, Q., Paley, S., Subhraveti, P., Weaver, D. S., Weerasinghe, D., Zhang, P., & Karp, P. D. (2014). The MetaCyc database of metabolic pathways and enzymes and the BioCyc collection of Pathway/Genome Databases. *Nucleic Acids Research*, *42*(D1), D459–D471. doi:10.1093/nar/gkt1103 PMID:24225315

Caspi, R., Dreher, K., & Karp, P. D. (2013). The challenge of constructing, classifying and representing metabolic pathways. *FEMS Microbiology Letters*, *345*(2), 85–93. doi:10.1111/1574-6968.12194 PMID:23746312

Cazzola, M., Mercuriali, F., & Brugnara, C. (1997). Use of recombinant human erythropoietin outside the setting of uremia. *Blood*, *89*(12), 4248–4267. doi:10.1182/blood.V89.12.4248 PMID:9192747

Cernak, P., Estrela, R., Poddar, S., Skerker, J. M., Cheng, Y.-F., Carlson, A. K., Chen, B., Glynn, V. M., Furlan, M., Ryan, O. W., Donnelly, M. K., Arkin, A. P., Taylor, J. W., & Cate, J. H. D. (2018). Engineering Kluyveromyces marxianus as a robust synthetic biology platform host. *mBio*, *9*(5), 10–1128. doi:10.1128/mBio.01410-18 PMID:30254120

Chamatidis, I., Katsika, A., & Spathoulas, G. (2017). Using deep learning neural net- works for ECG based authentication. In *2017 International Carnahan Conference on Security Technology (ICCST)* (pp. 1–6). IEEE.

Chanana, P., & Kumar, A. (2016). Possible involvement of nitric oxide modulatory mechanisms in the neuroprotective effect of *centella asiatica* against sleep deprivation induced anxiety like behaviour, oxidative damage and neuroinflammation. *Phytotherapy Research*, *30*(4), 671–680. doi:10.1002/ptr.5582 PMID:26848139

Chandak, S., & Rout, P. K. (2021). The implementation framework of a microgrid: A review. *International Journal of Energy Research*, *45*(3), 3523–3547. doi:10.1002/er.6064

Chandra, B., Sastry, C. S., Jana, S., & Patidar, S. (2017). Atrial fibrillation detection using convolutional neural networks. In 2017 Computing in cardiology (CINC) (pp. 1–4). IEEE.

Chandran, D., Copeland, W. B., Sleight, S. C., & Sauro, H. M. (2008). Mathematical modeling and synthetic biology. Drug Discovery Today: Disease Models, 5(4), 299-309. doi: 10.1016/j.ddmod.2009.07.002 PMID: 27840651. doi:10.1016/j.ddmod.2009.07.002

Chandrika, V., Sivakumar, A., Krishnan, T. S., Pradeep, J., Manikandan, S., & Boopathi, S. (2023). Theoretical Study on Power Distribution Systems for Electric Vehicles. In *Intelligent Engineering Applications and Applied Sciences for Sustainability* (pp. 1–19). IGI Global. doi:10.4018/979-8-3693-0044-2.ch001

Chang, R.J., & Kazer, R.R. (2014). Polycystic ovary syndrome. *The global library of women's medicine.*

Chang, A., Jeske, L., Ulbrich, S., Hofmann, J., Koblitz, J., Schomburg, I., Neumann-Schaal, M., Jahn, D., & Schomburg, D. (2021). BRENDA, the ELIXIR core data resource in 2021: New developments and updates. *Nucleic Acids Research*, *49*(D1), D498–D508. doi:10.1093/nar/gkaa1025 PMID:33211880

Chang, R. L., Ghamsari, L., Manichaikul, A., Hom, E. F. Y., Balaji, S., Fu, W., Shen, Y., Hao, T., Palsson, B., Salehi-Ashtiani, K., & Papin, J. A. (2011). Metabolic network reconstruction of *Chlamydomonas* offers insight into light-driven algal metabolism. *Molecular Systems Biology*, *7*(1), 518. doi:10.1038/msb.2011.52 PMID:21811229

Chao, R., Yuan, Y., & Zhao, H. (2014). Recent advances in DNA assembly technologies. *FEMS Yeast Research.* doi:10.1111/1567-1364.12171 PMID:24903193

Chaturvedi, P., Shukla, P., Giri, B. S., Chowdhary, P., Chandra, R., Gupta, P., & Pandey, A. (2021). Prevalence and hazardous impact of pharmaceutical and personal care products and antibiotics in environment: A review on emerging contaminants. *Environmental Research*, *194*, 110664. doi:10.1016/j.envres.2020.110664 PMID:33400949

Chaudhari, K. S., Tiwari, N. R., Tiwari, R. R., & Sharma, R. S. (2017). Neurocognitive effect of nootropic drug brahmi (*bacopa monnieri*) in alzheimer's disease. *Annals of Neurosciences*, *24*(2), 111–122. doi:10.1159/000475900 PMID:28588366

Checa, S. K., Zurbriggen, M. D., & Soncini, F. C. (2012). Bacterial signaling systems as platforms for rational design of new generations of biosensors. *Current Opinion in Biotechnology*, *23*(5), 766–772. doi:10.1016/j.copbio.2012.05.003 PMID:22658939

Chen, M., Wang, G., Xie, P., Sang, Z., Lv, T., Zhang, P., & Yang, H. (2018). Region aggregation network: Improving convolutional neural network for ECG characteristic detection. In *2018 40th Annual international conference of the ieee engineering in medicine and biology society (EMBC)* (pp. 2559–2562). IEEE.

Chen, C.-L., Tsai, W.-H., Chen, C.-J., & Pan, T.-M. (2016). *Centella asiatica* extract protects against amyloid β1–40-induced neurotoxicity in neuronal cells by activating the Antioxidative Defence System. *Journal of Traditional and Complementary Medicine*, *6*(4), 362–369. doi:10.1016/j.jtcme.2015.07.002 PMID:27774420

Chen, F., Yuan, L., Ding, S., Tian, Y., & Hu, Q. N. (2020). Data-driven rational biosynthesis design: From molecules to cell factories. *Briefings in Bioinformatics*, *21*(4), 1238–1248. doi:10.1093/bib/bbz065 PMID:31243440

Cheng, F., Yu, H., & Stephanopoulos, G. (2019). Engineering *Corynebacterium glutamicum* for high-titer biosynthesis of hyaluronic acid. *Metabolic Engineering*, *55*, 276–289. doi:10.1016/j.ymben.2019.07.003 PMID:31301358

Chen, L., & Kashina, A. (2021). Post-translational modifications of the protein termini. *Frontiers in Cell and Developmental Biology*, 9. PMID:34395449

Chenthamara, D., Subramaniam, S., Ramakrishnan, S. G., Krishnaswamy, S., Essa, M. M., Lin, F. H., & Qoronfleh, M. W. (2019, December). Therapeutic efficacy of nanoparticles and routes of administration. *Biomaterials Research*, *23*(1), 1–29. doi:10.118640824-019-0166-x PMID:31832232

Chen, W., Kong, Y., Li, J., Sun, Y., Min, J., & Hu, X. (2020). Enhanced biodegradation of crude oil by constructed bacterial consortium comprising salt-tolerant petroleum degraders and biosurfactant producers. *International Biodeterioration & Biodegradation*, *154*, 105047. doi:10.1016/j.ibiod.2020.105047

Chen, W., Zhang, Y., Zhang, Y., Pi, Y., Gu, T., Song, L., Wang, Y., & Ji, Q. (2018). CRISPR/Cas9-based genome editing in *Pseudomonas aeruginosa* and cytidine deaminase-mediated base editing in *Pseudomonas* species. *iScience*, *6*, 222–231. doi:10.1016/j.isci.2018.07.024 PMID:30240613

Chen, X., Sun, Q., & Lu, Y. (2020). Creating a locally crowded environment with nanoclay hydrogels for cell-free biosynthesis. *Soft Matter*, *16*(22), 5132–5138. doi:10.1039/D0SM00636J PMID:32478769

Che, S., & Men, Y. (2019). Synthetic microbial consortia for biosynthesis and biodegradation: Promises and challenges. *Journal of Industrial Microbiology & Biotechnology*, *46*(9-10), 1343–1358. doi:10.100710295-019-02211-4 PMID:31278525

Chiba, C. H., Knirsch, M. C., Azzoni, A. R., Moreira, A. R., & Stephano, M. A. (2021). Cell-free protein synthesis: Advances on production process for biopharmaceuticals and immunobiological products. *BioTechniques*, *70*(2), 126–133. doi:10.2144/btn-2020-0155 PMID:33467890

Chimezie, A., Kenneth, N., Olufunsho, A., Chinwe, N., Mohsen, A., & Herbert, C. (2011). Concentrations of heavy metals in some pharmaceutical effluents in Lagos, Nigeria. *Journal of Environmental Chemistry and Ecotoxicology*, *3*(2), 25–31.

Chi, N. T. L., Anto, S., Ahamed, T. S., Kumar, S. S., Shanmugam, S., Samuel, M. S., Mathimani, T., Brindhadevi, K., & Pugazhendhi, A. (2021). A review on biochar production techniques and biochar based catalyst for biofuel production from algae. *Fuel*, *287*, 119411. doi:10.1016/j.fuel.2020.119411

Ching, T., Himmelstein, D. S., Beaulieu-Jones, B. K., Kalinin, A. A., Do, B. T., Way, G. P., Ferrero, E., Agapow, P. M., Zietz, M., Hoffman, M. M., Xie, W., Rosen, G. L., Lengerich, B. J., Israeli, J., Lanchantin, J., Woloszynek, S., Carpenter, A. E., Shrikumar, A., Xu, J., ... Greene, C. S. (2018). Opportunities and obstacles for deep learning in biology and medicine. *Journal of the Royal Society, Interface*, *15*(141), 20170387. doi:10.1098/rsif.2017.0387 PMID:29618526

Chisti, Y. (2007). Biodiesel from microalgae. *Biotechnology Advances, 25*(3), 294–306. doi:10.1016/j.biotechadv.2007.02.001 PMID:17350212

Cho, E., & Lu, Y. (2020). Compartmentalizing cell-free systems: Toward creating life-like artificial cells and beyond. *ACS Synthetic Biology, 9*(11), 2881–2901. doi:10.1021/acssynbio.0c00433 PMID:33095011

Choi, D. H., Lee, W. S., Won, M., Park, M., Park, H. O., Kim, E., Lee, K. H., & Bae, J. (2010). The Apolipoprotein A-I level in downregulated in the grnualosa cells of patients with polycystic ovary syndrome and affects steroidogesis. *Journal of Proteome Research, 9*(9), 4329–4336. doi:10.1021/pr100008e PMID:20426491

Choudhary, S., & Parmar, N. (2013). Hazard assessment of liquid effluent treatment plant in pharmaceutical industry. *VSRD Int J Technol NonTechnol Res, 4*(9), 209–214.

Christen, V., Hickmann, S., Rechenberg, B., & Fent, K. (2010). Highly active human pharmaceuticals in aquatic systems: A concept for their identification based on their mode of action. *Aquatic Toxicology (Amsterdam, Netherlands), 96*(3), 167–181. doi:10.1016/j.aquatox.2009.11.021 PMID:20053463

Chung, B. K. S., Lakshmanan, M., Klement, M., Mohanty, B., & Lee, D. Y. (2013). Genome-scale in silico modeling and analysis for designing synthetic terpenoid-producing microbial cell factories. *Chemical Engineering Science, 103*, 100–108. doi:10.1016/j.ces.2012.09.006

Cid, R., & Bolívar, J. (2021, July 21). Platforms for production of protein-based vaccines: From classical to next-generation strategies. *Biomolecules, 11*(8), 1072. doi:10.3390/biom11081072 PMID:34439738

Condreay, J. P., & Kost, T. A. (2007). Baculovirus expression vectors for insect and mammalian cells. *Current Drug Targets, 8*(10), 1126–1131. doi:10.2174/138945007782151351 PMID:17979672

Cook-Deegan, R., & Heaney, C. (2010). Patents in genomics and human genetics. *Annual Review of Genomics and Human Genetics, 11*(1), 383–425. doi:10.1146/annurev-genom-082509-141811 PMID:20590431

Corchero, J. L., Gasser, B., Resina, D., Smith, W., Parrilli, E., Vázquez, F., Abasolo, I., Giuliani, M., Jäntti, J., Ferrer, P., Saloheimo, M., Mattanovich, D., Schwartz, S. Jr, Tutino, M. L., & Villaverde, A. (2013, March 1). Unconventional microbial systems for the cost-efficient production of high-quality protein therapeutics. *Biotechnology Advances, 31*(2), 140–153. doi:10.1016/j.biotechadv.2012.09.001 PMID:22985698

Cui, Y., Chen, X., Wang, Z., & Lu, Y. (2022). Cell-Free PURE System: Evolution and Achievements. *BioDesign Research, 9847014*. Advance online publication. doi:10.34133/2022/9847014

Currin, A., Parker, S., Robinson, C. J., Takano, E., Scrutton, N. S., & Breitling, R. (2021). The evolving art of creating genetic diversity: From directed evolution to synthetic biology. *Biotechnology Advances, 50*, 107762. doi:10.1016/j.biotechadv.2021.107762 PMID:34000294

CYP21A2 gene. (2010). https://medlineplus.gov/genetics/gene/cyp21a2/

Czajka, J. J., Oyetunde, T., & Tang, Y. J. (2021). Integrated knowledge mining, genome-scale modeling, and machine learning for predicting *Yarrowia lipolytica* bioproduction. *Metabolic Engineering, 67*, 227–236. doi:10.1016/j.ymben.2021.07.003 PMID:34242777

d'Espaux, L., Ghosh, A., Runguphan, W., Wehrs, M., Xu, F., Konzock, O., Dev, I., Nhan, M., Gin, J., Reider Apel, A., Petzold, C. J., Singh, S., Simmons, B. A., Mukhopadhyay, A., García Martín, H., & Keasling, J. D. (2017). Engineering high-level production of fatty alcohols by *Saccharomyces cerevisiae* from lignocellulosic feedstocks. *Metabolic Engineering, 42*, 115–125. doi:10.1016/j.ymben.2017.06.004 PMID:28606738

Dąbrowska, K., & Abedon, S. T. (2019, October 30). Pharmacologically aware phage therapy: Pharmacodynamic and pharmacokinetic obstacles to phage antibacterial action in animal and human bodies. *Microbiology and Molecular Biology Reviews, 83*(4), e00012–e00019. doi:10.1128/MMBR.00012-19 PMID:31666296

Dahabieh, M. S., Thevelein, J. M., & Gibson, B. (2020). Multimodal Microorganism Development: Integrating Top-Down Biological Engineering with Bottom-Up Rational Design. *Trends in Biotechnology, 38*(3), 241–253. doi:10.1016/j.tibtech.2019.09.006 PMID:31653446

Dai, G., & Lu, G. (2012). Different protein expression patterns associated with polycystic ovary syndrome in human follicular fluid during controlled ovarian hyperstimulation. *Reproduction, Fertility, and Development, 24*(7), 893–904. doi:10.1071/RD11201 PMID:22935150

Daletos, G., & Stephanopoulos, G. (2020). Protein engineering strategies for microbial production of isoprenoids. *Metabolic Engineering Communications, 11*, e00129. doi:10.1016/j.mec.2020.e00129 PMID:32612930

Dalipi, F., & Yayilgan, S. Y. (2016). Security and privacy considerations for IoT application on smart grids: Survey and research challenges. *Proceedings - 2016 4th International Conference on Future Internet of Things and Cloud Workshops, W-FiCloud 2016*, 63–68. 10.1109/W-FiCloud.2016.28

Damiati, S., Mhanna, R., Kodzius, R., & Ehmoser, E.-K. (2018). Cell-Free Approaches in Synthetic Biology Utilizing Microfluidics. *Genes, 9*(3), 144. Advance online publication. doi:10.3390/genes9030144 PMID:29509709

Danchin, A. (2022). In vivo, in vitro and in silico: An open space for the development of microbe-based applications of synthetic biology. *Microbial Biotechnology, 15*(1), 42–64. doi:10.1111/1751-7915.13937 PMID:34570957

De Castro, E., Sigrist, C. J., Gattiker, A., Bulliard, V., Langendijk-Genevaux, P. S., Gasteiger, E., & Hulo, N. (2006). Scan Prosite: Detection of PROSITE signature matches and ProRule-associated functional and structural residues in proteins. *Nucleic Acids Research, 34*(Web Server, suppl_2), W362–W365. doi:10.1093/nar/gkl124 PMID:16845026

de Lorenzo, V. (2022). Environmental Galenics: Large-scale fortification of extant microbiomes with engineered bioremediation agents. *Philosophical Transactions of the Royal Society B, 377*(1857), 20210395.

De Vet, S. J., & Rutgers, R. (2007). From waste to energy: First experimental bacterial fuel cells on board the International Space Station. *Microgravity Science and Technology, 19*(5-6), 225–229. doi:10.1007/BF02919487

Deheyab, A. O. A., Alwan, M. H., Rezzaqe, I. K. A., Mahmood, O. A., Hammadi, Y. I., Kareem, A. N., & Ibrahim, M. (2022). An overview of challenges in medical image processing. *ACM International Conference Proceeding Series*, 511–516. 10.1145/3584202.3584278

Desai, P. N., Shrivastava, N., & Padh, H. (2010, July 1). Production of heterologous proteins in plants: Strategies for optimal expression. *Biotechnology Advances, 28*(4), 427–435. doi:10.1016/j.biotechadv.2010.01.005 PMID:20152894

Deutsch, V. R., & Tomer, A. (2013, June). Advances in megakaryocytopoiesis and thrombopoiesis: From bench to bedside. *British Journal of Haematology, 161*(6), 778–793. doi:10.1111/bjh.12328 PMID:23594368

DeVries, J. K., & Zubay, G. (1967). DNA-directed peptide synthesis. II. The synthesis of the alpha-fragment of the enzyme beta-galactosidase. *Proceedings of the National Academy of Sciences of the United States of America, 57*(4), 1010–1012. doi:10.1073/pnas.57.4.1010 PMID:5342418

Dewangan, N. K., & Shukla, S. (2015). A survey on ECG signal feature extraction and analysis techniques. *International Journal of Innovative Research in Electrical, Electronics, Instrumentation and Control Engineering, 3*(6).

Dhanya, D., Kumar, S. S., Thilagavathy, A., Prasad, D., & Boopathi, S. (2023). Data Analytics and Artificial Intelligence in the Circular Economy: Case Studies. In Intelligent Engineering Applications and Applied Sciences for Sustainability (pp. 40–58). IGI Global.

Dhawan, B. N., & Singh, H. K. (1996). Pharmacological studies on *bacopa monniera*, an Ayurvedic nootropic agent. *European Neuropsychopharmacology*, *6*(1), 142–144. doi:10.1016/0924-977X(96)87969-7

Diamond, B. J., Shiflett, S. C., Feiwel, N., Matheis, R. J., Noskin, O., Richards, J. A., & Schoenberger, N. E. (2000). *Ginkgo biloba* extract: Mechanisms and clinical indications. *Archives of Physical Medicine and Rehabilitation*, *81*(5), 668–678. doi:10.1016/S0003-9993(00)90052-2 PMID:10807109

Didovyk, A., Tonooka, T., Tsimring, L., & Hasty, J. (2017). Rapid and Scalable Preparation of Bacterial Lysates for Cell-Free Gene Expression. *ACS Synthetic Biology*, *6*(12), 2198–2208. doi:10.1021/acssynbio.7b00253 PMID:28795570

Dinakarrao, S. M. P., Jantsch, A., & Shafique, M. (2019). Computer-aided arrhythmia diagnosis with bio-signal processing: A survey of trends and techniques. *ACM Computing Surveys*, *52*(2), 23.

Ding, X., Zheng, Z., Zhao, G., Wang, L., Wang, H., Yang, Q., Zhang, M., Li, L., & Wang, P. (2022). Bottom-up synthetic biology approach for improving the efficiency of menaquinone-7 synthesis in *Bacillus subtilis*. *Microbial Cell Factories*, *21*(1), 1–14. doi:10.118612934-022-01823-3 PMID:35643569

Dixon, C., Ma, X., & Marion, D. (1997). Effects of CDP-choline treatment on neurobehavioral deficits after TBI and on hippocampal and neocortical acetylcholine release. *Journal of Neurotrauma*, *14*(3), 161–169. doi:10.1089/neu.1997.14.161 PMID:9104933

Dolskiy, A. A., Grishchenko, I. V., & Yudkin, D. V. (2020). Cell Cultures for Virology: Usability, Advantages, and Prospects. *International Journal of Molecular Sciences*, *21*(21), 7978. Advance online publication. doi:10.3390/ijms21217978 PMID:33121109

Domakonda, V. K., Farooq, S., Chinthamreddy, S., Puviarasi, R., Sudhakar, M., & Boopathi, S. (2022). Sustainable Developments of Hybrid Floating Solar Power Plants: Photovoltaic System. In Human Agro-Energy Optimization for Business and Industry (pp. 148–167). IGI Global.

Domakonda, V. K., Farooq, S., Chinthamreddy, S., Puviarasi, R., Sudhakar, M., & Boopathi, S. (2023). Sustainable Developments of Hybrid Floating Solar Power Plants. In *Human Agro-Energy Optimization for Business and Industry* (pp. 148–167). IGI Global. doi:10.4018/978-1-6684-4118-3.ch008

Dong, O. X., & Ronald, P. C. (2019). Genetic Engineering for Disease Resistance in Plants: Recent Progress and Future Perspectives. *Plant Physiology*, *180*(1), 26–38. doi:10.1104/pp.18.01224 PMID:30867331

Donini, M., & Marusic, C. (2019, March). Current state-of-the-art in plant-based antibody production systems. *Biotechnology Letters*, *41*(3), 335–346. doi:10.100710529-019-02651-z PMID:30684155

Dormehi, I., & Jordaan, B. (1999). Spect monitoring of improved cerebral blood flow during long-term treatment of elderly patients with nootropic drugs. *Clinical Nuclear Medicine*, *24*(1), 29–34. doi:10.1097/00003072-199901000-00007 PMID:9890490

Doshi-Velez, F., & Kim, B. (2017). Towards a rigorous science of interpretable machine learning. *arXiv preprint arXiv.08608.*

Dua, D., & Graff, C. (2017). *UCI machine learning repository*. http://archive.ics.uci. edu/ml

Dumont, J., Euwart, D., Mei, B., Estes, S., & Kshirsagar, R. (2016). Human cell lines for biopharmaceutical manufacturing: History, status, and future perspectives. *Critical Reviews in Biotechnology*, *36*(6), 1110–1122. doi:10.3109/07388551.2015.1084266 PMID:26383226

Durairaj, M., Jayakumar, S., Karpagavalli, V., Maheswari, B. U., Boopathi, S., & ... (2023). Utilization of Digital Tools in the Indian Higher Education System During Health Crises. In *Multidisciplinary Approaches to Organizational Governance During Health Crises* (pp. 1–21). IGI Global. doi:10.4018/978-1-7998-9213-7.ch001

Durham, D. (1991). Dictionary of drugs—Chemical data, structures and bibliographies. Chapman & Hall.

Ebrahimi, S., Khanbabaei, H., Abbasi, S., Fani, M., Soltani, S., Zandi, M., & Najafimemar, Z. (2022). CRISPR-Cas System: A Promising Diagnostic Tool for Covid-19. *Avicenna Journal of Medical Biotechnology*, *14*(1), 3–9. doi:10.18502/ajmb.v14i1.8165 PMID:35509363

Eiras, M. C., Pinheiro, D. P., Mubarac Romcy, K. A., Ferriani, R. A., Reis, R. M., & Furtado, C. L. (2022). Polycystic ovary syndrome: The Epigenetic Behind the Disease. *Reproductive Sciences (Thousand Oaks, Calif.)*, *29*(3), 1–15. doi:10.100743032-021-00516-3 PMID:33826098

Ekor, M. (2014). The growing use of herbal medicines: Issues relating to adverse reactions and challenges in monitoring safety. *Frontiers in Pharmacology*, *177*(4), 35–38. doi:10.3389/fphar.2013.00177 PMID:24454289

Elani, Y. (2021). Interfacing living and synthetic cells as an emerging frontier in synthetic biology. *Angewandte Chemie*, *133*(11), 5662–5671. doi:10.1002/ange.202006941 PMID:32909663

Elekwachi, C. O. (2012). *Bioinformatics resources to support bioremediation research* [Doctoral dissertation]. University of Nottingham.

Elhardt, D., & Böck, A. (1982). An in vitro polypeptide synthesizing system from methanogenic bacteria: Sensitivity to antibiotics. *Molecular & General Genetics*, *188*(1), 128–134. doi:10.1007/BF00333006

Ellis, L. B., Roe, D., & Wackett, L. P. (2006). The University of Minnesota biocatalysis / biodegradation database: The first decade. *Nucleic Acids Research, 34*(suppl_1), D517-D521.

Endoh, T., Kanai, T., Sato, Y. T., Liu, D. V., Yoshikawa, K., Atomi, H., & Imanaka, T. (2006). Cell-free protein synthesis at high temperatures using the lysate of a hyperthermophile. *Journal of Biotechnology*, *126*(2), 186–195. doi:10.1016/j.jbiotec.2006.04.010 PMID:16730835

Endo, Y. (2021). Development of a cell-free protein synthesis system for practical use. *Proceedings of the Japan Academy. Series B, Physical and Biological Sciences*, *97*(5), 261–276. doi:10.2183/pjab.97.015 PMID:33980755

Enríquez, P. (2016). Genome editing and the jurisprudence of scientific empiricism. *Vand. J. Ent. & Tech. L.*, *19*, 603.

EPA ECOTOX database. (2013). https://www.epa.gov/ecotox/

Erdrich, P., Knoop, H., Steuer, R., & Klamt, S. (2014). Cyanobacterial biofuels: New insights and strain design strategies revealed by computational modeling. *Microbial Cell Factories*, *13*(1), 1–15. doi:10.118612934-014-0128-x PMID:25323065

Eroshenko, N., & Church, G. M. (2013). Mutants of Cre recombinase with improved accuracy. *Nature Communications*, *4*(1), 2509. Advance online publication. doi:10.1038/ncomms3509 PMID:24056590

Eskandar, K. (2023). Revolutionizing biotechnology and bioengineering: Unleashing the power of innovation. *J Appl Biotechnol Bioeng*, *10*(3), 81–88. doi:10.15406/jabb.2023.10.00332

Espinola, E. B., Dias, R. F., Mattei, R., & Carlini, E. A. (1997). Pharmacological activity of Guarana (*Paullinia Cupana*) in laboratory animals. *Journal of Ethnopharmacology*, *55*(3), 223–229. doi:10.1016/S0378-8741(96)01506-1 PMID:9080343

Eto, S., Matsumura, R., Shimane, Y., Fujimi, M., Berhanu, S., Kasama, T., & Kuruma, Y. (2022). Phospholipid synthesis inside phospholipid membrane vesicles. *Communications Biology*, *5*(1), 1016. doi:10.103842003-022-03999-1 PMID:36167778

Ewelina, L., Tomasz, K., Szymon, O., & Tomasz, S. (2016). Extraction and purification methods in downstream processing of plant-based recombinant proteins. *Protein Expression and Purification*. PMID:26742898

Ezure, T., Suzuki, T., Shikata, M., Ito, M., & Ando, E. (2010). A cell-free protein synthesis system from insect cells. *Methods in Molecular Biology (Clifton, N.J.)*, *607*, 31–42. doi:10.1007/978-1-60327-331-2_4 PMID:20204846

Falanga, A., Marchetti, M., & Vignoli, A. (2013, February). Coagulation and cancer: Biological and clinical aspects. *Journal of Thrombosis and Haemostasis*, *11*(2), 223–233. doi:10.1111/jth.12075 PMID:23279708

Fan, X., Yao, Q., Cai, Y., Miao, F., Sun, F., & Li, Y. (2018). Multiscaled fusion of deep convolutional neural networks for screening atrial fibrillation from single lead short ECG recordings. *IEEE Journal of Biomedical and Health Informatics*, *22*(6), 1744–1753. doi:10.1109/JBHI.2018.2858789 PMID:30106699

Faust, O., Shenfield, A., Kareem, M., San, T. R., Fujita, H., & Acharya, U. R. (2018). Au- tomated detection of atrial fibrillation using long short-term memory network with RR interval signals. *Computers in Biology and Medicine*, *102*, 327–335. doi:10.1016/j.compbiomed.2018.07.001 PMID:30031535

Feist, A. M., Herrgård, M. J., Thiele, I., Reed, J. L., & Palsson, B. (2009). Reconstruction of biochemical networks in microorganisms. *Nature Reviews Microbiology, 7*(2), 129–143. PMID: 19116616 doi:10.1038/nrmicro1949

Ferahtia, S., Djeroui, A., Rezk, H., Houari, A., Zeghlache, S., & Machmoum, M. (2022). Optimal control and implementation of energy management strategy for a DC microgrid. *Energy*, *238*, 121777. doi:10.1016/j.energy.2021.121777

Fernández-Robledo, J. A., & Vasta, G. R. (2010). Production of recombinant proteins from protozoan parasites. *Trends in Parasitology*, *26*(5), 244–254. doi:10.1016/j.pt.2010.02.004 PMID:20189877

Festa, F., Steel, J. C., Bian, X., & LaBaer, J. (2013). High-throughput cloning and expression library creation for functional proteomics. *Proteomics*, *13*(9), 1381–1399. doi:10.1002/pmic.201200456 PMID:23457047

Fogeron, M.-L., Lecoq, L., Cole, L., Harbers, M., & Böckmann, A. (2021). Easy synthesis of complex biomolecular assemblies: Wheat germ cell-free protein expression in structural biology. *Frontiers in Molecular Biosciences*, *8*, 639587. doi:10.3389/fmolb.2021.639587 PMID:33842544

Fowziya, S., Sivaranjani, S., Devi, N. L., Boopathi, S., Thakur, S., & Sailaja, J. M. (2023). Influences of nano-green lubricants in the friction-stir process of TiAlN coated alloys. *Materials Today: Proceedings*. Advance online publication. doi:10.1016/j.matpr.2023.06.446

Franks, S. (2002). Adult polycystic ovary syndrome begins in childhood. *Best Practice & Research. Clinical Endocrinology & Metabolism*, *16*(2), 263–272. doi:10.1053/beem.2002.0203 PMID:12064892

Freeze, H. H., & Elbein, A. D. (2009). Glycosylation precursors. Essentials of Glycobiology.

French, K. E. (2019). Harnessing synthetic biology for sustainable development. *Nature Sustainability*, *2*(4), 250–252. doi:10.103841893-019-0270-x

Frenzel, A., Hust, M., & Schirrmann, T. (2013, July 29). Expression of recombinant antibodies. *Frontiers in Immunology*, *4*, 217. doi:10.3389/fimmu.2013.00217 PMID:23908655

Froestl, W., Muhs, A., & Pfeifer, A. (2012). Cognitive enhancers (nootropics). part 1: Drugs interacting with receptors. *Journal of Alzheimer's Disease*, *32*(4), 793–887. doi:10.3233/JAD-2012-121186 PMID:22886028

Fuller, R. S., Kaguni, J. M., & Kornberg, A. (1981). Enzymatic replication of the origin of the Escherichia coli chromosome. *Proceedings of the National Academy of Sciences of the United States of America*, *78*(12), 7370–7374. doi:10.1073/pnas.78.12.7370 PMID:6278471

Fung, T. S., & Liu, D. X. (2018). Post-translational modifications of coronavirus proteins: Roles and function. *Future Virology*, *13*(6), 405–430. doi:10.2217/fvl-2018-0008 PMID:32201497

Fu, P. (2009). Genome-scale modeling of *Synechocystis* sp. PCC 6803 and prediction of pathway insertion. *Journal of Chemical Technology and Biotechnology*, *84*(4), 473–483. doi:10.1002/jctb.2065

Furuhama, A., Toida, T., Nishikawa, N., Aoki, Y., Yoshioka, Y., & Shiraishi, H. (2010). Development of an ecotoxicity QSAR model for the KAshinhou Tool for ecotoxicity (KATE) system, March 2009 version. *SAR and QSAR in Environmental Research*, *21*(5-6), 403–413. doi:10.1080/1062936X.2010.501815 PMID:20818579

Gadipelly, C., Perez-Gonzalez, A., Yadav, G. D., Ortiz, I., Ibáñez, R., Rathod, V. K., & Marathe, K. V. (2014). Pharmaceutical industry waste water: Review of the technologies for water treatment and reuse. *Industrial & Engineering Chemistry Research*, *53*(29), 11571–11592. doi:10.1021/ie501210j

Galambos, L. (2011, December 13). Pharmaceutical Firms and the Transition to Biotechnology: A Study in Strategic Innovation. *Business History Review*. https://www.cambridge.org/core/journals/business-history-review/article/abs/pharmaceutical-firms-and-the-transition-to-biotechnology-a-study-in-strategic-innovation/18D8F37FD43C813B-7984B06E58B733C0

Galazis, N., Pang, Y. L., Galazi, M., Haoula, Z., Layfield, R., & Atiomo, W. (2013). Proteomic biomarkers of endometrial cancer risk in women with polycystic ovary syndrome: A systematic review and biomarker database integration. *Gynecological Endocrinology*, *29*(7), 638–644. doi:10.3109/09513590.2013.777416 PMID:23527552

Garcia Serpa Osorio-de-Castro, C., Silva Miranda, E., Machado de Freitas, C., Rochel de Camargo, K. Jr, & Cranmer, H. H. (2017). The Zika Virus Outbreak in Brazil: Knowledge Gaps and Challenges for Risk Reduction. *American Journal of Public Health*, *107*(6), 960–965. doi:10.2105/AJPH.2017.303705 PMID:28426311

Garcia, C. B., Chapman, I. F., Chen, S. H., Lazear, E., Lentz, T. B., Williams, C. L., Dums, J. T., Goller, C. C., & Robertson, S. D. (2020). Integrating research into a molecular cloning course to address the evolving biotechnology landscape. *Biochemistry and Molecular Biology Education*, *49*(1), 115–128. doi:10.1002/bmb.21402 PMID:33176069

García-Granados, R., Lerma-Escalera, J. A., & Morones-Ramírez, J. R. (2019). Metabolic engineering and synthetic biology: Synergies, future, and challenges. *Frontiers in Bioengineering and Biotechnology*, *7*, 439905. doi:10.3389/fbioe.2019.00036 PMID:30886847

Garcia, S., Thompson, R. A., Giannone, R. J., Dash, S., Maranas, C. D., & Trinh, C. T. (2020). Development of a Genome-Scale Metabolic Model of Clostridium thermocellum and Its Applications for Integration of Multi-Omics Datasets and Computational Strain Design. *Frontiers in Bioengineering and Biotechnology*, *8*, 1–16. doi:10.3389/fbioe.2020.00772 PMID:32974289

Garenne, D., Haines, M. C., Romantseva, E. F., Freemont, P., Strychalski, E. A., & Noireaux, V. (2021). Cell-free gene expression. *Nature Reviews. Methods Primers*, *1*(1), 49. doi:10.103843586-021-00046-x

Gawand, P., Yang, L., Cluett, W. R., & Mahadevan, R. (2013). Metabolic model refinement using phenotypic microarray data. *Methods in Molecular Biology (Clifton, N.J.)*, *985*, 47–59. doi:10.1007/978-1-62703-299-5_3 PMID:23417798

Gellissen, G., & Hollenberg, C. P. (1997). Application of yeasts in gene expression studies: A comparison of Saccharomyces cerevisiae, Hansenula polymorpha and Kluyveromyces lactis- a review. *Gene*, *190*(1), 87–97. doi:10.1016/S0378-1119(97)00020-6 PMID:9185853

Gellissen, G., Kunze, G., Gaillardin, C., Cregg, J. M., Berardi, E., Veenhuis, M., & van der Klei, I. (2005). New yeast expression platforms based on methylotrophic Hansenula polymorpha and Pichia pastoris and on dimorphic Arxula adeninivorans and Yarrowia lipolytica– A comparison. *FEMS Yeast Research*, *5*(11), 1079–1096. doi:10.1016/j.femsyr.2005.06.004 PMID:16144775

Gerszberg, A., & Hnatuszko-Konka, K. (2022, March 17). Compendium on food crop plants as a platform for pharmaceutical protein production. *International Journal of Molecular Sciences*, *23*(6), 3236. doi:10.3390/ijms23063236 PMID:35328657

Gharani, N., Waterworth, D. M., Batty, S., White, D., Gilling-Smith, C., Conway, G. S., McCarthy, M., Franks, S., & Williamsom, R. (1997). Association of the steroid synthesis gene CYP11a with polycystic ovary syndrome and hyperandrogenism. *Human Molecular Genetics*, *6*(3), 397–402. doi:10.1093/hmg/6.3.397 PMID:9147642

Gharehbaghi, A., & Babic, A. (2018). Structural risk evaluation of a deep neural net- work and a Markov model in extracting medical information from phonocardio- graphy. *Studies in Health Technology and Informatics*, *251*, 157–160. PMID:29968626

Gharehbaghi, A., Lindén, M., & Babic, A. (2019b). An artificial intelligent-based model for detecting systolic pathological patterns of phonocardiogram based on time-growing neural network. *Applied Soft Computing*, *83*, 105615. doi:10.1016/j.asoc.2019.105615

Gharehbaghi, A., & Lindn, M. (2018). A deep machine learning method for classifying cyclic time series of biological signals using time-growing neural network. *IEEE Transactions on Neural Networks and Learning Systems*, *29*(9), 4102–4115. doi:10.1109/TNNLS.2017.2754294 PMID:29035230

Ghasempour, A. (2019). Internet of things in smart grid: Architecture, applications, services, key technologies, and challenges. *Inventions (Basel, Switzerland)*, *4*(1), 22. doi:10.3390/inventions4010022

Ghenai, C., Al-Mufti, O. A. A., Al-Isawi, O. A. M., Amirah, L. H. L., & Merabet, A. (2022). Short-term building electrical load forecasting using adaptive neuro-fuzzy inference system (ANFIS). *Journal of Building Engineering*, *52*, 104323. doi:10.1016/j.jobe.2022.104323

Ghiasi, S., Abdollahpur, M., Madani, N., Kiani, K., & Ghaffari, A. (2017). Atrial fibrillation detection using feature based algorithm and deep convolutional neural network. In 2017 Computing in cardiology (CINC) (pp. 1–4). IEEE.

Ghosh, A., Ando, D., Gin, J., Runguphan, W., Denby, C., Wang, G., Baidoo, E. E. K., Shymansky, C., Keasling, J. D., & Martín, H. G. (2016). 13C Metabolic Flux Analysis for Systematic Metabolic Engineering of S. cerevisiae for Overproduction of Fatty Acids. *Frontiers in Bioengineering and Biotechnology*, *4*, 76. doi:10.3389/fbioe.2016.00076 PMID:27761435

Giansily-Blaizot, M., & Schved, J. (2017). Recombinant human factor VIIa (rFVIIa) in hemophilia: Mode of action and evidence to date. *Therapeutic Advances in Hematology*, *8*(12), 345–352. doi:10.1177/2040620717737701 PMID:29204261

Gibson Assembly Cloning Guide (2nd ed.). (n.d.). BioCat. Retrieved August 2, 2022, from https://www.biocat.com/bc/files/Gibson_Guide_V2_101417_web_version_8.5_x_11_FINAL.pdf

Gibson, D. G. (2011). Enzymatic assembly of overlapping DNA fragments. In Methods in Enzymology (pp. 349–361). doi:10.1016/B978-0-12-385120-8.00015-2

Gibson, D. G., Benders, G. A., Andrews-Pfannkoch, C., Denisova, E. A., Baden-Tillson, H., Zaveri, J., Stockwell, T. B., Brownley, A., Thomas, D. W., Algire, M. A., Merryman, C., Young, L. S., Noskov, V. N., Glass, J. I., Venter, J. C., Hutchison, C. A. III, & Smith, H. O. (2008). Complete Chemical Synthesis, Assembly, and Cloning of a *Mycoplasma genitalium* Genome. *Science*, *319*(5867), 1215–1220. doi:10.1126cience.1151721 PMID:18218864

Gibson, D. G., Glass, J. I., Lartigue, C., Noskov, V. N., Chuang, R. Y., Algire, M. A., Benders, G. A., Montague, M. G., Ma, L., Moodie, M. M., Merryman, C., Vashee, S., Krishnakumar, R., Assad-Garcia, N., Andrews-Pfannkoch, C., Denisova, E. A., Young, L., Qi, Z. Q., Segall-Shapiro, T. H., ... Venter, J. C. (2010). Creation of a Bacterial Cell Controlled by a Chemically Synthesized Genome. *Science*, *329*(5987), 52–56. doi:10.1126cience.1190719 PMID:20488990

Gibson, D. G., Young, L. S., Chuang, R. Y., Venter, J. C., Hutchison, C. A. III, & Smith, H. O. (2009). Enzymatic assembly of DNA molecules up to several hundred kilobases. *Nature Methods*, *6*(5), 343–345. doi:10.1038/nmeth.1318 PMID:19363495

Gicevic, A., Hindija, L., & Karacic, A. (2019, May). Toxicity of azo dyes in pharmaceutical industry. In *International Conference on Medical and Biological Engineering* (pp. 581-587). Springer.

Giesbrecht, T., Rycroft, J. A., Rowson, M. J., & De Bruin, E. A. (2010). The combination of L-theanine and caffeine improves cognitive performance and increases subjective alertness. *Nutritional Neuroscience*, *13*(6), 283–290. doi:10.1179/147683010X12611460764840 PMID:21040626

Gimpel, J. A., & Mayfield, S. P. (2013). Analysis of heterologous regulatory and coding regions in algal chloroplasts. *Applied Microbiology and Biotechnology*, *97*(10), 4499–4510. doi:10.100700253-012-4580-4 PMID:23179624

Giurgea, C. (1973). The "nootropic" approach to the pharmacology of the integrative activity of the Brain 1, 2. *Conditional Reflex*, *8*(2), 108–115. doi:10.1007/BF03000311 PMID:4736348

Giurgea, C., & Salama, M. (1977). Nootropic drugs. *Progress in Neuro-Psychopharmacology*, *1*(34), 235–247. doi:10.1016/0364-7722(77)90046-7

Gleba, Y., Klimyuk, V., & Marillonnet, S. (2005). Magnifection- A new platform for expressing recombinant vaccines in plants. *Vaccine*, *23*(17-18), 23. doi:10.1016/j.vaccine.2005.01.006 PMID:15755568

Glont, M., Nguyen, T. V. N., Graesslin, M., Hälke, R., Ali, R., Schramm, J., Wimalaratne, S. M., Kothamachu, V. B., Rodriguez, N., Swat, M. J., Eils, J., Eils, R., Laibe, C., Malik-Sheriff, R. S., Chelliah, V., Le Novère, N., & Hermjakob, H. (2018). BioModels: Expanding horizons to include more modeling approaches and formats. *Nucleic Acids Research*, *46*(D1), D1248–D1253. doi:10.1093/nar/gkx1023 PMID:29106614

Gnanadass, S.A., Prabhu, Y.D., & Gopalakrishnan, A.V. (2021). Association of metabolic and inflammatory markers with polycystic ovarian syndrome (PCOS): An update. *Archives of Gynecology and obstestrics, 303*(3), 631-643.

Gnanaprakasam, C., Vankara, J., Sastry, A. S., Prajval, V., Gireesh, N., & Boopathi, S. (2023). Long-Range and Low-Power Automated Soil Irrigation System Using Internet of Things: An Experimental Study. In Contemporary Developments in Agricultural Cyber-Physical Systems (pp. 87–104). IGI Global.

Gohil, K. J., Patel, J. A., & Gajjar, A. K. (2010). Pharmacological review on Centella Asiatica: A potential herbal cure-all. *Indian Journal of Pharmaceutical Sciences*, *72*(5), 546. doi:10.4103/0250-474X.78519 PMID:21694984

Goldsmith, D. R., & Wellington, K. (2005). Naftidrofuryl. *Drugs & Aging*, *22*(11), 967–977. doi:10.2165/00002512-200522110-00006 PMID:16323973

Golfam, R., Kiarostami, K., Lohrasebi, T., Esfahani, K., & Ghazvini, H. (2021, June 1). Application of molecular farming based on barley plant (Hordeum vulgare L.) platform: A Review. *Research on Crops*, *22*(2).

Gomes de Oliveira Dal'Molin, C., Quek, L.-E., Palfreyman, R. W., & Nielsen, L. K. (2011). AlgaGEM - a genome-scale metabolic reconstruction of algae based on the Chlamydomonas reinhardtii genome. *BMC Genomics*, *12*(Suppl 4), 1–10. PMID: 22369158. doi:10.1186/1471-2164-12-S4-S5

Gomes, A.R., Byregowda, S.M., Veeregowda, B.M., & Balamurugan, V. (n.d.). *An overview of heterologous expression host systems for the production of recombinant proteins*. Academic Press.

Gonzales-Arimborgo, C., Yupanqui, I., Montero, E., Alarcón-Yaquetto, D., Zevallos-Concha, A., Caballero, L., Gasco, M., Zhao, J., Khan, I., & Gonzales, G. (2016). Acceptability, safety, and efficacy of oral administration of extracts of black or red maca (*lepidium meyenii*) in adult human subjects: A randomized, double-blind, placebo-controlled study. *Pharmaceuticals (Basel, Switzerland)*, *9*(3), 47–49. doi:10.3390/ph9030049 PMID:27548190

Goodwin, S., McPherson, J. D., & McCombie, W. R. (2016). Coming of age: Ten years of next-generation sequencing technologies. In Nature Reviews Genetics (Vol. 17, Issue 6, pp. 333–351). Nat Rev Genet. doi:10.1038/nrg.2016.49

Gootenberg, J. S., Abudayyeh, O. O., Lee, J. W., Essletzbichler, P., Dy, A. J., Joung, J., Verdine, V., Donghia, N., Daringer, N. M., Freije, C. A., Myhrvold, C., Bhattacharyya, R. P., Livny, J., Regev, A., Koonin, E. V., Hung, D. T., Sabeti, P. C., Collins, J. J., & Zhang, F. (2017). Nucleic acid detection with CRISPR-Cas13a/C2c2. *Science*, *356*(6336), 438–442. doi:10.1126cience.aam9321 PMID:28408723

Gordon & Ruddle. (1981). *Integration and stable germ line transmission of genes injected into mouse pronuclei*. doi:10.1126/science.6272397

Gowri, N. V., Dwivedi, J. N., Krishnaveni, K., Boopathi, S., Palaniappan, M., & Medikondu, N. R. (2023). Experimental investigation and multi-objective optimization of eco-friendly near-dry electrical discharge machining of shape memory alloy using Cu/SiC/Gr composite electrode. *Environmental Science and Pollution Research International*, 1–19. doi:10.100711356-023-26983-6 PMID:37126160

Gräslund, S., Nordlund, P., Weigelt, J., Hallberg, B. M., Bray, J., Gileadi, O., ... Gunsalus, K. C. (2008). Protein production and purification. *Nature Methods*, *5*(2), 135–146. doi:10.1038/nmeth.f.202 PMID:18235434

Graupp, M., Wehr, E., Schweighofer, N., Pieber, T. R., & Obermayer, B. (2011). Association of genetic variants in the two isoforms of 5alpha-reductase, SRD5A1 and SRD5A2, in lean patients with polycystic ovary syndrome. *European Journal of Obstetrics, Gynecology, and Reproductive Biology*, *157*(2), 175–179. doi:10.1016/j.ejogrb.2011.03.026 PMID:21530059

Green, A. A., Silver, P. A., Collins, J. J., & Yin, P. (2014). Toehold switches: De-novo-designed regulators of gene expression. *Cell*, *159*(4), 925–939. doi:10.1016/j.cell.2014.10.002 PMID:25417166

Greenberg, A. J., Hackett, S. R., Harshman, L. G., & Clark, A. G. (2011). Environmental and genetic perturbations reveal different networks of metabolic regulation. *Molecular Systems Biology*, *7*(1), 563. doi:10.1038/msb.2011.96 PMID:22186737

Gregorio, N. E., Levine, M. Z., & Oza, J. P. (2019). A User's Guide to Cell-Free Protein Synthesis. *Methods and Protocols*, *2*(1), 24. Advance online publication. doi:10.3390/mps2010024 PMID:31164605

Grue, B., Hamadache, S., Maddiboina, D., & Scott, B. (n.d.). *Invest in Canadian Synthetic Biology to Meet Commitments to Sustainable Development and Support Economic Recovery*. 10.38126/JSPG180106

Gu, C., Kim, G. B., Kim, W. J., Kim, H. U., & Lee, S. Y. (2019). Current status and applications of genome-scale metabolic models. *Genome Biology*, *20*(1), 1–18. doi:10.118613059-019-1730-3 PMID:31196170

Gunasekaran, K., & Boopathi, S. (2023). Artificial Intelligence in Water Treatments and Water Resource Assessments. In *Artificial Intelligence Applications in Water Treatment and Water Resource Management* (pp. 71–98). IGI Global. doi:10.4018/978-1-6684-6791-6.ch004

Gunasekaran, K., Boopathi, S., & Sureshkumar, M. (2022). Analysis of a Cryogenically Cooled Near-Dry WEDM Process using Different Dielectrics. *Materials Technology*, *56*(2), 179–186.

Gunasekaran, K., Boopathi, S., & Sureshkumar, M. (2022). Analysis of a Cryogenically Cooled Near-Dry WEDM Process using Different Dielectrics. *Mater-Tehnol.Si. Materials Technology*, *56*(2), 179–186.

Guo, G., Tian, F., Ding, K., Wang, L., Liu, T., & Yang, F. (2017). Effect of a bacterial consortium on the degradation of polycyclic aromatic hydrocarbons and bacterial community composition in Chinese soils. *International Biodeterioration & Biodegradation*, *123*, 56–62. doi:10.1016/j.ibiod.2017.04.022

Guo, Y., Su, L., Liu, Q., Zhu, Y., Dai, Z., & Wang, Q. (2022). Dissecting carbon metabolism of Yarrowia lipolytica type strain W29 using genome-scale metabolic modelling. *Computational and Structural Biotechnology Journal*, *20*, 2503–2511. doi:10.1016/j.csbj.2022.05.018 PMID:35664225

Gupta, S. K., & Shukla, P. (2018, December). Glycosylation control technologies for recombinant therapeutic proteins. *Applied Microbiology and Biotechnology*, *102*(24), 10457–10468. doi:10.100700253-018-9430-6 PMID:30334089

Gyori, B. M., & Bachman, J. A. (2021). From knowledge to models: Automated modeling in systems and synthetic biology. In *Current Opinion in Systems Biology* (Vol. 28, p. 100362). Elsevier. doi:10.1016/j.coisb.2021.100362

H., Ma, S., Cui, J., Xiao, W., & Liu, L. (2022, August 11). *Deep Learning-Based Data Augmentation and Model Fusion for Automatic Arrhythmia Identification and Classification Algorithms*. doi:10.1155/2022/1577778

Haider, S., Khaliq, S., & Haleem, D. (2007). Enhanced serotonergic neurotransmission in the hippocampus following tryptophan administration improves learning acquisition and memory consolidation in rats. *Pharmacological Reports*, *59*, 53–57. PMID:17377206

Haisma, H. J., & de Hon, O. (2006, May). Gene doping. *International Journal of Sports Medicine*, *27*(04), 257–266. doi:10.1055-2006-923986 PMID:16572366

Hammad, M., Iliyasu, A. M., Subasi, A., Ho, E. S. L., & El-Latif, A. A. A. (2021). A Multitier Deep Learning Model for Arrhythmia Detection. *IEEE Transactions on Instrumentation and Measurement*, *70*, 1–9. doi:10.1109/TIM.2020.3033072

Hang, L., & Kim, D. H. (2018). Enhanced model-based predictive control system based on fuzzy logic for maintaining thermal comfort in IoT smart space. *Applied Sciences (Basel, Switzerland)*, *8*(7), 1031. doi:10.3390/app8071031

Hansl, N. R., & Mead, B. T. (1978). PRL-8-53: Enhanced learning and subsequent retention in humans as a result of low oral doses of new psychotropic agent. *Psychopharmacology*, *56*(3), 249–253. doi:10.1007/BF00432846 PMID:418433

Hans, S., Kumar, N., Gohil, N., Khambhati, K., Bhattacharjee, G., Deb, S. S., Maurya, R., Kumar, V., Reshamwala, S. M. S., & Singh, V. (2022). Rebooting life: Engineering non-natural nucleic acids, proteins and metabolites in microorganisms. *Microbial Cell Factories*, *21*(1), 1–12. doi:10.118612934-022-01828-y PMID:35643549

Hanumanthakari, S., Gift, M. D. M., Kanimozhi, K. V., Bhavani, M. D., Bamane, K. D., & Boopathi, S. (2023a). Biomining Method to Extract Metal Components Using Computer-Printed Circuit Board E-Waste. In *Handbook of Research on Safe Disposal Methods of Municipal Solid Wastes for a Sustainable Environment* (pp. 123–141). IGI Global. doi:10.4018/978-1-6684-8117-2.ch010

Haribalaji, V., Venkatesan, G., Pandian, M., Subbiah, R., & Boopathi, S. (2022). Investigation on corrosion and tensile Characteristics: Friction stir welding of AA7075 and AA2014. *Materials Today: Proceedings*, *66*(3), 743–748. doi:10.1016/j.matpr.2022.04.037

Harikaran, M., Boopathi, S., Gokulakannan, S., & Poonguzhali, M. (2023). Study on the Source of E-Waste Management and Disposal Methods. In *Sustainable Approaches and Strategies for E-Waste Management and Utilization* (pp. 39–60). IGI Global. doi:10.4018/978-1-6684-7573-7.ch003

Harikaran, M., Boopathi, S., Rajkannan, M., & Gokulakannan, S. (2023). Impact analysis of ceramic tile powder aggregates on self-compacting concrete. *Engineering Research Express*, *5*(2), 025069. doi:10.1088/2631-8695/acde48

Harmsen, B., Robeyns, K., Wouters, J., & Leyssens, T. (2017). A study of Fasoracetam's Solid State forms: A potential Anti-Alzheimer Pharmaceutical. *Journal of Pharmaceutical Sciences*, *106*(5), 1317–1321. doi:10.1016/j.xphs.2017.01.016 PMID:28109793

Hartono, B. S., Budiyanto, & Setiabudy, R. (2013). Review of microgrid technology. *2013 International Conference on Quality in Research, QiR 2013 - In Conjunction with ICCS 2013: The 2nd International Conference on Civic Space*, 127–132. 10.1109/QiR.2013.6632550

Hase, A., Jung, S. E., & aan het Rot, M. (2015). Behavioral and cognitive effects of tyrosine intake in Healthy Human Adults. *Pharmacology, Biochemistry, and Behavior*, *133*, 1–6. doi:10.1016/j.pbb.2015.03.008 PMID:25797188

Heinemann, M., & Panke, S. (2006). Synthetic biology - Putting engineering into biology. *Bioinformatics (Oxford, England)*, *22*(22), 2790–2799. doi:10.1093/bioinformatics/btl469 PMID:16954140

Heirendt, L., Arreckx, S., Pfau, T., Mendoza, S. N., Richelle, A., Heinken, A., Haraldsdóttir, H. S., Wachowiak, J., Keating, S. M., Vlasov, V., Magnusdóttir, S., Ng, C. Y., Preciat, G., Žagare, A., Chan, S. H. J., Aurich, M. K., Clancy, C. M., Modamio, J., Sauls, J. T., ... Fleming, R. M. T. (2019). Creation and analysis of biochemical constraint-based models using the COBRA Toolbox v.3.0. *Nature Protocols*, *14*(3), 639–702. doi:10.103841596-018-0098-2 PMID:30787451

He, K., Zhang, X., Ren, S., & Sun, J. (2015). Spatial pyramid pooling in deep convolu- tional networks for visual recognition. *IEEE Transactions on Pattern Analysis and Machine Intelligence*, *37*(9), 1904–1916. Advance online publication. doi:10.1109/TPAMI.2015.2389824 PMID:26353135

Hema, N., Krishnamoorthy, N., Chavan, S. M., Kumar, N. M. G., Sabarimuthu, M., & Boopathi, S. (2023). A Study on an Internet of Things (IoT)-Enabled Smart Solar Grid System. In *Handbook of Research on Deep Learning Techniques for Cloud-Based Industrial IoT* (pp. 290–308). IGI Global. doi:10.4018/978-1-6684-8098-4.ch017

Hendry, J. I., Prasannan, C. B., Joshi, A., Dasgupta, S., & Wangikar, P. P. (2016). Metabolic model of Synechococcus sp. PCC 7002: Prediction of flux distribution and network modification for enhanced biofuel production. *Bioresource Technology*, *213*, 190–197. doi:10.1016/j.biortech.2016.02.128 PMID:27036328

Hérisson, J., Duigou, T., du Lac, M., Bazi-Kabbaj, K., Sabeti Azad, M., Buldum, G., Telle, O., El Moubayed, Y., Carbonell, P., Swainston, N., Zulkower, V., Kushwaha, M., Baldwin, G. S., & Faulon, J. L. (2022). The automated Galaxy-SynBioCAD pipeline for synthetic biology design and engineering. *Nature Communications*, *13*(1), 5082. doi:10.103841467-022-32661-x PMID:36038542

Hindmarch, I., Coleston, D. M., & Kerr, J. S. (1990). Psychopharmacological effects of pyritinol in normal volunteers. *Neuropsychobiology*, *24*(3), 159–164. doi:10.1159/000119478 PMID:2135070

Hinton, G. E., & Sejnowski, T. J. (1986). Learning and relearning in Boltzmann machines. *Parallel Distributed Processing: Explorations in the Microstructure of Cognition, 1*, 282-317.

Hinz, M., Stein, A., Neff, R., Weinberg, R., & Uncini, T. (2011). Treatment of attention deficit hyperactivity disorder with monoamine amino acid precursors and organic cation transporter assay interpretation. *Neuropsychiatric Disease and Treatment*, *32*(5), 29–31. doi:10.2147/NDT.S16270 PMID:21326653

Hoang, A. T., Ong, H. C., Fattah, I. R., Chong, C. T., Cheng, C. K., Sakthivel, R., & Ok, Y. S. (2021). Progress on the lignocellulosic biomass pyrolysis for biofuel production toward environmental sustainability. *Fuel Processing Technology, 223*, 106997. doi:10.1016/j.fuproc.2021.106997

Hoang, A. T., Varbanov, P. S., Nižetić, S., Sirohi, R., Pandey, A., Luque, R., Ng, K. H., & Pham, V. V. (2022). Perspective review on Municipal Solid Waste-to-energy route: Characteristics, management strategy, and role in circular economy. *Journal of Cleaner Production, 359*, 131897. doi:10.1016/j.jclepro.2022.131897

Hodgman, C. E., & Jewett, M. C. (2012). Cell-free synthetic biology: Thinking outside the cell. *Metabolic Engineering, 14*(3), 261–269. doi:10.1016/j.ymben.2011.09.002 PMID:21946161

Hong, S. H., & Fernandez-Baca, A. S. (2022). Cell-free synthetic biology as an emerging biotechnology. In *New Frontiers and applications of synthetic biology* (pp. 397–414). Elsevier. doi:10.1016/B978-0-12-824469-2.00016-6

Hong, S. H., Kwon, Y. C., & Jewett, M. C. (2014). Non-standard amino acid incorporation into proteins using Escherichia coli cell-free protein synthesis. *Frontiers in Chemistry, 2*, 34. doi:10.3389/fchem.2014.00034 PMID:24959531

Huang, A., Nguyen, P. Q., Stark, J. C., Takahashi, M. K., Donghia, N., Ferrante, T., Dy, A. J., Hsu, K. J., Dubner, R. S., Pardee, K., Jewett, M. C., & Collins, J. J. (2018). BioBits™ Explorer: A modular synthetic biology education kit. *Science Advances, 4*(8), eaat5105. Advance online publication. doi:10.1126ciadv.aat5105 PMID:30083608

Huang, C. Jr, Lowe, A. J., & Batt, C. A. (2010, June). Recombinant immunotherapeutics: Current state and perspectives regarding the feasibility and market. *Applied Microbiology and Biotechnology, 87*(2), 401–410. doi:10.100700253-010-2590-7 PMID:20422181

Huang, D., Li, S., Xia, M., Wen, J., & Jia, X. (2013). Genome-scale metabolic network guided engineering of Streptomyces tsukubaensis for FK506 production improvement. *Microbial Cell Factories, 12*(1), 1–18. doi:10.1186/1475-2859-12-52 PMID:23705993

Huang, D., Wen, J., Wang, G., Yu, G., Jia, X., & Chen, Y. (2012). In silico aided metabolic engineering of Streptomyces roseosporus for daptomycin yield improvement. *Applied Microbiology and Biotechnology, 94*(3), 637–649. doi:10.100700253-011-3773-6 PMID:22406858

Huang, J., & Koroteev, D. D. (2021). Artificial intelligence for planning of energy and waste management. *Sustainable Energy Technologies and Assessments, 47*, 101426. doi:10.1016/j.seta.2021.101426

Huang, Y. Y., Jian, X. X., Lv, Y. B., Nian, K. Q., Gao, Q., Chen, J., Wei, L. J., & Hua, Q. (2018). Enhanced squalene biosynthesis in Yarrowia lipolytica based on metabolically engineered acetyl-CoA metabolism. *Journal of Biotechnology, 281*, 106–114. doi:10.1016/j.jbiotec.2018.07.001 PMID:29986837

Huang, Y., Li, B., Liu, D., Xie, X., Zhang, H., Sun, H., Hu, X., & Zhang, S. (2020). Fundamental advances in biomass autothermal/oxidative pyrolysis: A review. *ACS Sustainable Chemistry & Engineering, 8*(32), 11888–11905. doi:10.1021/acssuschemeng.0c04196

Hua, S. (2019). Physiological and pharmaceutical considerations for rectal drug formulations. *Frontiers in Pharmacology, 10*, 1196. doi:10.3389/fphar.2019.01196 PMID:31680970

Huber, G., Bailly, Y., Martin, J. R., Mariani, J., & Brugg, B. (1997). Synaptic β-amyloid precursor proteins increase with learning capacity in rats. *Neuroscience, 80*(2), 313–320. doi:10.1016/S0306-4522(97)00120-6 PMID:9284337

Huber, S., Remberger, M., Kaj, L., Schlabach, M., Jörundsdóttir, H. O., Vester, J., Arnórsson, M., Mortensen, I., Schwartson, R., & Dam, M. (2016). A first screening and risk assessment of pharmaceuticals and additives in personal care products in waste water, sludge, recipient water and sediment from Faroe Islands, Iceland, and Greenland. *The Science of the Total Environment*, *562*, 13–25. doi:10.1016/j.scitotenv.2016.03.063 PMID:27092417

Huff, J. (2006). *ECG workout: Exercises in arrhythmia interpretation*. Lippincott Williams & Wilkins.

Hung, O., & Nelson, L. S. (1997). Toxnet - A Guide to clinical toxicology resources. *Journal of Toxicology. Clinical Toxicology*, *35*(6), 677–679. doi:10.3109/15563659709001254

Hunt, J. P., Zhao, E. L., Free, T. J., Soltani, M., Warr, C. A., Benedict, A. B., Takahashi, M. K., Griffitts, J. S., Pitt, W. G., & Bundy, B. C. (2022). Towards detection of SARS-CoV-2 RNA in human saliva: A paper-based cell-free toehold switch biosensor with a visual bioluminescent output. *New Biotechnology*, *66*, 53–60. doi:10.1016/j.nbt.2021.09.002 PMID:34555549

Hu, R., Wen, S., Zeng, Z., & Huang, T. (2017). A short-term power load forecasting model based on the generalized regression neural network with decreasing step fruit fly optimization algorithm. *Neurocomputing*, *221*, 24–31. doi:10.1016/j.neucom.2016.09.027

Hutchison, C. A. III, Chuang, R. Y., Noskov, V. N., Assad-Garcia, N., Deerinck, T. J., Ellisman, M. H., Gill, J., Kannan, K., Karas, B. J., Ma, L., Pelletier, J. F., Qi, Z. Q., Richter, R. A., Strychalski, E. A., Sun, L., Suzuki, Y., Tsvetanova, B., Wise, K. S., Smith, H. O., ... Venter, J. C. (2016a). Design and synthesis of a minimal bacterial genome. *Science*, *351*(6280), aad6253. Advance online publication. doi:10.1126cience.aad6253 PMID:27013737

Hu, W., Wang, Y., & Zhu, Z. (2006, January). A perspective on fish gonad manipulation for biotechnical applications. *Chinese Science Bulletin*, *51*(1), 1–6. doi:10.100711434-005-1055-3

Hu, X., Xie, C., He, S., Zhang, Y., Li, Y., & Jiang, L. (2013). Remifentanil post conditioning improves global cerebral ischemia-induced spatial learning and memory deficit in rats via inhibition of neuronal apoptosis through the PI3K signaling pathway. *Neurological Sciences*, *34*(11), 1955–1962. doi:10.100710072-013-1419-4 PMID:23609459

Hu, Y. C. (2005). Baculovirus as a highly efficient expression vector in insect and mammalian cells. *Acta Pharmacologica Sinica*, *26*(4), 405–416. doi:10.1111/j.1745-7254.2005.00078.x PMID:15780188

Ibrahim, M., Raajaraam, L., & Raman, K. (2021). Modeling microbial communities: Harnessing consortia for biotechnological applications. *Computational and Structural Biotechnology Journal*, *19*, 3892–3907. doi:10.1016/j.csbj.2021.06.048 PMID:34584635

Ihl, R., & Kretschmar, C. (1997). Zur Nootropikabewertung für die praxis. *Der Nervenarzt*, *68*(11), 853–861. doi:10.1007001150050207 PMID:9732727

Iizuka, N., & Sarnow, P. (1997). Translation-competent extracts from Saccharomyces cerevisiae: Effects of L-A RNA, 5′ cap, and 3′ poly(A) tail on translational efficiency of mRNAs. *Methods (San Diego, Calif.)*, *11*(4), 353–360. doi:10.1006/meth.1996.0433 PMID:9126550

Imam, S., Schäuble, S., Valenzuela, J., Lõpez García De Lomana, A., Carter, W., Price, N. D., & Baliga, N. S. (2015). A refined genome-scale reconstruction of Chlamydomonas metabolism provides a platform for systems-level analyses. *The Plant Journal*, *84*(6), 1239–1256. doi:10.1111/tpj.13059 PMID:26485611

Ishchuka, O. P., Domenzain, I., Sanchez, B. J., Muñiz-Paredes, F., Martínez, J. L., Nielsen, J., & Petranovic, D. (2022). Genome-scale modeling drives 70-fold improvement of intracellular heme production in Saccharomyces cerevisiae. *Proceedings of the National Academy of Sciences of the United States of America*, *119*(30), e2108245119. doi:10.1073/pnas.2108245119 PMID:35858410

Isin, A., & Ozdalili, S. (2017). Cardiac arrhythmia detection using deep learning. *Procedia Computer Science, 120*, 268–275. doi:10.1016/j.procs.2017.11.238

Iwamoto, T., & Nasu, M. (2001). Current bioremediation practice and perspective. *Journal of Bioscience and Bioengineering, 92*(1), 1–8. doi:10.1016/S1389-1723(01)80190-0 PMID:16233049

Izci, E., Ozdemir, M. A., Degirmenci, M., & Akan, A. (2019). Cardiac arrhythmia detection from 2d ecg images by using deep learning technique. *TIPTEKNO 2019 - Tip Teknolojileri Kongresi*, 1–4. doi:10.1109/TIPTEKNO.2019.8895011

Jagadevan, S., Banerjee, A., Banerjee, C., Guria, C., Tiwari, R., Baweja, M., & Shukla, P. (2018). Recent developments in synthetic biology and metabolic engineering in microalgae towards biofuel production. *Biotechnology for Biofuels, 11*(1), 1–21. doi:10.118613068-018-1181-1 PMID:29988523

Jäger, U., Barcellini, W., Broome, C. M., Gertz, M. A., Hill, A., Hill, Q. A., Jilma, B., Kuter, D. J., Michel, M., Montillo, M., Röth, A., Zeerleder, S. S., & Berentsen, S. (2020, May 1). Diagnosis and treatment of autoimmune hemolytic anemia in adults: Recommendations from the First International Consensus Meeting. *Blood Reviews, 41*, 100648. doi:10.1016/j.blre.2019.100648 PMID:31839434

Jain, K. K. (2020). An overview of drug delivery systems. *Drug Delivery System*, 1–54. PMID:31435914

Jaiswal, S., & Shukla, P. (2020). Alternative strategies for microbial remediation of pollutants via synthetic biology. *Frontiers in Microbiology, 11*, 808. doi:10.3389/fmicb.2020.00808 PMID:32508759

Janardhana, K., Singh, V., Singh, S. N., Babu, T. R., Bano, S., & Boopathi, S. (2023). Utilization Process for Electronic Waste in Eco-Friendly Concrete: Experimental Study. In Sustainable Approaches and Strategies for E-Waste Management and Utilization (pp. 204–223). IGI Global.

Janardhana, K., Singh, V., Singh, S. N., Babu, T. S. R., Bano, S., & Boopathi, S. (2023). Utilization Process for Electronic Waste in Eco-Friendly Concrete: Experimental Study. In Sustainable Approaches and Strategies for E-Waste Management and Utilization (pp. 204–223). IGI Global.

Javaid, M., & Haleem, A. (n.d.). *4D printing applications in medical field: A brief review*. Academic Press.

Javalkote, V. S., Kancharla, N., Bhadra, B., Shukla, M., Soni, B., Sapre, A., Goodin, M., Bandyopadhyay, A., & Dasgupta, S. (2022). CRISPR-based assays for rapid detection of SARS-CoV-2. *Methods (San Diego, Calif.), 203*, 594–603. doi:10.1016/j.ymeth.2020.10.003 PMID:33045362

Jeandet, P., Vasserot, Y., Chastang, T., & Courot, E. (2013). Engineering microbial cells for the biosynthesis of natural compounds of pharmaceutical significance. *BioMed Research International, 780145*, 1–13. Advance online publication. doi:10.1155/2013/780145 PMID:23710459

Jeevanantham, Y. A., A, S., V, V., J, S. Isaac., Boopathi, S., & Kumar, D. P. (2023). Implementation of Internet-of Things (IoT) in Soil Irrigation System. *IEEE Explore*, 1–5. doi:10.1109/ICPECTS56089.2022.10047185

Jeevanantham, Y. A., Saravanan, A., Vanitha, V., Boopathi, S., & Kumar, D. P. (2022). Implementation of Internet-of Things (IoT) in Soil Irrigation System. *IEEE Explore*, 1–5.

Jefferson, C., Lentzos, F., & Marris, C. (2014). Synthetic biology and biosecurity: Challenging the "myths". *Frontiers in Public Health, 2*, 115. doi:10.3389/fpubh.2014.00115 PMID:25191649

Jenkins, N. (2007). Modifications of therapeutic proteins: Challenges and prospects. *Cytotechnology, 53*(1-3), 121–125. doi:10.100710616-007-9075-2 PMID:19003198

Jeyakymar, M., Munusami, V., Shanmugam, P., Boopalan, N., Boopathi, S., & Sureshkumar, M. (2022). An investigation on wear loss and hardness of nitro-carburizing coated stainless-steel grade-316. *Materials Today: Proceedings*, *66*, 1398–1404. doi:10.1016/j.matpr.2022.05.223

Jiang, L., Zhao, J., Lian, J., & Xu, Z. (2018). Cell-free protein synthesis enabled rapid prototyping for metabolic engineering and synthetic biology. *Synthetic and Systems Biotechnology*, *3*(2), 90–96. doi:10.1016/j.synbio.2018.02.003 PMID:29900421

Jiang, W., Bikard, D., Cox, D., Zhang, F., & Marraffini, L. A. (2013). RNA-guided editing of bacterial genomes using CRISPR-Cas systems. *Nature Biotechnology*, *31*(3), 233–239. doi:10.1038/nbt.2508 PMID:23360965

Jiang, W., Zhao, X., Gabrieli, T., Lou, C., Ebenstein, Y., & Zhu, T. (2015). Cas9-Assisted Targeting of CHromosome segments CATCH enables one-step targeted cloning of large gene clusters. *Nature Communications*, *6*(1), 8101. Advance online publication. doi:10.1038/ncomms9101 PMID:26323354

Ji, J., Chen, X., Luo, C., & Li, P. (2018). A deep multi-task learning approach for ECG data analysis. In *2018 IEEE EMBS International conference on biomedical & health informatics (BHI)* (pp. 124–127). IEEE 10.1109/BHI.2018.8333385

Jiskoot, W., Randolph, T. W., Volkin, D. B., Middaugh, C. R., Schöneich, C., Winter, G., Friess, W., Crommelin, D. J., & Carpenter, J. F. (2012, March 1). Protein instability and immunogenicity: Roadblocks to clinical application of injectable protein delivery systems for sustained release. *Journal of Pharmaceutical Sciences*, *101*(3), 946–954. doi:10.1002/jps.23018 PMID:22170395

Joanna, T., & Luczak, M. (2021). Applying proteomics and integrative omics strategies to decipher the chronic kidney disease-Related Atherosclerosis. *International Journal of Molecular Sciences*, *22*(14), 1–19. PMID:34299112

Johanna, M., & George, P. L. (2016). Molecular pharming — VLPs made in plants. *Current Opinion in Biotechnology*, *37*, 201–206. doi:10.1016/j.copbio.2015.12.007 PMID:26773389

John, S. T. (2020). *First human efficacy study of a plant-derived influenza vaccine*. doi:10.1016/S0140-6736(20)32010-9Lancet

Joshi, A. J., Chandran, S., Jayaraman, V. K., & Kulkarni, B. D. (2009). *Hybrid SVM for multiclass arrhythmia classification. In 2009 IEEE international conference on bioinformatics and biomedicine*. IEEE.

Jouhten, P. (2012). Metabolic modeling in the development of cell factories by synthetic biology. *Computational and Structural Biotechnology Journal*, *3*(4), e201210009. doi:10.5936/csbj.201210009 PMID:24688669

Juhas, M. (2015). On the road to synthetic life: The minimal cell and genome-scale engineering. *Critical Reviews in Biotechnology*, 1–8. doi:10.3109/07388551.2014.989423 PMID:25578717

Jung, J. K., Alam, K. K., Verosloff, M. S., Capdevila, D. A., Desmau, M., Clauer, P. R., Lee, J. W., Nguyen, P. Q., Pastén, P. A., Matiasek, S. J., Gaillard, J.-F., Giedroc, D. P., Collins, J. J., & Lucks, J. B. (2020). Cell-free biosensors for rapid detection of water contaminants. *Nature Biotechnology*, *38*(12), 1451–1459. doi:10.103841587-020-0571-7 PMID:32632301

JunT. J.NguyenH. M.KangD.KimD.KimD.KimY.-H. (2018). *Ecg arrhythmia classification using a 2-d convolutional neural network*. arXiv:1804.06812.

Kagadis, G. C., Kloukinas, C., Moore, K., Philbin, J., Papadimitroulas, P., Alexakos, C., Nagy, P. G., Visvikis, D., & Hendee, W. R. (2013). Cloud computing in medical imaging. *Medical Physics*, *40*(7), 70901. doi:10.1118/1.4811272 PMID:23822402

Kai, L., Dötsch, V., Kaldenhoff, R., & Bernhard, F. (2013). Artificial environments for the co-translational stabilization of cell-free expressed proteins. *PLoS One*, *8*(2), e56637. doi:10.1371/journal.pone.0056637 PMID:23451062

Kamaleswaran, R., Mahajan, R., & Akbilgic, O. (2018). A robust deep convolutional neural network for the classification of abnormal cardiac rhythm using single lead electrocardiograms of variable length. *Physiological Measurement*, *39*(3), 035006. doi:10.1088/1361-6579/aaaa9d PMID:29369044

Kamerzell, T. J., Esfandiary, R., Joshi, S. B., Middaugh, C. R., & Volkin, D. B. (2011, October 1). Protein–excipient interactions: Mechanisms and biophysical characterization applied to protein formulation development. *Advanced Drug Delivery Reviews*, *63*(13), 1118–1159. doi:10.1016/j.addr.2011.07.006 PMID:21855584

Kaminski, M. M., Abudayyeh, O. O., Gootenberg, J. S., Zhang, F., & Collins, J. J. (2021). CRISPR-based diagnostics. *Nature Biomedical Engineering*, *5*(7), 643–656. doi:10.103841551-021-00760-7 PMID:34272525

Kampouropoulos, K., Andrade, F., Sala, E., Espinosa, A. G., & Romeral, L. (2018). Multiobjective optimization of multi-carrier energy system using a combination of ANFIS and genetic algorithms. *IEEE Transactions on Smart Grid*, *9*(3), 2276–2283. doi:10.1109/TSG.2016.2609740

Kanchiswamy, C. N., Maffei, M., Malnoy, M., Velasco, R., & Kim, J. S. (2016). Fine-tuning next-generation genome editing tools. *Trends in Biotechnology*, *34*(7), 562–574. doi:10.1016/j.tibtech.2016.03.007 PMID:27167723

Kanehisa, M. (2019). Toward understanding the origin and evolution of cellular organisms. *Protein Science*, *28*(11), 1947–1951. doi:10.1002/pro.3715 PMID:31441146

Kanehisa, M., Furumichi, M., Sato, Y., Kawashima, M., & Ishiguro-Watanabe, M. (2023). KEGG for taxonomy-based analysis of pathways and genomes. *Nucleic Acids Research*, *51*(D1), D587–D592. doi:10.1093/nar/gkac963 PMID:36300620

Kanehisa, M., & Goto, S. (2000). KEGG: Kyoto Encyclopedia of Genes and Genomes. *Nucleic Acids Research*, *28*(1), 27–30. doi:10.1093/nar/28.1.27 PMID:10592173

Kanehisa, M., Goto, S., Kawashima, S., Okuno, Y., & Hattori, M. (2004). The KEGG resource for deciphering the genome. *Nucleic Acids Research*, *32*(90001), D277–D280. doi:10.1093/nar/gkh063 PMID:14681412

Kang, M., Ko, E., & Mersha, T. B. (2022). A roadmap for multi-omics data integration using deep learning. *Briefings in Bioinformatics*, *23*(1), bbab454. Advance online publication. doi:10.1093/bib/bbab454 PMID:34791014

Kannan, E., Trabelsi, Y., Boopathi, S., & Alagesan, S. (2022). Influences of cryogenically treated work material on near-dry wire-cut electrical discharge machining process. *Surface Topography : Metrology and Properties*, *10*(1), 015027. doi:10.1088/2051-672X/ac53e1

Kannan, P., Venkataramanan, K., Sivaranjani, S., & Indhumathi, K. (2020)... *Analysis of ABC Algorithm Based ANFIS Controller for PV Integrated Low Voltage Weak Grid System.*, *3*(8), 196–201.

Karim, A. S., Dudley, Q. M., Juminaga, A., Yuan, Y., Crowe, S. A., Heggestad, J. T., Garg, S., Abdalla, T., Grubbe, W. S., Rasor, B. J., Coar, D. N., Torculas, M., Krein, M., Liew, F. M., Quattlebaum, A., Jensen, R. O., Stuart, J. A., Simpson, S. D., Köpke, M., & Jewett, M. C. (2020). In vitro prototyping and rapid optimization of biosynthetic enzymes for cell design. *Nature Chemical Biology*, *16*(8), 912–919. doi:10.103841589-020-0559-0 PMID:32541965

Karimi, B., Habibi, M., & Esvand, M. (2015). Biodegradation of naphthalene using *Pseudomonas aeruginosa* by up flow anoxic–aerobic continuous flow combined bioreactor. *Journal of Environmental Health Science & Engineering*, *13*(1), 1–10. doi:10.118640201-015-0175-1 PMID:25859393

Karp, P. D., Billington, R., Caspi, R., Fulcher, C. A., Latendresse, M., Kothari, A., Keseler, I. M., Krummenacker, M., Midford, P. E., Ong, Q., Ong, W. K., Paley, S. M., & Subhraveti, P. (2019). The BioCyc collection of microbial genomes and metabolic pathways. *Briefings in Bioinformatics*, *20*(4), 1085–1093. doi:10.1093/bib/bbx085 PMID:29447345

Karp, P. D., Riley, M., Saier, M., Paulsen, I. T., Paley, S. M., & Pellegrini-Toole, A. (2000). The ecocyc and metacyc databases. *Nucleic Acids Research*, *28*(1), 56–59. doi:10.1093/nar/28.1.56 PMID:10592180

Karthik, S., Hemalatha, R., Aruna, R., Deivakani, M., Reddy, R. V. K., & Boopathi, S. (2023). Study on Healthcare Security System-Integrated Internet of Things (IoT). In Perspectives and Considerations on the Evolution of Smart Systems (pp. 342–362). IGI Global.

Kasera, O. A., Owiso, M. O., & Mburu, B. (2021). Governing Emerging Technologies: A Systematic Exploration of Kenya's Biotechnology and LMO-Specific Policy Documents for Adoption and Implementation of Synthetic Biology. *International Journal of Innovation Science, Research and Technology*, 946–974.

Kasper, D. L., Fauci, A. S., Hauser, S. L., Longo, D. L., Jameson, J. L., & Loscalzo, J. (2018). Harrison's principles of internal medicine. Mc- Graw Hill Professional.

Kathleen H. (2017). Reconceptualizing cancer immunotherapy based on plant production systems. *Future Sci.*

Katyara, S., Shah, M. A., Chowdhary, B. S., Akhtar, F., & Lashari, G. A. (2019). Monitoring, Control and Energy Management of Smart Grid System via WSN Technology Through SCADA Applications. *Wireless Personal Communications*, *106*(4), 1951–1968. doi:10.100711277-018-5738-x

Kaur, H., & Sood, S. K. (2019). Adaptive Neuro Fuzzy Inference System (ANFIS) based wildfire risk assessment. *Journal of Experimental & Theoretical Artificial Intelligence*, *31*(4), 599–619. doi:10.1080/0952813X.2019.1591523

Kavallaris, M., & Marshall, G. M. (2005). Proteomics and disease: Opportunities and challenges. *The Medical Journal of Australia*, *182*(11), 575–579. doi:10.5694/j.1326-5377.2005.tb06817.x PMID:15938685

Kavitha, C., Geetha Malini, P. S., Charan Kantumuchu, V., Manoj Kumar, N., Verma, A., & Boopathi, S. (2023). An experimental study on the hardness and wear rate of carbonitride coated stainless steel. *Materials Today: Proceedings*, *74*, 595–601. doi:10.1016/j.matpr.2022.09.524

Kavšcek, M., Bhutada, G., Madl, T., & Natter, K. (2015). Optimization of lipid production with a genome-scale model of Yarrowia lipolytica. *BMC Systems Biology*, *9*(1), 1–13. doi:10.118612918-015-0217-4 PMID:26503450

Kellner, M. J., Koob, J. G., Gootenberg, J. S., Abudayyeh, O. O., & Zhang, F. (2019). SHERLOCK: Nucleic acid detection with CRISPR nucleases. *Nature Protocols*, *14*(10), 2986–3012. doi:10.103841596-019-0210-2 PMID:31548639

Kelwick, R. J. R., Webb, A. J., & Freemont, P. S. (2020). Biological Materials: The Next Frontier for Cell-Free Synthetic Biology. *Frontiers in Bioengineering and Biotechnology*, *8*, 399. doi:10.3389/fbioe.2020.00399 PMID:32478045

Kennedy, D. O., Haskell, C. F., Wesnes, K. A., & Scholey, A. B. (2004). Improved cognitive performance in human volunteers following administration of Guarana (*Paullinia Cupana*) extract: Comparison and interaction with panax ginseng. *Pharmacology, Biochemistry, and Behavior*, *79*(3), 401–411. doi:10.1016/j.pbb.2004.07.014 PMID:15582012

Kenney, K. L., Smith, W. A., Gresham, G. L., & Westover, T. L. (2013). Understanding biomass feedstock variability. *Biofuels*, *4*(1), 111–127. doi:10.4155/bfs.12.83

Keseler, I. M., Collado-Vides, J., Santos-Zavaleta, A., Peralta-Gil, M., Gama-Castro, S., Muniz-Rascado, L., Bonavides-Martinez, C., Paley, S., Krummenacker, M., Altman, T., Kaipa, P., Spaulding, A., Pacheco, J., Latendresse, M., Fulcher, C., Sarker, M., Shearer, A. G., Mackie, A., Paulsen, I., ... Karp, P. D. (2011). EcoCyc: A comprehensive database of Escherichia coli biology. *Nucleic Acids Research, 39*(Database), D583–D590. Advance online publication. doi:10.1093/nar/gkq1143 PMID:21097882

Keyvanrad, M. A., & Homayounpour, M. M. (2015). Deep belief network training im- provement using elite samples minimizing free energy. *International Journal of Pattern Recognition and Artificial Intelligence, 29*(05), 1551006. doi:10.1142/S0218001415510064

Khalid, R., & Javaid, N. (2020). A survey on hyperparameters optimization algorithms of forecasting models in smart grid. *Sustainable Cities and Society, 61*, 102275. doi:10.1016/j.scs.2020.102275

Khambhati, K., Bhattacharjee, G., Gohil, N., Braddick, D., Kulkarni, V., & Singh, V. (2019). Exploring the potential of cell-free protein synthesis for extending the abilities of biological systems. *Frontiers in Bioengineering and Biotechnology, 7*, 248. doi:10.3389/fbioe.2019.00248 PMID:31681738

Khan, A. H. (n.d.). *Arrhythmia Classification Techniques Using Deep Neural Network.* Www.hindawi.com

Khan, A. H., López-Maldonado, E. A., Alam, S. S., Khan, N. A., López, J. R. L., Herrera, P. F. M., Abutaleb, A., Ahmed, S., & Singh, L. (2022). Municipal solid waste generation and the current state of waste-to-energy potential: State of art review. *Energy Conversion and Management, 267*, 115905. doi:10.1016/j.enconman.2022.115905

Khan, A., & Rizwan, M. (2021). ANN and ANFIS Based Approach for Very Short Term Load Forecasting: A Step Towards Smart Energy Management System. *Proceedings of the 8th International Conference on Signal Processing and Integrated Networks, SPIN 2021*, 464–468. 10.1109/SPIN52536.2021.9566146

Khan, S., Ullah, M. W., Siddique, R., Nabi, G., Manan, S., Yousaf, M., & Hou, H. (2016). Role of Recombinant DNA Technology to Improve Life. *International Journal of Genomics, 2016*, 1–14. doi:10.1155/2016/2405954 PMID:28053975

Khattak, W. A., Ullah, M. W., Ul-Islam, M., Khan, S., Kim, M., Kim, Y., & Park, J. K. (2014). Developmental strategies and regulation of cell-free enzyme system for ethanol production: A molecular prospective. *Applied Microbiology and Biotechnology, 98*(23), 9561–9578. doi:10.100700253-014-6154-0 PMID:25359472

Khetan, S. K., & Collins, T. J. (2007). Human pharmaceuticals in the aquatic environment: A challenge to green chemistry. *Chemical Reviews, 107*(6), 2319–2364. doi:10.1021/cr020441w PMID:17530905

Khnouf, R., Beebe, D. J., & Fan, Z. H. (2009). Cell-free protein expression in a microchannel array with passive pumping. *Lab on a Chip, 9*(1), 56–61. doi:10.1039/B808034H PMID:19209336

Khnouf, R., Chapman, B. D., & Fan, Z. H. (2011). Fabrication optimization of a miniaturized array device for cell-free protein synthesis. *Electrophoresis, 32*(22), 3101–3107. doi:10.1002/elps.201100376 PMID:22038694

Khnouf, R., Olivero, D., Jin, S., & Fan, Z. H. (2010). Miniaturized fluid array for high-throughput protein expression. *Biotechnology Progress, 26*(6), 1590–1596. doi:10.1002/btpr.474 PMID:20661923

Khodarovich, Y. M., Goldman, I. L., Sadchikova, E. R., & Georgiev, P. G. (2013, December). Expression of eukaryotic recombinant proteins and deriving them from the milk of transgenic animals. *Applied Biochemistry and Microbiology, 49*(9), 711–722. doi:10.1134/S0003683813090020

Khoo, K. S., Chew, K. W., Yew, G. Y., Leong, W. H., Chai, Y. H., Show, P. L., & Chen, W. H. (2020). Recent advances in downstream processing of microalgae lipid recovery for biofuel production. *Bioresource Technology, 304*, 122996. doi:10.1016/j.biortech.2020.122996 PMID:32115347

Khuroo, M. S. (2020, September 1). Chloroquine and hydroxychloroquine in coronavirus disease 2019 (COVID-19). Facts, fiction and the hype: A critical appraisal. *International Journal of Antimicrobial Agents*, *56*(3), 106101. doi:10.1016/j.ijantimicag.2020.106101 PMID:32687949

Kim, D. M., & Choi, C. Y. (1996). A semicontinuous prokaryotic coupled transcription/translation system using a dialysis membrane. *Biotechnology Progress*, *12*(5), 645–649. doi:10.1021/bp960052l PMID:8879155

Kim, D. M., & Swartz, J. R. (1999). Prolonging cell-free protein synthesis with a novel ATP regeneration system. *Biotechnology and Bioengineering*, *66*(3), 180–188. doi:10.1002/(SICI)1097-0290(1999)66:3<180::AID-BIT6>3.0.CO;2-S PMID:10577472

Kim, D. M., & Swartz, J. R. (2001). Regeneration of adenosine triphosphate from glycolytic intermediates for cell-free protein synthesis. *Biotechnology and Bioengineering*, *74*(4), 309–316. doi:10.1002/bit.1121 PMID:11410855

Kim, D.-M., & Swartz, J. R. (2000). Prolonging Cell-Free Protein Synthesis by Selective Reagent Additions. *Biotechnology Progress*, *16*(3), 385–390. doi:10.1021/bp000031y PMID:10835240

Kim, H., Yoo, S. J., & Kang, H. A. (2015, February 1). Yeast synthetic biology for the production of recombinant therapeutic proteins. *FEMS Yeast Research*, *15*(1), 1–6. PMID:25130199

Kim, J., & Reed, J. L. (2010). OptORF: Optimal metabolic and regulatory perturbations for metabolic engineering of microbial strains. *BMC Systems Biology*, *4*(1), 53. doi:10.1186/1752-0509-4-53 PMID:20426856

Kim, M. S., Bang, J. H., Lee, J., Han, J. S., Baik, T. G., & Jeon, W. K. (2016). *Ginkgo biloba L.* Extract protects against chronic cerebral hypo perfusion by modulating neuroinflammation and the cholinergic system. *Phytomedicine*, *23*(12), 1356–1364. doi:10.1016/j.phymed.2016.07.013 PMID:27765355

Kim, M., Park, B. G., Kim, E. J., Kim, J., & Kim, B. G. (2019). In silico identification of metabolic engineering strategies for improved lipid production in yarrowia lipolytica by genome-scale metabolic modeling. *Biotechnology for Biofuels*, *12*(1), 1–14. doi:10.118613068-019-1518-4 PMID:31367232

Kim, T. W., Keum, J. W., Oh, I. S., Choi, C. Y., Park, C. G., & Kim, D. M. (2006). Simple procedures for the construction of a robust and cost-effective cell-free protein synthesis system. *Journal of Biotechnology*, *126*(4), 554–561. doi:10.1016/j.jbiotec.2006.05.014 PMID:16797767

King, Z. A., Lu, J., Dräger, A., Miller, P., Federowicz, S., Lerman, J. A., Ebrahim, A., Palsson, B. O., & Lewis, N. E. (2016). BiGG Models: A platform for integrating, standardizing and sharing genome-scale models. *Nucleic Acids Research*, *44*(D1), D515–D522. doi:10.1093/nar/gkv1049 PMID:26476456

Kishore, K., & Krishan, P. (2009). Pharmacology of recombinant or genetically engineered drugs. *Journal of Young Pharmacists : JYP*, *1*(2), 141. doi:10.4103/0975-1483.55747

Kitano, S., Lin, C., Foo, J. L., & Chang, M. W. (2023). Synthetic biology: Learning the way toward high-precision biological design. *PLoS Biology*, *21*(4), e3002116. doi:10.1371/journal.pbio.3002116 PMID:37099620

Klangsin, P., & Harding, A. K. (1998). Medical waste treatment and disposal methods used by hospitals in Oregon, Washington, and Idaho. *Journal of the Air & Waste Management Association*, *48*(6), 516–526.

Knoop, H., Gründel, M., Zilliges, Y., Lehmann, R., Hoffmann, S., Lockau, W., & Steuer, R. (2013). Flux Balance Analysis of Cyanobacterial Metabolism: The Metabolic Network of Synechocystis sp. PCC 6803. *PLoS Computational Biology*, *9*(6), e1003081. doi:10.1371/journal.pcbi.1003081 PMID:23843751

Knoop, H., Zilliges, Y., Lockau, W., & Steuer, R. (2010). The metabolic network of Synechocystis sp. PCC 6803: Systemic properties of autotrophic growth. *Plant Physiology*, *154*(1), 410–422. doi:10.1104/pp.110.157198 PMID:20616194

Kołodziej, B., Sugier, D., & Bielińska, E. (2013). The effect of leonardite application and various plantation modalities on yielding and quality of roseroot (*Rhodiola rosea L.*) and soil enzymatic activity. *Journal of Geochemical Exploration*, *129*(1), 64–69. doi:10.1016/j.gexplo.2012.10.014

Koncic, M., & Tomczyk, M. (2013). New insights into dietary supplements used in sport: Active substances, pharmacological and side effects. *Current Drug Targets*, *14*(9), 1079–1092. doi:10.2174/1389450111314090016 PMID:23574283

Koo, J., Yang, J., & Park, H. (2020). Cell-free systems: Recent advances and future outlook. *Biotechnology and Bioprocess Engineering; BBE*, *25*(6), 955–961. doi:10.100712257-020-0013-x

Koshariya, A. K., Kalaiyarasi, D., Jovith, A. A., Sivakami, T., Hasan, D. S., & Boopathi, S. (2023). AI-Enabled IoT and WSN-Integrated Smart Agriculture System. In *Artificial Intelligence Tools and Technologies for Smart Farming and Agriculture Practices* (pp. 200–218). IGI Global. doi:10.4018/978-1-6684-8516-3.ch011

Koshariya, A. K., Khatoon, S., Marathe, A. M., Suba, G. M., Baral, D., & Boopathi, S. (2023). Agricultural Waste Management Systems Using Artificial Intelligence Techniques. In *AI-Enabled Social Robotics in Human Care Services* (pp. 236–258). IGI Global. doi:10.4018/978-1-6684-8171-4.ch009

Kosjek, T., Heath, E., & Krbavcic, A. (2005). Determination of non-steroidal anti-inflammatory drug (NSAIDs) residues in water samples. *Environment International*, *31*(5), 679–685. doi:10.1016/j.envint.2004.12.001 PMID:15910964

Kovtun, O., Mureev, S., Jung, W., Kubala, M. H., Johnston, W., & Alexandrov, K. (2011). Leishmania cell-free protein expression system. *Methods (San Diego, Calif.)*, *55*(1), 58–64. doi:10.1016/j.ymeth.2011.06.006 PMID:21704167

Krämer, M., Bongaerts, J., Bovenberg, R., Kremer, S., Müller, U., Orf, S., Wubbolts, M., & Raeven, L. (2003). Metabolic engineering for microbial production of shikimic acid. *Metabolic Engineering*, *5*(4), 277–283. doi:10.1016/j.ymben.2003.09.001 PMID:14642355

Krizhevsky, A., Sutskever, I., & Hinton, G. E. (2012). ImageNet classification with deep convolutional neural networks. *Advances in Neural Information Processing Systems.*

Kruyer, N. S., Sugianto, W., Tickman, B. I., Alba Burbano, D., Noireaux, V., Carothers, J. M., & Peralta-Yahya, P. (2021). Membrane augmented cell-free systems: A new frontier in biotechnology. *ACS Synthetic Biology*, *10*(4), 670–681. doi:10.1021/acssynbio.0c00625 PMID:33749249

Kuepfer, L., Sauer, U., & Blank, L. M. (2005). Metabolic functions of duplicate genes in Saccharomyces cerevisiae. *Genome Research*, *15*(10), 1421–1430. doi:10.1101/gr.3992505 PMID:16204195

Kulkarni, R. (2016). Metabolic Engineering: Biological Art of Producing Useful Chemicals. *Resonance*, *21*(3), 233–237. doi:10.100712045-016-0318-4

Kumar, P., Sampath, B., Kumar, S., Babu, B. H., & Ahalya, N. (2023). Hydroponics, Aeroponics, and Aquaponics Technologies in Modern Agricultural Cultivation. In Trends, Paradigms, and Advances in Mechatronics Engineering (pp. 223–241). IGI Global.

Kumara, V., Mohanaprakash, T., Fairooz, S., Jamal, K., Babu, T., & Sampath, B. (2023). Experimental Study on a Reliable Smart Hydroponics System. In *Human Agro-Energy Optimization for Business and Industry* (pp. 27–45). IGI Global. doi:10.4018/978-1-6684-4118-3.ch002

Kumari, A., Kanchan, S., Sinha, R. P., & Kesheri, M. (2016). Applications of biomolecular databases in bioinformatics. In *Medical Imaging in Clinical Applications* (pp. 329–351). Springer. doi:10.1007/978-3-319-33793-7_15

Kumar, N. M., Muthukumaran, C., Sharmila, G., & Gurunathan, B. (2018). Genetically modified organisms and its impact on the enhancement of bioremediation. In S. J. Varjani, A. K. Agarwal, E. Gnansounou, & B. Gurunathan (Eds.), *Bioremediation: Applications for environmental protection and management* (pp. 53–76). Springer. doi:10.1007/978-981-10-7485-1_4

Kumar, S., & Kumar, A. (2015). Role of genetic engineering in agriculture. *Plant Archives*, *15*(1), 1–6. http://plantarchives.org/PDF%2015%20-%201/1-6%20(2950).pdf

Kwon, G., Bhatnagar, A., Wang, H., Kwon, E. E., & Song, H. (2020). A review of recent advancements in utilization of biomass and industrial wastes into engineered biochar. *Journal of Hazardous Materials*, *400*, 123242. doi:10.1016/j.jhazmat.2020.123242 PMID:32585525

Kwon, S. H., & Dong, L. (2022). Flexible sensors and machine learning for heart monitoring. *Nano Energy*, *102*, 107632. doi:10.1016/j.nanoen.2022.107632

Labati, R. D., Muñoz, E., Piuri, V., Sassi, R., & Scotti, F. (2018). Deep-ECG: Convolu- tional neural networks for ECG biometric recognition. *Pattern Recognition Letters*.

Ladd, S. L., Sommer, S. A., LaBerge, S., & Toscano, W. (1993). Brief report. *Clinical Neuropharmacology*, *16*(6), 540–549. doi:10.1097/00002826-199312000-00007 PMID:9377589

Lajoie, M. J., Kosuri, S., Mosberg, J. A., Gregg, C., Zhang, D., & Church, G. M. (2013). Probing the limits of genetic recoding in essential genes. *Science*, *342*(6156), 361–363. doi:10.1126cience.1241460 PMID:24136967

Lakshmanan, M., Kok, Y. J., Lee, A. P., Kyriakopoulos, S., Lim, H. L., Teo, G., Poh, S. L., Tang, W. Q., Hong, J., Tan, A. H.-M., Bi, X., Ho, Y. S., Zhang, P., Ng, S. K., & Lee, D. Y. (2019). Multi-omics profiling of CHO parental hosts reveals cell line-specific variations in bioprocessing traits. *Biotechnology and Bioengineering*, *116*(9), 2117–2129. doi:10.1002/bit.27014 PMID:31066037

Lamnabhi-Lagarrigue, F., Annaswamy, A., Engell, S., Isaksson, A., Khargonekar, P., Murray, R. M., Nijmeijer, H., Samad, T., Tilbury, D., & Van den Hof, P. (2017, January 1). Systems & control for the future of humanity, research agenda: Current and future roles, impact and grand challenges. *Annual Reviews in Control*, *43*, 1–64. doi:10.1016/j.arcontrol.2017.04.001

Lampariello, L. R., Cortelazzo, A., Guerranti, R., Sticozzi, C., & Valacchi, G. (2012). The magic velvet bean of *mucuna pruriens*. *Journal of Traditional and Complementary Medicine*, *2*(4), 331–339. doi:10.1016/S2225-4110(16)30119-5 PMID:24716148

Landon, S., Rees-Garbutt, J., Marucci, L., & Grierson, C. (2019). Genome-driven cell engineering review: In vivo and in silico metabolic and genome engineering. *Essays in Biochemistry*, *63*(2), 267–284. doi:10.1042/EBC20180045 PMID:31243142

Laohakunakorn, N. (2020). Cell-Free Systems: A Proving Ground for Rational Biodesign. *Frontiers in Bioengineering and Biotechnology*, *8*, 788. Advance online publication. doi:10.3389/fbioe.2020.00788 PMID:32793570

Laohakunakorn, N., Grasemann, L., Lavickova, B., Michielin, G., Shahein, A., Swank, Z., & Maerkl, S. J. (2020). Bottom-up construction of complex biomolecular systems with cell-free synthetic biology. *Frontiers in Bioengineering and Biotechnology*, *8*, 213. doi:10.3389/fbioe.2020.00213 PMID:32266240

Lawson, C. E., Martí, J. M., Radivojevic, T., Jonnalagadda, S. V. R., Gentz, R., Hillson, N. J., Peisert, S., Kim, J., Simmons, B. A., Petzold, C. J., Singer, S. W., Mukhopadhyay, A., Tanjore, D., Dunn, J. G., & Garcia Martin, H. (2021). Machine learning for metabolic engineering: A review. *Metabolic Engineering*, *63*, 34–60. doi:10.1016/j.ymben.2020.10.005 PMID:33221420

LeCun, Y., Bottou, L., Bengio, Y., & Haffner, P. (1998). Gradient-based learning applied to document recognition. *Proceedings of the IEEE*. https://doi.org/. 72679110.1109/5

Lecun, Y., Bengio, Y., & Hinton, G. (2015). Deep learning. *Nature*, *521*(7553), 436–444. doi:10.1038/nature14539 PMID:26017442

Lee, C. R., & Benfield, P. (1994). Aniracetam. *Drugs & Aging*, *4*(3), 257–273. doi:10.2165/00002512-199404030-00007 PMID:8199398

Lee, K. Y., Park, J. M., Kim, T. Y., Yun, H., & Lee, S. Y. (2010). The genome-scale metabolic network analysis of Zymomonas mobilis ZM4 explains physiological features and suggests ethanol and succinic acid production strategies. *Microbial Cell Factories*, *9*(1), 94. doi:10.1186/1475-2859-9-94 PMID:21092328

Lee, S. K., Chou, H., Ham, T. S., Lee, T. S., & Keasling, J. D. (2008). Metabolic engineering of microorganisms for biofuels production: From bugs to synthetic biology to fuels. *Current Opinion in Biotechnology*, *19*(6), 556–563. doi:10.1016/j.copbio.2008.10.014 PMID:18996194

Lee, U., Han, J., & Wang, M. (2017). Evaluation of landfill gas emissions from municipal solid waste landfills for the life-cycle analysis of waste-to-energy pathways. *Journal of Cleaner Production*, *166*, 335–342. doi:10.1016/j.jclepro.2017.08.016

Lei, L., Jiang, W., & Lu, Y. (2018). A Modified Gibson Assembly Method for Cloning Large DNA Fragments with High GC Contents. In Methods in molecular biology (pp. 203–209). doi:10.1007/978-1-4939-7295-1_13

Leisle, L., Valiyaveetil, F., Mehl, R. A., & Ahern, C. A. (2015). Incorporation of non-canonical amino acids. *Novel Chemical Tools to Study Ion Channel Biology*, 119-151.

Lei, Y., Zhang, S., Fang, L., Akash, M. S., Shi, W., Sun, K., Xu, Y., & Chen, S. (2013). A sensitive and specific enzyme immunoassay for detecting tartrazine in human urinary samples. *Analytical Methods*, *5*(4), 925–930. doi:10.1039/c2ay26121a

Leonard, P. J., & Miller, W. L. (1987). Cloning and sequence of the human gene for P450c17 (steroid 17α-hydroxylase/17,20lyase): Similarity with the gene for P450. *DNA (Mary Ann Liebert, Inc.)*, *6*(5), 439–448. doi:10.1089/dna.1987.6.439 PMID:3500022

Lewis, J. E., & Kemp, M. L. (2021). Integration of machine learning and genome-scale metabolic modeling identifies multi-omics biomarkers for radiation resistance. *Nature Communications*, *12*(1), 1–14. doi:10.103841467-021-22989-1 PMID:33976213

Le, X. T., Pham, H. T., Do, P. T., Fujiwara, H., Tanaka, K., Li, F., Van Nguyen, T., Nguyen, K. M., & Matsumoto, K. (2013). *Bacopa monnieri* ameliorates memory deficits in olfactory bulbectomized mice: Possible involvement of glutamatergic and cholinergic systems. *Neurochemical Research*, *38*(10), 2201–2215. doi:10.100711064-013-1129-6 PMID:23949198

Li, D., Yin, H., & Zheng, B. (2018). An Annual Load Forecasting Model Based on Generalized Regression Neural Network With Multi-Swarm Fruit Fly Optimization Algorithm. *Dianwang Jishu/Power System Technology*, *42*(2), 585–590. doi:10.13335/j.1000-3673.pst.2017.1403

Liang, J., Luo, Y., & Zhao, H. (2011). Synthetic biology: Putting synthesis into biology. *Wiley Interdisciplinary Reviews. Systems Biology and Medicine*, *3*(1), 7–20. doi:10.1002/wsbm.104 PMID:21064036

Liang, Y. (2020, January). Deep Learning Algorithm Classifies Heartbeat Events Based On Electrocardiogram Signals. *Frontiers*, *1*. Www.frontiersin.org PMID:33117191

Liang, Y., Cui, Y., Trushenski, J., & Blackburn, J. W. (2010). Converting crude glycerol derived from yellow grease to lipids through yeast fermentation. *Bioresource Technology, 101*(19), 7581–7586. doi:10.1016/j.biortech.2010.04.061 PMID:20478702

Li, C. T., Yelsky, J., Chen, Y., Zuñiga, C., Eng, R., Jiang, L., Shapiro, A., Huang, K. W., Zengler, K., & Betenbaugh, M. J. (2019). Utilizing genome-scale models to optimize nutrient supply for sustained algal growth and lipid productivity. *NPJ Systems Biology and Applications, 5*(1), 33. Advance online publication. doi:10.103841540-019-0110-7 PMID:31583115

Licona-Cassani, C., Marcellin, E., Quek, L. E., Jacob, S., & Nielsen, L. K. (2012). Reconstruction of the Saccharopolyspora erythraea genome-scale model and its use for enhancing erythromycin production. *Antonie van Leeuwenhoek. Antonie van Leeuwenhoek, 102*(3), 493–502. doi:10.100710482-012-9783-2 PMID:22847261

Li, F., Wang, L., Liu, C., Wu, D., & Song, H. (2018). Engineering exoelectrogens by synthetic biology strategies. *Current Opinion in Electrochemistry, 10*, 37–45. doi:10.1016/j.coelec.2018.03.030

Li, H., & d'Anjou, M. (2009). Pharmacological significance of glycosylation in therapeutic proteins. *Current Opinion in Biotechnology, 20*(6), 678–684. doi:10.1016/j.copbio.2009.10.009 PMID:19892545

Li, J., Castano, I., Liu, W.-Q., Oza, J. P., & Black, M. W., Jew, K., Smith, P. E., So, B., Kasman, J., & Oza, J. P. (2023). Cell-Free. *Synthetic Biology, II*, 87.

Li, J., Kwon, Y.-C., Lu, Y., & Moore, S. J. (2021). Cell-Free Synthetic Biology. *Frontiers in Bioengineering and Biotechnology, 9*, 799122. doi:10.3389/fbioe.2021.799122 PMID:34912793

Li, K., Pan, W., Li, Y., Jiang, Q., & Liu, G. (2018a). A method to detect sleep apnea based on deep neural network and hidden Markov model using single-lead ECG signal. *Neurocomputing, 294*, 94–101. doi:10.1016/j.neucom.2018.03.011

Li, L., Feng, Q., Ye, M., He, Y., Yao, A., & Shi, K. (2017). Metabolic effect of obesity on polycystic ovary syndrome in adolescents: A meta-analysis. *Journal of Obstetrics & Gynaecology, 37*(8), 1036–1047. doi:10.1080/01443615.2017.1318840 PMID:28657375

Limam, M., & Precioso, F. (2017). Atrial fibrillation detection and ECG classification based on convolutional recurrent neural network. In 2017 computing in cardiology (CINC) (pp. 1–4). IEEE.

Lin, D. M., Koskella, B., & Lin, H. C. (2017). Phage therapy: An alternative to antibiotics in the age of multi-drug resistance. *World Journal of Gastrointestinal Pharmacology and Therapeutics, 8*(3), 162. doi:10.4292/wjgpt.v8.i3.162 PMID:28828194

Lin, G.-M., Warden-Rothman, R., & Voigt, C. A. (2019). Retrosynthetic design of metabolic pathways to chemicals not found in nature. *Current Opinion in Systems Biology, 14*, 82–107. doi:10.1016/j.coisb.2019.04.004

Li, P., Sakuragi, K., & Makino, H. (2019). Extraction techniques in sustainable biofuel production: A concise review. *Fuel Processing Technology, 193*, 295–303. doi:10.1016/j.fuproc.2019.05.009

Li, S., Huang, D., Li, Y., Wen, J., & Jia, X. (2012). Rational improvement of the engineered isobutanol-producing Bacillus subtilis by elementary mode analysis. *Microbial Cell Factories, 11*(1), 1–12. doi:10.1186/1475-2859-11-101 PMID:22862776

Liu, M., & Kim, Y. (2018). Classification of heart diseases based on ECG signals using long short-term memory. In *2018 40th Annual international conference of the IEEE engineering in medicine and biology society (EMBC)* (pp. 2707–2710). IEEE. 10.1109/EMBC.2018.8512761

Liu, H. Y., Hopping, G. C., Vaidyanathan, U., Ronquillo, Y. C., Hoopes, P. C., & Moshirfar, M. (2019a). Polymerase Chain Reaction and Its Application in the Diagnosis of Infectious Keratitis. *Medical Hypothesis, Discovery and Innovation in Ophthalmology*, *8*(3), 152–155. PMID:31598517

Liu, L., Agren, R., Bordel, S., & Nielsen, J. (2010). Use of genome-scale metabolic models for understanding microbial physiology. *FEBS Letters*, *584*(12), 2556–2564. doi:10.1016/j.febslet.2010.04.052 PMID:20420838

Liu, Q., Xie, Y. J., Qu, L. H., Zhang, M. X., & Mo, Z. C. (2019). Dyslipidemia involvement in the development of polycystic ovary syndrome. *Taiwanese Journal of Obstetrics & Gynecology*, *58*(4), 447–453. doi:10.1016/j.tjog.2019.05.003 PMID:31307731

Liu, R., Bassalo, M. C., Zeitoun, R. I., & Gill, R. T. (2015). Genome scale engineering techniques for metabolic engineering. *Metabolic Engineering*, *32*, 143–154. doi:10.1016/j.ymben.2015.09.013 PMID:26453944

Liu, S., Sanchez, D. J., Aliyari, R., Sun, L., & Cheng, G. (2012). Systematic identification of type I and type II interferon-induced antiviral factors. *Proceedings of the National Academy of Sciences of the United States of America*, *109*(11), 4239–4244. doi:10.1073/pnas.1114981109 PMID:22371602

Liu, S., Yu, J. M., Gan, Y. C., Qiu, X. Z., Gao, Z. C., Wang, H., Chen, S.-X., Xiong, Y., Liu, G.-H., Lin, S.-E., McCarthy, A., John, J. V., Wei, D.-X., & Hou, H. H. (2023). Biomimetic natural biomaterials for tissue engineering and regenerative medicine: New biosynthesis methods, recent advances, and emerging applications. *Military Medical Research*, *10*(1), 16. doi:10.118640779-023-00448-w PMID:36978167

Liu, W., Huang, Q., Chang, S., Wang, H., & He, J. (2018). Multiple-feature-branch conolutional neural network for myocardial infarction diagnosis using electrocardiogram. *Biomedical Signal Processing and Control*, *45*, 22–32. doi:10.1016/j.bspc.2018.05.013

Liu, W.-Q., Zhang, L., Chen, M., & Li, J. (2019b). Cell-free protein synthesis: Recent advances in bacterial extract sources and expanded applications. *Biochemical Engineering Journal*, *141*, 182–189. doi:10.1016/j.bej.2018.10.023

Liu, W., Yuan, J. S., & Stewart, C. N. Jr. (2013). Advanced genetic tools for plant biotechnology. *Nature Reviews. Genetics*, *14*(11), 781–793. doi:10.1038/nrg3583 PMID:24105275

Li, Y., Pang, Y., Wang, J., & Li, X. (2018b). Patient-specific ECG classification by deeper CNN from generic to dedicated. *Neurocomputing*, *314*, 336–346. doi:10.1016/j.neucom.2018.06.068

Li, Y., Xing, B., Ding, Y., Han, X., & Wang, S. (2020). A critical review of the production and advanced utilization of biochar via selective pyrolysis of lignocellulosic biomass. *Bioresource Technology*, *312*, 123614. doi:10.1016/j.biortech.2020.123614 PMID:32517889

Li, Z., Li, Y., Lin, X., Cui, Y., Wang, T., Dong, J., & Lu, Y. (2022). Supramolecular protein assembly in cell-free protein synthesis system. *Bioresources and Bioprocessing*, *9*(1), 28. doi:10.118640643-022-00520-8

Li, Z., Zhang, X., Müller, H., & Zhang, S. (2018). Large-scale retrieval for medical image analytics: A comprehensive review. *Medical Image Analysis*, *43*, 66–84. doi:10.1016/j.media.2017.09.007 PMID:29031831

Loira, N., Dulermo, T., Nicaud, J. M., & Sherman, D. J. (2012). A genome-scale metabolic model of the lipid-accumulating yeast Yarrowia lipolytica. *BMC Systems Biology*, *6*(1), 1–9. doi:10.1186/1752-0509-6-35 PMID:22558935

Loira, N., Mendoza, S., Paz Cortés, M., Rojas, N., Travisany, D., Di Genova, A., Gajardo, N., Ehrenfeld, N., & Maass, A. (2017). Reconstruction of the microalga Nannochloropsis salina genome-scale metabolic model with applications to lipid production. *BMC Systems Biology*, *11*(1), 1–17. doi:10.118612918-017-0441-1 PMID:28676050

Lomedico, P. T. (1982). Use of recombinant DNA technology to program eukaryotic cells to synthesize rat proinsulin: A rapid expression assay for cloned genes. *Proceedings of the National Academy of Sciences of the United States of America, 79*(19), 5798–5802. doi:10.1073/pnas.79.19.5798 PMID:6764530

Londei, P., Altamura, S., Cammarano, P., & Petrucci, L. (1986). Differential features of ribosomes and of poly(U)-programmed cell-free systems derived from sulphur-dependent archaebacterial species. *European Journal of Biochemistry, 157*(3), 455–462. doi:10.1111/j.1432-1033.1986.tb09689.x PMID:3087750

Long, M. R., Ong, W. K., & Reed, J. L. (2015). Computational methods in metabolic engineering for strain design. *Current Opinion in Biotechnology, 34*, 135–141. doi:10.1016/j.copbio.2014.12.019 PMID:25576846

Loni, M., Sinaei, S., Zoljodi, A., Daneshtalab, M., & Sjödin, M. (2020). Deepmaker: A multi-objective optimization framework for deep neural networks in embedded systems. *Microprocessors and Microsystems, 73*, 102989. doi:10.1016/j.micpro.2020.102989

Loni, M., Zoljodi, A., Sinaei, S., Daneshtalab, M., & Sjödin, M. (2019). Neu- ropower: Designing energy efficient convolutional neural network architecture for embedded systems. In *International conference on artificial neural networks* (pp. 208–222). Springer.

Lopes, H., & Rocha, I. (2017). Genome-scale modeling of yeast: Chronology, applications and critical perspectives. *FEMS Yeast Research, 17*(5), 1–14. doi:10.1093/femsyr/fox050 PMID:28899034

Lu, J., Wu, Y., Deng, C., Liu, Y., Lv, X., Li, J., Du, G., & Liu, L. (2022). Model-based dynamic engineering of Escherichia coli for N-acetylglucosamine overproduction. *Biotechnology Notes, 3*, 15–24. doi:10.1016/j.biotno.2022.02.001

Lui, H. W., & Chow, K. L. (2018). Multiclass classification of myocardial infarction with convolutional and recurrent neural networks for portable ECG devices. *Informatics in Medicine Unlocked, 13*, 26–33. doi:10.1016/j.imu.2018.08.002

Lu, J., Bi, X., Liu, Y., Lv, X., Li, J., Du, G., & Liu, L. (2022). In silico cell factory design driven by comprehensive genome-scale metabolic models: Development and challenges. *Systems Microbiology and Biomanufacturing, 3*(2), 207–222. doi:10.100743393-022-00117-4

Luk, F. C., Johnson, T. M., & Beckers, C. J. J. (2008). N-linked glycosylation of proteins in the protozoan parasite Toxoplasma gondii. *Molecular and Biochemical Parasitology, 157*(2), 169–178. doi:10.1016/j.molbiopara.2007.10.012 PMID:18096254

Luo, L., Dong, L., Yan, Q., Cao,S., Wen, X., Huang, Y., Huang, X., Wu, R., & Ping Ma, X. (2014). *Research progress in applying proteomics technology to explore early diagnosis biomarkers of breast cancer, lung cancer and ovarian cancer*. Academic Press.

Luo, K., Li, J., Wang, Z., & Cuschieri, A. (2017). Patient-specific deep architectural model for ECGclassification. *Journal of Healthcare Engineering*.

Luo, Y., Zhang, L., Barton, K. W., & Zhao, H. (2015). Systematic identification of a panel of strong constitutive promoters from Streptomyces albus. *ACS Synthetic Biology, 4*(9), 1001–1010. doi:10.1021/acssynbio.5b00016 PMID:25924180

Lu, T. K., & Collins, J. J. (2007). Dispersing biofilms with engineered enzymatic bacteriophage. *Proceedings of the National Academy of Sciences of the United States of America, 104*(27), 11197–11202. doi:10.1073/pnas.0704624104 PMID:17592147

Lu, Y. (2020). *Cell-Free Synthetic Biology*. Springer. doi:10.1007/978-981-13-1171-0

Lu, Y. (2022). Textile-embedded cell-free biosensors. *Nature Biomedical Engineering, 6*(3), 225–226. doi:10.103841551-022-00869-3 PMID:35314799

Madan, P., Singh, V., Singh, D. P., Diwakar, M., Pant, B., & Kishor, A. (2022, April 2). A *Hybrid Deep Learning Approach for ECG-Based Arrhythmia Classification.* MDPI. doi:10.3390/bioengineering9040152

Magnusson. (2010). Selective vulnerabilities of N-methyl-D-aspartate (NMDA) receptors during brain aging. *Frontiers in Aging Neuroscience, 12*(3), 8-14.

Magoffin, D. A. (2006). Ovarian enzyme activities in women with polycystic ovary syndrome. *Fertility and Sterility, 86*(1), 9–11. doi:10.1016/j.fertnstert.2006.03.015 PMID:16798289

Maguluri, L. P., Ananth, J., Hariram, S., Geetha, C., Bhaskar, A., & Boopathi, S. (2023). Smart Vehicle-Emissions Monitoring System Using Internet of Things (IoT). In Handbook of Research on Safe Disposal Methods of Municipal Solid Wastes for a Sustainable Environment (pp. 191–211). IGI Global.

Mahadevan, R., & Schilling, C. H. (2003). The effects of alternate optimal solutions in constraint-based genome-scale metabolic models. *Metabolic Engineering, 5*(4), 264–276. doi:10.1016/j.ymben.2003.09.002 PMID:14642354

Mahady, G. B., Gyllenhaal, C., Fong, H. H. S., & Farnsworth, N. R. (2000). Ginsengs: A Review of Safety and Efficacy. *Nutrition in Clinical Care, 3*(2), 90–101. doi:10.1046/j.1523-5408.2000.00020.x

Maheswari, B. U., Imambi, S. S., Hasan, D., Meenakshi, S., Pratheep, V., & Boopathi, S. (2023). Internet of Things and Machine Learning-Integrated Smart Robotics. In Global Perspectives on Robotics and Autonomous Systems: Development and Applications (pp. 240–258). IGI Global. doi:10.4018/978-1-6684-7791-5.ch010

Mahla, S. K., Ardebili, S. M. S., Sharma, H., Dhir, A., Goga, G., & Solmaz, H. (2021). Determination and utilization of optimal diesel/n-butanol/biogas derivation for small utility dual fuel diesel engine. *Fuel, 289*, 119913. doi:10.1016/j.fuel.2020.119913

Maier, A., Syben, C., Lasser, T., & Riess, C. (2019). A gentle introduction to deep learning in medical image processing. *Zeitschrift für Medizinische Physik, 29*(2), 86–101. doi:10.1016/j.zemedi.2018.12.003 PMID:30686613

Majumdar, A., & Ward, R. (2017). Robust greedy deep dictionary learning for ECGar- rhythmia classification. In *2017 International joint conference on neural networks (IJCNN)* (pp. 4 400–4 407). IEEE.

Majumder, S. (n.d.). *Engineering Platforms for Mimicking Cellular Processes using Cell-Free Expression Systems* [Doctoral dissertation].

Maknickas, V., & Maknickas, A. (2017). Atrial fibrillation classification using QRS complex features and LSTM. In 2017 computing in cardiology (CINC) (pp. 1–4). IEEE. doi:10.22489/CinC.2017.350-114

Makrides, S. C. (1996). Strategies for achieving high-level expression of genes in Escherichia coli. *Microbiological Reviews, 60*(3), 512–538. doi:10.1128/mr.60.3.512-538.1996 PMID:8840785

Mali, H., Shah, C., Patel, D. H., Trivedi, U., & Subramanian, R. B. (2022). Biocatalytic system of metallohydrolases for remediation of neurotoxin organophosphates and applications with a future vision. *Journal of Inorganic Biochemistry, 231*, 111771. doi:10.1016/j.jinorgbio.2022.111771 PMID:35278760

Malik, R., Sangwan, A., Saihgal, R., Paul Jindal, D., & Piplani, P. (2007). Towards better brain management: Nootropics. *Current Medicinal Chemistry, 14*(2), 123–131. doi:10.2174/092986707779313408 PMID:17266573

Malik-Sheriff, R. S., Glont, M., Nguyen, T. V. N., Tiwari, K., Roberts, M. G., Xavier, A., Vu, M. T., Men, J., Maire, M., Kananathan, S., Fairbanks, E. L., Meyer, J. P., Arankalle, C., Varusai, T. M., Knight-Schrijver, V., Li, L., Dueñas-Roca, C., Dass, G., Keating, S. M., ... Hermjakob, H. (2020). BioModels-15 years of sharing computational models in life science. *Nucleic Acids Research, 48*(D1), D407–D415. doi:10.1093/nar/gkz1055 PMID:31701150

Manichaikul, A., Ghamsari, L., Hom, E. F. Y., Lin, C., Murray, R. R., Chang, R. L., Balaji, S., Hao, T., Shen, Y., Chavali, A. K., Thiele, I., Yang, X., Fan, C., Mello, E., Hill, D. E., Vidal, M., Salehi-Ashtiani, K., & Papin, J. A. (2009). Metabolic network analysis integrated with transcript verification for sequenced genomes. *Nature Methods, 6*(8), 589–592. doi:10.1038/nmeth.1348

Manu, M., Soni, T., & Prabhakar, P.K. (2021). Pathophysiology of polycystic ovarian Syndrome. *Intech Open.*

Marcer, D., & Hopkins, S. A. N. D. R. A. M. (1977). The differential effects of Meclofenoxate on memory loss in the elderly. *Age and Ageing*, *6*(2), 123–131. doi:10.1093/ageing/6.2.123 PMID:329662

Marques, L. L., Klein, T., & de Mello, J. C. (2019). Guarana. *Nonvitamin and Nonmineral Nutritional*, (Supplements), 283–288.

Marques, R., Lacerda, R., & Romão, L. (2022). Internal Ribosome Entry Site (IRES)-Mediated Translation and Its Potential for Novel mRNA-Based Therapy Development. *Biomedicines*, *10*(8), 1865. Advance online publication. doi:10.3390/biomedicines10081865 PMID:36009412

Marshall, R., Maxwell, C. S., Collins, S. P., Jacobsen, T., Luo, M. L., Begemann, M. B., Gray, B. N., January, E., Singer, A., He, Y., Beisel, C. L., & Noireaux, V. (2018). Rapid and scalable characterization of CRISPR technologies using an *E-coli* cell-free transcription-translation system. *Molecular Cell*, *69*(1), 146–157. doi:10.1016/j.molcel.2017.12.007 PMID:29304331

Martin-Pascual, M., Batianis, C., Bruinsma, L., Asin-Garcia, E., Garcia-Morales, L., Weusthuis, R. A., van Kranenburg, R., & Dos Santos, V. A. M. (2021). A navigation guide of synthetic biology tools for Pseudomonas putida. *Biotechnology Advances*, *49*, 107732. doi:10.1016/j.biotechadv.2021.107732 PMID:33785373

Masteikova, R., Muselik, J., Bernatonienė, J., & Bernatonienė, R. (2007). Antioxidative activity of *ginkgo*, echinacea, and ginseng tinctures. *Medicina*, *43*(4), 306. doi:10.3390/medicina43040038 PMID:17485958

Mathews, S. M., Kambhamettu, C., & Barner, K. E. (2018). A novel application of deep learning for single-lead ECG classification. *Computers in Biology and Medicine*, *99*, 53–62. doi:10.1016/j.compbiomed.2018.05.013 PMID:29886261

Mcpherson, J. D., Marra, M. A., Hillier, L. D. W., Waterston, R. H., Chinwalla, A. T., Wallis, J. W., Sekhon, M., Wylie, K. M., Mardis, E. R., Wilson, R. K., Fulton, R. S., Kucaba, T. A., Wagner-McPherson, C., Barbazuk, W. B., Gregory, S., Humphray, S., French, L., Evans, R. S., Bethel, G., ... Asakawa, S. (2001). A physical map of the human genome. *Nature*, *409*(6822), 934–941. doi:10.1038/35057157 PMID:11237014

Meadows, A. L., Hawkins, K. M., Tsegaye, Y., Antipov, E., Kim, Y., Raetz, L., Dahl, R. H., Tai, A., Mahatdejkul-Meadows, T., Xu, L., Zhao, L., Dasika, M. S., Murarka, A., Lenihan, J., Eng, D., Leng, J. S., Liu, C. L., Wenger, J. W., Jiang, H., ... Tsong, A. E. (2016). Rewriting yeast central carbon metabolism for industrial isoprenoid production. *Nature*, *537*(7622), 694–697. doi:10.1038/nature19769 PMID:27654918

Medlock, G. L., & Papin, J. A. (2020). Guiding the Refinement of Biochemical Knowledgebases with Ensembles of Metabolic Networks and Machine Learning. *Cell Systems*, *10*(1), 109–119.e3. doi:10.1016/j.cels.2019.11.006 PMID:31926940

Megchelenbrink, W., Huynen, M., & Marchiori, E. (2014). optGpSampler: An Improved Tool for Uniformly Sampling the Solution-Space of Genome-Scale Metabolic Networks. *PLoS One*, *9*(2), e86587. doi:10.1371/journal.pone.0086587 PMID:24551039

Meher, S. K., Das, B., Panda, P., Bhuyan, G. C., & Rao, M. M. (2016). Uses of *withania somnifera* (linn) Dunal (ashwagandha) in ayurveda and its pharmacological evidences. *Research Journal of Pharmacology and Pharmacodynamics*, *8*(1), 20–23. doi:10.5958/2321-5836.2016.00006.9

Mehrotra, P. (2016). Biosensors and their applications - A review. *Journal of Oral Biology and Craniofacial Research*, *6*(2), 153–159. doi:10.1016/j.jobcr.2015.12.002 PMID:27195214

Mei, Q., Khnouf, R., Simon, A., & Fan, Z. H. (2010). Protein synthesis in a device with nanoporous membranes and microchannels. *Lab on a Chip*, *10*(19), 2541–2545. doi:10.1039/c005233g PMID:20730191

Mellit, A., & Kalogirou, S. A. (2011). ANFIS-based modelling for photovoltaic power supply system: A case study. *Renewable Energy*, *36*(1), 250–258. doi:10.1016/j.renene.2010.06.028

Mendoza, S. N., Olivier, B. G., Molenaar, D., & Teusink, B. (2019). A systematic assessment of current genome-scale metabolic reconstruction tools. *Genome Biology*, *20*(1), 1–20. doi:10.118613059-019-1769-1 PMID:31391098

Michael, P., Inger, B. O., Jon, M. B., & Henrik, B. P. (2022). *Modulation of Barely (Hordeum vulgare L.)* Grain Protein Sink-Source Relations Towards Human Epidermal Growth Factor Instead of B-hordein Storage Protein. Springer Nature.

Michel, C., Jean, M., Coulon, S., Dictor, M. C., Delorme, F., Morin, D., & Garrido, F. (2007). Biofilms of As (III)-oxidizing bacteria: Formation and activity studies for bioremediation process development. *Applied Microbiology and Biotechnology*, *77*(2), 457–467. doi:10.100700253-007-1169-4 PMID:17846760

Mihaela-O, D., Catalin, L., Lisa, P., Andre, V. E., Jihong, L. C., Catalin, T., Luliana, C., Sonya, C., Adrian, O., Vlad, T., Alexandru, B., Stephan, U., Crina, S., & Norica, B.-N. (2018). Oral administration of a chimeric Hepatitis B Virus S/preS1 antigen produced in lettuce triggers infection-neutralizing antibodies in mice. *Vaccine*, *36*(38), 5789–5795. Advance online publication. doi:10.1016/j.vaccine.2018.07.072 PMID:30082163

Miller, W. L. (1988). Molecular biology of steroid hormone synthesis. *Endocrine Reviews*, *9*(3), 295–318. doi:10.1210/edrv-9-3-295 PMID:3061784

Minxiang. (n.d.). Feature selection and extraction in sequence labeling for arrhythmia detection. *Fourth International Balkan Conference on Communications and Networking*.

Mitragotri, S., Burke, P. A., & Langer, R. (2014, September). Overcoming the challenges in administering biopharmaceuticals: Formulation and delivery strategies. *Nature Reviews. Drug Discovery*, *13*(9), 655–672. doi:10.1038/nrd4363 PMID:25103255

Moen, E., Bannon, D., Kudo, T., Graf, W., Covert, M., & Van Valen, D. (2019). Deep learning for cellular image analysis. *Nature Methods*, *16*(12), 1233–1246. doi:10.103841592-019-0403-1 PMID:31133758

Moerk, K. (2000). The development of major depressive episodes during the course of dysthymic and Episodic Major Depressive Disorders: A retrospective examination of life events. *Journal of Affective Disorders*, *58*(2), 117–123. doi:10.1016/S0165-0327(99)00102-0 PMID:10781701

Mohammad Bilal, M. Z. M. T. A. B. K. B. & O. K. (2022). Analysis and Validation of an Intelligent Energy Monitoring Framework for a Yacht Powered System Based on Adaptive Neuro-Fuzzy Inference System (Anfis). *International Journal of Robotics Research and Development*, *12*(1), 1–10.

Mohammadinejad, R., Shavandi, A., Raie, D. S., Sangeetha, J., Soleimani, M., Hajibehzad, S. S., Thangadurai, D., Hospet, R., Popoola, J. O., Arzani, A., & Gómez-Lim, M. A. (2019). Plant molecular farming: Production of metallic nanoparticles and therapeutic proteins using green factories. *Green Chemistry*, *21*(8), 1845–1865. doi:10.1039/C9GC00335E

Mohammadi, R., Fallah-Mehrabadi, J., Bidkhori, G., Zahiri, J., Javad Niroomand, M., & Masoudi-Nejad, A. (2016). A systems biology approach to reconcile metabolic network models with application to: Synechocystis sp. PCC 6803 for biofuel production. *Molecular BioSystems*, *12*(8), 2552–2561. doi:10.1039/C6MB00119J PMID:27265370

Mohanty, A., Venkateswaran, N., Ranjit, P. S., Tripathi, M. A., & Boopathi, S. (2023). Innovative Strategy for Profitable Automobile Industries: Working Capital Management. In Handbook of Research on Designing Sustainable Supply Chains to Achieve a Circular Economy (pp. 412–428). IGI Global.

Mohanty, A., Venkateswaran, N., Ranjit, P., Tripathi, M. A., & Boopathi, S. (2023). Innovative Strategy for Profitable Automobile Industries: Working Capital Management. In Handbook of Research on Designing Sustainable Supply Chains to Achieve a Circular Economy (pp. 412–428). IGI Global.

Mohanty, A., Jothi, B., Jeyasudha, J., Ranjit, P., Isaac, J. S., & Boopathi, S. (2023). Additive Manufacturing Using Robotic Programming. In *AI-Enabled Social Robotics in Human Care Services* (pp. 259–282). IGI Global. doi:10.4018/978-1-6684-8171-4.ch010

Mo, M. L., Palsson, B., & Herrgård, M. J. (2009). Connecting extracellular metabolomic measurements to intracellular flux states in yeast. *BMC Systems Biology*, *3*(1), 1–17. doi:10.1186/1752-0509-3-37 PMID:19321003

Monlau, F., Suarez-Alvarez, S., Lallement, A., Vaca-Medina, G., Giacinti, G., Munarriz, M., Urreta, I., Raynaud, C., Ferrer, C., & Castañón, S. (2021). A cascade biorefinery for the valorization of microalgal biomass: Biodiesel, biogas, fertilizers and high valuable compounds. *Algal Research*, *59*, 102433. doi:10.1016/j.algal.2021.102433

Moreno-Paz, S., Schmitz, J., Martins dos Santos, V. A. P., & Suarez-Diez, M. (2022). Enzyme-constrained models predict the dynamics of Saccharomyces cerevisiae growth in continuous, batch and fed-batch bioreactors. *Microbial Biotechnology*, *15*(5), 1434–1445. doi:10.1111/1751-7915.13995 PMID:35048533

Moriguchi, S., Tanaka, T., Narahashi, T., & Fukunaga, K. (2013). Novel nootropic drug Sunifiram enhances hippocampal synaptic efficacy via glycine-binding site of *n*-methyl-d-aspartate receptor. *Hippocampus*, *23*(10), 942–951. doi:10.1002/hipo.22150 PMID:23733502

Morreale, H. E., Pazos, F., Potau, N., Garcia-Robles, R., Sancho, J. M., & Varela, C. (1994). Ovarian suppression with triptorelin and adrenal stimulation with adrenocorticotropin in functional hyperadrogenism: Role of adrenal and ovarian cytochrome P450c17 alpha. *Fertility and Sterility*, *62*(3), 521–530. doi:10.1016/S0015-0282(16)56940-4 PMID:8062947

Mrid, R. B., Bouchmaa, N., Ainani, H., Fatimy, R. E., Malka, G., & Mazini, L. (2022). Anti-rheumatoid drugs advancements: New insights into the molecular treatment of rheumatoid arthritis. *Biomedicine and Pharmacotherapy*, *151*, 113126. doi:10.1016/j.biopha.2022.113126 PMID:35643074

Mudeppa, D. G., Pang, C. K. T., Tsuboi, T., Endo, Y., Buckner, F. S., Varani, G., & Rathod, P. K. (2007). Cell-free production of functional Plasmodium falciparum dihydrofolate reductase-thymi-dylate synthase. *Molecular and Biochemical Parasitology*, *151*(2), 216–219. doi:10.1016/j.molbiopara.2006.10.016 PMID:17157934

Mukherjee, A., Bisht, B., Dutta, S., & Paul, M. K. (2022, April 4). Current advances in the use of exosomes, liposomes, and bioengineered hybrid nanovesicles in cancer detection and therapy. *Acta Pharmacologica Sinica*, *43*(11), 1–8. doi:10.103841401-022-00902-w PMID:35379933

Mukherjee, S., Joardar, N., Sengupta, S., & Babu, S. P. (2018, November 1). Gut microbes as future therapeutics in treating inflammatory and infectious diseases: Lessons from recent findings. *The Journal of Nutritional Biochemistry*, *61*, 111–128. doi:10.1016/j.jnutbio.2018.07.010 PMID:30196243

Müller, W. E., Koch, S., Scheuer, K., Rostock, A., & Bartsch, R. (1997). Effects of piracetam on membrane fluidity in the aged mouse, rat, and human brain. *Biochemical Pharmacology*, *53*(2), 135–140. doi:10.1016/S0006-2952(96)00463-7 PMID:9037245

Murat, F., Yildirim, O., Talo, M., Baloglu, U. B., Demir, Y., & Acharya, U. R. (2020). Application of deep learning techniques for heartbeats detection using ECG signals-analysis and review. *Computers in Biology and Medicine*, *120*, 103726. doi:10.1016/j.compbiomed.2020.103726 PMID:32421643

Myhrvold, C., Freije, C. A., Gootenberg, J. S., Abudayyeh, O. O., Metsky, H. C., Durbin, A. F., Kellner, M. J., Tan, A. L., Paul, L. M., Parham, L. A., Garcia, K. F., Barnes, K. G., Chak, B., Mondini, A., Nogueira, M. L., Isern, S., Michael, S. F., Lorenzana, I., Yozwiak, N. L., ... Sabeti, P. C. (2018). Field-deployable viral diagnostics using CRISPR-Cas13. *Science*, *360*(6387), 444–448. doi:10.1126cience.aas8836 PMID:29700266

Myilsamy, S., & Sampath, B. (2017). Grey Relational Optimization of Powder Mixed Near-Dry Wire Cut Electrical Discharge Machining of Inconel 718 Alloy. *Asian Journal of Research in Social Sciences and Humanities*, *7*(3), 18–25. doi:10.5958/2249-7315.2017.00157.5

Nachbar, F., Korting, H. C., & Vogl, T. (1993). Erythema multiforme-like eruption in association with severe headache following pyritinol. *Dermatology (Basel, Switzerland)*, *187*(1), 42–46. doi:10.1159/000247196 PMID:8324277

Nadif, M., & Role, F. (2021). Unsupervised and self-supervised deep learning approaches for biomedical text mining. *Briefings in Bioinformatics*, *22*(2), 1592–1603. doi:10.1093/bib/bbab016 PMID:33569575

Nagasawa, H., Kogure, K., Kawashima, K., Ido, T., Itoh, M., & Hatazawa, J. (1990). Effects of co-dergocrine mesylate (Hydergine) in multi-infarct dementia as evaluated by positron emission tomography. *The Tohoku Journal of Experimental Medicine*, *162*(3), 225–233. doi:10.1620/tjem.162.225 PMID:2091324

Naidu, P. S., Singh, A., & Kulkarni, S. K. (2006). Effect of *withania somnifera* root extract on reserpine-induced orofacial dyskinesia and cognitive dysfunction. *Phytotherapy Research*, *20*(2), 140–146. doi:10.1002/ptr.1823 PMID:16444668

Napoletano, F., Schifano, F., Corkery, J. M., Guirguis, A., Arillotta, D., Zangani, C., & Vento, A. (2020). The psychonauts' world of cognitive enhancers. *Frontiers in Psychiatry*, •••, 11. PMID:33024436

Naresh, V., & Lee, N. (2021). A review on biosensors and recent development of nanostructured materials-enabled biosensors. *Sensors (Basel)*, *21*(4), 1109. doi:10.339021041109 PMID:33562639

Nathan, P., Lu, K., Gray, M., & Oliver, C. (2006). The neuropharmacology of *L*-theanine (N-ethyl-L-glutamine). *Journal of Herbal Pharmacotherapy*, *6*(2), 21–30. PMID:17182482

Naushin, M., Saraswat, A., & Abhishek, K. (2022). Deep Learning Approach to Deal with E-Waste. In *Lecture Notes in Electrical Engineering* (Vol. 858, pp. 123–136). Springer. doi:10.1007/978-981-19-0840-8_9

Naveenkumar, R., Iyyappan, J., Pravin, R., Kadry, S., Han, J., Sindhu, R., Awasthi, M. K., Rokhum, S. L., & Baskar, G. (2023). A strategic review on sustainable approaches in municipal solid waste management and energy recovery: Role of artificial intelligence, economic stability and life cycle assessment. *Bioresource Technology*, *379*, 129044. doi:10.1016/j.biortech.2023.129044 PMID:37044151

Nazir, S., Dickson, D. M., & Akram, M. U. (2023). Survey of explainable artificial intelligence techniques for biomedical imaging with deep neural networks. *Computers in Biology and Medicine*, *156*, 106668. doi:10.1016/j.compbiomed.2023.106668 PMID:36863192

Nguyen, B., Neidle, S., & Wilson, W. D. (2009). A role for water molecules in DNA-ligand minor groove recognition. *Accounts of Chemical Research*, *42*(1), 11–21. doi:10.1021/ar800016q PMID:18798655

Nguyen, M. T., Van Nguyen, B., & Kim, K. (2018). Deep feature learning for sudden cardiac arrest detection in automated external defibrillators. *Scientific Reports*, *8*(1), 1–12. doi:10.103841598-018-33424-9 PMID:30464177

Nguyen, P. Q., Huang, X., Collins, D. S., Collins, J. J., & Lu, T. (2023). Harnessing synthetic biology to enhance ocean health. *Trends in Biotechnology*, *41*(7), 860–874. Advance online publication. doi:10.1016/j.tibtech.2022.12.015 PMID:36669947

Nguyen, P. Q., Soenksen, L. R., Donghia, N. M., Angenent-Mari, N. M., de Puig, H., Huang, A., Lee, R., Slomovic, S., Galbersanini, T., Lansberry, G., Sallum, H. M., Zhao, E. M., Niemi, J. B., & Collins, J. J. (2021). Wearable materials with embedded synthetic biology sensors for biomolecule detection. *Nature Biotechnology*, *39*(11), 1366–1374. doi:10.103841587-021-00950-3 PMID:34183860

Nielsen, J., & Keasling, J. D. (2016). Engineering Cellular Metabolism. *Cell*, *164*(6), 1185–1197. doi:10.1016/j.cell.2016.02.004 PMID:26967285

Nirenberg, M. W., & Matthaei, J. H. (1961). The dependence of cell-free protein synthesis in E. coli upon naturally occurring or synthetic polyribonucleotides. *Proceedings of the National Academy of Sciences of the United States of America*, *47*(10), 1588–1602. doi:10.1073/pnas.47.10.1588 PMID:14479932

Nishanth, J., Deshmukh, M. A., Kushwah, R., Kushwaha, K. K., Balaji, S., & Sampath, B. (2023). Particle Swarm Optimization of Hybrid Renewable Energy Systems. In *Intelligent Engineering Applications and Applied Sciences for Sustainability* (pp. 291–308). IGI Global. doi:10.4018/979-8-3693-0044-2.ch016

Nogales, J., Gudmundsson, S., Knight, E. M., Palsson, B. O., & Thiele, I. (2012). Detailing the optimality of photosynthesis in cyanobacteria through systems biology analysis. *Proceedings of the National Academy of Sciences of the United States of America*, *109*(7), 2678–2683. doi:10.1073/pnas.1117907109 PMID:22308420

Nogales, J., Mueller, J., Gudmundsson, S., Canalejo, F. J., Duque, E., Monk, J., Feist, A. M., Ramos, J. L., Niu, W., & Palsson, B. O. (2020). High-quality genome-scale metabolic modeling of *Pseudomonas putida* highlights its broad metabolic capabilities. *Environmental Microbiology*, *22*(1), 255–269. doi:10.1111/1462-2920.14843 PMID:31657101

Noireaux, V., & Libchaber, A. (2004). A vesicle bioreactor as a step toward an artificial cell assembly. *Proceedings of the National Academy of Sciences of the United States of America*, *101*(51), 17669–17674. doi:10.1073/pnas.0408236101 PMID:15591347

Nowak, K., Błażej, P., Wnetrzak, M., Mackiewicz, D., & Mackiewicz, P. (2021). Some theoretical aspects of reprogramming the standard genetic code. *Genetics*, *218*(1), iyab040. doi:10.1093/genetics/iyab040 PMID:33711098

O'Brien, E. J., Monk, J. M., & Palsson, B. O. (2015). Using Genome-scale Models to Predict Biological Capabilities. *Cell*, *161*(5), 971–987. doi:10.1016/j.cell.2015.05.019 PMID:26000478

O'Flaherty, R., Bergin, A., Flampouri, E., Mota, L. M., Obaidi, I., Quigley, A., Xie, Y., & Butler, M. (2020, November 1). Mammalian cell culture for production of recombinant proteins: A review of the critical steps in their biomanufacturing. *Biotechnology Advances*, *43*, 107552. doi:10.1016/j.biotechadv.2020.107552 PMID:32416132

Ohashi, H., Kanamori, T., Shimizu, Y., & Ueda, T. (2010). A highly controllable reconstituted cell-free system—A breakthrough in protein synthesis research. *Current Pharmaceutical Biotechnology*, *11*(3), 267–271. doi:10.2174/138920110791111889 PMID:20210745

Oh, S. L., Ng, E. Y., San Tan, R., & Acharya, U. R. (2018). Automated diagnosis of arrhythmia using combination of cnn and lstm techniques with variable length heart beats. *Computers in Biology and Medicine*, *102*, 278–287. doi:10.1016/j.compbiomed.2018.06.002 PMID:29903630

Oh, Y. K., Palsson, B. O., Park, S. M., Schilling, C. H., & Mahadevan, R. (2007). Genome-scale reconstruction of metabolic network in Bacillus subtilis based on high-throughput phenotyping and gene essentiality data. *The Journal of Biological Chemistry*, *282*(39), 28791–28799. doi:10.1074/jbc.M703759200 PMID:17573341

Olbrich, G. (2016). Energy management and smart cities. *Gis. Business (Atlanta, Ga.)*, *1*(4), 3.

Oliveira, A. G., & Fiorotto, R. (2021). Novel approach to liver disease diagnosis and modelling. *Translational Gastroenterology and Hepatology*, *6*(19), 1–14. PMID:33824923

Ong, H. C., Chen, W.-H., Singh, Y., Gan, Y. Y., Chen, C.-Y., & Show, P. L. (2020). A state-of-the-art review on thermochemical conversion of biomass for biofuel production: A TG-FTIR approach. *Energy Conversion and Management*, *209*, 112634. doi:10.1016/j.enconman.2020.112634

Opgenorth, P., Costello, Z., Okada, T., Goyal, G., Chen, Y., Gin, J., Benites, V., De Raad, M., Northen, T. R., Deng, K., Deutsch, S., Baidoo, E. E. K., Petzold, C. J., Hillson, N. J., Garcia Martin, H., & Beller, H. R. (2019). Lessons from Two Design-Build-Test-Learn Cycles of Dodecanol Production in Escherichia coli Aided by Machine Learning. *ACS Synthetic Biology*, *8*(6), 1337–1351. doi:10.1021/acssynbio.9b00020 PMID:31072100

Orth, J. D., Conrad, T. M., Na, J., Lerman, J. A., Nam, H., Feist, A. M., & Palsson, B. (2011). A comprehensive genome-scale reconstruction of Escherichia coli metabolism—2011. *Molecular Systems Biology*, *7*(1), 535. Advance online publication. doi:10.1038/msb.2011.65 PMID:21988831

Owczarek, B., Gerszberg, A., & Hnatuszko-Konka, K. (2019, January 8). A brief reminder of systems of production and chromatography-based recovery of recombinant protein biopharmaceuticals. *BioMed Research International*, *2019*, 2019. doi:10.1155/2019/4216060 PMID:30729123

Pabo, C. O., & Nekludova, L. (2000). Geometric analysis and comparison of protein-DNA interfaces: Why is there no simple code for recognition? 1. *Journal of Molecular Biology*, *301*(3), 597–624. doi:10.1006/jmbi.2000.3918 PMID:10966773

Padmanaban, S., Priyadarshi, N., Bhaskar, M. S., Holm-Nielsen, J. B., Ramachandaramurthy, V. K., & Hossain, E. (2019). A Hybrid ANFIS-ABC Based MPPT Controller for PV System with Anti-Islanding Grid Protection: Experimental Realization. *IEEE Access : Practical Innovations, Open Solutions*, *7*, 103377–103389. doi:10.1109/ACCESS.2019.2931547

Padmaperuma, G., Butler, T. O., Shuhaili, F. A. A., Almalki, W. J., & Vaidyanathan, S. (2020). Microbial consortia: Concept and application in fruit crop management. In *Fruit crops* (pp. 353–366). Elsevier. doi:10.1016/B978-0-12-818732-6.00025-3

Palaniappan, M., Tirlangi, S., Mohamed, M. J. S., Moorthy, R. M. S., Valeti, S. V., & Boopathi, S. (2023a). Fused deposition modelling of polylactic acid (PLA)-based polymer composites: A case study. In Development, Properties, and Industrial Applications of 3D Printed Polymer Composites (pp. 66–85). IGI Global. doi:10.4018/978-1-6684-6009-2.ch005

Palaniappan, M., Tirlangi, S., Mohamed, M. J. S., Moorthy, R. S., Valeti, S. V., & Boopathi, S. (2023). Fused Deposition Modelling of Polylactic Acid (PLA)-Based Polymer Composites: A Case Study. In Development, Properties, and Industrial Applications of 3D Printed Polymer Composites (pp. 66–85). IGI Global.

Palaniappan, M., Tirlangi, S., Mohamed, M. J. S., Moorthy, R. S., Valeti, S. V., & Boopathi, S. (2023b). Fused Deposition Modelling of Polylactic Acid (PLA)-Based Polymer Composites: A Case Study. In Development, Properties, and Industrial Applications of 3D Printed Polymer Composites (pp. 66–85). IGI Global.

Pan, P., & Hua, Q. (2012). Reconstruction and In Silico Analysis of Metabolic Network for an Oleaginous Yeast, Yarrowia lipolytica. *PLoS One*, *7*(12), e51535. doi:10.1371/journal.pone.0051535 PMID:23236514

Paramasivan, K., Kumar, H. N. P., & Mutturi, S. (2019). Systems-based Saccharomyces cerevisiae strain design for improved squalene synthesis. *Biochemical Engineering Journal*, *148*(April), 37–45. doi:10.1016/j.bej.2019.04.025

Pardee, K. (2018). Perspective: Solidifying the impact of cell-free synthetic biology through lyophilization. *Biochemical Engineering Journal*, *138*, 91–97. doi:10.1016/j.bej.2018.07.008 PMID:30740032

Park, J. H., & Lee, S. Y. (2008). Towards systems metabolic engineering of microorganisms for amino acid production. *Current Opinion in Biotechnology*, *19*(5), 454–460. doi:10.1016/j.copbio.2008.08.007 PMID:18760356

Parvaneh, S., Rubin, J., Babaeizadeh, S., & Xu-Wilson, M. (2019). Cardiac arrhythmia detection using deep learning: A review. *Journal of Electrocardiology*, *57*, S70–S74. doi:10.1016/j.jelectrocard.2019.08.004 PMID:31416598

Paszke, A., Gross, S., Massa, F., Lerer, A., Bradbury, J., Chanan, G., . . . Antiga, L. (2019). Pytorch: An imperative style, high-performance deep learning library. In Advances in neural information processing systems (pp. 8024–8035). Academic Press.

Pater, C. (2005). Methodological considerations in the design of trials for safety as- sessment of new drugs and chemical entities. *Current Controlled Trials in Cardiovascular Medicine*, *6*(1), 1. doi:10.1186/1468-6708-6-1 PMID:15691384

Pathania, R., Srivastava, A., Srivastava, S., & Shukla, P. (2022). Metabolic systems biology and multi-omics of cyanobacteria: Perspectives and future directions. *Bioresource Technology*, *343*, 126007. doi:10.1016/j.biortech.2021.126007

Patil, K. R., Rocha, I., Förster, J., & Nielsen, J. (2005). Evolutionary programming as a platform for in silico metabolic engineering. *BMC Bioinformatics*, *6*(1), 1–12. doi:10.1186/1471-2105-6-308 PMID:16375763

Paul Berg. (2022, March 29). *Science History Institute*. https://www.sciencehistory.org/historical-profile/paul-berg

Paul Divakar, P., Vijaya Lakshmi, G., & Devi, L. (2020). Applications of Internet of things on Smart Grid. *IOP Conference Series. Materials Science and Engineering*, *906*(1), 13–17. doi:10.1088/1757-899X/906/1/012015

Paul, G., Roy, S., Sarkar, S., Hanumaiah, N., & Mitra, S. (2013). Investigations on influence of process variables on crater dimensions in micro-EDM of γ-titanium aluminide alloy in dry and oil dielectric media. *International Journal of Advanced Manufacturing Technology*, *65*(5–8), 1009–1017. doi:10.100700170-012-4235-8

Paulina, N. N., & Albertha, R. (2020, December 6). Immunogenicity of Plant-Produced Human Papillomavirus (HPV) Virus-Like Particles (VLPs). *Vaccines*, *8*(4), 740. Advance online publication. doi:10.3390/vaccines8040740

Paul, O., Jasu, A., Lahiri, D., Nag, M., & Ray, R. R. (2021). in Situ and Ex Situ Bioremediation of Heavy Metals: The Present Scenario. *Journal of Environmental Engineering and Landscape Management*, *29*(4), 454–469. doi:10.3846/jeelm.2021.15447

Pawar, P., TarunKumar, M., & Vittal K, P. (2020). An IoT based Intelligent Smart Energy Management System with accurate forecasting and load strategy for renewable generation. *Measurement*, *152*, 107187. doi:10.1016/j.measurement.2019.107187

Pawar, S., & Kumar, A. (2002). Issues in the formulation of drugs for oral use in children. *Paediatric Drugs*, *4*(6), 371–379. doi:10.2165/00128072-200204060-00004 PMID:12038873

Payal, S. (2016, October 20). *7 Main Stages of Recombinant DNA Technology*. Biology Discussion. https://www.biologydiscussion.com/dna/recombinant-dna-technology/7-main-stages-of-recombinant-dna-technology/56320

Pazos, F., Guijas, D., Valencia, A., & De Lorenzo, V. (2005). MetaRouter: Bioinformatics for bioremediation. *Nucleic Acids Research*, *33*(Database issue, suppl_1), D588–D592. doi:10.1093/nar/gki068 PMID:15608267

Pearcy, N., Hu, Y., Baker, M., Maciel-Guerra, A., Xue, N., Wang, W., Kaler, J., Peng, Z., Li, F., & Dottorini, T. (2021). Genome-Scale Metabolic Models and Machine Learning Reveal Genetic Determinants of Antibiotic Resistance in Escherichia coli and Unravel the Underlying Metabolic Adaptation Mechanisms. *mSystems*, *6*(4), e00913-20. Advance online publication. doi:10.1128/mSystems.00913-20 PMID:34342537

Pere, M. A., Richard,, M. T., Derry,, A., & Pedro,, C. B., Merce, B., Paul, C., & Teresa, C. (2019). A simplified techno-economic model for the molecular pharming of antibodies. *Biotechnology and Bioengineering*. PMID:31209856

Perez, J. G., Stark, J. C., & Jewett, M. C. (2016). Cell-free synthetic biology: Engineering beyond the cell. *Cold Spring Harbor Perspectives in Biology*, *8*(12), a023853. doi:10.1101/cshperspect.a023853 PMID:27742731

Peruzzi, J., Steinkuehler, J., Vu, T., Lu, P., Baker, D., & Kamat, N. (2022). Organizing cell-free expressed membrane proteins in synthetic membranes using lipid-protein interactions. *Biophysical Journal*, *121*(3), 313a. doi:10.1016/j.bpj.2021.11.1194

Peter, C. S., Ben, M. M. D., Barbara, S., Theo, C. V., Peter, J. M. V. D. E., & André, H. (1990). *Production of Correctly Processed Human Serum Albumin in Transgenic Plants*. Academic Press.

Peters, C., & Brown, S. (2015). Antibody–drug conjugates as novel anti-cancer chemotherapeutics. *Bioscience Reports*, *35*(4), e00225. doi:10.1042/BSR20150089 PMID:26182432

Pham, N., Reijnders, M., Suarez-Diez, M., Nijsse, B., Springer, J., Eggink, G., & Schaap, P. J. (2021). Genome-scale metabolic modeling underscores the potential of Cutaneotrichosporon oleaginosus ATCC 20509 as a cell factory for biofuel production. *Biotechnology for Biofuels*, *14*(1), 1–17. doi:10.118613068-020-01838-1 PMID:33407779

Pharkya, P., Burgard, A. P., & Maranas, C. D. (2004). OptStrain: A computational framework for redesign of microbial production systems. *Genome Research*, *14*(11), 2367–2376. doi:10.1101/gr.2872004 PMID:15520298

Plesinger, F., Nejedly, P., Viscor, I., Halamek, J., & Jurak, P. (2017). Automatic detec- tion of atrial fibrillation and other arrhythmias in holter ECG recordings using rhythm features and neural networks. In 2017 Computing in cardiology (CINC) (pp. 1–4). IEEE.

Pogue, G. P., Lindbo, J. A., Garger, S. J., & Fitzmaurice, W. P. (2002, September). Making an ally from an enemy: Plant virology and the new agriculture. *Annual Review of Phytopathology*, *40*(1), 45–74. doi:10.1146/annurev.phyto.40.021102.150133 PMID:12147754

Poh, M.-Z., Poh, Y. C., Chan, P.-H., Wong, C.-K., Pun, L., Leung, W. W.-C., Wong, Y.-F., Wong, M. M.-Y., Chu, D. W.-S., & Siu, C.-W. (2018). Diagnostic assessment of a deep learning system for detecting atrial fibrillation in pulse waveforms. *Heart (British Cardiac Society)*, *104*(23), 1921–1928. doi:10.1136/heartjnl-2018-313147 PMID:29853485

Pontrelli, S., Chiu, T. Y., Lan, E. I., Chen, F. Y. H., Chang, P., & Liao, J. C. (2018). *Escherichia coli* as a host for metabolic engineering. *Metabolic Engineering*, *50*, 16–46. doi:10.1016/j.ymben.2018.04.008 PMID:29689382

Poonam, S., Varsha, S., Anju, N., Joycee, J., Vandana, G., Ajay, R., & Smita, B. (2018). Plantibodies as biopharmaceuticals: A review. *Journal of Pharmacognosy and Phytochemistry*.

Posner, J., Barrington, P., Brier, T., & Datta-Mannan, A. (2019). Monoclonal antibodies: past, present and future. *Concepts and Principles of Pharmacology*, 81-141.

Pourbabaee, B., Roshtkhari, M. J., & Khorasani, K. (2017). Deep convolutional neu- ral networks and learning ECG-features for screening paroxysmal atrial fibrilla- tion patients. *IEEE Transactions on Systems, Man, and Cybernetics. Systems*, *48*(12), 2095–2104. doi:10.1109/TSMC.2017.2705582

Prado, M. S., Bendtzen, K., & Andrade, L. E. (2017, September 2). Biological anti-TNF drugs: Immunogenicity underlying treatment failure and adverse events. *Expert Opinion on Drug Metabolism & Toxicology*, *13*(9), 985–995. doi:10.1080/17425255.2017.1360280 PMID:28772079

Pramila, P., Amudha, S., Saravanan, T., Sankar, S. R., Poongothai, E., & Boopathi, S. (2023). Design and Development of Robots for Medical Assistance: An Architectural Approach. In Contemporary Applications of Data Fusion for Advanced Healthcare Informatics (pp. 260–282). IGI Global.

Prapas, N., Karkanaki, A., Prapas, I., Kalogiannidis, I., Katsikis, I., & Panidis, D. (2009). Genetics of Polycystic ovary syndrome. *Hippokratia*, *13*(4), 216–223. PMID:20011085

Preiss, T., & Hentze, M. W. (1998). Dual function of the messenger RNA cap structure in poly(A)-tail-promoted translation in yeast. *Nature*, *392*(6675), 516–520. doi:10.1038/33192 PMID:9548259

Premkumar, S., & Sigappi, A. N. (2021). Functional framework for edge-based agricultural system. In AI, Edge and IoT-based Smart Agriculture (pp. 71–100). Elsevier. doi:10.1016/B978-0-12-823694-9.00029-3

Presta, L. G. (2008, August 1). Molecular engineering and design of therapeutic antibodies. *Current Opinion in Immunology*, *20*(4), 460–470. doi:10.1016/j.coi.2008.06.012 PMID:18656541

Purnick, P. E., & Weiss, R. (2009). The second wave of synthetic biology: From modules to systems. *Nature Reviews. Molecular Cell Biology*, *10*(6), 410–422. doi:10.1038/nrm2698 PMID:19461664

Pyhtila, B. (2016, March 1). *Plasmids 101: Gibson Assembly and Other Long-Homology Based Cloning Methods.* Addgene. https://blog.addgene.org/plasmids-101-gibson-assembly

Qiao, J., Wang, G., Li, W., & Li, X. (2018). A deep belief network with PLSR for non- linear system modeling. *Neural Networks*, *104*, 68–79. doi:10.1016/j.neunet.2017.10.006 PMID:29729561

Rahamathunnisa, U., Sudhakar, K., Murugan, T. K., Thivaharan, S., Rajkumar, M., & Boopathi, S. (2023). *Cloud Computing Principles for Optimizing Robot Task Offloading Processes.* doi:10.4018/978-1-6684-8171-4.ch007

Ramudu, K., Mohan, V. M., Jyothirmai, D., Prasad, D., Agrawal, R., & Boopathi, S. (2023). Machine Learning and Artificial Intelligence in Disease Prediction: Applications, Challenges, Limitations, Case Studies, and Future Directions. In Contemporary Applications of Data Fusion for Advanced Healthcare Informatics (pp. 297–318). IGI Global.

Ranganathan, S., Suthers, P. F., & Maranas, C. D. (2010). OptForce: An Optimization Procedure for Identifying All Genetic Manipulations Leading to Targeted Overproductions. *PLoS Computational Biology*, *6*(4), e1000744. doi:10.1371/journal.pcbi.1000744 PMID:20419153

Ravisankar, A., Sampath, B., & Asif, M. M. (2023). Economic Studies on Automobile Management: Working Capital and Investment Analysis. In Multidisciplinary Approaches to Organizational Governance During Health Crises (pp. 169–198). IGI Global.

Reddy, M. A., Gaurav, A., Ushasukhanya, S., Rao, V. C. S., Bhattacharya, S., & Boopathi, S. (2023). Bio-Medical Wastes Handling Strategies During the COVID-19 Pandemic. In Multidisciplinary Approaches to Organizational Governance During Health Crises (pp. 90–111). IGI Global. doi:10.4018/978-1-7998-9213-7.ch006

Reddy, K. R., Deepika, M. L. N., Supriya, K., Latha, K. P., Lakshmana Rao, S. S., Usha Rani, V., & Jahan, P. (2014). CYP11A1 microsatellite (tttta)n polymorphism in PCOS women from South India. *Journal of Assisted Reproduction and Genetics*, *31*(7), 857–863. doi:10.100710815-014-0236-x PMID:24793009

Reddy, M. A., Reddy, B. M., Mukund, C., Venneti, K., Preethi, D., & Boopathi, S. (2023). Social Health Protection During the COVID-Pandemic Using IoT. In *The COVID-19 Pandemic and the Digitalization of Diplomacy* (pp. 204–235). IGI Global. doi:10.4018/978-1-7998-8394-4.ch009

Reed, J. L., Vo, T. D., Schilling, C. H., & Palsson, B. O. (2003). An expanded genome-scale model of Escherichia coli K-12 (iJR904 GSM/GPR). *Genome Biology*, *4*(9), 1–12. doi:10.1186/gb-2003-4-9-r54 PMID:12952533

Remer, T., Maser-Gluth, C., & Wudy, S. A. (2008). Glucocorticoid measurements in health and disease-metabolic implications and the potential of 24-h urine analyses. *Mini-Reviews in Medicinal Chemistry*, *8*(2), 153–170. doi:10.2174/138955708783498096 PMID:18289099

Renoux, J. M., & Rohmer, M. (1986). Enzymatic cyclization of all-trans pentaprenyl and hexaprenyl methyl ethers by a cell-free system from the protozoon Tetrahymena pyriformis: The biosynthesis of scalarane and polycyclohexaprenyl derivatives. *European Journal of Biochemistry*, *155*(1), 125–132. doi:10.1111/j.1432-1033.1986.tb09467.x PMID:3081345

Reynders, E., Foulquier, F., Annaert, W., & Matthijs, G. (2011). How Golgi glycosylation meets and needs trafficking: The case of the COG complex. *Glycobiology*, *21*(7), 853–863. doi:10.1093/glycob/cwq179 PMID:21112967

Roberts, M. A. J., Cranenburgh, R. M., Stevens, M. P., & Oyston, P. C. F. (2013). Synthetic biology: Biology by design. *Microbiology*, *159*(Pt 7), 1219–1220. doi:10.1099/mic.0.069724-0 PMID:23744901

Roberts, R. J., Belfort, M., Bestor, T. H., Bhagwat, A. S., Bickle, T. A., Bitinaite, J., Blumenthal, R., Degtyarev, S. K., Dryden, D. T. F., Dybvig, K., Firman, K., Gromova, E. S., Gumport, R. I., Halford, S. E., Hattman, S., Heitman, J., Hornby, D., Janulaitis, A., Jeltsch, A., ... Xu, S. (2003). A nomenclature for restriction enzymes, DNA methyltransferases, homing endonucleases and their genes. *Nucleic Acids Research*, *31*(7), 1805–1812. doi:10.1093/nar/gkg274 PMID:12654995

Roberts, S. B., Gowen, C. M., Brooks, J. P., & Fong, S. S. (2010). Genome-scale metabolic analysis of Clostridium thermocellum for bioethanol production. *BMC Systems Biology*, *4*(1), 31. Advance online publication. doi:10.1186/1752-0509-4-31 PMID:20307315

Robinson, J. L., Kocabaş, P., Wang, H., Cholley, P. E., Cook, D., Nilsson, A., Anton, M., Ferreira, R., Domenzain, I., Billa, V., Limeta, A., Hedin, A., Gustafsson, J., Kerkhoven, E. J., Svensson, L. T., Palsson, B. O., Mardinoglu, A., Hansson, L., Uhlén, M., & Nielsen, J. (2020). An atlas of human metabolism. *Science Signaling*, *13*(624), eaaz1482. Advance online publication. doi:10.1126cisignal.aaz1482 PMID:32209698

Robishaw, J. D., Alter, S. M., Solano, J. J., Shih, R. D., DeMets, D. L., Maki, D. G., & Hennekens, C. H. (2021). Genomic surveillance to combat COVID-19: Challenges and opportunities. *The Lancet Microbe*, *2*(9), e481–e484. doi:10.1016/S2666-5247(21)00121-X PMID:34337584

Rocher Medical Center. (n.d.). *Intercity digital electrocardiogram Alliance' ideal database* (e-hol-03-0202-003). http://thew-project.org/Database/E-HOL-03-0202-003.html

Rodionova, M. V., Bozieva, A. M., Zharmukhamedov, S. K., Leong, Y. K., Lan, J. C.-W., Veziroglu, A., Veziroglu, T. N., Tomo, T., Chang, J.-S., & Allakhverdiev, S. I. (2022). A comprehensive review on lignocellulosic biomass biorefinery for sustainable biofuel production. *International Journal of Hydrogen Energy*, *47*(3), 1481–1498. doi:10.1016/j.ijhydene.2021.10.122

Rodriguez, E. A., Ramirez, D., Balcazar, J. L., & Jimenez, J. N. (2021). Metagenomic analysis of urban waste water resistome and mobilome: A support for antimicrobial resistance surveillance in an endemic country. *Environmental Pollution*, *276*, 116736. doi:10.1016/j.envpol.2021.116736 PMID:33618114

Roldão, A., Cox, M., Alves, P., Carrondo, M., & Vicente, T. (2014). *Industrial large scale of suspension culture of insect cells.* Academic Press.

Rolf, J., Rosenthal, K., & Lütz, S. J. C. (2019). Application of cell-free protein synthesis for faster biocatalyst development. *Catalysts*, *9*(2), 190. doi:10.3390/catal9020190

Rollié, S., Mangold, M., & Sundmacher, K. (2012). Designing biological systems: Systems Engineering meets Synthetic Biology. *Chemical Engineering Science*, *69*(1), 1–29. doi:10.1016/j.ces.2011.10.068

Romero, M., Gallego, D., Blaz, J., Lechuga, A., Martínez, J. F., Barajas, H. R., & Alcaraz, L. D. (2019). Rhizosphere metagenomics of mine tailings colonizing plants: Assembling and selecting synthetic bacterial communities to enhance *in-situ* bioremediation. BioRxiv, 664805. doi:10.1101/664805

Rosenblum, G., & Cooperman, B. S. (2014). Engine out of the chassis: Cell-free protein synthesis and its uses. *FEBS Letters*, *588*(2), 261–268. doi:10.1016/j.febslet.2013.10.016 PMID:24161673

Roy, K., & Kar, S. (2016). In-silico models for ecotoxicity of pharmaceuticals. In *In-silico methods for predicting drug toxicity* (pp. 237–304). Humana Press. doi:10.1007/978-1-4939-3609-0_12

Rubin, B. E., Diamond, S., Cress, B. F., Crits-Christoph, A., He, C., Xu, M., & Doudna, J. A. (2020). Targeted genome editing of bacteria within microbial communities. bioRxiv. doi:10.1101/2020.07.17.209189

Rubin, J., Parvaneh, S., Rahman, A., Conroy, B., & Babaeizadeh, S. (2018). Densely connected convolutional networks for detection of atrial fibrillation from short single-lead ECG recordings. *Journal of Electrocardiology*, *51*(6), S18–S21. doi:10.1016/j.jelectrocard.2018.08.008 PMID:30122456

Ruggero, D., Creti, R., & Londei, P. (1993). In vitro translation of archaeal natural mRNAs at high temperature. *FEMS Microbiology Letters*, *107*(1), 89–94. doi:10.1111/j.1574-6968.1993.tb06009.x

Rugnetta, M. (2020, April 15). Cloning. *Encyclopedia Britannica*. https://www.britannica.com/science/cloning

Sadrawi, M., Lin, C.-H., Lin, Y.-T., Hsieh, Y., Kuo, C.-C., Chien, J. C., Haraikawa, K., Abbod, M., & Shieh, J.-S. (2017). Arrhythmia evaluation in wearable ECG devices. *Sensors (Basel)*, *17*(11), 2445. doi:10.339017112445 PMID:29068369

Sahota, N. K., & Sharma, R. (2020). Bioremediation: Harnessing Natural Forces for Solid Waste Management. Handbook of Solid Waste Management: Sustainability through Circular Economy, 1–32.

Saini, D. K., Rai, A., Devi, A., Pabbi, S., Chhabra, D., Chang, J. S., & Shukla, P. (2021). A multi-objective hybrid machine learning approach-based optimization for enhanced biomass and bioactive phycobiliproteins production in Nostoc sp. CCC-403. *Bioresource Technology*, *329*, 124908. doi:10.1016/j.biortech.2021.124908 PMID:33690058

Saitua, F., Torres, P., Pérez-Correa, J. R., & Agosin, E. (2017). Dynamic genome-scale metabolic modeling of the yeast Pichia pastoris. *BMC Systems Biology*, *11*(1), 1–21. doi:10.118612918-017-0408-2 PMID:28222737

Sajeewani, R. K., & Imalka, T. (2022). Molecular Farming: Implication for Future Pharmaceutical products Sri Lankan. *Journal of Health Sciences*.

Sakarika, M., & Kornaros, M. (2019). Chlorella vulgaris as a green biofuel factory: Comparison between biodiesel, biogas and combustible biomass production. *Bioresource Technology*, *273*, 237–243. doi:10.1016/j.biortech.2018.11.017 PMID:30447625

Saleem, M. U., Usman, M. R., & Shakir, M. (2021). Design, Implementation, and Deployment of an IoT Based Smart Energy Management System. *IEEE Access : Practical Innovations, Open Solutions*, *9*, 59649–59664. doi:10.1109/ACCESS.2021.3070960

Salguero, D. A. M., Fernández-Niño, M., Serrano-Bermúdez, L. M., Melo, D. O. P., Winck, F. V., Caldana, C., & Barrios, A. F. G. (2018). Development of a Chlamydomonas reinhardtii metabolic network dynamic model to describe distinct phenotypes occurring at different CO2 levels. *PeerJ*, *6*(9), e5528. Advance online publication. doi:10.7717/peerj.5528 PMID:30202653

Samikannu, R., Koshariya, A. K., Poornima, E., Ramesh, S., Kumar, A., & Boopathi, S. (2022). Sustainable Development in Modern Aquaponics Cultivation Systems Using IoT Technologies. In *Human Agro-Energy Optimization for Business and Industry* (pp. 105–127). IGI Global.

Sampath, B. (2021). *Sustainable Eco-Friendly Wire-Cut Electrical Discharge Machining: Gas Emission Analysis*. Academic Press.

Sampath, B., Sasikumar, C., & Myilsamy, S. (2023). Application of TOPSIS Optimization Technique in the Micro-Machining Process. In Trends, Paradigms, and Advances in Mechatronics Engineering (pp. 162–187). IGI Global.

Sampath, B., Pandian, M., Deepa, D., & Subbiah, R. (2022). Operating parameters prediction of liquefied petroleum gas refrigerator using simulated annealing algorithm. *AIP Conference Proceedings*, *2460*(1), 070003. doi:10.1063/5.0095601

Sampath, B., Yuvaraj, T., & Velmurugan, D. (2022). Parametric analysis of mould sand properties for flange coupling casting. *AIP Conference Proceedings*, *2460*(1), 070002. doi:10.1063/5.0095599

San León, D., & Nogales, J. (2022). Toward merging bottom–up and top–down model-based designing of synthetic microbial communities. *Current Opinion in Microbiology*, *69*, 102169. doi:10.1016/j.mib.2022.102169 PMID:35763963

Sánchez-Costa, M., & López-Gallego, F. (2023). *Solid-Phase Cell-Free Protein Synthesis and Its Applications in Biotechnology*. Springer. doi:10.1007/10_2023_226

Sankar, K. M., Booba, B., & Boopathi, S. (2023). Smart Agriculture Irrigation Monitoring System Using Internet of Things. In *Contemporary Developments in Agricultural Cyber-Physical Systems* (pp. 105–121). IGI Global. doi:10.4018/978-1-6684-7879-0.ch006

Sannino, G., & De Pietro, G. (2018). A deep learning approach for ECG-based heartbeat classification for arrhythmia detection. *Future Generation Computer Systems*, *86*, 446–455. doi:10.1016/j.future.2018.03.057

Sapundzhi, F. I., & Dzimbova, T. A. (2022). The importance of biological databases in modeling of structure-activity relationship. *Interpretation (Tulsa)*, *7*, 11.

Saratale, R. G., Ponnusamy, V. K., Jeyakumar, R. B., Sirohi, R., Piechota, G., Shobana, S., Dharmaraja, J., Lay, C., Saratale, G. D., Shin, H. S., & ... (2022). Microalgae cultivation strategies using cost–effective nutrient sources: Recent updates and progress towards biofuel production. *Bioresource Technology*, *361*, 127691. doi:10.1016/j.biortech.2022.127691 PMID:35926554

Saravanan, A., Deivayanai, V., Kumar, P. S., Rangasamy, G., & Varjani, S. (2022). CO2 bio-mitigation using genetically modified algae and biofuel production towards a carbon net-zero society. *Bioresource Technology*, *363*, 127982. doi:10.1016/j.biortech.2022.127982 PMID:36126842

Saravanan, A., Kumar, P. S., Jeevanantham, S., Karishma, S., & Vo, D.-V. N. (2022). Recent advances and sustainable development of biofuels production from lignocellulosic biomass. *Bioresource Technology*, *344*, 126203. doi:10.1016/j.biortech.2021.126203 PMID:34710606

Saravanan, A., Venkatasubramanian, R., Khare, R., Surakasi, R., Boopathi, S., Ray, S., & Sudhakar, M. (2022). Policy trends of renewable energy and non. *Renewable Energy*.

Saravanan, M., Vasanth, M., Boopathi, S., Sureshkumar, M., & Haribalaji, V. (2022). Optimization of Quench Polish Quench (QPQ) Coating Process Using Taguchi Method. *Key Engineering Materials*, *935*, 83–91. doi:10.4028/p-z569vy

Sarkar, A., Umar, M. F., Fatima, U., Hossain, K., & Rafatullah, M. (2022). Modification of Bioanodes from Different Nanocomposite Materials for Wastewater Bioremediation through Microbial Fuel Cells. In Biorenewable Nanocomposite Materials, Vol. 2: Desalination and Wastewater Remediation (pp. 113–140). ACS Publications. doi:10.1021/bk-2022-1411.ch005

Sathish, T., Sunagar, P., Singh, V., Boopathi, S., Sathyamurthy, R., Al-Enizi, A. M., Pandit, B., Gupta, M., & Sehgal, S. S. (2023). Characteristics estimation of natural fibre reinforced plastic composites using deep multi-layer perceptron (MLP) technique. *Chemosphere*, *337*(June), 139346. doi:10.1016/j.chemosphere.2023.139346 PMID:37379988

Sathiyaseelan, G., Srinath, P., Melwyn, J. G., Bhagyanathan, C., Socrates, S., & Boopathi, S. (2023). Tribology and hardness measurement of biosilica dispersed AA7075 metal matrix composites. *Materials Today: Proceedings*. Advance online publication. doi:10.1016/j.matpr.2023.02.440

Savalia, S., & Emamian, V. (2018). Cardiac arrhythmia classification by multi-layer perceptron and convolution neural networks. *Bioengineering (Basel, Switzerland)*, *5*(2), 35. doi:10.3390/bioengineering5020035 PMID:29734666

Saveen, G., Prudhvi Raju, P., Manikanta, D. V., & Satya Praveen, M. (2018). Design and implementation of energy management system with fuzzy control for multiple microgrid. *Proceedings of the 2nd International Conference on Inventive Systems and Control, ICISC 2018*, *28*(4), 1239–1244. 10.1109/ICISC.2018.8399003

Sawhney, N. S., Anousheh, R., Chen, W. C., & Feld, G. K. (2009). *Diagnosis and management of typical atrial flutter*. . doi:10.1016/j.ccl.2008.09.010

Sayantan, G., Kien, P., & Kadambari, K. (2018). Classification of ECG beats using deep belief network and active learning. *Medical & Biological Engineering & Computing*, *56*(10), 1887–1898. doi:10.100711517-018-1815-2 PMID:29651694

Scheba, P., Scebba, G. C., Zhang, J., Delai, M., & Karlen, W. (2017). Beat by beat: Classifying cardiac arrhythmias with recurrent neural networks. In 2017 computing in cardiology (CINC) (pp. 1–4). IEEE.

Schellekens, H. (2005, September 1). Immunologic mechanisms of EPO-associated pure red cell aplasia. *Best Practice & Research. Clinical Haematology*, *18*(3), 473–480. doi:10.1016/j.beha.2005.01.016 PMID:15792922

Schellenberger, J., Que, R., Fleming, R. M. T., Thiele, I., Orth, J. D., Feist, A. M., Zielinski, D. C., Bordbar, A., Lewis, N. E., Rahmanian, S., Kang, J., Hyduke, D. R., & Palsson, B. (2011). Quantitative prediction of cellular metabolism with constraint-based models: The COBRA Toolbox v2.0. *Nature Protocols*, *6*(9), 1290–1307. doi:10.1038/nprot.2011.308 PMID:21886097

Schinn, S. M., Morrison, C., Wei, W., Zhang, L., & Lewis, N. E. (2021). A genome-scale metabolic network model and machine learning predict amino acid concentrations in Chinese Hamster Ovary cell cultures. *Biotechnology and Bioengineering*, *118*(5), 2118–2123. doi:10.1002/bit.27714 PMID:33580712

Schloßhauer, J. L., Cavak, N., Zemella, A., Thoring, L., & Kubick, S. (2022). Cell Engineering and Cultivation of Chinese Hamster Ovary Cells for the Development of Orthogonal Eukaryotic Cell-free Translation Systems. *Frontiers in Molecular Biosciences*, *9*, 832379. doi:10.3389/fmolb.2022.832379 PMID:35586195

Schmid-Dannert, C., & López-Gallego, F. (2019). Advances and opportunities for the design of self-sufficient and spatially organized cell-free biocatalytic systems. *Current Opinion in Chemical Biology*, *49*, 97–104. doi:10.1016/j.cbpa.2018.11.021 PMID:30551093

Schmidt, M., & Pei, L. (2011). Synthetic toxicology: Where engineering meets biology and Toxicology? *Toxicological Sciences*, *120*(Supplement 1), S204–S224. doi:10.1093/toxsci/kfq339 PMID:21068213

Schomburg, I., Chang, A., Hofmann, O., Ebeling, C., Ehrentreich, F., & Schomburg, D. (2002). BRENDA: A resource for enzyme data and metabolic information. *Trends in Biochemical Sciences*, *27*(1), 54–56. doi:10.1016/S0968-0004(01)02027-8 PMID:11796225

Schuergers, N., Werlang, C., Ajo-Franklin, C. M., & Boghossian, A. A. (2017a). A synthetic biology approach to engineering living photovoltaics. *Energy & Environmental Science*, *10*(5), 1102–1115. doi:10.1039/C7EE00282C PMID:28694844

Schuster, M., & Paliwal, K. K. (1997). Bidirectional recurrent neural networks. *IEEE Transactions on Signal Processing*, *45*(11), 2673–2681. doi:10.1109/78.650093

Scossa, F., & Fernie, A. R. (2020). The evolution of metabolism: How to test evolutionary hypotheses at the genomic level. *Computational and Structural Biotechnology Journal*, *18*, 482–500. doi:10.1016/j.csbj.2020.02.009 PMID:32180906

Scott, A., Noga, M. J., de Graaf, P., Westerlaken, I., Yildirim, E., & Danelon, C. (2016). Cell-free phospholipid biosynthesis by gene-encoded enzymes reconstituted in liposomes. *PLoS One*, *11*(10), e0163058. doi:10.1371/journal.pone.0163058 PMID:27711229

Seaver, S. M. D., Liu, F., Zhang, Q., Jeffryes, J., Faria, J. P., Edirisinghe, J. N., Mundy, M., Chia, N., Noor, E., Beber, M. E., Best, A. A., DeJongh, M., Kimbrel, J. A., D'haeseleer, P., McCorkle, S. R., Bolton, J. R., Pearson, E., Canon, S., Wood-Charlson, E. M., ... Henry, C. S. (2021). The ModelSEED Biochemistry Database for the integration of metabolic annotations and the reconstruction, comparison and analysis of metabolic models for plants, fungi and microbes. *Nucleic Acids Research*, *49*(D1), D575–D588. doi:10.1093/nar/gkaa746 PMID:32986834

Seifert, R., Weber, M., Kocakavuk, E., Rischpler, C., & Kersting, D. (2021). Artificial Intelligence and Machine Learning in Nuclear Medicine: Future Perspectives. *Seminars in Nuclear Medicine*, *51*(2), 170–177. doi:10.1053/j.semnuclmed.2020.08.003 PMID:33509373

Selvakumar, S., Adithe, S., Isaac, J. S., Pradhan, R., Venkatesh, V., & Sampath, B. (2023). A Study of the Printed Circuit Board (PCB) E-Waste Recycling Process. In Sustainable Approaches and Strategies for E-Waste Management and Utilization (pp. 159–184). IGI Global.

Selvakumar, S., Shankar, R., Ranjit, P., Bhattacharya, S., Gupta, A. S. G., & Boopathi, S. (2023). E-Waste Recovery and Utilization Processes for Mobile Phone Waste. In *Handbook of Research on Safe Disposal Methods of Municipal Solid Wastes for a Sustainable Environment* (pp. 222–240). IGI Global. doi:10.4018/978-1-6684-8117-2.ch016

Sengeni, D., Padmapriya, G., Imambi, S. S., Suganthi, D., Suri, A., & Boopathi, S. (2023). Biomedical Waste Handling Method Using Artificial Intelligence Techniques. In *Handbook of Research on Safe Disposal Methods of Municipal Solid Wastes for a Sustainable Environment* (pp. 306–323). IGI Global. doi:10.4018/978-1-6684-8117-2.ch022

Senthil, T. S. R. Ohmsakthi vel, Puviyarasan, M., Babu, S. R., Surakasi, R., & Sampath, B. (2023). Industrial Robot-Integrated Fused Deposition Modelling for the 3D Printing Process. In Development, Properties, and Industrial Applications of 3D Printed Polymer Composites (pp. 188–210). IGI Global. doi:10.4018/978-1-6684-6009-2.ch011

Senthil, T., Puviyarasan, M., Babu, S. R., Surakasi, R., Sampath, B., & Associates. (2023). Industrial Robot-Integrated Fused Deposition Modelling for the 3D Printing Process. In Development, Properties, and Industrial Applications of 3D Printed Polymer Composites (pp. 188–210). IGI Global.

Shabestary, K., & Hudson, E. P. (2016). Computational metabolic engineering strategies for growth-coupled biofuel production by Synechocystis. *Metabolic Engineering Communications*, *3*, 216–226. doi:10.1016/j.meteno.2016.07.003 PMID:29468126

Shadish, J. A., & DeForest, C. A. (2020). Site-Selective Protein Modification: From Functionalized Proteins to Functional Biomaterials. *Matter*, *2*(1), 50–77. doi:10.1016/j.matt.2019.11.011

Shah, A. R., Ahmad, A., Srivastava, S., & Jaffar Ali, B. M. (2017). Reconstruction and analysis of a genome-scale metabolic model of Nannochloropsis gaditana. *Algal Research*, *26*, 354–364. doi:10.1016/j.algal.2017.08.014

Shahkouhi, A. M., & Motamedian, E. (2020). Reconstruction of a regulated two-cell metabolic model to study biohydrogen production in a diazotrophic cyanobacterium Anabaena variabilis ATCC 29413. *PLoS One*, *15*(1), 1–18. doi:10.1371/journal.pone.0227977 PMID:31978122

Shah, M. V., Nazem-Bokaee, H., Antoney, J., Kang, S. W., Jackson, C. J., & Scott, C. (2021). Improved production of the non-native cofactor F420 in Escherichia coli. *Scientific Reports*, *11*(1), 1–16. doi:10.103841598-021-01224-3 PMID:34741069

Shalini, D., Prodipta, R., & Upasana, B. (2021). Molecular pharming for the betterment of crop yielding. *International Research Journal of Plant Science*.

Sharma, A., Singh, Y., Ansari, N. A., Pal, A., & Lalhriatpuia, S. (2020). Experimental investigation of the behaviour of a DI diesel engine fuelled with biodiesel/diesel blends having effect of raw biogas at different operating responses. *Fuel*, *279*, 118460. doi:10.1016/j.fuel.2020.118460

Sharma, B., & Shukla, P. (2022). Futuristic avenues of metabolic engineering techniques in bioremediation. *Biotechnology and Applied Biochemistry*, *69*(1), 51–60. doi:10.1002/bab.2080 PMID:33242354

Sharpes, C. E., McManus, J. B., Blum, S. M., Mgboji, G. E., & Lux, M. W. (2021). Assessment of Colorimetric Reporter Enzymes in the PURE System. *ACS Synthetic Biology*, *10*(11), 3205–3208. doi:10.1021/acssynbio.1c00360 PMID:34723497

Shashikumar, S. P., Shah, A. J., Clifford, G. D., & Nemati, S. (2018). Detection of paroxysmal atrial fibrillation using attention-based bidirectional recurrent neural networks. In *Proceedings of the 24th ACM SIGKDD international conference on knowledge discovery & data mining* (pp. 715–723). ACM. 10.1145/3219819.3219912

Sheridan, C. (2005). The business of making vaccines. *Nature Biotechnology*, *23*(11), 1359–1366. doi:10.1038/nbt1105-1359 PMID:16273061

Shih, H. H. (2012). *Discovery process for antibody-based therapeutics. InDevelopment of Antibody-Based Therapeutics*. Springer.

Shimizu, Y., Inoue, A., Tomari, Y., Suzuki, T., Yokogawa, T., Nishikawa, K., & Ueda, T. (2001). Cell-free translation reconstituted with purified components. *Nature Biotechnology*, *19*(8), 751–755. doi:10.1038/90802 PMID:11479568

Shimizu, Y., Kuruma, Y., Ying, B.-W., Umekage, S., & Ueda, T. (2006). Cell-free translation systems for protein engineering. *The FEBS Journal*, *273*(18), 4133–4140. doi:10.1111/j.1742-4658.2006.05431.x PMID:16930131

Shin, J., & Noireaux, V. (2010). Efficient cell-free expression with the endogenous E. Coli RNA polymerase and sigma factor 70. *Journal of Biological Engineering*, *4*(1), 8. doi:10.1186/1754-1611-4-8 PMID:20576148

Sindhu, R., Binod, P., Pandey, A., Ankaram, S., Duan, Y., & Awasthi, M. K. (2019). Biofuel production from biomass: Toward sustainable development. In *Current developments in biotechnology and bioengineering* (pp. 79–92). Elsevier. doi:10.1016/B978-0-444-64083-3.00005-1

Singh Grewal, G., & Parkash Dhiman, D. (2019). Effect of deep cryogenic treatment on copper electrode for non-traditional electric discharge machining (EDM). *Mechanical Sciences*, *10*(2), 413–427. doi:10.5194/ms-10-413-2019

Singh, S. (2018). Synthetic biology: Omics tools and their applications. In *Synthetic Biology*. Omics Tools and Their Applications., doi:10.1007/978-981-10-8693-9

Singh, S., Pandey, S. K., Pawar, U., & Janghel, R. R. (2018). Classification of ECG Arrhythmia using Recurrent Neural Networks. *Procedia Computer Science*, *132*, 1290–1297. doi:10.1016/j.procs.2018.05.045

Singh, V. (2014). Recent advancements in synthetic biology: Current status and challenges. *Gene*, *535*(1), 1–11. doi:10.1016/j.gene.2013.11.025 PMID:24269673

Sistla, S., & Rao, D. N. (2004). S-Adenosyl-L-methionine–Dependent Restriction Enzymes. *Critical Reviews in Biochemistry and Molecular Biology*, *39*(1), 1–19. doi:10.1080/10409230490440532 PMID:15121719

Sivadasan, D., Sultan, M. H., Madkhali, O., Almoshari, Y., & Thangavel, N. (2021, August 18). Polymeric Lipid Hybrid Nanoparticles (PLNs) as Emerging Drug Delivery Platform—A Comprehensive Review of Their Properties, Preparation Methods, and Therapeutic Applications. *Pharmaceutics*, *13*(8), 1291. doi:10.3390/pharmaceutics13081291 PMID:34452251

Smolskaya, S., Logashina, Y. A., & Andreev, Y. A. (2020). Escherichia coli Extract-Based Cell-Free Expression System as an Alternative for Difficult-to-Obtain Protein Biosynthesis. *International Journal of Molecular Sciences*, *21*(3), 928. Advance online publication. doi:10.3390/ijms21030928 PMID:32023820

Soh, J. H., Balleza, E., Abdul Rahim, M. N., Chan, H.-M., Mohd Ali, S., Chuah, J. K. C., Edris, S., Atef, A., Bahieldin, A., Ying, J. Y., & Sabir, J. S. M. (2022). CRISPR-based systems for sensitive and rapid on-site COVID-19 diagnostics. *Trends in Biotechnology*, *40*(11), 1346–1360. doi:10.1016/j.tibtech.2022.06.002 PMID:35871983

Sohn, S. B., Graf, A. B., Kim, T. Y., Gasser, B., Maurer, M., Ferrer, P., Mattanovich, D., & Lee, S. Y. (2010). Genome-scale metabolic model of methylotrophic yeast Pichia pastoris and its use for in silico analysis of heterologous protein production. *Biotechnology Journal*, *5*(7), 705–715. doi:10.1002/biot.201000078 PMID:20503221

Soni, T., Vivekanand, V., & Pareek, N. (2022). Biosensors as an effective tool for detection of emerging water and waste-water pollutants. In I. Haq, A. S. Kalamdhad, & M. P. Shah (Eds.), *Biodegradation and Detoxification of Micropollutants in Industrial Wastewater* (pp. 39–54). Elsevier. doi:10.1016/B978-0-323-88507-2.00010-5

Spencer, A. C., Torre, P., & Mansy, S. S. (2013). The encapsulation of cell-free transcription and translation machinery in vesicles for the construction of cellular mimics. *Journal of Visualized Experiments*, (80), e51304. PMID:24192867

Sproles, A. E., Fields, F. J., Smalley, T. N., Le, C. H., Badary, A., & Mayfield, S. P. (2021, March 1). Recent advancements in the genetic engineering of microalgae. *Algal Research*, *53*, 102158. doi:10.1016/j.algal.2020.102158

Stalidzans, E., Seiman, A., Peebo, K., Komasilovs, V., & Pentjuss, A. (2018). Model-based metabolism design: Constraints for kinetic and stoichiometric models. *Biochemical Society Transactions*, *46*(2), 261–267. doi:10.1042/BST20170263 PMID:29472367

Stark, J. C., Huang, A., Nguyen, P. Q., Dubner, R. S., Hsu, K. J., Ferrante, T. C., Anderson, M., Kanapskyte, A., Mucha, Q., Packett, J. S., Patel, P., Patel, R., Qaq, D., Zondor, T., Burke, J., Martinez, T., Miller-Berry, A., Puppala, A., Reichert, K., ... Jewett, M. C. (2018). BioBits™ Bright: A fluorescent synthetic biology education kit. *Science Advances*, *4*(8), eaat5107. Advance online publication. doi:10.1126ciadv.aat5107 PMID:30083609

Stefan, S., Nicole, R., Holger, S., Stefan, R., & Matthias, B. (2019). *Critical Analysis of the Commercial Potential of Plants for the Production of Recombinant Proteins hypothesis and theory*. Academic Press.

Steinberg, F. (1998, May 1). *Biotech pharmaceuticals and biotherapy: An overview*. Semantic Scholar. https://www.semanticscholar.org/paper/Biotech-pharmaceuticals-and-biotherapy%3A-an-Steinberg-Raso/93e6672461b240dc57fe4548d48c1e785d6d7df7

Stephanopoulos, G. N., Aristidou, A. A., & Nielsen, J. (1998). Metabolic Engineering: Principles and Methodologies. In Metabolic Engineering (Vol. 54). Academic Press.

Stevenson, J., Krycer, J. R., Phan, L., & Brown, A. (2013). A practical comparison of Ligation-Independent cloning techniques. *PLoS One*, *8*(12), e83888. doi:10.1371/journal.pone.0083888 PMID:24376768

Straathof, A. J. J., Wahl, S. A., Benjamin, K. R., Takors, R., Wierckx, N., & Noorman, H. J. (2019). Grand Research Challenges for Sustainable Industrial Biotechnology. *Trends in Biotechnology*, *37*(10), 1042–1050. doi:10.1016/j.tibtech.2019.04.002 PMID:31054854

Strohl, W.R. (2009). *Therapeutic monoclonal antibodies: Past, present, and future*. Academic Press.

Subha, S., Inbamalar, T. M., R, K. C., Suresh, L. R., Boopathi, S., & Alaskar, K. (2023, February). A Remote Health Care Monitoring system using internet of medical things (IoMT). *IEEE Explore*. doi:10.1109/ICIPTM57143.2023.10118103

Subha, S., Inbamalar, T., Komala, C., Suresh, L. R., Boopathi, S., & Alaskar, K. (2023). A Remote Health Care Monitoring system using internet of medical things (IoMT). *IEEE Explore*, 1–6.

Suganthi, L., Iniyan, S., & Samuel, A. A. (2015). Applications of fuzzy logic in renewable energy systems—A review. *Renewable & Sustainable Energy Reviews*, *48*, 585–607. doi:10.1016/j.rser.2015.04.037

Suhail, M., Akhtar, I., Kirmani, S., & Jameel, M. (2021). Development of Progressive Fuzzy Logic and ANFIS Control for Energy Management of Plug-In Hybrid Electric Vehicle. *IEEE Access : Practical Innovations, Open Solutions*, *9*, 62219–62231. doi:10.1109/ACCESS.2021.3073862

Sun, X., Xu, C., Wu, G., Ye, Q., & Wang, C. (2017, June 1). Poly (lactic-co-glycolic acid): Applications and future prospects for periodontal tissue regeneration. *Polymers*, *9*(6), 189. doi:10.3390/polym9060189 PMID:30970881

Sun, Z. Z., Hayes, C. A., Shin, J., Caschera, F., Murray, R. M., & Noireaux, V. (2013). Protocols for implementing an Escherichia coli based TX-TL cell-free expression system for synthetic biology. *Journal of Visualized Experiments*, *79*(79), e50762. doi:10.3791/50762 PMID:24084388

Swain, N., Cirenza, C., & Samanta, L. (2016). Role of proteomics in female infertility. *Proteomics in Human Reproduction*, 45-63.

Syamala, M., Komala, C., Pramila, P., Dash, S., Meenakshi, S., & Boopathi, S. (2023). Machine Learning-Integrated IoT-Based Smart Home Energy Management System. In *Handbook of Research on Deep Learning Techniques for Cloud-Based Industrial IoT* (pp. 219–235). IGI Global. doi:10.4018/978-1-6684-8098-4.ch013

Szklarczyk, D., Gable, A. L., Nastou, K. C., Lyon, D., Kirsch, R., Pyysalo, S., Doncheva, N. T., Legeay, M., Fang, T., Bork, P., Jensen, L. J., & von Mering, C. (2021). The STRING database in 2021: Customizable protein-protein networks, and functional characterization of user-uploaded gene/measurement sets. *Nucleic Acids Research*, *49*(D1), D605–D612. doi:10.1093/nar/gkaa1074 PMID:33237311

Tanveer, T., Shaheen, K., Parveen, S., Misbah, Z. T., Babar, M. M., & Gul, A. (2018). Omics-based bioengineering in environmental biotechnology. In D. Barh & V. Azevedo (Eds.), *Omics Technologies and Bio-Engineering* (pp. 353–364). Academic Press. doi:10.1016/B978-0-12-815870-8.00019-X

Tayal, V., & Kalra, B. S. (2008, January 28). Cytokines and anti-cytokines as therapeutics—An update. *European Journal of Pharmacology*, *579*(1-3), 1–2. doi:10.1016/j.ejphar.2007.10.049 PMID:18021769

Tay, P. K. R., Nguyen, P. Q., & Joshi, N. S. (2017). A synthetic circuit for mercury bioremediation using self-assembling functional amyloids. *ACS Synthetic Biology*, *6*(10), 1841–1850. doi:10.1021/acssynbio.7b00137 PMID:28737385

Tecon, R., & Van der Meer, J. R. (2008). Bacterial biosensors for measuring availability of environmental pollutants. *Sensors (Basel)*, *8*(7), 4062–4080. doi:10.33908074062 PMID:27879922

Thakare, S., Shriyan, A., Thale, V., Yasarp, P., & Unni, K. (2017). Implementation of an energy monitoring and control device based on IoT. *2016 IEEE Annual India Conference, INDICON 2016*, 1–6. 10.1109/INDICON.2016.7839066

Thiele, I., & Palsson, B. (2010). A protocol for generating a high-quality genome-scale metabolic reconstruction. *Nature Protocols*, *5*(1), 93–121. doi:10.1038/nprot.2009.203 PMID:20057383

Thomas, S. R., Maynard, N. D., & Gill, J. (2015). DNA library construction using Gibson Assembly®. *Nature Methods*, *12*(11), i–ii. doi:10.1038/nmeth.f.384

Thombre, R. S., Kaur, K., Jagtap, S. S., Dixit, J., & Vaishampayan, P. V. (2022). *Microbial Life in Space*. New Frontiers in Astrobiology. doi:10.1016/B978-0-12-824162-2.00013-0

Thoring, L., Dondapati, S. K., Stech, M., Wüstenhagen, D. A., & Kubick, S. (2017). High-yield production of "difficult-to-express" proteins in a continuous exchange cell-free system based on CHO cell lysates. *Scientific Reports*, *7*(1), 11710. doi:10.103841598-017-12188-8 PMID:28916746

Thoring, L., Wüstenhagen, D. A., Borowiak, M., Stech, M., Sonnabend, A., & Kubick, S. (2016). Cell-Free Systems Based on CHO Cell Lysates: Optimization Strategies, Synthesis of "Difficult-to-Express" Proteins and Future Perspectives. *PLoS One*, *11*(9), e0163670. doi:10.1371/journal.pone.0163670 PMID:27684475

Tinafar, A., Jaenes, K., & Pardee, K. (2019). Synthetic Biology Goes Cell-Free. *BMC Biology*, *17*(1), 64. doi:10.118612915-019-0685-x PMID:31395057

Tiongco, R. E., Rivera, N., Clemente, B., Dizon, D., Salita, C., & Pineda-Cortel, M. R. (2019). Serum Ferritin as a candidate diagnostic biomarker of polycystic ovarian syndrome: A meta-analysis. *Biomarkers*, *24*(5), 484–491. doi:10.1080/1354750X.2019.1620335 PMID:31096807

Torchen, L. C., Idkowiak, J., Fogel, N. R., O'Neil, D. M., Shackleton, C. L., Wiebke, A., & Dunaif, A. (2016). Evidence for Increased 5alpha-Reductase Activity During Early Childhood in Daughters of Women with Polycystic Ovary Syndrome. *The Journal of Clinical Endocrinology and Metabolism*, *101*(5), 2069–2075. doi:10.1210/jc.2015-3926 PMID:26990942

Torres, N. V., & Voit, E. O. (2002). Pathway Analysis and Optimization in Metabolic Engineering. In *Pathway Analysis and Optimization in Metabolic Engineering*. Cambridge University Press., doi:10.1017/CBO9780511546334

Trachsel, H., Ranu, R. S., & London, I. M. (1978). Regulation of protein synthesis in rabbit reticulocyte lysates: Purification and characterization of heme-reversible translational inhibitor. *Proceedings of the National Academy of Sciences of the United States of America*, *75*(8), 3654–3658. doi:10.1073/pnas.75.8.3654 PMID:278981

Trojovský, P., Dhasarathan, V., & Boopathi, S. (2023). Experimental investigations on cryogenic friction-stir welding of similar ZE42 magnesium alloys. *Alexandria Engineering Journal*, *66*(1), 1–14. doi:10.1016/j.aej.2022.12.007

Tsuji, T., Shiozaki, A., Kohno, R., Yoshizato, K., & Shimohama, S. (2002) Article. *Neurochemical Research*, *27*(10), 1245–1253. doi:10.1023/A:1020941929414 PMID:12462422

Uçar, A., Demir, Y., & Güzeliş, C. (2017). Object recognition and detection with deep learning for autonomous driving applications. *Simulation*, *93*(9), 759–769. doi:10.1177/0037549717709932

Uchida, M., LaFrance, B., Broomell, C. C., Prevelige, P. E. Jr, & Douglas, T. (2015). Higher order assembly of virus-like particles (VLPs) mediated by multi-valent protein linkers. *Small*, *11*(13), 1562–1570. doi:10.1002mll.201402067 PMID:25641768

Ulislam, M. (2014, January 1). *Yeast cell-free enzyme system for bio-ethanol production at elevated temperatures.* Academia. https://www.academia.edu/60123976/Yeast_cell_free_enzyme_system_for_bio_ethanol_production_at_elevated_temperatures?from_sitemaps=true&version=2

Ullah, M. W., Khattak, W. A., Ul-Islam, M., Khan, S., & Park, J. K. (2014). Bio-ethanol production through simultaneous saccharification and fermentation using an encapsulated reconstituted cell-free enzyme system. *Biochemical Engineering Journal*, *91*, 110–119. doi:10.1016/j.bej.2014.08.006

Ullah, M. W., Khattak, W. A., Ul-Islam, M., Khan, S., & Park, J. K. (2015). Encapsulated yeast cell-free system: A strategy for cost-effective and sustainable production of bio-ethanol in consecutive batches. *Biotechnology and Bioprocess Engineering; BBE*, *20*(3), 561–575. doi:10.100712257-014-0855-1

Unluturk, U., Harmanci, A., Kocaefe, C., & Yildiz, B. O. (2007). The genetic basis of the polycystic ovary syndrome: A literature review including discussion of PPAR-gamma. *PPAR Research*, *2007*, 1–23. doi:10.1155/2007/49109 PMID:17389770

Uribe-Echeverry, P. T., & Lopez-Gartner, G. A. (2017, June 1). Fungal immunomodulatory proteins in the context of biomedicine. *Frontiers in Bioscience (Elite Edition)*, *9*(2), 286–306. doi:10.2741/e803 PMID:28410152

US Energy Information Administration (EIA). (2021). *International energy outlook 2021: With projections to 2050*. Author.

Usmani, S. S., Bedi, G., Samuel, J. S., Singh, S., Kalra, S., Kumar, P., Ahuja, A. A., Sharma, M., Gautam, A., & Raghava, G. P. (2017, July 31). THPdb: Database of FDA-approved peptide and protein therapeutics. *PLoS One*, *12*(7), e0181748. doi:10.1371/journal.pone.0181748 PMID:28759605

Utturkar, S. M., Bollmann, A., Brzoska, R. M., Klingeman, D. M., Epstein, S. E., Palumbo, A. V., & Brown, S. D. (2013). Draft genome sequence for *Ralstonia* sp. *strain OR*214, a bacterium with potential for bioremediation. *Genome Announcements*, *1*(3), e00321–e13. doi:10.1128/genomeA.00321-13 PMID:23792748

Vallecillos, L., Sadef, Y., Borrull, F., Pocurull, E., & Bester, K. (2017). Degradation of synthetic fragrances by laccase-mediated system. *Journal of Hazardous Materials*, *334*, 233–243. doi:10.1016/j.jhazmat.2017.04.003 PMID:28415001

Van Aken, J., & Hammond, E. (2003). Genetic engineering and biological weapons: New technologies, desires and threats from biological research. *EMBO Reports*, *4*(S1), S57–S60. doi:10.1038j.embor.embor860 PMID:12789409

Vandana, U. K., Gulzar, A., Laskar, I. H., Meitei, L. R., & Mazumder, P. (2021). Role of microbes in bioremediation of radioactive waste. *Microbial Rejuvenation of Polluted Environment*, *1*, 329–352. doi:10.1007/978-981-15-7447-4_13

Vanitha, S., Radhika, K., & Boopathi, S. (2023). Artificial Intelligence Techniques in Water Purification and Utilization. In *Human Agro-Energy Optimization for Business and Industry* (pp. 202–218). IGI Global. doi:10.4018/978-1-6684-4118-3.ch010

Vazana, Y., Barak, Y., Unger, T., Peleg, Y., Shamshoum, M., Ben-Yehezkel, T., Mazor, Y., Shapiro, E., Lamed, R., & Bayer, E. A. (2013). A synthetic biology approach for evaluating the functional contribution of designer cellulosome components to deconstruction of cellulosic substrates. *Biotechnology for Biofuels*, *6*(1), 1–18. doi:10.1186/1754-6834-6-182 PMID:24341331

Veeranjaneyulu, R., Boopathi, S., Kumari, R. K., Vidyarthi, A., Isaac, J. S., & Jaiganesh, V. (2023). Air Quality Improvement and Optimisation Using Machine Learning Technique. *IEEE Explore*, 1–6.

Veeranjaneyulu, R., Boopathi, S., Narasimharao, J., Gupta, K. K., Reddy, R. V. K., & Ambika, R. (2023). Identification of Heart Diseases using Novel Machine Learning Method. *IEEE Explore*, 1–6.

Venkateswaran, N., Vidhya, K., Ayyannan, M., Chavan, S. M., Sekar, K., & Boopathi, S. (2023). A Study on Smart Energy Management Framework Using Cloud Computing. In 5G, Artificial Intelligence, and Next Generation Internet of Things: Digital Innovation for Green and Sustainable Economies (pp. 189–212). IGI Global. doi:10.4018/978-1-6684-8634-4.ch009

Venkateswaran, N., Vidhya, R., Naik, D. A., Raj, T. F. M., Munjal, N., & Boopathi, S. (2023). Study on Sentence and Question Formation Using Deep Learning Techniques. In *Digital Natives as a Disruptive Force in Asian Businesses and Societies* (pp. 252–273). IGI Global. doi:10.4018/978-1-6684-6782-4.ch015

Vennila, T., Karuna, M., Srivastava, B. K., Venugopal, J., Surakasi, R., & Sampath, B. (2022). New Strategies in Treatment and Enzymatic Processes: Ethanol Production From Sugarcane Bagasse. In Human Agro-Energy Optimization for Business and Industry (pp. 219–240). IGI Global.

Venter, M. (2007). Synthetic promoters: Genetic control through cis engineering. *Trends in Plant Science, 12*(3), 118–124. doi:10.1016/j.tplants.2007.01.002 PMID:17292658

Ventola, C. L. (2015). The antibiotic resistance crisis: part 1: causes and threats. *P&T, 40*(4), 277. PMID:25859123

Ventorim, R. Z., Ferreira, M. A. de M., de Almeida, E. L. M., Kerkhoven, E. J., & da Silveira, W. B. (2022). Genome-scale metabolic model of oleaginous yeast Papiliotrema laurentii. *Biochemical Engineering Journal, 180*, 108353. doi:10.1016/j.bej.2022.108353

Verma, J., Kumar, D., Singh, N., Katti, S. S., & Shah, Y. T. (2021). Electricigens and microbial fuel cells for bioremediation and bioenergy production: A review. *Environmental Chemistry Letters, 19*(3), 2091–2126. doi:10.100710311-021-01199-7

Vidali, M. (2001). Bioremediation: An overview. *Pure and Applied Chemistry, 73*(7), 1163–1172. doi:10.1351/pac200173071163

Vidic, J., Manzano, M., Chang, C.-M., & Jaffrezic-Renault, N. (2017). Advanced biosensors for detection of pathogens related to livestock and poultry. *Veterinary Research, 48*(1), 11. doi:10.118613567-017-0418-5 PMID:28222780

Vidigal, J. A., & Ventura, A. (2015). Rapid and efficient one-step generation of paired gRNA CRISPR-Cas9 libraries. *Nature Communications, 6*(1), 8083. Advance online publication. doi:10.1038/ncomms9083 PMID:26278926

Vilkhovoy, M., Adhikari, A., Vadhin, S., & Varner, J. D. (2020). The Evolution of Cell Free Biomanufacturing. *Processes (Basel, Switzerland), 8*(6), 675. Advance online publication. doi:10.3390/pr8060675

Vilkhovoy, M., Dai, D., Vadhin, S., Adhikari, A., & Varner, J. D. (2019). Absolute Quantification of Cell-Free Protein Synthesis Metabolism by Reversed-Phase Liquid Chromatography-Mass Spectrometry. *Journal of Visualized Experiments*, (152). Advance online publication. doi:10.3791/60329-v PMID:31710042

Vinaya, R. (2022). *Metabolic engineering of plants – Molecular Pharming. Science – Policy Brief for the Multistakeholder Forum on Science, Technology and Innovation for the SDGs.* Academic Press.

Vojvoda, E.J., Burrington, L.R., & Oza, J.P. (2022). Development of next-generation diagnostic tools using synthetic biology. In *New Frontiers and Applications of Synthetic Biology* (pp. 287-330). Academic Press.

Vu, N. T., & Oh, C.-S. (2020). Bacteriophage usage for bacterial disease management and diagnosis in plants. *The Plant Pathology Journal, 36*(3), 204–217. doi:10.5423/PPJ.RW.04.2020.0074 PMID:32547337

Wahlang, I., Sharma, P., Sanyal, S., Saha, G., & Maji, A. K. (2020). Deep learning techniques for classification of brain MRI. *International Journal of Intelligent Systems Technologies and Applications, 19*(6), 571–588. doi:10.1504/IJISTA.2020.112441

Waller, T. C., Berg, J. A., Lex, A., Chapman, B. E., & Rutter, J. (2020). Compartment and hub definitions tune metabolic networks for metabolomic interpretations. *GigaScience, 9*(1), 1–18. doi:10.1093/gigascience/giz137 PMID:31972021

Wang, F., & Zhang, W. (2019). Synthetic biology: Recent progress, biosafety and biosecurity concerns, and possible solutions. *Journal of Biosafety and Biosecurity, 1*(1), 22–30. doi:10.1016/j.jobb.2018.12.003

Wang, G., Dai, Y., Yang, H., Xiong, Q., Wang, K., Zhou, J., Li, Y., & Wang, S. (2020). A review of recent advances in biomass pyrolysis. *Energy & Fuels, 34*(12), 15557–15578. doi:10.1021/acs.energyfuels.0c03107

Wang, G., Kong, Q., Wang, D., & Asmi, F. (2023). Ethical and social insights into synthetic biology: Predicting research fronts in the post-COVID-19 era. *Frontiers in Bioengineering and Biotechnology, 11*, 1085797. doi:10.3389/fbioe.2023.1085797 PMID:37274167

Wang, H., Chen, X., Huang, Z., Zhou, B., Jia, G., Liu, G., & Zhao, H. (2014). Expression and purification of porcine PID1 gene in Escherichia coli. *Turkish Journal of Biology, 38*(4), 523–527. doi:10.3906/biy-1403-38

Wang, H., Marcišauskas, S., Sánchez, B. J., Domenzain, I., Hermansson, D., Agren, R., Nielsen, J., & Kerkhoven, E. J. (2018). RAVEN 2.0: A versatile toolbox for metabolic network reconstruction and a case study on Streptomyces coelicolor. *PLoS Computational Biology, 14*(10), e1006541. doi:10.1371/journal.pcbi.1006541 PMID:30335785

Wang, J. W., Wang, A., Li, K., Wang, B., Jin, S., Reiser, M., & Lockey, R. F. (2015). CRISPR/Cas9 nuclease cleavage combined with Gibson assembly for seamless cloning. *BioTechniques, 58*(4), 161–170. doi:10.2144/000114261 PMID:25861928

Weidong, Q., Tingting, G., Xinxin, L., Jinnan, G., Muhammad, N., Yaying, L., & Yongxin, G. (2020, August 19). Molecular Pharming of the Recombinant Protein hEGF-hEGF Concatenated with Oleosin Using Transgenic *Arabidopsis. Genes, 11*(9), 959. Advance online publication. doi:10.3390/genes11090959

Weissmann, F., Petzold, G., VanderLinden, R. T., Veld, P. J. H. I., Brown, N. G., Lampert, F., Westermann, S., Stark, H., Schulman, B. A., & Peters, J. (2016). biGBac enables rapid gene assembly for the expression of large multisubunit protein complexes. *Proceedings of the National Academy of Sciences of the United States of America, 113*(19). 10.1073/pnas.1604935113

White, A., Light, A., Hughen, R., Bateman, L., Martins, T., Hill, H., & Light, K. (2010). Severity of symptom flare after moderate exercise is linked to cytokine activity in chronic fatigue syndrome. *Psychophysiology, 47*, 615–624. doi:10.1111/j.1469-8986.2010.00978.x PMID:20230500

Whitman, E. (2011, September 21). *Corporate Profits Trumping Public Health*. Global Issues. https://www.globalissues.org/news/2011/09/21/11262

Wickenheisser, J. K., Nelson-DeGrave, V. L., & McAllister, J. M. (2005). Dysregulation of cytochrome P450 17α-hydroxylase messenger ribonucleic acid stability in theca cells isolated from women with polycystic ovary syndrome. *The Journal of Clinical Endocrinology and Metabolism, 90*(3), 1720–1727. doi:10.1210/jc.2004-1860 PMID:15598676

Widiastuti, H., Kim, J. Y., Selvarasu, S., Karimi, I. A., Kim, H., Seo, J. S., & Lee, D. Y. (2011). Genome-scale modeling and in silico analysis of ethanologenic bacteria Zymomonas mobilis. *Biotechnology and Bioengineering, 108*(3), 655–665. doi:10.1002/bit.22965 PMID:20967753

Williams, H. F., Layfield, H. J., Vallance, T., Patel, K., Bicknell, A. B., Trim, S. A., & Vaiyapuri, S. (2019, June 20). The urgent need to develop novel strategies for the diagnosis and treatment of snakebites. *Toxins, 11*(6), 363. doi:10.3390/toxins11060363 PMID:31226842

Wisniewski, A., Zimmerman, M., Crews, T. Jr, Haulbrook, A., Fitzgerald, D. C., & Sistino, J. J. (2020). Reducing the impact of perfusion medical waste on the environment. *The Journal of Extra-Corporeal Technology, 52*(2), 135–141. doi:10.1051/ject/202052135 PMID:32669740

Wisselink, H. W., Cipollina, C., Oud, B., Crimi, B., Heijnen, J. J., Pronk, J. T., & van Maris, A. J. A. (2010). Metabolome, transcriptome and metabolic flux analysis of arabinose fermentation by engineered Saccharomyces cerevisiae. *Metabolic Engineering, 12*(6), 537–551. doi:10.1016/j.ymben.2010.08.003 PMID:20816840

Witchel, S. F., & Aston, C. E. (2000). The role of heterozygosity for CYP21 in the polycystic ovary syndrome. *Journal of Pediatric Endocrinology & Metabolism, 13*(5), 1315–1317. PMID:11117678

Witchel, S. F., Lee, P. A., Suda-Hartman, M., & Hoffman, E. P. (1997). Hyperandrogenism and manifesting heterozygotes for 21-hydroxylase deficiency. *Biochemical and Molecular Medicine*, *62*(2), 151–158. doi:10.1006/bmme.1997.2632 PMID:9441866

Witchel, S. F., Oberfield, S. E., & Pena, A. S. (2019). Polycystic ovary syndrome: Pathophysiology presentation and treatment with emphasis on adolescent girls. *Journal of the Endocrine Society*, *3*(8), 1545–1573. doi:10.1210/js.2019-00078 PMID:31384717

Woo, C. H., Jang, S., Shin, G., Jung, G. Y., & Lee, J. W. (2020). Sensitive fluorescence detection of SARS-CoV-2 RNA in clinical samples via one-pot isothermal ligation and transcription. *Nature Biomedical Engineering*, *4*(12), 1168–1179. doi:10.103841551-020-00617-5 PMID:32948855

Woolston, B. M., Edgar, S., & Stephanopoulos, G. (2013). Metabolic engineering: Past and future. *Annual Review of Chemical and Biomolecular Engineering*, *4*(1), 259–288. doi:10.1146/annurev-chembioeng-061312-103312 PMID:23540289

Wu, S., Snajdrova, R., Moore, J. C., Baldenius, K., & Bornscheuer, U. T. (2021). Biocatalysis: Enzymatic Synthesis for Industrial Applications. *Angewandte Chemie International Edition*, *60*(1), 88–119. doi:10.1002/anie.202006648 PMID:32558088

Xavier, J. C., Patil, K. R., & Rocha, I. (2017). Integration of Biomass Formulations of Genome-Scale Metabolic Models with Experimental Data Reveals Universally Essential Cofactors in Prokaryotes. *Metabolic Engineering*, *39*, 200–208. doi:10.1016/j.ymben.2016.12.002 PMID:27939572

Xu, C., Liu, L., Zhang, Z., Jin, D., Qiu, J., & Chen, M. (2013). Genome-scale metabolic model in guiding metabolic engineering of microbial improvement. *Applied Microbiology and Biotechnology*, *97*(2), 519–539. doi:10.100700253-012-4543-9 PMID:23188456

Xu, G., Zhao, S., Liu, J., & He, J. (2023). Bioremediation of organohalide pollutants: Progress, microbial ecology, and emerging computational tools. *Current Opinion in Environmental Science & Health*, *32*, 100452. doi:10.1016/j.coesh.2023.100452

Xu, P., Bhan, N., & Koffas, M. A. (2013). Engineering plant metabolism into microbes: From systems biology to synthetic biology. *Current Opinion in Biotechnology*, *24*(2), 291–299. doi:10.1016/j.copbio.2012.08.010 PMID:22985679

Xu, Y., Lee, J., Tran, C., Heibeck, T. H., Wang, W. D., Yang, J., Stafford, R. L., Steiner, A. R., Sato, A. K., Hallam, T. J., & Yin, G. (2015). Production of bispecific antibodies in "knobs-into-holes" using a cell-free expression system. *mAbs*, *7*(1), 231–242. doi:10.4161/19420862.2015.989013 PMID:25427258

Xu, Z., Zheng, P., Sun, J., & Ma, Y. (2013). ReacKnock: Identifying Reaction Deletion Strategies for Microbial Strain Optimization Based on Genome-Scale Metabolic Network. *PLoS One*, *8*(12), e72150. doi:10.1371/journal.pone.0072150 PMID:24348984

Yaashikaa, P. R., Devi, M. K., & Kumar, P. S. (2022). Engineering microbes for enhancing the degradation of environmental pollutants: A detailed review on synthetic biology. *Environmental Research*, *214*, 113868. doi:10.1016/j.envres.2022.113868 PMID:35835162

Yaashikaa, P., Devi, M. K., & Kumar, P. S. (2022). Algal biofuels: Technological perspective on cultivation, fuel extraction and engineering genetic pathway for enhancing productivity. *Fuel*, *320*, 123814. doi:10.1016/j.fuel.2022.123814

Yang, F., Ruan, Y. C., Yang, Y., Wang, K., Liang, S., Han, Y., Teng, X.-M., & Yang, J.-Z. (2015). Follicular hyperandrogenism downregulates aromatase in luteinizing granulosa cells in polycystic ovary syndrome women. *Reproduction (Cambridge, England)*, *150*(4), 289–296. doi:10.1530/REP-15-0044 PMID:26199450

Yang, S., Fei, Q., Zhang, Y., Contreras, L. M., Utturkar, S. M., Brown, S. D., Himmel, M. E., & Zhang, M. (2016). Zymomonas mobilis as a model system for production of biofuels and biochemicals. *Microbial Biotechnology*, *9*(6), 699–717. doi:10.1111/1751-7915.12408 PMID:27629544

Yano, D., Funadani, R., Uda, K., Matsuoka, T., & Suzuki, T. (2020). Role of arginine kinase in Paramecium tetraurelia (Ciliophora, Peniculida): Subcellular localization of AK3 and phosphoarginine shuttle system in cilia. *European Journal of Protistology*, *74*, 125705. doi:10.1016/j.ejop.2020.125705 PMID:32464434

Yatika, G., Kumar, S. R., Patel, P. B., Singh, D. P., Rajkamal, M., & Boopathi, S. (2023). Experimental investigation of RHA biochar and tamarind fibre epoxy composite. *Materials Today: Proceedings*. Advance online publication. doi:10.1016/j.matpr.2023.02.439

Ye, C., Wei, X., Shi, T., Sun, X., Xu, N., Gao, C., & Zou, W. (2022). Genome-scale metabolic network models: From first-generation to next-generation. *Applied Microbiology and Biotechnology*, *106*(13), 4907–4920. doi:10.100700253-022-12066-y PMID:35829788

Yim, S. K., Jung, H. C., Pan, J. G., Kang, H. S., Ahn, T., & Yun, C. H. (2006). Functional expression of mammalian NADPH-cytochrome P450 oxido-reductase on the cell surface of *Escherichia coli*. *Protein Expression and Purification*, *49*(2), 292–298. doi:10.1016/j.pep.2006.05.013 PMID:16829136

Yue, K., Chen, J., Li, Y., & Kai, L. (2023). Advancing synthetic biology through cell-free protein synthesis. *Computational and Structural Biotechnology Journal*, *21*, 2899–2908. doi:10.1016/j.csbj.2023.05.003 PMID:37216017

Yupapin, P., Trabelsi, Y., Nattappan, A., & Boopathi, S. (2023). Performance improvement of wire-cut electrical discharge machining process using cryogenically treated super-conductive state of Monel-K500 alloy. *Iranian Journal of Science and Technology. Transaction of Mechanical Engineering*, *47*(1), 267–283. doi:10.100740997-022-00513-0

Zamecnik, P. C., & Frantz, I. D. Jr. (1948). Incorporation in vitro of radioactive carbon from carboxyl-labeled dl-alanine and glycine into proteins of normal and malignant rat livers. *The Journal of Biological Chemistry*, *175*(1), 299–314. doi:10.1016/S0021-9258(18)57260-4 PMID:18873305

Zekrifa, D. M. S., Kulkarni, M., Bhagyalakshmi, A., Devireddy, N., Gupta, S., & Boopathi, S. (2023). Integrating Machine Learning and AI for Improved Hydrological Modeling and Water Resource Management. In *Artificial Intelligence Applications in Water Treatment and Water Resource Management* (pp. 46–70). IGI Global. doi:10.4018/978-1-6684-6791-6.ch003

Zemella, A., Thoring, L., Hoffmeister, C., & Kubick, S. (2015). Cell-Free Protein Synthesis: Pros and Cons of Prokaryotic and Eukaryotic Systems. *ChemBioChem*, *16*(17), 2420–2431. doi:10.1002/cbic.201500340 PMID:26478227

Zhang, J., Zhang, Q., & Wei, X. (2018). Classification of ECG signals based on 1d convolution neural network. In *2017 IEEE 19th international conference on e-health networking, applications and services (healthcom)* (pp. 1–6). IEEE.

Zhang, F., Sun, G., Zheng, B., & Dong, L. (2021). Design and implementation of energy management system based on spring boot framework. *Information (Basel)*, *12*(11), 457. doi:10.3390/info12110457

Zhang, L., Guo, W., & Lu, Y. (2020). Advances in Cell-Free Biosensors: Principle, Mechanism, and Applications. *Biotechnology Journal*, *15*(9), 2000187. doi:10.1002/biot.202000187 PMID:32667120

Zheng, Y., & Sriram, G. (2010). Mathematical modeling: Bridging the gap between concept and realization in synthetic biology. *BioMed Research International*, *541609*, 1–16. Advance online publication. doi:10.1155/2010/541609 PMID:20589069

Zhou, W., Wang, Y., Zhang, J., Zhao, M., Tang, M., Zhou, W., & Gong, Z. (2021). A metabolic model of Lipomyces starkeyi for predicting lipogenesis potential from diverse low-cost substrates. *Biotechnology for Biofuels*, *14*(1), 1–17. doi:10.118613068-021-01997-9 PMID:34210354

Zhu, J. (2012, September 1). Mammalian cell protein expression for biopharmaceutical production. *Biotechnology Advances*, *30*(5), 1158–1170. doi:10.1016/j.biotechadv.2011.08.022 PMID:21968146

Zhu, Y., Guo, X., Liu, J., Li, F., & Yang, D. (2020). Emerging advances of cell-free systems toward artificial cells. *Small Methods*, *4*(10), 2000406. doi:10.1002mtd.202000406

Zubay, G. (1973). In vitro synthesis of protein in microbial systems. *Annual Review of Genetics*, *7*(1), 267–287. doi:10.1146/annurev.ge.07.120173.001411 PMID:4593305

Zuñiga, C., Levering, J., Antoniewicz, M. R., Guarnieri, M. T., Betenbaugh, M. J., & Zengler, K. (2018). Predicting Dynamic Metabolic Demands in the Photosynthetic Eukaryote Chlorella vulgaris. *Plant Physiology*, *176*(1), 450–462. doi:10.1104/pp.17.00605 PMID:28951490

About the Contributors

Muhammad Arshad is an Associate Professor at the International Islamic University, Islamabad. Dr. Arshad completed his M. Phil. from the Centre of Excellence in Molecular Biology at the University of Punjab. Recognizing his potential, he pursued a Ph.D. in Cell Biology from Tsinghua University, one of China's most prestigious institutions. To further enhance his knowledge and expertise, he also undertook post-doctoral studies at the renowned Chinese Academy of Sciences in Beijing. In recognition of his scholarly achievements, Dr. Arshad has received the Young Scientist Post-Doctoral Fellowship from the Government of China and the Best Teacher Award from Pakistan International School, Taif. He is also affiliated with the Safe Blood Transfusion Program and with his strong commitment, contributes to various scientific societies. Throughout his academic journey, Dr. Arshad has been actively engaged in research activities. He has authored 75+ research articles and review papers as well as three scientific handbooks, which have contributed to the advancement of knowledge in his field. His research focuses on molecular biotechnology, infectious diseases, and exploring viral genomics.

* * *

Fatma Ece Altinisik Kaya, in 2014, started her master education in the Department of Bioengineering at Marmara University. In February 2016, she started to work as a teaching assistant in Marmara University Bioengineering Department. In 2017, she completed her master's degree with the thesis titled "Understanding alkaliphilic adaptation of Bacillus marmarensis sp. nov. using omic tools". In 2018, she started her PhD studies at Marmara University, Department of Bioengineering, and in March 2019, she passed the PhD qualification exam and started her thesis studies.

Fatma Gizem Avci received her B.Sc. (2010) and M.Sc. (2013) degrees from the Department of Bioengineering, Ege University. During her M.Sc. studies, she focused on the production of 1,3-propanediol (1,3-PDO) using a cell recycle bioreactor. She participated in a project on the production of 1,3-PDO using the metabolic engineering approach for 3 months at the Centre for Industrial Biotechnology and Biocatalysis, Ghent University, Belgium. She completed her Ph.D. at the Department of Bioengineering, Marmara University in 2018. Her research during her Ph.D. was mainly focused on antibiotic resistance, investigation of antimicrobial candidates specifically from plant-derived compounds to tackle this issue, and understanding their mechanism of action. With support from of FEMS Research Fellowship, she also carried out studies on the characterization of bacterial membrane proteins at the National Scientific Research Center (CNRS-Centre National de la Recherche Scientifique) in Lyon, France. Fatma Gizem Avci is currently an Assistant Professor at Uskudar University, Bioengineering Department, Turkey. Her research areas include antibiotic resistance, understanding antimicrobial mechanisms, omic technologies, and industrial/environmental biotechnology.

Alka Bali is Professor of Pharmaceutical Chemistry with a teaching and research experience of 28 years. Her research interests include synthetic medicinal chemistry and pharmaceutical analysis.

Shahid Bashir is a neuroscientist whose main research interest is in brain stimulation and the molecular mechanisms behind these brain-stimulating techniques. Born in Lahore, Pakistan, Dr. Shahid Bashir completed his BSc and MSc education at the University of Punjab, Lahore, Pakistan. Dr. Shahid got his Ph.D. degree in Physiology from the University of Fribourg, Switzerland. He is a postdoctoral fellow at Harvard Medical School in the United States. He also remains affiliated with the University of Lausanne, the University of Bern, and the University of Zurich, Switzerland. Dr. Shahid was also awarded grants from HEC Pakistan and the Swiss Society of Neuroscience. He has published more than 250 scientific papers in high-impact and peer-reviewed journals. In addition, he serves as a reviewer and editor for major journals in neurology, neuroscience, psychiatry, and neurophysiology. Currently, he is serving as a neurologist at King Fahad Specialist.

M. Beulah Viji Christiana, Professor, Department of Management Studies, Panimalar Engineering College, Chennai, Tamil Nadu 600123, India, has got a rich teaching experience which spans over 23 years. She is an innovative and student focused professional and mentor with a proven track record of excellent teaching skills.

Sampath Boopathi is an accomplished individual with a strong academic background and extensive research experience. He completed his undergraduate studies in Mechanical Engineering and pursued his postgraduate studies in the field of Computer-Aided Design. Dr. Boopathi obtained his Ph.D. from Anna University, focusing his research on Manufacturing and optimization. Throughout his career, Dr. Boopathi has made significant contributions to the field of engineering. He has authored and published over 145 research articles in internationally peer-reviewed journals, highlighting his expertise and dedication to advancing knowledge in his area of specialization. His research output demonstrates his commitment to conducting rigorous and impactful research. In addition to his research publications, Dr. Boopathi has also been granted one patent and has three published patents to his name. This indicates his innovative thinking and ability to develop practical solutions to real-world engineering challenges. With 17 years of academic and research experience, Dr. Boopathi has enriched the engineering community through his teaching and mentorship roles. He has served in various engineering colleges in Tamilnadu, India, where he has imparted knowledge, guided students, and contributed to the overall academic development of the institutions. Dr. Sampath Boopathi's diverse background, ranging from mechanical engineering to computer-aided design, along with his specialization in manufacturing and optimization, positions him as a valuable asset in the field of engineering. His research contributions, patents, and extensive teaching experience exemplify his expertise and dedication to advancing engineering knowledge and fostering innovation.

Manisha Jankiram Gadekar is Assistant Professor in Computer science with 18 years of teaching and 5 years of research experience. She is author of 4 textbooks. Her research expertise is in developing software related to life science.

Aparna Gunjal is Assistant Professor, Department of Microbiology, Dr. D. Y. Patil, Arts, Commerce & Science College, Pimpri, Pune, Maharashtra, India. She has 9 years teaching and 12 years research experience. Aparna Gunjal has 170 publications to her credit which includes research papers, books, book chapters, review articles, research papers in Proceedings, articles in magazines and newsletters, Report at International level and one German patent which is granted. She has 15 National and International awards to her credit. She is editor and reviewer for many journals. She is member of Biotech Research Society India, Indian Women Scientist Association, etc.

Nishika Kapoor is graduate in pharmacy (B. Pharm.) and currently pursuing her M. Pharmacy degree in Pharmacology from Panjab University Chandigarh.

Dipali Deepak Mali is Assistant Professor with 16 years of teaching and 5 years of research experience. She has keen interest in developing informatic tools and cyber security. She is author of 7 text books.

Manish Manoria was born on 12 Jan 1969. He have completed his Ph.D from RGTU Bhopal in 2007. He have 28 Year of Experience of academics and Research of different Institution Like SATI Vidisha(Autonomous Under RGPV Bhopal), Truba College, SIRT, TIET, Bhopal,etc. Currently he is working as a Director Bansal Institute of Research Technology and sciences, Bhopal. He has 78 international research publication and 5 Books in field of computer science and engineering. He have guided 25 PG and 02 Ph.d students. He have received 15 + Lac Research grant from different Govt. Agency. He is a good Enterprise Academic Leader with rich experience of educational institutions through innovations and streamlining operation.

Jay Prakash Maurya was born in India on June, 1987. He received the Engineering degree, B.Tech. in Computer Science and Information Technology from Institute of Engineering & Technology (IET), Mahatma Jyotiba Phule Rohilkhand University, Bareilly(U.P), India in 2008. He has completed a completed post graduate diploma in System database Administration from CDAC, Noida(U.P), India. He have completed his Post Graduate degree M. Tech. in Computer Science and engineering from Bansal Institute of Science & Technology, Bhopal (M.P), India in 2013. He is pursuing Ph.D from SATI(Autonomous), RGPV, Bhopal (M.P), and India. He is currently working as a Assistant Professor in the Department of Computing Science Engineering VIT Bhopal, University Madhya Pradesh, India. He has 11 years of experience in teaching and research. He have guided 09 P.G Dissertation under him. He has published more than 35 scientific papers in International and National reputed Journals and conference proceedings (Including SCOPUS index and UGC Paper),04 Book Chapter in the field of data mining, image processing, and Network security. His current research interests include Data Mining, Machine Learning, Soft Computing, and IoT. He also has the experience of conducting workshops with collaboration of IIT-Bombay for the project "SMART CLASSROOM" and "BODHI TREE", & Spoken Tutorial Project. Nominated as expert, reviewer and jury in different event like Smart India Hackathon, Toycathon, MPCST Project, and many more.

Jatin Phogat is a graduate at Panjab University, Chandigarh. He completed his Bachelors of pharmacy from the University Institute of Pharmaceutical Sciences, Panjab University, Chandigarh. He writes and presents widely on the topics related to Pharmaceutical Sciences and Medical information.

Hammad Qaiser is an emerging molecular biologist who is currently pursuing his research work at the Human Molecular Lab, IIUI. His upcoming scientific work will be based on genetics, molecular biology, and neurological sciences. Having been a Managing Director and Founder of Biologia360°, Hammad has experience in management consultation, team building, professional development, strategic implementation, and company collaboration. Hammad loves science, so he majored in biotechnology at International Islamic University Islamabad and is motivating other fellows regarding biological sciences and applications, including molecular biology, genetics, personalized medicine, and potential therapeutic options.

R. E. Ugandar is Professor and Head, Department of Pharmacy Practice Santhiram College of Pharmacy, Department of Pharmacy Practice, Santhiram College of Pharmacy, NH-40. Nerawada (V&P)-518112. Panyam (M). Nandyal District Andhrapradesh-India. Having 25 years of experience in Academic Teaching and Research at National & International Universities in India & Malaysia. Public Health, Natural Polymers, Pharmaceutical Formulations, Pharmaceutical Technology, Pharmaceutical Engineering, and Pharmacogenomics. Forty-Six (46) Research papers are published in SCI, Scopus, and UGC-indexed journals and also hold several Indian Patent Publications, Patent Designs, Copyrights, authored Books, and Book Chapters. Most Sporting employee of Asia Metropolitan University-Malaysia. Innovator in Natural Polymer Science/Santhiram College of Pharmacy-India. A life member of many scientific, Health, and Professional Associations such as the Association of Pharmaceutical Teachers of India (APTI), Indian Red Cross Society, Indian Public Health (IPH), Epidemiology Foundation of India (EFI), Karnataka State Pharmacy Council (KSPC)-India, and also involved in Pharmacy Practice, Public health activities, and Community Services. Certified in Diabetic Pharmacotherapy, Cardiology Series-1 with Basic Cardio Vascular Diseases and Clinical Pharmacogenomics.

Mohammad Uzair is graduated in Biotechnology with additional two years of research at different labs, where he skilled in molecular biology techniques. Uzair is an upcoming researcher in the field of biological sciences who believes in hard work in an organized way that leads to success. Uzair has excellent scientific writing skills and published various scientific papers in impact factor journals. His main research interest in neurology, molecular biology, and genetics. He is also working on different projects of green synthesis of nanoparticles and their biological applications. Alongside, Uzair is also the CEO & Founder of Biologia360°. Uzair is a powerful force in the workplace and uses his positive attitude and tireless energy to encourage others to work hard and succeed.

N. Venkateswaran, working as Professor at Panimalar Engineering College, having 26 years of teaching experience published more papers in International journals and also member of editorial board in more reputed journals. Published books on International Business Management, HR Analytics, Entrepreneurship Development, Professional Ethics and Human Values, Fundamentals of Management. Area of expertise is on Business Analytics, Artificial Intelligence and Supply Chain Management" Publications.

Meghmala Waghmode has 15 years of teaching experience and 7 years of research experience. She has 30 research papers, 2 books and 9 book chapters to her credit.

Index

E

F

G

H

I

L

M

N

O

P

Q

R

S

T

U

V

W

Y

www.ingramcontent.com/pod-product-compliance
Lightning Source LLC
LaVergne TN
LVHW060256011026
849161LV00011B/2728